Advances in Intelligent Systems and Computing

Volume 264

Series editor

Janusz Kacprzyk, Polish Academy of Sciences, Warsaw, Poland
e-mail: kacprzyk@ibspan.waw.pl

For further volumes:
http://www.springer.com/series/11156

About this Series

The series "Advances in Intelligent Systems and Computing" contains publications on theory, applications, and design methods of Intelligent Systems and Intelligent Computing. Virtually all disciplines such as engineering, natural sciences, computer and information science, ICT, economics, business, e-commerce, environment, healthcare, life science are covered. The list of topics spans all the areas of modern intelligent systems and computing.

The publications within "Advances in Intelligent Systems and Computing" are primarily textbooks and proceedings of important conferences, symposia and congresses. They cover significant recent developments in the field, both of a foundational and applicable character. An important characteristic feature of the series is the short publication time and world-wide distribution. This permits a rapid and broad dissemination of research results.

Advisory Board

Chairman

Nikhil R. Pal, Indian Statistical Institute, Kolkata, India
e-mail: nikhil@isical.ac.in

Members

Emilio S. Corchado, University of Salamanca, Salamanca, Spain
e-mail: escorchado@usal.es

Hani Hagras, University of Essex, Colchester, UK
e-mail: hani@essex.ac.uk

László T. Kóczy, Széchenyi István University, Győr, Hungary
e-mail: koczy@sze.hu

Vladik Kreinovich, University of Texas at El Paso, El Paso, USA
e-mail: vladik@utep.edu

Chin-Teng Lin, National Chiao Tung University, Hsinchu, Taiwan
e-mail: ctlin@mail.nctu.edu.tw

Jie Lu, University of Technology, Sydney, Australia
e-mail: Jie.Lu@uts.edu.au

Patricia Melin, Tijuana Institute of Technology, Tijuana, Mexico
e-mail: epmelin@hafsamx.org

Nadia Nedjah, State University of Rio de Janeiro, Rio de Janeiro, Brazil
e-mail: nadia@eng.uerj.br

Ngoc Thanh Nguyen, Wroclaw University of Technology, Wroclaw, Poland
e-mail: Ngoc-Thanh.Nguyen@pwr.edu.pl

Jun Wang, The Chinese University of Hong Kong, Shatin, Hong Kong
e-mail: jwang@mae.cuhk.edu.hk

Sabu M. Thampi · Alexander Gelbukh
Jayanta Mukhopadhyay

Editors

Advances in Signal Processing and Intelligent Recognition Systems

 Springer

Editors
Sabu M. Thampi
Indian Inst. of Information Technology
    and Management - Kerala (IIITMK)
Technopark Campus
Trivandrum
India

Jayanta Mukhopadhyay
Department of Computer Science
    and Engineering
Indian Institute of Technology
Kharagpur
India

Alexander Gelbukh
National Polytechnic Institute
Mexico City
Mexico

ISSN 2194-5357   ISSN 2194-5365   (electronic)
ISBN 978-3-319-04959-5   ISBN 978-3-319-04960-1   (eBook)
DOI 10.1007/978-3-319-04960-1
Springer Cham Heidelberg New York Dordrecht London

Library of Congress Control Number: 2014931505

Printed on acid-free paper

Springer is part of Springer Science+Business Media (www.springer.com)

Preface

This Edited Volume contains a selection of refereed and revised papers originally presented at the International Symposium on Signal Processing and Intelligent Recognition Systems (SIRS-2014), March 13–15, 2014, Trivandrum, India. SIRS-2014 provided a forum for the sharing, exchange, presentation and discussion of original research results in both methodological issues and different application areas of signal processing and pattern recognition.

Credit for the quality of the symposium proceedings goes first and foremost to the authors. They contributed a great deal of effort and creativity to produce this work, and we are very thankful that they chose SIRS-2014 as the place to present it. All the authors who submitted papers, both accepted and rejected, are responsible for keeping the SIRS program vital. The program committee received 134 submissions from 11 countries: India, United Kingdom, Iran, Saudi Arabia, Macao, Kazakhstan, Algeria, Egypt, France and USA. The program committee had a very challenging task of choosing high quality submissions. Each paper was peer re-viewed by at least three or more independent referees of the program committee and the papers were selected based on the referee recommendations. The technical program of SIRS'14 comprises of 52 papers including 6 short papers. The selected papers offer stimulating insights into signal processing and intelligent recognition systems.

The success of such an event is mainly due to the hard work and dedication of a number of people and the collaboration of several institutions. We are grateful to the members of the program committee for reviewing and selecting papers in a very short period of time. Many thanks to all the Chairs and their involvement and support have added greatly to the quality of the symposium. We also wish to thank all the members of the Steering Committee, whose work and commitment were invaluable. We would like to express our sincere gratitude to local organizing committees that has made this event a success. Our special thanks also to the keynote speakers and tutorial presenters for their effort in preparing the lectures. The EDAS conference system proved very helpful during the submission, review, and editing phases.

We wish to express our sincere thanks to Thomas Ditzinger, Senior Editor, Engineering/AppliedSciences Springer-Verlag and Janusz Kacprzyk, Series Editor for their help and cooperation.

Finally, we hope that you will find this edited book to be a valuable resource in your professional, research, and educational activities whether you are a student, academic, researcher, or a practicing professional.

Alexander Gelbukh
Sabu M. Thampi
Jayanta Mukhopadhyay

Organized by

Indian Institute of Information Technology and Management-Kerala (IIITM-K),
Trivandrum, India

http://www.iiitmk.ac.in

Organization

Committee

Steering Committee

Rajasree M.S.	Director IIITM-K, India
Janusz Kacprzyk	Polish Academy of Sciences, Poland
Achuthsankar S. Nair	Centre of Excellence in Bioinformatics, University of Kerala, India
Axel Sikora	University of Applied Sciences Offenburg, Germany
Francesco Masulli	University of Genoa, Italy
Juan Manuel Corchado Rodriguez	University of Salamanca, Spain
Ronald R. Yager	Machine Intelligence Insitute, Iona College, USA
Sankar Kumar Pal	Indian Statistical Institute, Kolkata, India
Soura Dasgupta	The University of Iowa, USA
Suash Deb	President Intl. Neural Network Society (INNS), India Regional Chapter
Subir Biswas	Michigan State University, USA

Technical Program Committee (TPC)

General Chairs

Sabu M. Thampi	IIITM-K, India
Adel M. Alimi	University of Sfax, Tunisia
Alexander Gelbukh	Mexican Academy of Science, Mexico

Program Chairs

Jayanta Mukhopadhyay	Indian Institute of Technology, Kharagpur, India
Kuan-Ching Li	Providence University, Taiwan

TPC Members

Abdallah Kassem   Notre Dame University, Lebanon
Abduladhem Ali   University of Basrah, Iraq
Abul Bashar   Prince Mohammad Bin Fahd University,
                                       Saudi Arabia
Aditya Vempaty   Syracuse University, USA
Agilandeeswari Loganathan   VIT University, India
Ahmed Abdelgawad   Central Michigan University, USA
Ajey Kumar   Symbiosis Centre for Information Technology,
                                       India
Akash Singh   IBM, USA
Alex James   Nazarbayev University, Kazakhstan
Alexandre Ramos   Federal University of Itajubá, Brazil
Alexandru Lavric   Stefan cel Mare University of Suceava, Romania
Ali Al-Sherbaz   The University of Northampton,
                                       United Kingdom
Ananthram Swami   Army Research Lab., USA
André Almeida   Federal University of Ceará, Brazil
Andrew Mercer   Mississippi State University, USA
Andrzej Borys   University of Technology and Life Sciences,
                                       Poland
Andy Peng   Lockheed Martin, USA
Ang Chen   The Hong Kong Polytechnic University,
                                       Hong Kong
Anil Yekkala   Forus Health Pvt Ltd., India
Ankit Chaudhary   MUM, USA
Anna Antonyová   University of Prešov in Prešov, Slovakia
Anna Bartkowiak   University of Wroclaw, Poland
António Trigo   ISCAC - Coimbra Business School, Portugal
Antony Chung   Lancaster University, United Kingdom
Aravind Kailas   LLC, USA
Arpan Kar   Indian Institute of Management, Rohtak, India
Ashutosh Tripathi   Infosys India Pvt Ltd., India
B. Harish   S.J. College of Engineering, India
Balaji Raghothaman   InterDigital, USA
Bao Rong Chang   National University of Kaohsiung, Taiwan
Beatriz Sainz   University of Valladolid, Spain
Belal Amro   Hebron University, Palestine
Biju Issac   Teesside University, Middlesbrough,
                                       United Kingdom
Björn Schuller   Imperial College London, Germany
Brian Sadler   Army Research Laboratory, USA
Carlos Gonzalez Morcillo   University of Castilla-La Mancha, Spain
Carlos Regis   University Federal of Campina Grande, Brazil

César Cárdenas	Tecnológico de Monterrey - Campus Querétaro, Mexico
Christophe Jego	IMS CNRS Laboratory, France
Christopher Hollitt	Victoria University of Wellington, New Zealand
Daisuke Umehara	Kyoto Institute of Technology, Japan
Dalila Boughaci	University of Sciences and Technology USTHB, Algeria
Damon Chandler	Oklahoma State University, USA
Daniel Benevides da Costa	Federal University of Ceara (UFC), Brazil
Daniele Toti	University of Salerno, Italy
Deepak Choudhary	LPU, India
Dhiya Al-Jumeily	Liverpool John Moores University, United Kingdom
Djalma Carvalho Filho	Federal University of Campina Grande-UFCG, Brazil
E. George Prakash Raj	Bharathidasan University - Trichy - South India, India
Eduard Babulak	Sungkyunkwan University, Korea
Elpida Tzafestas	University of Athens, Greece
Emilio Jiménez Macías	University of La Rioja, Spain
Erik Markert	Chemnitz University of Technology, Germany
Erwin Daculan	University of San Carlos, Philippines
Faheem Ahmed	Thompson Rivers University, Canada
Fakhrul Alam	Massey University, New Zealand
Farid Naït-Abdesselam	University of Paris Descartes, France
Felix Albu	Valahia University of Targoviste, Romania
Fuad Alnajjar	CUNY - City College of New York, USA
Gancho Vachkov	The University of the South Pacific (USP), Fiji
George Mastorakis	Technological Educational Institute of Crete, Greece
George Papakostas	TEI of Eastern Macedonia & Thrace, Greece
George Tambouratzis	Institute for Language & Speech Processing, Greece
Georgios Sirakoulis	Democritus University of Thrace, Greece
Gianluigi Ferrari	University of Parma, Italy
Giovanni Barroso	Universidade Federal do Ceará, Brazil
Giovanni Iacca	INCAS3, The Netherlands
Girija Chetty	University of Canberra, Australia
Grzegorz Debita	Wroclaw University of Technology, Poland
Hamed Mojallali	University of Guilan, Iran
Hendrik Richter	HTWK Leipzig-University of Applied Sciences, Germany
Hikmat Farhat	Notre Dame University, Lebanon
Houcem Gazzah	University of Sharjah, UAE
Hugh Kennedy	University of South Australia, Australia

Michal Kopcek   Slovak University of Technology, Slovakia
Mohamed Dahmane   University of Montreal, Canada
Mohammad Firoj Mithani   Telstra Corporation, Australia
Mohammad Mozumdar   California State University, Long Beach, USA
Mumtaz Ahmad   INRIA-Nancy, France
Muthukumar Subramanyam   National Institute of Technology, Puducherry,
                                     India
Najett Neji   Ecole Supérieure d'Electricité (SUPELEC),
                                     France
Nor Hayati Saad   UiTM, Malaysia
Nordholm Sven   Curtin University of Technology, Australia
Ola Amayri   Concordia University, Canada
Osamu Ono   Meiji University, Japan
Oscar Castillo   Tijuana Institute of Technology, Mexico
Paolo Crippa   Università Politecnica delle Marche, Italy
Pascal Lorenz   University of Haute Alsace, France
Patrick Siarry   University of Paris XII, France
Paul Fergus   Liverpool John Moores University,
                                     United Kingdom
Philip Moore   Birmingham City University, United Kingdom
Prashant More   CDAC Mumbai (Formerly NCST), India
Praveen Srivastava   Indian Institute of Management (IIM), India
Prem Jain   Shiv Nadar University, India
Priya Ranjan   Templecity Institute of Technology and
                                     Engineering, USA
Qiang Yang   Zhejiang University, P.R. China
R. Agrawal   Jawaharlal Nehru University, New Delhi, India
R. Prasad Kodaypak   Senior Member IEEE, USA
Radu Vasiu   Politehnica University of Timisoara, Romania
Rafael Lopes   Instituto Federal do Maranhão, Brazil
Rajan Ma   Tata Consultancy Services, India
Ralph Turner   Eastern Kentucky University, USA
Ramesh Babu   DSCE, Bangalore, India
Ramesh Rayudu   Victoria University of Wellington, New Zealand
Ramin Halavati   Sharif University of Technology, Iran
Ramiro Barbosa   Institute of Engineering of Porto, Portugal
Rashid Saeed   Telekom Malaysia, R&D Innovation Center,
                                     Malaysia
Ravi Subban   Pondicherry University, Pondicherry, India
Reyer Zwiggelaar   Aberystwyth University, United Kingdom
Ricardo Rodriguez   Technologic University of Ciudad Juarez,
                                     Mexico
Robinson Pino   ICF International, USA
Rohollah Dosthoosseini   Yazd University, Iran
Rosaura Palma-Orozco   Instituto Politécnico Nacional, Mexico

Ruqiang Yan	Southeast University, P.R. China
Ryosuke Ando	Toyota Transportation Research Institute (TTRI), Japan
Saad Harous	UAE University, UAE
Salah Bourennane	Ecole Centrale Marseille, France
Sameera Abar	University College Dublin/IBM-Ireland, Ireland
Sandeep Paul	Dayalbagh Educational Institute, India
Santoso Wibowo	CQUniversity Melbourne, Malaysia
Senthil kumar Thangavel	Amrita School of Engineering, India
Seyed (Reza) Zekavat	Michigan Technological University, USA
Shawn Kraut	MIT Lincon Laboratory, USA
Shikha Tripathi	Amrita School of Engineering, India
Silvana Costa	Instituto Federal de Educação, Ciência e Tecnologiada Paraíba, Brazil
Simon Fong	University of Macau, Macao
Spyridon Mouroutsos	Democritus University of Thrace, Greece
Stephane Senecal	Orange Labs, France
Steven Guan	Xian Jiatong-Liverpool University, Australia
Suma N.	B.M.S. College of Engineering, India
Sung-Bae Cho	Yonsei University, Korea
Sunil Kumar Kopparapu	Tata Consultancy Services, India
T. Manjunath	HKBK College of Engineering Bangalore Karnataka, India
Tarek Bejaoui	University of Paris-Sud 11, France
Tiago de Carvalho	Federal Rural University of Pernambuco, Brazil
Tzung-Pei Hong	National University of Kaohsiung, Taiwan
Ugo Dias	University of Brasilia, Brazil
Valentina Balas	Aurel Vlaicu University of Arad, Romania
Valerio Scordamaglia	University of Reggio Calabria, Italy
Varun Mittal	Singtel, India
Wali Ali	University of Sfax, Tunisia
Waslon Lopes	UFCG - Federal University of Campina Grande, Brazil
Wei Zhan	Texas A&M University, USA
Xianghua Xie	Swansea University, United Kingdom
Ye Kyaw Thu	National Institute of Information and Communications Technology, Japan
Yingqiong Gu	University of Notre Dame, USA
Yordan Chervenkov	Naval Academy - Varna, Bulgaria
Yoshikazu Miyanaga	Hokkaido University, Japan
Yun Huoy Choo	Universiti Teknikal Malaysia Melaka, Malaysia
Zebin Chen	Microsoft, USA
Zeeshan Ahmed	University of Wuerzburg Germany, Germany
Zsofia Lendek	Technical University of Cluj-Napoca, Romania

Additional Reviewers

Abdul Talib Din	Universiti Teknikal Malaysia Melaka, Malaysia
Afaf Merazi	Djillali Liabès University of Sidi Bel-Abbès, Algeria
Ahmad Usman	Georgia Institute of Technology - Atlanta, USA
Andreas Buschermoehle	University of Osnabrueck, Germany
Bhaskar Belavadi	BGS Health & Education city, Bangalore, India
Bo Han	Aalborg University, Denmark
Carlos Oliveira	IFRJ, Brazil
Davood Mohammadi souran	Shiraz University, Iran
Francesco Maiorana	University of Catania, Italy
G. Deka	DGE&T, India
Hrudya P.	Amrita Cenetr for Cyber Security, India
Jirapong Manit	King Mongkut's University of Technology Thonburi, Thailand
K. Mahantesh	SJBIT, India
K. Hemant Reddy	BPUT, Rourkela, ORISSA, India
Karthikeyan Ramasamy	Teclever Solutions Pvt Ltd, India
Kuan-Chieh Huang	National Cheng Kung University, Taiwan
M. Udin Harun Al Rasyid	Polytechnic Institute of Surabaya, Indonesia
Mahdi Mahmoodi Vaneghi	Shahed University, Iran
Md. Whaiduzzaman	Jahangirnagar University, Bangladesh
Mohammad Nasiruddin	Université de Grenoble, France
Phuc Nguyen	Asian Institute of Technology and Management, Vietnam
Pourya Hoseini	Urmia University, Iran
Rajesh Siddavatam	Apex University, Jaipur, India
Rajiv Singh	University of Allahabad, India
Ramalingeswara Rao K.V.	Dravidian University, India
Raza Hasan	Middle East College, Oman
S. Vijaykumar	6th Sense Advanced Research Foundation, India
Sayed Mostafa Taheri	University College London (UCL), United Kingdom
Thiang Hwang Liong Hoat	Petra Christian University, Indonesia
Vijayaratnam Ganeshkumar	Creative Technology Solutions PTE, Sri Lanka
Vikrant Bhateja	Shri Ramswaroop Professional Colleges, India
Vishnu Pendyala	Santa Clara University, USA
Wan Hussain Wan Ishak	Universiti Utara Malaysia, Malaysia

Contents

Pattern Recognition, Machine Learning and Knowledge-Based Systems

Signal and Speech Processing

Image and Video Processing

Mobile Computing and Applications

International Workshop on Advances in Image Processing, Computer Vision, and Pattern Recognition (IWICP-2014)

Short Papers

Ear Recognition Using Texture Features – A Novel Approach

Lija Jacob and G. Raju

Abstract. Ear is a new class of relatively stable biometric that is invariant from childhood to old age. It is not affected with facial expressions, cosmetics and eye glasses. Human ear is one of the representative human biometrics with uniqueness and stability. Ear Recognition for Personal Identification using 2-D ear from a side face image is a challenging problem. This paper analyzes the efficiency of using texture features such as Gray Level Co-occurrence Matrix (GLCM), Local Binary Pattern (LBP) and Gabor Filter for the recognition of ears. The combination of three feature vectors was experimented with. It is found that the combination gives better results compared to when the features were used in isolation. Further, it is found that the recognition accuracy improves by extracting local texture features extracted from sub-images. The proposed technique is tested using an ear database which contains 442 ear images of 221 subjects and obtained 94.12% recognition accuracy.

Keywords: Ear recognition, texture features, co-occurrence matrix, Local Binary Pattern, Gabor Filter.

1   Introduction

There are many established human traits that can be used as a biometric like fingerprint, face, voice and iris. Despite extensive research, many problems in these methods remain largely unsolved. In security situations, active identification methods like fingerprint or iris recognition may not be suitable. Hence over the last

Lija Jacob
Saintgits College of Engineering, Kottayam
e-mail: lija.jacob@saintgits.org
G. Raju
Department of Information Technology, Kannur University, Kannur
e-mail: kurupgraju@gmail.com

S.M. Thampi, A. Gelbukh, and J. Mukhopadhyay (eds.), *Advances in Signal Processing and Intelligent Recognition Systems*, Advances in Intelligent Systems and Computing 264,
DOI: 10.1007/978-3-319-04960-1_1, © Springer International Publishing Switzerland 2014

few decades, ear biometrics has been widely paid attention due its promising characteristics like friendliness, uniqueness and stability etc. Human ear is a perfect
data for physiological, passive person identification, which can be applied to provide security in the public places [1] [2] [3]. Human ears have been used as a major feature in the forensic science for many years. Ear prints found on the crime
scene have been used as a proof in over hundred cases in many countries especially United States and Netherlands [4].

Ear satisfies all the properties that should be possessed by a biometric such as
universality, uniqueness and permanence [6]. It has a rich and firm structure that
changes little with age and does not suffer from changes in facial expression or
make-up effects. It is firmly fixed in the middle side of the head so that the immediate background is predictable. The ear is large, compared with iris, retina and
fingerprint that make it more easily captured at distance with or without the knowledge of examined person; thus unlikely to cause anxiety as that may happen with
the iris and retina measurements.

Human ear contains large amount of precise and unique features that allows for
human individualization. The characteristic appearance of the human outer ear (or
pinna) is formed by several morphological components such as the outer helix, the
antihelix, the lobe, the tragus, the antitragus, and the concha. The external ear flap,
known as the pinna, of a human is shown in Fig 1. It has been proved through studies that finding two ears which are wholly identical are almost unfeasible. Even
though the left and the right ear of the same individual show some similarities,
they are not symmetric [6]. Consequently, ear biometrics is suitable for security,
surveillance, access control and monitoring applications.

Ear has proved its competency in multimodal recognition. A multimodal system where face features are combined with ear is found more accurate than a
unimodal system with face or ear features taken separately [19] [20] [21].

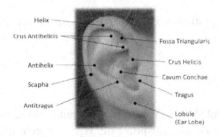

Fig. 1 The human outer ear (Pinna)

As in every biometric method, the steps involved in ear biometrics are
Image acquisition (side face image), Image preprocessing, Shape Segmentation /
Localization (cropping the ear images either manually or automatic), Ear detec-
tion, Feature Extraction and Ear recognition/Verification.

Automatic ear recognition has become one of the challenging areas of research
in ear biometrics. In any recognition phase, the key task is to explore discriminative information and extract features accordingly. Major works in automatic ear

recognition includes structure-based ear recognition and statistic-based ear recognition methods [13]. Structure based methods focus on extracting feature vectors with geometrical structure of the ear [25] [28]. This method is susceptible to light and imaging angles, resulting in poor robustness. The statistic-based methods use global features of ear image for recognition. It includes principal component analysis, independent component analysis, kernel principal component analysis, and moment invariants [5] [13]. The statistic-based method is easy to extract and has better performance in response to noise. However, the computational complexity becomes very high when the dimension of statistical features increases.

The texture analysis represents a kind of local and structural features including not only statistical information, but also structural information. Hence, to describe ear images, the texture descriptors suits better than the structural or statistical approach [12].

In this paper we have implemented some of the standard feature extraction methods for ear recognition, including geometrical, texture and transform features. We carried out a performance analysis by considering the feature extraction methods in isolation and also by combining some of the prominent texture feature extraction methods.

The paper proposes a block segmentation method of the ear images for more accurate results. In order to verify the effectiveness of the block segmentation method, texture feature extraction methods such as Gabor Filter, LBP and GLCM in isolation and combination were applied both on non blocked images (full images) and on blocked images .The experimental results of the methods indicate that the recognition rate is significantly increased in the block segmentation method.

2 A Review on Automatic 2D Ear Recognition Techniques

Ear features are potentially a promising biometric for use in human identification [1][2]. There exist many methods to perform ear recognition. Mark Burge and Wilhelm Burger reported the first attempt to automate the ear recognition process in 1997 [1]. They used a mathematical graph model to represent and match the curves and edges in a 2D ear image. Later Belen Moreno et.al [9] described a fully automated ear recognition system based on features such as ear shape and wrinkles. Since then, researchers have proposed numerous feature extraction and matching schemes, based on computer vision and image processing algorithms, for ear recognition. This ranges from simple appearance-based methods such as principal component analysis and independent component analysis to more sophisticated techniques based on scale-invariant feature transforms, local binary patterns, wavelet transforms, and force fields [6].

Some ear recognition methods were based on geometric features [25] [28] and force field transformation method developed by Hurley et al. [8] but were very sensitive to poses. Some are based on algebraic features like "eigen-ear" approach [1][2].Algebraic features are extracted based on statistics theory which does not give good results in all scenarios. Lu Lu et.al [7] have experimented on ear shape features for recognition. Active Shape Models (ASMs) were applied to model the shape and local appearance of the ear in a statistical manner. Steerable features from the ear images, which could predetermine rich discriminate information of the local structural texture and shape location, were also extracted in the work.

Texture analysis represents a kind of local and structural features which includes both statistical and structural information. In the early 1970s, Haralick et al. has proposed to use co-occurrence statistics to describe texture features [16], but these features were not much appreciated in direct human vision. Until recently, Gabor filter method has been considered state-of-the-art in texture analysis by many researchers [17].This approach has shown very good performance because of the combination of spatial frequencies and local edge information. Although theoretically simple, Gabor filtering tends to be computationally very demanding, especially with large mask sizes. It is also affected by varying lighting conditions.

Local Binary Pattern (LBP) has proved its efficiency in texture analysis [18]. LBPs described the comparative change of gray values between the samples in the circular local neighborhood as texture structural features and made a statistical analysis for these texture features using texture spectrum histogram. LBP operator with small radius cannot describe the texture information of images effectively because of the presence of structural information on a larger scale. Hence it is necessary to use multi-resolution versions. Ojala [26] developed original LBP and made it a more flexible operator which is in circular form with discretionary radius and neighborhoods. This makes it possible to acquire different resolution LBP operators.

3   Theoretical Background

Each ear recognition/verification system consists of feature extraction algorithm and feature vector comparison using classification methods. Fig 2 depicts major components in an Ear recognition system.

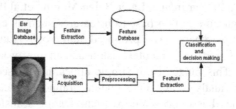

Fig. 2 Components of an Ear Recognition System

A concise description of the three major texture based feature extraction method is explained below.

3.1 Feature Extraction Methods

In pattern recognition, feature extraction is a special form of dimensionality reduction. The large input data will be transformed into a reduced representation set of features (also named features vector). A better-quality feature extraction algorithm simplifies the amount of resources required to describe the large set of data and ensures accuracy. The major features experimented in this work includes Gray level co-occurrence matrix (GLCM), Local Binary Pattern (LBP) and Gabor Filter.

Gray Level Co-occurrence MATRIX (GLCM)
Gray Level Co-occurrence Matrices (GLCM) is one of the prominent techniques used for image texture analysis. A co-occurrence matrix of an image is the distribution of co-occurring values at a given offset. Gray Level Co-occurrence Matrix, GLCM (also called the Gray Tone Spatial Dependency Matrix) is a tabulation of how often different combinations of pixel brightness values (gray levels) occur in an image.

GLCM estimates image properties related to second-order statistics. A GLCM is a matrix where the number of rows and columns is equal to the number of gray levels, N, in the image. The matrix element P(i, j | d, θ) is the relative frequency with which two pixels, separated by distance d, and in direction specified by the particular angle (θ) , one with intensity i and the other with intensity j. Each entry (i,j) in GLCM corresponds to the number of occurrences of the pair of gray levels i and j which are at a distance d apart in original image. The distance parameter can be selected as one or higher. In our work, d value is set to 1. Haralick [16] proposed 14 statistical features for characterizing the co-occurrence matrix contents and estimating the similarity. To reduce the computational complexity, only four relevant features that are widely used in literature were chosen. The features computed are shown in the Table 1.

Table 1 GLCM Properties

Property	Formula
Contrast	$\displaystyle\sum_{i,j=0}^{N-1} P(i,j)(i-j)^2$
Entropy	$\displaystyle\sum_{i,j=0}^{N-1} P(i,j)(-lnP(i,j))$
Energy	$\displaystyle\sum_{i,j=0}^{N-1} P(i,j)^2$
Homogeneity	$\displaystyle\sum_{i,j=0}^{N-1} \frac{P_{i,j}}{1+(i-j)^2}$

Local Binary Pattern (LBP)

The LBP operator is a theoretically simple yet one of the best performing texture descriptors. It has been widely used in various applications [15][23]. It has proven to be highly discriminative and the key advantages, namely, its invariance to monotonic gray-level changes and computational efficiency, make it suitable for demanding image analysis tasks [5]. The operator labels the pixels by comparing the neighborhoods with the center value. The result is recorded as a binary number. The binary derivatives are used to form a short code to describe the pixel neighborhood which can be used as a texture descriptor.

The LBP, by definition, is invariant to monotonic changes in gray scale; it was supplemented by an independent measure of local contrast (C). Fig. 3 shows how the contrast measure (C) is derived. The average of the gray levels below the center pixel is subtracted from that of the gray levels above (or equal to) the center pixel. Two-dimensional distributions of the LBP and local contrast measures can be used as features [23].

In the proposed work, the multiscale LBP is calculated where the arbitrary circular neighborhoods are considered [26]. Fig 4 shows the multiscale circularly symmetric neighbor sets of LBP with different values of P and R where P is the number of neighbors considered and R the radius.

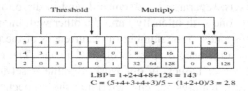

$$LBP = 1+2+4+8+128 = 143$$
$$C = (5+4+3+4+3)/5 - (1+2+0)/3 = 2.8$$

Fig. 3 Calculating the original LBP code and Local contrast measure

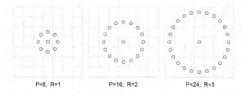

P=8, R=1   P=16, R=2   P=24, R=3

Fig. 4 Multiscale circularly symmetric neighbor sets of LBP

The value of LBP code of a pixel (x_c, y_c) is given by (1) where g_c corresponds to the gray value of the center pixel of a local neighborhood. g_p (p = 0,...,P -1) correspond to the gray values of P equally spaced pixels on a circle of radius R (R> 0) that form a circularly symmetric set of neighbors. The function LBP is defined as [23]

$$LBP = \sum_{p=0}^{P-1} s(g_p - g_c)2^p$$

where

$$g_c = (0,0)$$

$$g_p = \left(\left(x_c + R\cos\left(\frac{2\pi p}{P}\right), y_c + R\sin\left(\frac{2\pi p}{P}\right) \right) \right) \tag{1}$$

$$p = \{0,1,\dots P-1\}$$

$$s(x) = \{1 \text{ when } x > 0$$

$$0 \text{ when } x < 0\} .$$

Gabor Filter

Gabor is a technique that extracts texture information from an image. Gabor filters have the ability to perform multi-resolution decomposition due to its optimal localization as per uncertainty principle in both the spatial and spatial-frequency domain. As texture segmentation requires simultaneous measurements in both the spatial and the spatial-frequency domains along with the accurate localization of texture boundaries, Gabor filters are well suited for texture segmentation problem and ensures more accurate results [24].

The 2D Gabor filter can be represented as a complex sinusoidal signal modulated by Gaussian function as

$$g(x,y) = s(x,y) * w_r(x,y) . \tag{2}$$

where $s(x, y)$ is a complex sinusoid, known as the carrier, and $w_r(x, y)$ is a 2-D Gaussian-shaped function, known as the envelope [27].

The complex sinusoid is defined as

$$s(x,y) = \exp\big(j(2\pi(u_0 x + v_0 y) + P)\big) . \tag{3}$$

where (u_0, v_0) and P define the spatial frequency and the phase of the sinusoid respectively. The real part and the imaginary part of this sinusoid are

$$Re\,(s(x,y)) = \cos\,(2\pi(u_0 x + v_0 y) + P) . \tag{4}$$

$$Im\,(s(x,y)) = \sin\,(2\pi(u_0 x + v_0 y) + P) . \tag{5}$$

The parameters u0 and v0 define the spatial frequency of the sinusoid in Cartesian coordinates.

Gabor filters can be highly selective in both position and frequency, thus resulting in sharper texture boundary detection. Gabor filters related feature extraction
or segmentation paradigm is based on filter bank model in which several filters are
applied simultaneously to an input image. The filters focus on particular range of
frequencies [14]. Gabor values spectra for various orientations are considered as
features. This spectrum identifies the texture element more exactly.

3.2   Extraction of Local Features from Image Blocks

The block division method is a simple approach that relies on sub-images to address the spatial properties of images [11][12][13]. The original image matrix I of
size m x n is divided into sub-images (blocks) of size p x q and texture features are
extracted from each sub-image independently. In this work, the GLCM features,
LBP distributions, statistical features such as mean and standard deviation of each
block is calculated and the feature values are combined into a single vector
representing the image.

3.3   Classification and Decision Making

Classification includes a broad range of decision-theoretic approaches to the image recognition or identification. Classification algorithms analyze the various image features and organize data into categories. The feature vectors of the images
from the database are extracted using feature extraction algorithms and are stored
in the feature library. These vectors are used for classification. In the recognition
phase, preprocessing and feature extraction steps are applied on the query image.
Then the feature vector of the query image is matched with the feature vectors in
feature library.

For classification we have used similarity measure called Euclidean Distance.
Euclidean Distance measure is one of the simplest distance classifier. Let u denote
an image in ear image database F. T(u) denote the feature vector of the image u, v
denote a queried image, T(v) denote the feature vector of the image v. The Euclidean measure is defined as:

$$\text{Dist}(T(u), T(v)) = \sqrt{\sum (T(u(k)) - T(v(k)))^2}. \qquad (6)$$

The distance between feature vector of each image in the feature library and the
query image is calculated. The minimum of Dist (T (u), T (v)) is considered as a
match for the input query image.

4 Experiments and Results

4.1 Dataset

The ear images are taken from IIT Delhi ear database [10] .The ear images of 221 subjects with 3 images per person is chosen for the work. Fig 5 depicts a few samples of images from the dataset. The feature library is constructed by taking two ear images each of the 221 subjects. The rest (one per subject) is used for testing.

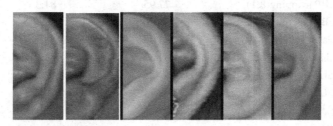

Fig. 5 Typical ear image samples

4.2 Feature Extraction

We have implemented the feature extraction methods GLCM, LBP and Gabor filter. In addition to the texture features we used shape[10], edge [28] and HoG [22] feature extraction methods for comparison.

Experiment 1

Each image of size 256 * 256 in the database is preprocessed using histogram equalization for adjusting image intensities. The Shape, Edge, HoG, GLCM, LBP and Gabor features are extracted from image blocks of size 8x8 and 16x16.

The distances to the edge pixels from a common point is evaluated to obtain the shape feature. The HoG technique counts occurrences of gradient orientation in localized portions of the image. For each block the gradients of the chosen region are evaluated. The returned features encode local shape information from regions within an image. For LBP, two different neighborhoods, (8, 1) and (16, 2) were employed. The recognition rate is obtained for each type of features in isolation and in combination of different types of features. The results obtained are summarized in Table 2.

As it is evident, image block size of 16 is found to give better recognition rate. Gabor filter is found to give highest accuracy when used in isolation. The highest recognition rate is obtained when the entire three prominent feature types are combined. This increased accuracy is at the cost of time, as the extraction of the three different types of features as well as distance calculation with large sized feature vectors is expensive.

Table 2 Comparison of performance of different features in isolation and in combination

Features Considered	Recognition Rate (%)	
	Image Block Size 8	Image Block Size 16
Shape	57.25	60.63
Edge	39.85	43.23
HoG	85.22	88.6
GLCM	88.48	91.86
LBP mapping(8,1)	77.95	81.3
LBP mapping(16,2)	84.26	87.64
Gabor Filter	87.13	90.51
Shape+Hog+Gabor	85.85	89.23
GLCM+Shape +HoG	89.23	92.61
GLCM + LBP(16,2) + Gabor Filter	90.74	94.12

Experiment 2

Here we compare the performance of feature extraction from the whole image and from sub images. Only GLCM, LBP and Gabor filter features are employed for the study. Based on experiment 1, we choose a block size of 16x16. The results obtained are given in Table 3. The results clearly show the advantage of block based feature extraction over whole image based.

Table 3 Comparison of Block based and Whole image based Feature Extraction

Features Considered	Recognition Rate(%)		
	Feature Extracted from Non Blocked Image	Feature Extracted from Blocked Image	Time for execution (Blocked Images)(in sec)
GLCM	58.13	91.86	48.5
LBP(8,1)	47.6	81.33	123.3
LBP(16,2)	53.91	87.64	32.2
Gabor Filter	56.78	90.51	97.12
GLCM + LBP(16,2) + Gabor Filter	60.39	94.12	199.42

4.3 Discussion

In this work, distance calculation for classification is carried out without considering the variations in the feature values. When features are combined, better results can be obtained with normalized feature vector. Feature reduction methods like PCA can also be applied on combined features. Improved results with block based images can be attributed to the fact that local features are more discriminating than global features of non - blocked images. In this work we have used a few number of samples as well as relatively small number of classes. Hence Euclidean distance measure is used for comparison. The performance of other popular classifiers like NN, SVM are to be investigated.

5 Conclusion

In this paper a novel approach for ear recognition is presented. In the proposed approach a combination of three texture features specifically Gabor filter, GLCM and LBP are employed. Also, the features are extracted from image blocks rather than from the whole image. The extraction of features from blocks is found to outperform that from whole images. Also, we found that a combination of the three features gave the best recognition rate. The combination of features results in more complex and time consuming recognition system. A detailed investigation incorporating simple features as well as employing feature selection methods is being carried out.

References

1. Burge, M.: Ear Biometrics for Computer Vision. In: Proc. 21st Workshop Austrian Assoc. for Pattern Recognition, 1997, pp. 275–282 (2000)
2. Burge, W.: Burger: Using Ear Biometrics for Passive Identification. Published by Chapman & Hall. IP (1998)
3. Jitendra, B., Anjali, S.: Ear Based Attendance Monitoring System. In: Proceedings of ICETECT (2011)
4. Meijerman, L., Thean, A., Maat, G.J.R.: Earprints In Forensic Investigations. Forensic Science, Medicine and Pathology 1(4), 247–256 (2005)
5. Guo, Y., Xu, Z.: Ear recognition using a new local matching approach. In: Image Processing, ICIP 2008, pp. 289–292 (2008)
6. Pug, A., Busch, C.: Ear Biometrics - A Survey of Detection, Feature Extraction and Recognition Methods. IET Biometrics Journal 1(2), 114–129 (2012) ISSN 2047-4938
7. Lu, L., Xiaoxun, Z., Youdong, Z., Yunde, J.: Ear Recognition Based on Statistical Shape Model. CICIC (3), 353–356 (2006)
8. Hurley, D.J., Nixon, M.S., Carter, J.N.: Force Field Feature Extraction for Ear Biometrics. Computer Vision and Image Understanding, 491–512 (June 2005)
9. Moreno, B., Sanchez, A., Velez, J.F.: On the use of outer ear images for personal identification in security applications. In: Proceedings of IEEE Conference on Security Technology, pp. 469–476 (1999)
10. Kumar, A., Wu, C.: Automated human identification using ear imaging. Pattern Recognition 45(3), 956–968
11. Takala, V., Ahonen, T., Pietikäinen, M.: Block-Based Methods for Image Retrieval Using Local Binary Patterns. In: Proc. 14th Scandinavian Conference, SCIA, pp. 882–891 (2005)

12. Wang, Y., Mu, Z.-C., Zeng, H.: Block-based and Multi-resolution Methods for Ear Recognition Using Wavelet Transform and Uniform Local Binary Patterns. In: Pattern Recognition, ICPR 2008, pp. 1–4 (2008)
13. Xiaoyun, W., Weiqi, Y.: Human Ear Recognition Based on Block Segmentation Cyber-Enabled Distributed Computing and Knowledge Discovery, pp. 262–266 (2009)
14. Jawale, S.: Wavelet Transform and Co-occurance matrix based texture features for CBIR. International Journal of Engineering Research and Applications (IJERA) ISSN: 2248-9622
15. Ahonen, T., Hadid, A., Pietikanen, M.: Face Description with Local Binary Patterns: Application to Face Recognition. IEEE, IEEE Transactions on Pattern Analysis And Machine Intelligence 28(12) (December 2006)
16. Haralick, R.M., Shanmugam, K., Dinstein, I.: Textural Features for Image Classification. IEEE Transactions on Systems, Man and Cybernetics SMC-3(6), 610–621 (1973)
17. Manjunath, B., Ma, W.: Texture features for browsing and retrieval of image data. IEEE Transactions on Pattern Analysis and Machine Intelligence 18(8), 837–842 (1996)
18. Mäenpää, T., Pietikäinen, M.: Texture Analysis with Local Binary Patterns. In: Chen, C.H., Wang, P.S.P. (eds.) Handbook of Pattern Recognition and Computer Vision, 3rd edn., pp. 197–216. World Scientific (2005)
19. Ross, A., Jain, A.K.: Multimodal Biometrics: an overview. In: Proceedings of the European Signal Processing Conference, pp. 1221–1224 (2004)
20. Cadavid, S., Mahoor, M.H., Abdel-Mottaleb, M.: Multi-modal Biometric Modeling and Recognition of the Human Face and Ear. Cambridge University Press (2011)
21. Yuan, L., Mu, Z.-C.: Multimodal recognition using ear and face. In: 5th International Conference on Wavelet Analysis and Pattern Recognition, Beijing (2007)
22. Deniz, O., Bueno, G., Salido, J., De la Torre, F.: Face Recognition using Histogram of Oriented Gradients. Pattern Recognition Letters 32(12) (September 1, 2011)
23. Maenpaa, T., Pietikainen, M.: Texture Analysis with local binary patterns
24. Hammouda, K., Jernigan, E.: Texture Segmentation Using Gabor Filters, University of Waterloo, Ontario, Canada
25. Choras, M.: Ear Biometrics based on Geometrical Feature Extraction. Electronic Letters on Computer Vision and Image Analysis 5(3), 84–95 (2005)
26. Ojala, T., Pietikainen, M., Maenpaa, T.: Multiresolution gray scale and rotation invariant texture analysis with local binary patterns. IEEE Transactions on Pattern Analysis and Machine Intelligence 24, 971–987 (2002)
27. Movellan, J.: Tutorial on Gabor Filters, Technical report, MPLab Tutorials, Univ. of California, San Diego (2005)
28. Shailaja, D., Gupta, P.: A Simple Geometric Approach for Ear Recognition. In: 9th International Conference on Information Technology (ICIT 2006) (2006)

Aggregate Metric Networks for Nearest Neighbour Classifiers

Dhanya Alex and Alex Pappachen James

Abstract. A common idea prevailing in distance or similarity measures is the use of aggregate operators in localised and point-wise differences or similarity calculation between two patterns. We test the impact of aggregate operations such as min, max, average, sum, product, sum of products and median on distance measures and similarity measures for nearest neighbour classification. The point-wise differences or similarities extends the idea of distance measurements from the decision space to feature space for the extraction of inter-feature dependencies in high dimensional patterns such as images. Inter-feature spatial differences are extracted using the gradient functions across various directions and then applied on aggregate function, to result in a fused feature set. The initial study is conducted on Iris flower and verified using AR face database. The resulting method shows an accuracy of 92% on face recognition task using the standard AR database.

1 Introduction

Similarity and distance calculations is fundamental to the design of pattern classification and retrieval techniques [10, 7, 16]. Similarity measurements find a wide variety of applications in the broad areas of psychology[16], taxonomy[15, 9] and molecular biology[6], while distance measurements often find applications in the fields of fuzzy set theory[12], trigonometry[7, 6] and graph theory[14]. The concept of similarity has found its way into various areas like anthropology, biology, chemistry, computer science, ecology, information theory, geology, mathematics, physics, psychology, and statistics [18, 15, 12, 14]. Similarity is used for pattern matching,

Dhanya Alex
Enview R&D labs
e-mail: dhanya.alex@gmail.com
www.enviewres.com

Alex Pappachen James
Nazarbayev University
e-mail: apj@ieee.org

S.M. Thampi, A. Gelbukh, and J. Mukhopadhyay (eds.), *Advances in Signal Processing and Intelligent Recognition Systems*, Advances in Intelligent Systems and Computing 264,
DOI: 10.1007/978-3-319-04960-1_2, © Springer International Publishing Switzerland 2014

classification, information retrieval, clustering, and decision making[18, 8]. One of the oldest and simplest ways of calculating similarity is by inversely relating it to the distance between the two objects[17]. The assumption here is that if two objects are similar or closely related, then the difference or distance between them will be smaller than two objects which are dissimilar.

Though the concept of similarity seems simple, it is quite difficult to measure its accuracy. Similarity can be considered as a numerical quantity that measures the strength of relationship between two objects or characteristics or features. A numerical value of 1 reflects perfect similarity between the two objects while, a numerical value of 0, reflects zero similarity. Similarity therefore is an indication of proximity between two objects. Dissimilarity or zero similarity indicates disorder between the two objects and can be measured through distance measures. Higher the distance between two objects, higher will be their dissimilarity. A large number of methods has been proposed to calculate similarity and distance measure between two objects under comparison. However, what method to use and whether there is any need for pre-processing the data depends on the purpose of the study.

The idea of similarity and distance measurements are interchangeably used in nearest neighbour classifiers with often limited importance given on the selection of measure. However, it is often observed that the selection of these measures is tightly linked with the performance of the classification. In addition, the approach to fuse the feature level distances and similarities to form global distances or similarities defines the global definition of these measures.

We propose the use of linear and non-linear aggregate functions for localised inter-feature dependency calculations for improving the performance of nearest neighbour classifiers. The system was implemented using both Euclidian distance measures and Shepherds similarity measure to understand the differences, if any, in the two approaches on the system performance.

2 Proposed Method

A multi-layered moving window aggregate operator for similarity/distance measure is the primary methodological advancement in this research. Figure 1 shows an illustrative example of the proposed similarity aggregator network. Consider N set of training samples and an equal number of non-overlapping test samples. For each training sample T_i and test sample G_i, pair-wise local distance/similarity measure is calculated to obtain the distance/similarity measures between the two pairs. The distance measures so obtained are then put through a smoothing filter operation using the local aggregator networks. The aggregator operation shown in Fig. 1 is representative of linear operators such as averages, median, sum, and product or non-linear operators such as min, max, sigmoid, and threshold. The aggregates so obtained at different levels is then again put through a global aggregator to obtain the distance/similarity measures.

The working of the system is illustrated through a simple classification problem. For testing the system during the initial phase, Iris[5, 4] dataset from UCI Machine

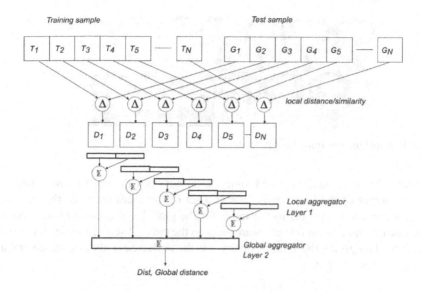

Fig. 1 Aggregate Similarity Network

Learning Repository[11] is used. Three classes (Fig 2), each referring to a type of type of iris plant forms the dataset.

(a) Iris Setosa   (b) Iris Versicolour (c) Iris Virginica

Fig. 2 Members of the Iris database

For classification purpose, the Iris data set is randomly divided into non- overlapping equal sized training and testing set. We get a total of 75 training samples, made up of 25 instances from each class, and the remaining 75 forms the test samples. In the first stage, local distance or similarity is calculated using Euclidian distance measure or Shepards distance based similarity measure respectively. The Euclidian distance $d(p,q) = \sum |p-q|$ is used for calculating distance , while the Shepherds similarity $s(p,q) = e^{d(p,q)}$ is used for similarity calculations.

These measures form the input for the aggregator stage of the proposed two layer aggregate network. Layer 1 of the proposed system referred to as local aggregator layer performs a smoothing operation. Simulations were run using a window size of 3 distance measures for each class. Both linear and non-linear operators are used to analyse the performance of the local aggregate network. Linear operators such as *averages*, *sum* and *product* obey the superposition conditions whereas non-linear

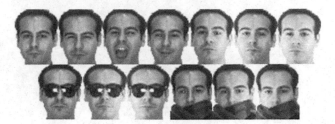

Fig. 3 Example images from the AR database

operators like *min*, *max*, threshold logic etc do not obey the superposition principle. The output of the local aggregator is a set of distances obtained through the smoothing method. This forms the input for the global aggregate network, where again linear or non-linear operators are used on the entire distance measures in each class. The class giving the shortest distance is then selected as the one most similar to the given test sample.

2.1 Example of Face Recognition

The absence of deformities or irregularities in iris database makes it an easier classification problem, making it necessary to test the proposed system on a high dimensional classification problem such as face recognition in images. One of the most widely recognized AR face database [13] is used for this purpose. The database consists of over 4000 face images in color of nearly 126 people. Each participant were asked to come in for two sessions separated by 2 weeks time. In each session 13 photos were taken under the following conditions: Neutral expression, Smile, Anger, Scream, left light on, right light on, all side lights on, with sun glasses, with scarf, with sun glasses and left light on, with sun glasses and right light on, with scarf and left light on, and with scarf and right light on.

2.1.1 Data Representation in the AR Image Set

The AR face database is given in Fig 3. The images are stored as RGB RAW files as pixel information. Each image are of size 768 x 576 pixel and 24 bits in depth. Male images are stored in the format of M-xx-yy.raw and female images are stored as F-xx-yy.raw. Here M and F correspond to male and female respectively. 'xx' is a unique person identifier and ranges from 00 to 70 for men and from 00 to 56 for women. 'yy' ranges from 00 to 26 and indicates the 13 features given above Numbers 00 to 13 corresponds to the first section and 14 to 26 corresponds to the second session taken 2 weeks later. Data in the AR face database are pixel information in RGB model[3, 2, 1].

2.1.2 Feature Extraction from Images

The images in the AR dataset are coloured images ranging in intensity values between 0 to 255. For classification purposes it is easier to bring the range within [0, 1], and for this we use feature extraction techniques. Normalization in image processing is the process of bringing the original pixel intensity value of images into the range specified by the user. For bringing the values into the desirable range, we first obtain a range of threshold values. The image is then extracted depending on whether the pixel information at each point is above or below this defined threshold value. This is achieved through the following equation:

$$\theta_n = n\frac{R}{N},\tag{1}$$

where n=[1 to N-1], R is the pixel intensity value and N is the total number of threshold points defined. For each theta value, we then obtain n different image vectors,such that each pixel value is defined by

$$x_i = \begin{cases} 1 & \forall x_i > \theta_i \\ 0 & \text{otherwise} \end{cases}\tag{2}$$

The image so obtained is again put through a feature extraction process. The gradient of the image across the horizontal and vertical directions is essentially a difference function represented as:

$$\Delta f = \frac{\delta f}{\delta x}\hat{x} + \frac{\delta f}{\delta y}\hat{y}\tag{3}$$

where: $\frac{\delta f}{\delta x}$ is the gradient in the x direction, and $\frac{\delta f}{\delta y}$ is the gradient in the y direction of the image plane. The gradients are quantised and added together to form the gradient threshold feature vector. An example of the gradient threshold features for 13 images from AR database is shown in Fig. 4.

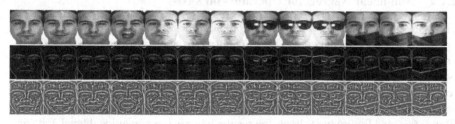

Fig. 4 The images show the original images, gradient threshold images and spatial change images from the AR database

The magnitude of gradients in the images can vary from first order functions as in Eq. (3) to second or third order. This necessitate that the magnitude be normalised using any normalisation function. We apply local window mean sigma normalisation as:

$$I_{norm} = \frac{|\Delta f - \mu \Delta f|}{3\sigma \Delta f} + 0.5 \tag{4}$$

The example results of applying the spatial change detection over the gradient threshold feature vector is shown in Fig 4.

2.1.3 Linear Aggregator Operators on k-NN

Figure 5 shows the result of using different linear operators at local and global aggregate levels of the network. These operators are used with both distance and similarity measures to understand if there is any difference in operation of the two concepts of difference and similarity. Figure 5(a) and Figure 5(b) shows the recognition accuracies when using sum operators at local and global aggregator levels. As can be seen, a higher recognition rate is obtained when using distance measure as opposed to similarity measure. However, the similarity measure provides for a higher instances of k values (between 6 and 8 gives optimum results) values making it to perform better in noisy environments as opposed to a k value of only 7 in difference measure. Figure 5(c) and Figure 5(d) shows the results when using mean at local level and sum at global aggregator level. Here also, a higher recognition rate is obtained when using distance measure as opposed to similarity measure. However, the similarity measure offer a higher number of k values making it more suitable in noisy environments. Figure 5(e) and Figure 5(f) shows the results when using product measure at local level and sum at global aggregator level. Here, difference measure results in a slightly better recognition rates and similarity measure still provides for larger window size in k value. Figure 5(g) and Figure 5(h) shows the results when using product measure at local level and mean at global aggregator level. The change in operator from sigma to mean has not provided for any changes in the results.

2.1.4 Non-linear Aggregator Operators on k-NN

Figure 6 shows the result of using different non-linear operators at local and global aggregate levels of the network. These operators are again repeated for both distance and similarity measures to understand if there is any difference in operation of the two concepts of difference and similarity. Figure 6(a) and Figure 6(b) shows the results when using *min* operators at local and global aggregator levels. As can be seen, there is no difference in performance with distance or similarity measures. Figure 6(c) and Figure 6(d) shows the results when using *min* at local level and *sum* at global aggregator level. Here also, no difference in observed in performance with distance or similarity measures. Figure 6(e) and Figure 6(f) shows the results when using *min* measure at local level followed by *mean* at global aggregator level. In this case however, the distance measure clearly outperforms the similarity measure both in terms of recognition accuracy, in the size and range of k values. Figure 6(g) and Figure 6(h) shows the results when using *sum* measure at local level and *min* at global aggregator level. Here the operators in the second case is reversed and as can be seen from the figures there is a change in performance when using distance and similarity measures. In this case also, the distance measure clearly outperforms the similarity measure both in terms of recognition accuracy, in the size and range of k values. The results suggest that when using non-linear operator *min* in the aggregator levels, the distance measures are found to work better, whereas when we use linear operators at the aggregator levels similarity measures provide for higher stability.

2.1.5 Results

AR database is used in our experiments for face recognition. In order to perform the face recognition experiments, for each class, the 13 images in the session 1 of the AR database is used as training while the 13 images recorded in the second session is used for testing. Images are converted to the feature vectors using the feature extraction process followed by using a localised binary aggregate distance function. The number of correctly matched test images to that in the training images reflect the recognition accuracy. Overall, using the feature extraction followed by the agrregator nearest neighbour network, we achived a recognition accuracy of 92%, in contrast with global distance nearest neighbour classifier with raw features resulted in 70% recognition accuracy.

3   Conclusion and Future Work

We proposed a new approach for calculating similarity between two subjects or patterns by looking at the localised differences and calculating the global similarity using the local aggregates. The designed classifier works in three layers. First the differences or similarities between the two subjects are calculated. This is followed by calculation of the localised differences using a simple moving average technique. This step is followed by a calculation of the global similarity of the test subject with the existing subject group. The aggregator network was successfully used for classification of the samples in the iris dataset. The subject sample in this case was of quantitative nature with no much irregularities or noise interference. The method was able to give a slightly better recognition accuracy of 92% as compared with the 91.6% accuracy obtained by applying k-nearest neighbour method directly. The method was also successfully demonstrated to solve complex pattern recognition problems on 2-D image problems through the testing of the AR face database. The method generated an accuracy of 92% on the AR database as opposed to 70% achieved with using the global distance nearest neighbour classifier. We also establish that distance and similarity measures generate different outcomes and that the similarity measures was observed to work better than distance measure when used with linear operators. The proposed system can be used for solving datasets with

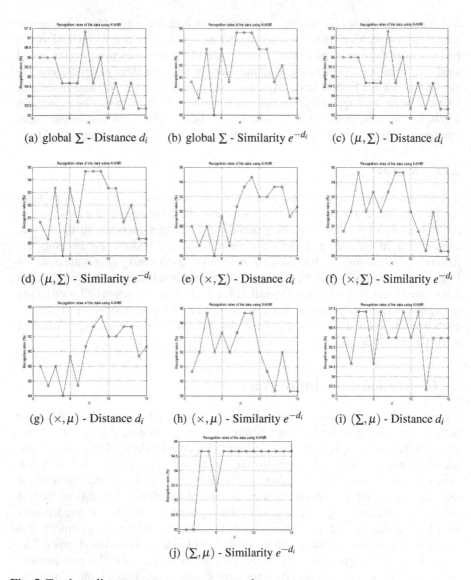

(a) global Σ - Distance d_i (b) global Σ - Similarity e^{-d_i} (c) (μ, Σ) - Distance d_i

(d) (μ, Σ) - Similarity e^{-d_i} (e) $(\times, \Sigma)$ - Distance d_i (f) $(\times, \Sigma)$ - Similarity e^{-d_i}

(g) $(\times, \mu)$ - Distance d_i (h) $(\times, \mu)$ - Similarity e^{-d_i} (i) (Σ, μ) - Distance d_i

(j) (Σ, μ) - Similarity e^{-d_i}

Fig. 5 Two layer linear aggregate operator networks

nominal or ordinal feature types. In addition to applications such as identification and classification, the proposed method can also be used in clustering problems. It is most suitable for situations where the data is sparse and the conventional classification methods fail to work. The ability to make decisions at local level, which is one of main attribute of the proposed similarity classifier, can be used in large data analysis and can be extended to cloud based solutions. The ability of the system to process the system in parallel makes the proposed system practically realisable in

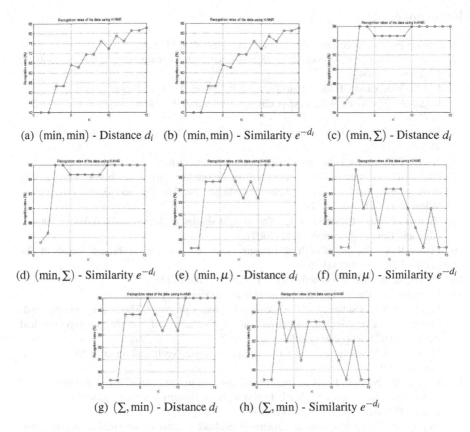

(a) $(\min, \min)$ - Distance d_i (b) $(\min, \min)$ - Similarity e^{-d_i} (c) $(\min, \Sigma)$ - Distance d_i

(d) $(\min, \Sigma)$ - Similarity e^{-d_i} (e) $(\min, \mu)$ - Distance d_i (f) $(\min, \mu)$ - Similarity e^{-d_i}

(g) $(\Sigma, \min)$ - Distance d_i (h) $(\Sigma, \min)$ - Similarity e^{-d_i}

Fig. 6 Two layer non-linear aggregate operator networks

real-time hardware. The system also needs some degree of normalization to be done to convert the data into a form that can be applied on the system. The alignment of the features is important to the proper comparison of the features. For example, in the face recognition task the alignment of the spatial features between the images are essential to make a factual comparison between the images. In the case of problems that involve missing data, the alignment can be challenging as the feature normalisation becomes error prone, and cumulative to the similarity calculation. The classification system was tested only on conditions where deformities and occlusions where at a minimum. The classification system needs to be tested on real-time images where the amount of noise will be much higher and where images might not be very clear or precise. Future works could include studying the impact of outliers on the system. The classification system can also be tested on other databases for assessing the stability of the system.

References

1. Digital Picture Processing. Academic Press, New York (1982)
2. The Reproduction of Colour. WileyIS and T Series in Imaging Science and Technology, Chichester (2004)
3. Digital Video and HDTV: Algorithms and Interfaces, p. 203. Morgan Kaufmann
4. Edgar, A.: The irises of the gasp peninsula. Bulletin of the American Iris Society 59, 2–5 (1935)
5. Edgar, A.: The species problem in iris. Annals of the Missouri Botanical Garden 23(3), 457–509 (1936)
6. Ashby, F.G., Ennis, D.M.: Similarity measures. Scholarpedia 2(12), 4116 (2007)
7. Cha, S.-H.: Comprehensive survey on distance/similarity measures between probability density functions. International Journal of Mathematical Models and Methods in Applied Sciences 1, 300–307 (2007)
8. Vitanyi., P.M.B., Cilibrasi., R.L.: The google similarity distance. IEEE Trans. Knowledge and Data Engineering 19(3) (2007)
9. Fisher, R.A.: The use of multiple measurements in taxonomic problems. Annals of Eugenics 7(2), 179–188 (1936)
10. Hamilton, W.: Websters dictionary @ONLINE (June 1913)
11. Lichman, M., Bache, K.: UCI machine learning repository (2013)
12. Landauer, S.T., Dumais, T.K.: A solution to plato's problem: The latent semantic analysis theory of acquisition, induction, and representation of knowledge. Psychological Review 104(2), 211–240 (1997)
13. Martinez, A.M., Benavente, R.: The ar face database. Technical report, CVC Technical Report (1998)
14. Lapata, M., Navigli, R.: An experimental study of graph connectivity for unsupervised word sense disambiguation. IEEE Transactions on Pattern Analysis and Machine Intelligence (TPAMI) 32(4), 678–692 (2010)
15. Resnik, P.: Using information content to evaluate semantic similarity in a taxonomy. In: Proceedings of the 14th International Joint Conference on Artificial Intelligence, pp. 448–453 (1995)
16. Tappert, C.C., Choi, S., Cha, S.: A survey of binary similarity and distance measures. Journal of Systemics, Cybernetics and Informatics 8(1), 43–48 (2010)
17. Shepard, R.N.: Toward a universal law of generalization for psychological science. Science 237, 1317–1323 (1987)
18. Phillips, P.J., Rosenfeld, A., Zhao, W., Chellappa, R.: Face recognition: A literature survey. ACM Computing Surveys 35(4), 399–458 (2003)

A Comparative Study of Linear Discriminant and Linear Regression Based Methods for Expression Invariant Face Recognition

Nitin Kumar, R.K. Agrawal, and Ajay Jaiswal

Abstract. In the literature, the performance of Fisher's Linear Discriminant (FLD), Linear Regression (LR) and their variants is found to be satisfactory for face recognition under illumination variation. However, face recognition under expression variation is also a challenging problem and has received little attention. To determine suitable method for expression invariant face recognition, in this paper, we have investigated several methods which are variants of FLD or LR. Extensive experiments are performed on three publicly available datasets namely ORL, JAFFE and FEEDTUM with varying number of training images per person. The performance is evaluated in terms of average classification accuracy. Experimental results demonstrate superior performance of Enhanced FLD (EFLD) method in comparison to other methods on all the three datasets. Statistical ranking used for comparison of methods strengthen the empirical findings.

Keywords: scatter, expression, comparison, ranking.

1 Introduction

Face recognition has drawn significant attention from the research community in past few decades. It has several applications including biometrics, surveillance

Nitin Kumar
Department of Computer Science and Engineering,
National Institute of Technology, Uttarakhand - 246174, India
R.K. Agrawal
School of Computer & System Sciences,
Jawaharlal Nehru University, New Delhi -110067, India
Ajay Jaiswal
S. S. College of Business Studies,
University of Delhi, Vivek Vihar - 110095, India
{nitin2689,rkajnu}@gmail.com, a_ajayjaiswal@yahoo.com

S.M. Thampi, A. Gelbukh, and J. Mukhopadhyay (eds.), *Advances in Signal Processing and Intelligent Recognition Systems*, Advances in Intelligent Systems and Computing 264, DOI: 10.1007/978-3-319-04960-1_3, © Springer International Publishing Switzerland 2014

systems, security such as ATM, access control and so on [12]. Automatic face recognition is carried out with the help of digital facial images which pose certain challenges such as varying illumination, pose, expression and occlusion *etc.* [5]. Expression variation is a challenging problem in which local appearance of face image gets modified due to different facial expressions. These expressions are grouped into seven basic categories [6] *i.e.* happy, sad, angry, neutral, disgust, surprise and fear. Expression invariant face recognition has received little attention in the past as only few methods have been proposed in literature to address this challenge. Tsai and Jan [14] proposed benchmark subspace methods such as Principal Component Analysis (PCA) [15] and Fisher Linear Discriminant (FLD) [3, 7] for feature extraction. An and Ruan [1] suggested modified FLD method to address expression variation called Enhanced Fisher Linear Discriminant (EFLD). Recently, Heish and Lai [8] have proposed an approach which exploits constrained optical flow warping to solve the problem. In this method, facial feature points are marked manually which are vulnerable to error and is also time consuming. Hence, this approach is not suitable for automatic face recognition.

Zhang *et al.* [17] have proposed Exponential Discriminant Analysis (EDA) which employs matrix exponential for feature extraction. It has been shown that EDA outperforms popular techniques such as PCA, FLD and others on a variety of datasets under illumination variation. In all these methods, the facial images of the identities are transformed to another space and recognition is performed in the transformed space. In contrast, face recognition is also formulated as a Linear Regression (LR) problem in which facial image of a person is generated (called virtual views) with the help of training images. In this direction, Chai *et al.* [4] have proposed techniques called Global LR (GLR) and Local LR (LLR) for pose invariant face recognition. The identity of a person is classified according to the best approximation of the test image by the available training images of a person. Similar to this method, Naseem *et al.* [10] have proposed Linear Regression Classification (LRC) algorithm to address illumination, expression and pose. This algorithm is evaluated on various datasets and outperforms several other methods.

FLD, LR and their variants have shown better performance in comparison to other methods across challenges such as variation in illumination and pose. Further, these approaches do not involve manual intervention which is essential for automatic face recognition. To the best of our knowledge, no comparative research work has been done to determine suitable method for expression invariant face recognition. To investigate this, in this paper, we have performed extensive experiments on three publicly available datasets with varying number of training images per person. The rest of the paper is organized as follows: Sect. 2 provides an overview of the popular FLD and its variants methods. Linear Regression and its variants are briefly described in Sect. 3. Experimental set up and results are presented in Sect. 4 while some concluding remarks and future work are given in Sect. 5.

2   Fisher's Linear Discriminant and its Variants

Given training data, $\mathbf{X} = [\mathbf{x}_1, \mathbf{x}_2, ..., \mathbf{x}_N]$ where $\mathbf{x}_i \in \Re^n$ (i = 1, 2, ..., N) represents the image of a person as a column vector and N is the total number of images. Let N_i denotes the number of images in class i (i = 1, 2,..., C) such that $N = \sum N_i$. FLD based methods involve between-Class ($\mathbf{S_b}$) and within-Class scatter ($\mathbf{S_w}$) which are defined as:

$$\mathbf{S_b} = \sum_{i=1}^{C} \mathbf{N_i}(\mathbf{m_i} - \mathbf{m})(\mathbf{m_i} - \mathbf{m})^T, \tag{1}$$

$$\mathbf{S_w} = \sum_{i=1}^{C} \sum_{\mathbf{x_j} \in c_i} (\mathbf{x_j} - \mathbf{m_i})(\mathbf{x_j} - \mathbf{m_i})^T \tag{2}$$

where $\mathbf{m} = (1/N) \sum_{i=1}^{N} \mathbf{x}_i$ is the mean of the dataset and $\mathbf{m}_i$ is the mean of class i. The objective of FLD is to find the direction in which between-Class scatter ($\mathbf{S_b}$) is maximized and within-Class scatter ($\mathbf{S_w}$) is minimized as given in (3).

$$\mathbf{J}_{fld} = \frac{\arg\max}{\mathbf{W}} \frac{\left| \mathbf{W}^T \mathbf{S_b} \mathbf{W} \right|}{\left| \mathbf{W}^T \mathbf{S_w} \mathbf{W} \right|} \tag{3}$$

But in practice, due to few available samples and high dimensionality, $\mathbf{S_w}$ becomes singular. To overcome this limitation, some variants of FLD are proposed in literature as discussed next.

Fisherfaces

Fisherfaces [3] is a popular variant of FLD method. In this method, first the dimensionality of the data samples is reduced using PCA and then FLD criterion is applied in the reduced dimension for feature extraction. Thus the optimal transformation $\mathbf{W}_{opt}$ in Fisherfaces is a combination of two transformations i.e. $\mathbf{W}_{fld}$ and $\mathbf{W}_{pca}$ as given below:

$$\mathbf{W}_{opt} = \mathbf{W}_{pca} \mathbf{W}_{fld} \tag{4}$$

where $\mathbf{W}_{pca} = \frac{\arg\max}{\mathbf{W}} \left| \mathbf{W}^T \mathbf{S}_t \mathbf{W} \right|$ and $\mathbf{W}_{fld}$ is found as given in (3).

Here, $\mathbf{S_t}$ represents the total-Scatter such that $\mathbf{S_t} = \mathbf{S_b} + \mathbf{S_w}$.

Enhanced Fisher Linear Discriminant (EFLD)

In the original formulation of FLD, all the features for S_b and S_w were treated equally. However, the features possess different scales and importance which are useful for classification. Keeping this in mind, An and Ruan [1] proposed Enhanced FLD (EFLD) method in which the scale of features Φ_j ($j = 1, 2,\ldots, n$) was used to find their importance and is defined as:

$$\Phi_j = \sqrt{\frac{1}{N-1}\sum_{i=1}^{N-1}(\mathbf{x}_i^{(j)} - \boldsymbol{\mu}^{(j)})^2} \tag{5}$$

Here $\boldsymbol{\mu}^{(j)}$ is mean of feature j. Thus, Fisher's criterion as given in (3) is modified as:

$$\mathbf{J}_{efld} = \underset{\mathbf{W}}{\arg\max}\left|\frac{\mathbf{W}^T\boldsymbol{\Lambda}^{-\alpha}\mathbf{S}_b(\boldsymbol{\Lambda}^{-\alpha})^T\mathbf{W}}{\mathbf{W}^T\boldsymbol{\Lambda}^{-\alpha}\mathbf{S}_w(\boldsymbol{\Lambda}^{-\alpha})^T\mathbf{W}}\right| \tag{6}$$

where $\boldsymbol{\Lambda} = \mathrm{diag}(\Phi_1,\Phi_2,\ldots,\Phi_n)$ and α is a scale parameter which controls the effect of features and ranges between $(0,\infty)$. S_w being singular due to limited samples, the optimal transformation is found by first reducing the dimension using Enhanced PCA (EPCA) [1] and then applying EFLD as follows:

$$\mathbf{W}_{opt} = \mathbf{W}_{epca}\mathbf{W}_{efld} \tag{7}$$

where $\mathbf{W}_{epca} = \underset{\mathbf{W}}{\arg\max}\left|\mathbf{W}^T\boldsymbol{\Lambda}^{-\alpha}\mathbf{S}_t(\boldsymbol{\Lambda}^{-\alpha})^T\mathbf{W}\right|$ and

$$\mathbf{W}_{efld} = \underset{\mathbf{W}}{\arg\max}\left|\frac{\mathbf{W}^T\boldsymbol{\Lambda}^{-\alpha}\mathbf{S}_b(\boldsymbol{\Lambda}^{-\alpha})^T\mathbf{W}}{\mathbf{W}^T\boldsymbol{\Lambda}^{-\alpha}\mathbf{S}_w(\boldsymbol{\Lambda}^{-\alpha})^T\mathbf{W}}\right|$$

Exponential Discriminant Analysis (EDA)

Zhang et al. [17] have proposed EDA to address the singularity problem of FLD. In contrast to FLD which is based on second-Order mixed central moment, this technique is able to capture a linear combination of the mixed central moments including second-order mixed central moments. This helps in more useful representation of data resulting in better classification. In EDA, the exponential of matrix is used to transform the input data into a new space and FLD criterion is applied as given below [17]:

$$J_{eda} = \frac{\arg\max}{W} \left| \frac{W^T \exp(S_b)W}{W^T \exp(S_w)W} \right| \tag{8}$$

3 Linear Regression and Its Variants

Chai *et al.* [4] proposed face recognition as regression problem for pose invariant face recognition. In this approach, the training images of a person are used to generate virtual views and the identity is classified according to minimum reconstruction error. LR can be applied on the whole facial image called Global LR (GLR) or on smaller image patches called Local LR (LLR). In LLR, the image is represented in terms of patches which can be without overlap (LLRWO) or with overlap (LLRO). The regression is performed separately on individual patches and the virtual views are constructed by combining the non-overlapping/overlapping patches. Recently, Naseem *et al.* [10] suggested using LR for face recognition across illumination, expression and occlusion. The outline of the proposed algorithm (LRC) is reproduced below [10]:

Algorithm: Linear Regression Classification (LRC)
Inputs: Class models $X_i \in \mathfrak{R}^{n \times N_i}$; $i = 1, 2, \ldots, C$ and a test image $y \in \mathfrak{R}^n$
Output: Class label of y

Step 1: $\hat{\delta}_i \in \mathfrak{R}^n$ is computed for each class model as $\hat{\delta}_i = (X_i^T X_i)X_i^T y$

Step 2: $\hat{y}_i$ is computed for each $\hat{\delta}_i$ as $\hat{y}_i = X_i \hat{\delta}_i$; $i = 1, 2, \ldots, C$

Step 3: Compute distance between original and predicted response variables $d_i(y) = \|y - \hat{y}_i\|$, $i = 1, 2, \ldots, C$

Step 4: Decision is made in favour of the class with minimum distance $d_i(y)$.

Ridge Regression
The regression coefficients in step 2 of LRC algorithm are found as a solution to the ordinary least squares method. But due to availability of few samples, these coefficients are not reliable. To better estimate these coefficients, Ridge Regression is proposed by An *et al.* [2]. In this technique, a penalizing term (λ) is added in the coefficient estimation equation as given below:

$$\hat{\delta}_i = (X_i^T X_i + \lambda I)X_i^T y \tag{9}$$

The optimum value of the parameter λ is determined by cross validation on the face dataset.

Robust Regression

Robust Regression [11] plays an important role when the issues of contamination of test images by variations such as expression variation in facial images of different people are present. In this technique, the coefficients are determined by the following equation [11]:

$$\hat{\delta}_i = \underset{\hat{\delta}_i \in \Re^{N_i}}{\arg\min} \{F(\hat{\delta}) \equiv \sum_{j=1}^{n} \rho(r_j(\hat{\delta}_i))\}, i = 1,2,...,C \qquad (10)$$

where $r_j(\hat{\delta}_i)$ is the j-th component of the residual $i.e.$ $r_j(\hat{\delta}_i) = \mathbf{y} - \mathbf{X}_i\hat{\delta}_i$, $i = 1, 2,$ ..., C and ρ is a symmetric function having minimum at zero. Once the regression coefficients are determined, the rest of the method follows similar procedure as LRC.

4　Experimental Setup and Results

To investigate the performance of various methods based on FLD and LR approaches for expression invariant face recognition, we have conducted extensive experiments on three publicly available datasets $i.e.$ ORL [13], JAFFE [9] and FEEDTUM [16]. The details of the datasets are summarized in Table 1. JAFFE and FEEDTUM datasets contain all facial expressions as mentioned in Sect. 1 while ORL have most but not all expressions. The images of JAFFE dataset were manually cropped. Further, images of all the datasets were resized to 50×50. Sample images from all the datasets are shown in Fig. 1. The experiments on all the datasets are performed by varying the number of images per person from three to nine as given in Tables 2-4. Each dataset is randomly partitioned into training set and testing set. The model is learnt from the training set and classification accuracy is obtained using a testing set with nearest neighbor classifier. This process is repeated ten times for each dataset and their average is computed. In LLRWO, face image is divided into 4 patches of size 25×25 each and in EFLD method; the optimal value of parameter α is computed from the set [0.05, 0.6].

Table 1 Summary of datasets used in Experiments

Dataset	Number of Identities	Images per Person	Total Images
ORL	40	10	400
JAFFE	10	21(Approx.)	213
FEEDTUM	18	21	378

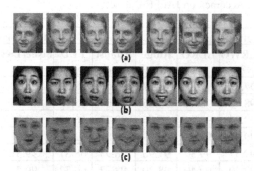

Fig. 1 Sample faces from datasets (a) ORL (b) JAFFE and (c) FEEDTUM

The classification accuracy on ORL dataset is given in Table 2. It is observed that with the increase in number of training images per person, the classification accuracy of all the methods increase except GLR, FLD and EFLD. While EDA gives worst performance, EFLD clearly outperforms other techniques over different training images per person with one exception. The experimental results on JAFFE dataset are given in Table 3. The best performance is again obtained with EFLD. The performance of EDA increases consistently while this increase is inconsistent for all other methods. Further, EFLD is able to achieve 100 percent classification accuracy with 9 training images per person. The classification accuracy on FEEDTUM dataset is given in Table 4. The classification accuracy of all the methods increases with increase in the number of training images except FLD, Fisherfaces and EFLD. Again, EFLD reports the best performance with two exceptions (in row 4 and 7 where LLRWO performs best). It is also observed that the performance of LR based methods and EDA increases with increase in training data. The performance of various methods is compared based on the average percentage increase (P_j; j =1,2, ...,9) in classification accuracy with respect to the method with worst performance. For this purpose we define a variable p_{ij}^k which denotes the percentage increase in classification of k-th datasets with i-th instance and j-th method as given below:

Table 2 Classification Accuracy on ORL Dataset

Training	Regression based methods					FLD based methods			
	GLR	LLRWO	LLRO	Ridge Reg.	Robust Reg.	FLD	Fisherfaces	EFLD	EDA
3	88.93	89.36	87.68	89.93	86.14	31.75	88.75	**93.57**	73.93
4	92.38	92.13	90.25	93.46	90.75	32.88	89.75	**95.50**	79.63
5	94.85	94.05	92.70	95.35	92.95	30.90	91.70	**96.80**	84.30
6	95.31	94.50	93.75	96.06	93.94	31.25	92.38	**96.69**	86.81
7	97.42	96.58	96.58	97.75	96.33	28.58	94.17	**97.92**	88.83
8	97.25	97.13	96.88	**98.38**	97.50	29.75	94.88	98.25	90.50
9	97.75	98.25	98.25	**98.75**	**98.75**	30.25	95.75	98.00	94.75

Table 3 Classification Accuracy on JAFFE Dataset

Training	Regression based methods					FLD based methods			
	GLR	LLRWO	LLRO	Ridge Reg.	Robust Reg.	FLD	Fisherfaces	EFLD	EDA
3	97.38	97.32	97.32	93.17	96.61	58.31	98.20	**98.25**	92.90
4	98.50	98.38	98.32	95.66	98.04	65.55	98.96	**99.19**	97.51
5	99.08	98.77	98.90	96.75	98.77	66.87	99.75	**99.82**	98.40
6	99.15	99.02	99.02	96.99	98.89	59.48	99.61	**99.74**	98.69
7	99.30	99.37	99.30	97.90	98.95	68.81	99.51	**99.72**	99.30
8	99.25	99.40	99.40	98.20	98.87	66.32	99.77	**99.85**	99.70
9	99.35	99.35	99.51	98.54	99.11	66.26	99.92	**100**	99.84

Table 4 Classification Accuracy on FEEDTUM Dataset

Training	Regression based methods					FLD based methods			
	GLR	LLRWO	LLRO	Ridge Reg.	Robust Reg.	FLD	Fisherfaces	EFLD	EDA
3	97.84	97.19	96.67	97.01	96.64	60.31	98.06	**98.61**	96.02
4	98.37	98.33	98.04	97.45	97.39	64.90	98.66	**98.89**	97.35
5	98.51	98.54	98.23	97.57	97.29	64.27	98.65	**98.92**	97.71
6	98.70	**98.96**	98.59	97.78	97.30	65.07	98.74	98.83	98.26
7	98.81	99.05	98.93	97.98	97.90	62.30	99.09	**99.33**	98.37
8	99.06	99.15	99.02	97.99	97.86	62.14	99.10	**99.27**	98.38
9	99.12	**99.17**	99.03	98.38	98.29	63.84	99.12	99.12	98.43

Table 5 Ranking of the methods

Method	GLR	LLRWO	LLRO	Ridge Reg.	Robust Reg.	FLD	Fisherfaces	EFLD	EDA
P_i	106.28	105.95	104.93	105.58	104.35	0.00	104.06	108.78	95.38
Rank	2	3	5	4	6	9	7	1	8

$$\mathbf{P}_j = \frac{1}{n_k \times n_i} \sum_{k=1}^{n_k} \sum_{i=1}^{n_i} p_{ij}^k \qquad (11)$$

where n_k is the number of datasets and n_i is the number of instances compared (n_k = 3 and n_i = 7 in our experiments). The methods investigated here are ranked based on the value of $\mathbf{P}_j$ ($j = 1,2,\ldots,9$) where rank 1 is given to the method which performs the best among all the method and rank 9 is given to the method performing worst as given in Table 5.

We observe the following from Tables 2-5:

1. EFLD performs best among all the methods on all the datasets.
2. The performance of various methods in general increases or remains same with the increase in number of training images per person.
3. There is less variation in classification accuracy among LR, Fisherfaces and their variants on JAFFE and FEEDTUM datasets.
4. The performance of FLD is worst for all the datasets.
5. The variation in performance of LR based methods is less in comparison to FLD based methods.

5 Conclusion

In literature, FLD, LR and their many variants are proposed for face recognition under illumination and pose variations. In this paper, we investigated the performance of these methods for expression invariant face recognition. Extensive experiments on three well known publicly available face datasets show that EFLD performed best on all the datasets while FLD performed worst. The performance of all the methods in general increases or remains the same with the increase in number of training images. The variation in performance of LR based methods is less in comparison to FLD based methods. In future work, we will explore nonlinear methods in contrast to the linear methods to determine their suitability for face recognition under expression variation.

References

1. An, G.Y., Ruan, Q.Q.: A novel mathematical model for enhanced Fishers linear discriminant and its application to face recognition. In: The 18th Int. Conf. on Patt. Rec (ICPR 2006), pp. 524–527 (2006)
2. An, S., Liu, W., Venkatesh, S.: Face Recognition Using Kernel Ridge Regression. In: IEEE Conference on Computer Vision and Pattern Recognition, CVPR 2007, pp. 1–7 (2007)
3. Belhumeur, P.N., Hespanha, J.P., Kriegman, D.J.: Eigenfaces vs. Fisherfaces: recognition using class specific linear projection. IEEE Trans. Pattern Anal. Mach. Intell. 19, 711–720 (1997)
4. Chai, X., Shan, S., Chen, X., Gao, W.: Locally linear regression for pose invariant face recognition. IEEE Trans. Image Proc. 16, 1716–1725 (2007)
5. Cordiner, A.: Illumination invariant face detection. University of Wollongong, MComSc thesis (2009)
6. Eisert, P., Girod, B.: Analysing facial expressions for virtual conferencing. IEEE Trans. Comp. Graph. and Appl. 18, 70–78 (1998)
7. Fisher, R.A.: The use of multiple measures in taxonomic problems. Ann. Eugenics 7, 179–188 (1936)
8. Hsieh, C.K., Lai, S.H.: Expression invariant face recognition with constrained optical flow warping. IEEE Trans. Multimedia 11, 600–610 (2009)

 9. Lyons, M.L., Akamatsu, S., Kamachi, M., Gyoba, J.: Coding facial expressions with gabor wavelets. In: IEEE Int. Conf. Auto. Face and Gest. Rec., Japan, pp. 200–205 (1998)
10. Naseem, I., Togneri, I., Bennamoun, M.: Linear regression for face recognition. IEEE Tran. Pattern Anal. Machine Intell. 32, 2106–2112 (2010)
11. Naseem, I., Togneri, I., Bennamoun, M.: Robust regression for face recognition. Pattern Recognition 45, 104–118 (2012)
12. Prabhakar, S., Pankanti, S., Jain, A.K.: Biometric recognition: security and privacy concerns. IEEE Security and Privacy 1, 33–42 (2003)
13. Samaria, F., Harter, A.: Parameterization of a stochastic model for human face identification. In: Proc. 2nd IEEE Workshop on Appl. of Comp. Vision, pp. 138–142 (1994)
14. Tsai, P.H., Jan, T.: Expression-invariant face recognition system using subspace model analysis. IEEE Int. Conf. on Systems, Man and Cyber. 2, 1712–1717 (2005)
15. Turk, M., Pentland, A.: Eigenfaces for recognition. J. Cogn. Neuro. 3, 71–86 (1991)
16. Wallhoff, F.: Facial expressions and emotion database. Technische Universitt Mnchen (2006), http://www.mmk.ei.tum.de/waf/fgnet/feedtum.html
17. Zhang, T., Fang, B., Tang, Y.Y., Shang, Z., Xu, B.: Generalized discriminant analysis: a matrix exponential approach. IEEE Trans. Systems, Man, and Cybernetics, Part B: Cybernetics 40, 186–197 (2010)

Real-Time Video Based Face Identification Using Fast Sparse Representation and Incremental Learning

Selvakumar Karuppusamy and Jovitha Jerome

Abstract. Video based face identification is a challenging problem as it needs to learn a robust model to account for face appearance change caused by pose, scale, expression and illumination variations. This paper proposes a novel video based face identification framework by combining fast sparse representation and incremental learning. For robust face identification, we proposed class specific subspace model based sparse representation which gives dense target coefficients and sparse error coefficients. Each subspace model is learned by using Principal Component Analysis (PCA). The test face is identified by using residual errors obtained for each model. For video based face identification, to harness the temporal information we integrated incremental face learning with our proposed face identification algorithm. By using reconstruction error and sparse error coefficients we have formulated new decision rules using rejection ratio and occlusion ratio respectively which helps in effective subspace model update in online. Numerous experiments using static and video datasets show that our method performs efficiently.

1 Introduction

Video based face recognition in non controlled environment is very crucial in public security, human machine interaction and entertainment applications. The specific problems and recent developments involved in video based face recognition can be found in Chellappa et al., (2012), Barr et al., (2012). The conventional still based frontal to half profile face recognition algorithms like Eigenfaces and Fisherfaces with nearest neighbor (Turk and Pentland, 1991 and Belhumeur et al., 1997) and Support Vector Machine (SVM) (Bernd et al., 2003) perform well in non controlled settings. But they do not have inbuilt mechanisms to handle large

Selvakumar Karuppusamy · Jovitha Jerome
PSG College of Technology, India
e-mail: {kskumareee,jovithajerome}@gmail.com

S.M. Thampi, A. Gelbukh, and J. Mukhopadhyay (eds.), *Advances in Signal Processing and Intelligent Recognition Systems*, Advances in Intelligent Systems and Computing 264, DOI: 10.1007/978-3-319-04960-1_4, © Springer International Publishing Switzerland 2014

contiguous occlusions like glasses, scarf etc., which are commonly encountered in videos. They also require large number of training images for better recognition performance which is not possible for real world applications. In recent face recognition developments, linear regression based classification (LRC) (Naseem et al., 2010) and sparse representation based classification (SRC) (Wright et al., 2009) are considered as most representative methods. In LRC, the test image is represented using class specific linear subspace models by least squares estimation method. The face is identified based on distance between test image and reconstructed image. To handle contiguous occlusion, partition scheme based LRC is proposed in Naseem et al., (2010), which divides test image into multiple non-overlapping parts. However, if the contiguous occlusion appears on many parts, this method fails. On other hand, SRC based method is proven to be very robust against noise and occlusion. Entire theory on sparse representation and learning for vision applications can be found in Cheng et al., (2013). Along with class specific linear models, this method uses large trivial template matrix to code occlusion and noises at the cost of huge computation.

To address the poses variation issue in videos, Chen et al., (2012) proposed joint sparse representation method using pose specific sub dictionaries for each class. However, learning too many poses is not possible in training stage and also that will increases the computational complexity. Very recently, Ptucha et al., (2013) proposed manifold learning based sparse representation to handle nonlinear appearance change. But all the above said SRC methods work fast only in low dimension feature vectors (e.g., 12 x 15 patches in Mei and Ling, (2009)) which may not capture the discriminating details in high resolution and large size face images in videos. Specifically, when video captured in long shot (i.e. working distance is more than 5 m), it contains very small faces which cannot be down sampled further to get low dimension feature vector.

Contributions of this paper are summarized as follows. First, we have proposed real-time face identification algorithm using class specific linear subspaces learned through PCA and l_1 regularized sparse representation method (Wang et al., 2013). This method uses simple least squares estimation for face representation and imposes l_1 regularization to obtain sparse noise term. The dense face coefficients are used to reconstruct face using corresponding subspace models. Second, the sparse noise coefficients are effectively used to reconstruct occlusion map which is used to handle large contiguous block occlusion. The class of the test face is identified based on minimum residual error obtained using test face, reconstructed face and occlusion map. Third, to the reject invalid faces detected in video, residual errors are utilized to formulate rejection rule. To avoid the occluded faces used for subspace model learning, we have constructed occlusion ratio by using occlusion map. After rejection of invalid and occluded faces, only genuine faces are used to update the class specific models. Finally, in video based face identification algorithm, we have integrated Singular Value Decomposition (SVD) based incremental learning approach is used to update the models in online.

The rest of this paper is organized as follows: Section 2 explains class specific appearance modelling using PCA and fast l_1 regularization for sparse representation. It also gives the proposed face identification algorithm. In section 3, comprehensive video based face identification using incremental learning approach is explained. The formulation of decision rules are also given in this section. Detailed experimental analysis is given in section 4. This paper is concluded in section 5.

2   Appearance Modeling Framework

In most of the face identification approaches, it is assumed that the faces belong to one subject will lie on a linear subspace (Chen et al., 2011). Using that principle, the proposed method uses PCA based linear subspace learning method to learn model for each subject. The training faces of each subject are rescaled into same size feature vector $f \in \Re^d$ to learn the appearance model. Let N be the total number of subjects and D_i be the dictionary contains face images of particular subject as given below;

$$D_i = \{f_1,...,f_{n_i}\} \in \Re^{d \times n_i}; i = 1,....,N; \tag{1}$$

where n_i be number of training images for each subject. Using PCA, linear subspace model for each D_i is learned as explained below. All the dictionaries are centered to eliminate the influence caused by the mean;

$$\bar{D}_i = D_i - m_i 1_{1 \times n_i} \tag{2}$$

where $m_i \in \Re^{d \times 1}$ is the mean of the D_i Using $\bar{D}_i$, subspace model of the each subject is obtained with the help of projection matrix

$$U_i = \{u_1,....,u_{p_i}\} \in \Re^{d \times p_i}; \tag{3}$$

Projection matrix can be calculated by directly decomposing the $\bar{D}_i$ using SVD to get the eigenvectors of the p_i largest singular values. p_i is the number of principal components and it has been chosen such that

$$\frac{\sum \xi_j}{\sum \xi_k} \geq T; j = 1,..p_i; k = 1,..,n_i; \tag{4}$$

where ξ is the eigenvalue, T is the variance threshold. A larger T will lead to larger p_i and corresponding subspace will preserve the most variation of the $\bar{D}_i$. All the chosen principal components (u_i) are the basis vectors of the corresponding subspace and all are orthogonal to each other. The learned subspace model of each subject can be represented by the m_i and the projection matrix U_i . Each subject is approximated using this linear subspace models and it can be denoted by tiny dictionaries $A_i = \{A_1,...,A_N\}$; where $A_i = \{U_i,m_i\}$. Given test face image (y) can be represented as given as;

$$y = Uz + e \tag{5}$$

where e is the noise and z indicates corresponding coefficient vector which can be estimated as $z = U^T y$ (6)

In PCA, the error vector e is assumed to be Gaussian distributed small dense noise. The reconstruction error can by approximated by $\left\| y - UU^T y \right\|_2^2$. However this assumption is not valid when the given test face is with contiguous block occlusion as the noise term cannot be modeled. Hence to estimate the sparse noise term successfully, identity matrix is used in SRC based face recognition. So we have used sparse representation framework to estimate the noise term using linear subspaces U_i as given below.

$$y = [U_i \, I] \begin{bmatrix} \sigma \\ e \end{bmatrix} = Bc \tag{7}$$

where U is the matrix which contains basis vectors for i^{th} subject, $I \in \Re^{d \times d}$ is the identity matrix and σ and e are the coefficients for face and sparse noise of the test image respectively. Generally to solve this underdetermined system, the conventional sparse representation methods like Orthogonal Matching Pursuit (OMP) or Least Shrinkage Selection Absolute Operator (LASSO) are used. However these methods take huge computation time to obtain sparse coefficients which is not suitable for identifying faces in real-time videos. Also, these methods work well only for test faces with spatially correlated noises, not for contiguous block occlusion.

To model the sparse noise term effectively with less computation time, the fast l_1 regularization method is used in our proposed face identification algorithm. Using this method, equation (7) can be solved via following optimization formulation,

$$\min \frac{1}{2} \left\| y - Bc \right\|_2^2 + \lambda \left\| c \right\|_1 \tag{8}$$

wh(8) $\left\| . \right\|_1$ and $\left\| . \right\|_2$ denote the l_1 and l_2 norms respectively. Equation (8) can be rewritten as given below;

$$\min_{\sigma, e} \frac{1}{2} \left\| y - U\sigma - e \right\|_2^2 + \lambda \left\| e \right\|_1 \tag{9}$$

(9)In Lagrangian form,
and the objective function is defined as

$$\min_{\sigma, e} L(\sigma, e) \text{ s.t. } U^T U = I \tag{10}$$

λ is l_1 regularization parameter. As there is no analytical solution for equation (10), σ_{opt} and e_{opt} are solved using iterative techniques (Wang et al., 2013). The face and noise factorization of a typical test image is shown in figure 1.

Fig. 1 Face and noise factorization

The obtained representation coefficients σ_{opt} and e_{opt} for each tiny dictionary are used to calculate the residual error of the test image as given below;

$$r_i(y) = \left\| y - U_i \sigma_i - e_i \right\|_2^2 \tag{11}$$

Using r_i, the class of the test face is identified as

$$identity(y) = \arg\min(r_i(y)) \tag{12}$$

Large contiguous block occlusion can be handled by partitioning approach as proposed in Naseem et al., (2010). The face image is divided into number of non-overlapped sub images and each sub-image is processed individually. In the proposed method, for each class number of subspace models is learned using corresponding sub-images. The test image also divided into same manner and for each partition residual error is obtained using (11). Decision is taken based on minimum residual error among all class specific partitioned subspace models.

3 Incremental Face Learning Using Rejection Ratio and Occlusion Detection

In video based face recognition, face appearance change has to be learned dynamically to combat the gradual variations of pose, expression, illumination and scale. Conventional frontal face identification methods use fixed appearance models which are learned in offline which may not work well for test face with appearance change in videos. In our proposed approach, to update the class specific subspace models $A_i = \{U_i, m_i\}$, we have used incremental PCA (David et al., 2008), which learns the appearance changes of faces in online effectively. However, there are few important factors which decide effective online face learning in videos as given below. First, invalid face images should not be used to learn class specific models. Second, wrong identifications must be rejected otherwise this may lead to updating wrong model. Finally, face models should not be updated during occlusion. To handle first two issues, we have constructed an index called rejection ratio (ϕ) which can reject wrong identifications and invalid faces. The ϕ is constructed using residual error obtained by (11) for each model for a given test image. Here we have assumed that, for a given test face image, the genuine class residual error is much lower than average of all class residual errors. For a invalid face image, residual errors of all classes are equally high as shown in figure. With those presumptions, we calculate ϕ as given below,

$$mean(r(y)) = \frac{\sum_{i=1}^{N} r_i(y)}{N} \qquad \phi(y) = \frac{mean(r(y)) - \min(r(y))}{mean(r(y))}$$

It can be observed that, $\phi(y)$ value is high for genuine test faces and very low for invalid test images. To show the validity of our assumption, we have illustrated rejection ratio value of different kinds of example test images in figure 2. Based on cross validation, we formulated a decision rule by setting threshold tr_1 to decide whether to accept or reject the identity of a given test face.

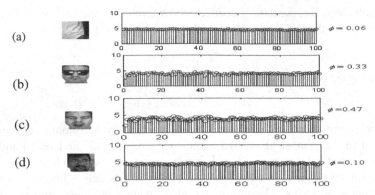

Fig. 2 Illustration of rejection ratio for various test images (a) Non face (b) Valid face with occlusion (c) Valid face with different expression and (d) New face which is not in gallery

If $th_1 > \phi$, test image identity is accepted otherwise that will be rejected and system will start to process next frame. Once test face identity is accepted, that will be scrutinized for subspace model update by setting strong threshold (th_2) i.e. $\phi > th_2 > th_1$. After that, we have to find whether test face is occluded or not to avoid partially occluded faces used for corresponding model update. To detect the occlusion on test face, we have used sparse error coefficients (e_{opt}) of corresponding identity model. Any nonzero coefficient in e_{opt} shows that, corresponding pixel in the test image has been corrupted. Using e_{opt} coefficients, we have defined occlusion ratio η, as ratio of number of non-zero coefficients in e_{opt} to total number of coefficients in e_{opt}. Based on cross validation, we set threshold th_3, and if $\eta > th_3$, the test face will be considered as occluded and if $\eta \le th_3$, the test face is not occluded as illustrated in figure 3. Only those test faces passed the above evaluation criteria's are used to update the corresponding subspace model in online.

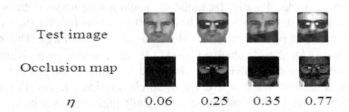

Fig. 3 Illustration of occlusion ratio for various test faces

Based on rejection ratio and occlusion ratio, the proposed face identification algorithm is integrated with incremental learning approach. The flowchart of the proposed comprehensive video based face identification algorithm is shown in figure 4.

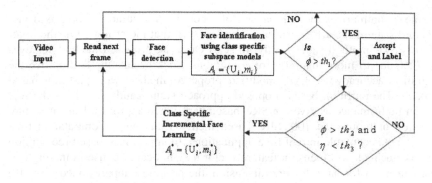

Fig. 4 Flow chart of Comprehensive video based face identification algorithm

4 Experimental Setup and Results

We will first examine the robustness of the proposed face identification algorithm against contiguous block occlusion using AR and Extended Yale B datasets. To compare the recognition performance and computation time, we have used another state of the art method SRC. Finally, we demonstrate video based face identification using incremental learning technique. All methods are implemented using MATLAB and executed on Intel Core 2 Duo processor at 2.53 GHz. To solve l_1 minimization for SRC method we have used SPArse Modeling Software (SPAMS) package (Mairal et al., 2010).

4.1 AR Dataset

In this section, we evaluated our proposed face identification method on AR dataset (Martinez and Kak, 2001) which contains 126 subjects. We have used randomly selected 100 subjects with 50 males and 50 females. Each subject contains 26 images which incorporates expression variation, illumination variation and partial occlusions. Evaluation is done by using 8 non occluded training images to identify the 4 occluded images as shown in figure 5.a and 5.b. All experiments are done by down sampling 165×120 cropped images to an order of 50×40. First, the proposed method is evaluated without partition. To handle large contiguous occlusion, we also proposed classical partition based approach. In partitioned approach, each sub image (in this experiment number of partition is 2) is down samples to 20×30 as shown in figure 5.c.

(a)   (b)   (c)

Fig. 5 Sample (a) training and (b) testing images of a typical subject (c) partitioned images

Table 1 summarizes the comparison of recognition accuracy of proposed methods and state of the art algorithms. It can be seen that the proposed method outperforms SRC method significantly. In scarf case, the proposed method achieves only 72% due to more than 40% of contiguous block occlusion. To improve the recognition accuracy we have modified proposed method with partition based approach. The partition based proposed approach significantly improves the recognition performance as given in SRC based partitioned approach. But for feature dimension of more than 100, SRC based methods take huge computation time which is not acceptable in real time applications. In appearance based recognition methods, such a low dimension feature vector may not capture discriminating features. But, even for large feature dimension, the proposed unpartitioned (50×40) and partitioned (20×30) approaches take only 312ms and 203ms respectively.

Table 1 Experimental results on AR Dataset with contiguous occlusion

Method	Recognition Accuracy (%)	
	Sunglasses	Scarf
SRC (Wright et al., 2009)	87	59.5
SRC +partition (Wright et al., 2009)	97.5	93.5
Proposed	**100**	**72**
Proposed + partition	**100**	**95**

4.2   Extended Yale B Dataset

The extended Yale B dataset contains 2414 fontal face images of 38 subjects taken at various lighting conditions. Each subject consists of 64 images. As given in Georghiades et al., (2001) and Lee et al., (2005), we divided the dataset into five subsets based on normal to severe light variations. Subset 1 consists of 266 images (seven images per subject) which are used for training. Subsets 2, 3, 4 and 5 each consists of 12, 12, 14, and 19 images per subject respectively. Subset 2 and 3 are used for testing with 10% to 60% occlusion by replacing a square block of each test image with a baboon image. The example 10% to 60% occluded images of subset 2 is shown in figure 6. Subset 4 and 5 are not considered for this experiment due to the assumptions of sufficient lighting condition under indoor surveillance environment and more focus on contiguous block occlusion analysis. All experiments are done by downsampling 192×168 images to an order of 20×20.

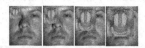

Fig. 6 Test images under varying level of contiguous occlusion from 10% to 60% of Extended Yale B dataset.

Table 2 shows the recognition accuracy and computation time of various algorithms using subset 1 and 2. The proposed method outperforms the other methods for all level of occlusion. In subset 2, up to 40% of occlusion the proposed method correctly identified 100% of test faces. Even at 60%, above 85% recognition rate is achieved. Even LRC partitioned and SRC methods achieved 100% recognition accuracy for up to 20% occlusion. In subset 3, up to 20% of occlusion, the proposed method correctly identified about 100% of test faces. At 40% occlusion, above 84% recognition rate is achieved. The LRC partitioned method achieved recognition accuracy on par with proposed method from 0% to 60% due to location of the block occlusion. If the Baboon block image is placed in center, its performance will be degraded as occlusion is shared by all partitions and also it will huge computation time. It can be observed that SRC does not perform well in higher level of block occlusions and also it takes more computation time when dimension of feature vector increases. The computation time of LRC is very less when compare to all other methods but recognition accuracy is very poor for 20% to 60% occlusion. Figure 7 shows recognition rate across varying level of contiguous occlusion for various algorithms using subset 1 and subset 2.

Table 2 Experimental results of Extended Yale B dataset with varying level of contiguous occlusion

Method	Recognition Accuracy (%)										Computation Time per image (ms)
	Subset2					Subset3					
	0%	10%	20%	40%	60%	0%	10%	20%	40%	60%	
LRC	100	99.5	96	70	21	100	97.3	91.2	48.2	12.7	**17.3**
LRC-Partition	100	100	100	92	51.5	100	98	97.8	**87.9**	**46.2**	140
SRC	100	100	100	71.9	36.1	100	99.1	96.7	38.3	16	106
Proposed	**100**	**100**	**100**	**100**	**85.7**	**100**	**100**	**99.5**	84.6	43.6	34.5

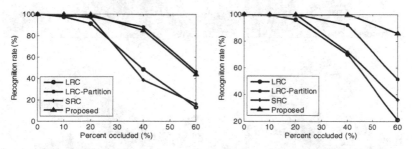

Fig. 7 The recognition rate against varying level of occlusion for various algorithms using (a) subset and (b) subset 2

4.3   Honda/UCSD Video Dataset

The third set of experiments is conducted on Honda/UCSD video dataset (Lee et al., 2005). This dataset consists of 20 subject's training video sequences which incorporates pose variation, illumination variation and scale variation and resolution is 640×480. In each video, the person rotates and turns his/her head in his/her own preferred order and speed, and typically in about 15 seconds, the individual is able to provide a wide range of different poses. The detected faces (Viola and Michael, 2004) are down sampled into 32×32. To obtain the subspace model of each subject initially, 10 randomly selected sample faces are used which are obtained from first 30 frames of each video sequence. Out of all other test frames, in few frames, face could not be detected by detection module due extreme to face pose variation. The recognition rates is computed by taking ratio of number of correct iendentifications to total number of detected faces from video. For this experiment, the evaluation criteria thresholds th_1, th_2 and th_3 are set as 0.2, 0.5, and 0.2 respectively.

In incremental learning approach, gradual pose, scale and expression variations are incorporated on subspace models in online which are not possible in static face identification methods. For incremental face learning following requirements should be followed; (i) False identifications should be strictly rejected which may lead to false learning on another subject model (ii)Invalid faces must be rejected (iii) Occluded faces may be identified correctly but that should not be used for learning. To meet above said requirements, the formulated decision rules by using rejection ratio (ϕ) and occlusion ratio (η) are utilized very effectively. Figure 8 shows correct identification and rejection results for different test video sequences.

Fig. 8 Correct face identification during pose, expression variations and invalid faces

Table 3 Results of proposed video based face identification method on Honda/UCSD dataset

Method	Recognition accuracy (%)
Proposed	69.56
Proposed with incremental learning	77.07

It can be seen in table 3 that proposed incremental learning method significantly outperforms the proposed static face identification method. It is due to the incremental learning of nonlinear appearance variations of face which are not available during training. But, whereas in conventional static face identification method, face models are learned only once in training stage. In few cases like extreme pose variation and drastic illumination variation, the proposed method fails to identify or reject the faces as shown in figure 9. This can be improved by using illumination invariant face descriptors at the cost of computation.

Fig. 9 False identification and rejection during extreme pose and illumination variation

5   Conclusion

In this paper, we have presented robust face identification algorithm under severe occlusion using subspace learning and fast sparse representation. We also integrated our proposed face identification algorithm with incremental face learning approach to extend for video based face identification. The newly formulated decision rules using residual errors and sparse noise coefficients are proved to be very effective in incremental subspace model update in online. In this, different kinds of face appearance are learned for each subspace model in online. The efficacy of the proposed methods is demonstrated using various static face datasets and videos. It remains an interesting topic for future work to implement the proposed framework in embedded hardware and test in real-time.

References

1. Chellappa, R., Ni, J., Patel, V.M.: Remote identification of faces: Problems, pros pects, and progress. International Journal of Pattern Recognition Letters 33, 1849–1859 (2012)
2. Barr, J.R., Bowyer, K.W., Flynn, J., Biswas, S.: Face Recognition from Video: A Review. International Journal of Pattern Recognition and Artificial Intelligence 26(5) (2012)

3. Turk, M., Pentland, A.: Eigen faces for face recognition. Journal of Cognitive Neuroscience 3, 71–86 (1991)
4. Belhumeur, P.N., Hespanha, J.P., Kriengman, D.J.: Eigenfaces vs. fisherfaces: Recognition using class specific linear projection. IEEE Transactions on Pattern Recognition and Machine Intelligence 19, 711–720 (1997)
5. Bernd, H., Purdy, H., Tomaso, P.: Face recognition: Component based versus global approaches. Journal of Computer Vision and Image Understanding 91, 6–21 (2003)
6. Naseem, I., Togneri, R., Bennamoun, M.: 'Linear Regression for Face Recognition'. IEEE Transactions on Pattern Recognition and Machine Intelligence 32(11), 2106–2112 (2010)
7. Wright, J., Yang, A.Y., Ganesh, A., Sastry, S.S., Ma, Y.: Robust Face Recognition via Sparse Representation. IEEE Transactions on Pattern Analysis and Machine Intelligence 31(2), 210–227 (2009)
8. Cheng, H., Liu, Z., Yang, L., Chen, X.: Sparse Representation and learning in visual recognition: Theory and Applications. International Journal on Signal Processing 93, 1408–1425 (2013)
9. Chen, Y.-C., Patel, V.M., Phillips, P.J., Chellappa, R.: Dictionary-based Face Recognition from Video. In: Fitzgibbon, A., Lazebnik, S., Perona, P., Sato, Y., Schmid, C. (eds.) ECCV 2012, Part VI. LNCS, vol. 7577, pp. 766–779. Springer, Heidelberg (2012)
10. Ptucha, R., Savakis, A.: Manifold based sparse representation for facial understanding in natural images. International Journal on Image and Vision Computing 31, 365–378 (2013)
11. Mei, X., Ling, H.: Robust visual tracking using minimization. In: Proceedings of IEEE Conference on Computer Vision, pp. 1436–1443 (2009)
12. Wang, D., Lu, H., Yang, M.H.: Online Object Tracking with Sparse Prototypes. IEEE Transactions on Image Processing 22(1), 314–325 (2013)
13. Viola, P., Michael, J.: Robust Real time Face Detection. International Journal of Computer Vision 57(2), 137–154 (2004)
14. David, A.R., Lim, J., Lin, R.S., Yang, M.H.: Incremental Learning for Robust Visual Tracking. International Journal of Computer Vision 77(1-3), 125–141 (2008)
15. Chen, F., Wang, Q., Wang, S., Zhang, W., Xu, W.: 'Object tracking via appearance modeling and sparse representation'. International Journal on Image and Vision Computing 29, 787–796 (2011)
16. Hale, E.T., Yin, W., Zhang, Y.: Fixed point continuation for l1 minimization: Methodology and convergence. SIAM Journal on Optimization 19(3), 1107–1130 (2008)
17. Mairal, J., Bach, F., Ponce, J., Sapiro, G.: Online Learning for Matrix Factorization and Sparse Coding. Journal of Machine Learning Research 11, 19–60 (2010)
18. Martinez, A.M., Kak, A.C.: PCA versus LDA. IEEE Transactions on Pattern Analysis and Machine Intelligence 23(2), 228–233 (2001)
19. Georghiades, A., Belhumeur, P., Kriegman, D.: From Few to Many: Illumination cone models for face recognition under variable lighting and pose. IEEE Transactions on Pattern Analysis and Machine Intelligence 23(6), 643–660 (2001)
20. Lee, K.C., Ho, J., Kriegman, D.: Acquiring linear subspaces for Face Recognition under Variable Lighting. IEEE Transactions on Pattern Analysis and Machine Intelligence 27(5), 684–698 (2005)
21. Lee, K.C., Ho, J., Yang, M.H., Kriegman, D.: Visual tracking and recognition using probabilistic appearance models. International Journal of Computer Vision and Image Understanding 99(3), 303–331 (2005)

Face Recognition Using MPCA-EMFDA Based Features Under Illumination and Expression Variations in Effect of Different Classifiers

Chandan Tripathi, Aditya Kumar, and Punit Mittal

Abstract. The paper proposes a new method for feature extraction using tensor based Each Mode Fisher Discriminant Analysis(EMFDA) over Multilinear Principle Components (MPCA) in effect of different classifiers while changing feature size. Initially the face datasets have been mapped into curvilinear tensor space and features have been extracted using Multilinear Principal Component Analysis (MPCA) followed by Fisher Discriminant Analysis, in each mode of tensor space. The ORL and YALE databases have been used, without any pre-processing, in order to test the effect of classifier in real time environment.

1 Introduction

Face Recognition is one of the highly researched areas of the pattern recognition that just opened great application opportunities as basic as class-room attendance management till high end automated robotic surveillance and many more[1]. Among all, Principle Component Analysis (PCA) [2] was the first method that interested many of the researches of this field. Further, 2-D PCA [7, 8] was introduced to improve the recognition accuracy. After the success of 2-D PCA, K.N. Plataniotis introduced Multilinear Principle Component Analysis (MPCA) over tensor objects first in face recognition [11] and then in gait recognition [9]. While increasing of the dimension size, all the above methods were only relied over eigen vectors having higher eigen values, in each dimension eigen space. It restricted the use of more relevant eigen vectors with other non-zero eigen values that were able to increase the recognition rate. A class variance based method, named as Fisher Discriminant Analysis (FDA), [3, 4, 6] aimed that issue while selecting the eigen vectors with non-zero eigen values. Furthermore, a Two Dimensional Linear Discriminant Analysis(2D-LDA) [12] based approach tried to resolve the same issue in two dimensional space.

Chandan Tripathi · Aditya Kumar · Punit Mittal
Department of Computer Science, VKP, India
e-mail: {ctripathi007,aditya.brt,punit.mittal06}@gmail.com

S.M. Thampi, A. Gelbukh, and J. Mukhopadhyay (eds.), *Advances in Signal Processing and Intelligent Recognition Systems*, Advances in Intelligent Systems and Computing 264, DOI: 10.1007/978-3-319-04960-1_5, © Springer International Publishing Switzerland 2014

In process of classification the size of feature dimension of above methods were leading towards higher computational time. Keeping above views [5, 16] a new adaptive method, named as Multidimensional Discriminant Analysis (MLDA), has been proposed with optimal reduction in feature dimension without compromising with both, recognition accuracy as well as computational time.

Further experiments were more focused to find robust classification methods rather than more experimenting in feature extraction method. Among all supervised learning based classifiers, the Support Vector Machine (SVM) [14, 15] was able to separate two classes more accurately than the other classification methods. Albeit, the requirement of the classification depends on the characteristics of the features. This paper focuses more to test the effect of different classifiers using combination of two different feature extraction algorithms MPCA and FDA in each mode. This feature extraction method is envisioned to utilize the strength of both the MPCA in tensor space and followed by more discriminating features in each mode. Furthermore, a comparative study with MPCA has also been included to test effectiveness of used method while changing the count of gallery and probe images. Moreover, different classification methods such as different kind of KNN and SVM functions with a comparative study have also been tested.

2 Multilinear Principle Component Analysis

This method works on tensor representation [11] of the faces in order to preserve the relationship among neighboring pixels as is absent in PCA. A tensor object is a higher dimensional matrix of order N stands as $TNS \in R^{(T_1 \times T_2 \times \cdots \times T_N)}$, where T_i represents each mode of that tensor. For P number of images where image is symbolized as F_i , can be projected into a third order tensor $TNS \in R^{T_1 \times t_2 \times t_3}$. In TNS the T_1 stands as row size (height) of image as mode-1 of tensor TNS, T_2 is column size (width) of image as mode-2 of tensor TNS and T_3 is number of images P used in training as mode-3 of tensor TNS. The MPCA algorithm is explained below:

1. *Obtain the mean Image value of all training images F_{Mean} as:*

$$F_{Mean} = \frac{1}{T_3} \sum_{t=1}^{T_3} F_{M \times N}^t \tag{1}$$

   where $T_3 = P$

2. *Center the tensor object representation of training images:*

$$\widehat{TNS} = \left[F_{M \times N}^1 - F_{Mean}, F_{M \times N}^2 - F_{Mean}, \ldots, F_{M \times N}^P - F_{Mean} \right] \tag{2}$$

3. *Unfold the tensor for different n-modes into matrix for different modes. Unfolding of tensor TNS along the n-mode can be defined as $TNS(n) \in R^{(T_n \times (T_1 \times T_2 \times \cdots \times T_N))}$*

i.e.; for $(\widehat{TNS}_n)_{index\_val} = f_{(t_1,t_2,t_3)}$, *where f if the value present at* $(t_1,t_2,t_3)^{th}$ *index value. Further, index value selection can be done as:*

$$index\_val = [t_n, \sum_{m=3}^{n+1} (t_m-1)(\prod_{q=m}^{n+1} T_q)(\prod_{q=n-1}^{1} T_q) + \sum_{m=n-1}^{1} (t_m-1)(\prod_{q=m-1}^{1} T_q)] \quad (3)$$

4. *Evaluate the co-variance matrix mentioned as* C_n *and* $C_n = \widehat{TNS}_n(\widehat{TNS}_n)^{Transpose}$ *of each n-mode and find the set of eigenvectors* $X_n = [x_1,.....x_{k_{(n)}}]$ *founded on the selected eigenvectors according to the largest k (n) Eigenvalues.*

   One can obtain optimal features set using any of the subsequent feature selection process:

   (i) *Using* $Y_t = ((F_{m\times n})^{Input} - F_{Mean} \times_n X_{(n)}^{Transpose}$ *which will give* $k_{(1)} \times T_2 + k_{(2)} \times T_1$ *features.*
   (ii) *Using* $Y_t = X_{(1)}^T ranspose(AB_{(M\times N)_{Input}}^t - AB_{Mean})X_{(2)}$ *different value for n (for different modes), which will give* $k_{(1)} \times k_{(2)}$ *features.*

5. *Test the classifier with obtained feature set* Y_t.

3   Fisher Discriminant Analysis

In this paper we used a Fisher discriminant analysis that is orthogonal in nature. The FDA [3,4,6] aims to search the projection P_{FDA}, a set of basis vectors which are normal vectors to discriminant hyper-plane of feature vector y_m that can be obtained using any of the feature selection method. FDA is envisioned to use Gaussian information present in image data while using class information. For an mth training sample image having class C_m FDA aims to maximize the ratio of the $S_{between\_class}$ (between-class scattering matrix) and S_{within_class} (within-class scattering matrix), where

$$S_{within\ class} = \sum_{m=1}^{M} (y_m - \bar{y}_{C_m})(y_m - \bar{y}_{C_m})^{Transpose} \quad (4)$$

where $\bar{y}_{C_m} = \frac{1}{N_c} \Sigma_{m,C_m=C} y_m$ and

$$S_{between\_class} = \sum_{C_m=1}^{C_M} (y_{C_m} - \bar{y})(y_{C_m} - \bar{y})^{Transpose} \quad (5)$$

where $\bar{y} = \frac{1}{M} \Sigma_m y_m$

The purpose of DFA is to find the maximized set of basis vector P as

$$P_{DFA} = argmaxP \frac{|(P^T S_{between\_class} P)|}{|(P^T S_{within\_class} P)|} \quad (6)$$

The outcome of above equation (6) can be generalized as $P_{FDA} = [p_1, p_2......p_{hz}]$. Here $p_{hz}, h_Z = 1,2,.....,H_Z$ is the set of eigen vectors of $S_{between\_class}$ and $S_{between\_class}$ for $H_Z C - 1$ largest generalized eigenvalues. The discriminant feature vector for m_{th} training images can be obtained as $yd_m = P_{FDA}^T y_m$.

4   K-Nearest Neighborhood Classifier

K-nearest neighborhood (KNN) classifier is a well-known and simplest most classifier among all the other unsupervised machine learning algorithms. We have tested
two different math functions for classification separately. The first and commonly
used function is euclidian based, here named as KNN_1, and the second one is using
cosine measurement, here named as KNN_2.

5   Support Vector Machine

Support vector machine is a well-known supervised machine learning classifier that
try to find an optimal separating hyper plane between two given classes $y \in (+1, 1)$
using training datasets [14, 15]. The hyper plane can be expressed as:

$$w.x + b = 0 \tag{7}$$

where x is a probe face image vector projection and w is an angular orientation of the
hyper plane for each binary class separation with b distance from hyperspace origin.
Albeit SVM is originally designed for two class classification problems[15] but
most of the classification problems encourage the multi-class classification system.
Principally there are two methods that enables binary SVM applicable for multiclass
classification as One-Against-One (OAO) classification method and One-Against-
All (OAA) classification method. Suppose a sample has N number of classes then
for OAO method $N \times (N-1)/2$ number of SVM classifiers will be needed and in
OAA method only N number of SVM classifiers required. In this paper we used
OAA method. Two different kind of kernel functions has been tested over the datasets. First function is Gaussian type redial basis kernel function (RBF), named as
SVM_1, defined as

$$F(x_i, x_j) = e^{-\gamma \times (x_i - x_j)^2} \tag{8}$$

whereas the second function is order 3 heterogeneous polynomial kernel function,
named as:

$$F(x_i, x_j) = (a \times (x_i - x_j) + b)^c \tag{9}$$

The above used symbols have following meanings:

x_i= database point
x_j= test vector point of input vectors(support vector
  samples) and $\gamma > 0$
a = constant multiplier in a polynomial function
b = trade-off between training error and margin
c = power of polynomial kernel function

6   Result and Discussions

Proposed method has been tested over windows platform of 32-bit based machine
having Intel core-2 duo 2.4 GHz and 2GB RAM. The code has been simulated

using Matlab-2011. The experimented algorithm has been tested over two bench-mark databases: AT&T and YALE databases. The AT&T contains sample face vary-ing in poses and size, whereas YALE database contains varying facial and lightning conditions. The performance of EMFDA has been tested with above mentioned four classifiers KNN_1, KNN_2, SVM_1 and SVM_2. One more study has been done using these classifiers with change in count of gallery and probe images while the perfor-mance has been compared with MPCA method.

6.1   Experiment on At&T Face Dataset

The database, as shown in Fig. 1, contains 400 images of size 112×92 equally di-vided into 40 classes with varying expressions, controlled facial rotations, with/without white lens glasses and with same background information. In this ex-periment we randomly elected the gallery and probe images with varying ratio from 2/8 to 5/5 without any pre-processing.

Fig. 1 Sample images of original AT&T data-set of size 112×92

6.1.1   Results in Effect of KNN Classifiers

The KNN classifiers (KNN_1 and KNN_2) have been tested with changing the value of number of nearest neighbors count k from 1 to maximum accuracy pursuing value.

In experiment it is found that after some value of k (as after 40 in case of 5/5 sample ratio) the recognition accuracy stopped changing as most of the neighbors were belonging to the expected class. While after some values of k (as after 130 in case of 5/5 sample ratio) the accuracy started decreasing. It can be said that, based on experimental results, in training samples the different persons faces, which originally were so much close in some dimension, had come up with false class information and therefore increases the most common class majority ratio while comparing true class majority.

One more experiment has been done with increasing the feature dimension while keeping maximum but constant accuracy providing value of k (as 40 in case of 5/5 sample ratio). In this experiment it is found that after some feature size the accuracy does not further increase in case of both the MPCA and EMFDA as shown in Fig. 2. The results of classifiers with change in gallery and probe images ratio, has been shown in Table-1.

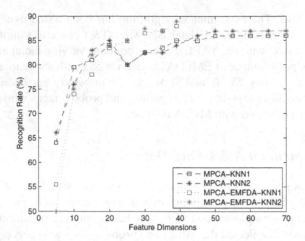

Fig. 2 Effect of KNN classifiers with increase in size of feature vectors over AT&T data-set with 3/7 Gallery/Probe sample ratio

Table 1 Results along with their respective dimensions based on AT&T Dataset in effect of KNN classifiers

Sample Ratio (Training/Testing)			2/8	3/7	4/6	5/5
MPCA	KNN_1	Features	65	65	65	65
		Accuracy(%)	76.88	85.71	87.08	87.00
	KNN_2	Features	65	65	65	65
		Accuracy(%)	78.13	83.93	87.08	88.00
MPCA-EMFDA	KNN_1	Features	39	39	39	39
		Accuracy(%)	78.13	87.86	89.16	93.00
	KNN_2	Features	39	39	39	39
		Accuracy(%)	78.13	88.93	92.08	94.00

6.1.2   Results in Effect of SVM Classifiers

The SVM based classifiers, named as SVM_1 and SVM_2, have been tested over the MPCA-EMFDA features and the effects were compared with MPCA based features using following values as given below:

In SVM_1

Gamma value in redial basis kernel =0.000009

In SVM_2

Value of parameter a in Polynomial kernel=0.00018

Value of parameter b in Polynomial kernel=0.000009

Dimension c of Polynomial function=3

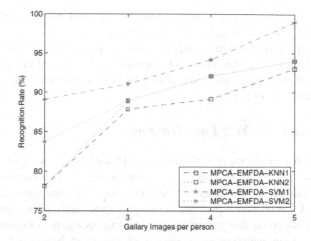

Fig. 3 Effect of all classifiers over AT&T dataset with changing Gallery Images from 2 to 5

The above values of gamma, a, b and c have been obtained based on maximum accuracy achieved programmatically. The program was designed to pass these parameters until the recognition further stopped changing. Initially it found that the program always stuck in infinite loop because the accuracy was either increasing or decreasing in alternate fashion while increasing the iteration. In order to control the iteration the accuracy has been checked with 10 more iteration after the accuracy further stopped increasing or decreasing (with +0.50 or -0.50) in five consecutive iterations. Albeit the values of gamma, a, b and c were the best experimented optimal values that can further be polished.

Here the separation between kernel function hyper-plane and corresponding support vector containing hyper-plane plays a major role in classification of the input face image feature vector. The results ,as shown in Table-2 , indicate that the redial basis hyper-plane separates the training data points more accurately than third order polynomial hyper-plane. As in case of redial basis kernel function we are able to find

Table 2 Results along with their respective dimensions based on AT&T Dataset in effect of SVM classifiers

	Sample Ratio (Training/Testing)		2/8	3/7	4/6	5/5
MPCA	SVM_1	Features	65	65	65	65
		Accuracy(%)	85.94	87.86	90.00	94.00
	SVM_2	Features	65	65	65	65
		Accuracy(%)	81.88	87.86	89.16	88.00
MPCA-EMFDA	SVM_1	Features	39	39	39	39
		Accuracy(%)	86.06	91.07	94.17	99.00
	SVM_2	Features	39	39	39	39
		Accuracy(%)	83.75	88.93	92.08	94.00

more nearer data points of a class as a support vectors, when compared with third order polynomial. Now it can be said that the normal vector distance of support vectors from hyper-plane is more uniformly distributed in redial basis hyper-plane than third degree polynomial hyper-plane. The results of SVM based classifiers, when compared with KNN based classifiers, shows higher recognition accuracy. The results have been shown in Fig. 3.

6.2 Experiment on Yale Face Database

The YALE database consists of 165 images of 15 individuals with 11 images per person with 243×320 as original size, shown in Fig. 4. The reason using this database is to understand the effect of the classifiers with change in illumination variation with different directional focus of light. This database is more challenging than AT&T database due to presence of varying background information with different facial appearance (normal, sleepy, wink, surprised, happy and sad).

Fig. 4 Example images of YALE data-set with size 243×320 as used in experiment

6.2.1 Results in Effect of KNN Classifiers

In this experiment KNN_1 and KNN_2 classifications have been verified in same fashion while varying feature dimension as in case of AT&T data-set as shown in Fig. 5.

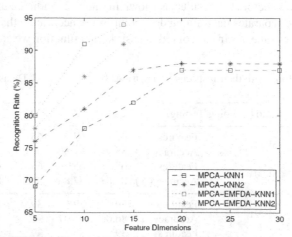

Fig. 5 Effect of KNN classifiers with increase in size of feature vectors over YALE data-set with 3/7 Gallery/Probe sample ratio

The results of classifiers with change in gallery and probe images ratio, has been shown in Table-3. The experiment over YALE data-set shows that the size of features greatly affected by the data-set size while comparing the well performing feature dimensions of AT&T data-set. One more observation came in experiment that the shadow effect appearing at background was greatly affecting the Gaussian nature of the classes. It resulted a slight degradation in recognition accuracy. This effect can be seen in table-3 as when those images were being introduced in k.

Table 3 Results along with their respective dimensions based on YALE Dataset in effect of KNN classifiers

Sample Ratio (Training/Testing)			2/8	3/7	4/6	5/5
MPCA	KNN_1	Features	43	43	43	43
		Accuracy(%)	62.96	70.83	79.05	85.56
	KNN_2	Features	43	43	43	43
		Accuracy(%)	79.26	86.67	89.52	91.11
MPCA-EMFDA	KNN_1	Features	14	14	14	14
		Accuracy(%)	68.15	74.17	77.14	82.22
	KNN_2	Features	14	14	14	14
		Accuracy(%)	80.00	82.96	84.76	87.78

6.2.2 Results in Effect of SVM Classifiers

The SVM based classifiers named as SVM_1 and SVM_2 have been tested over the MPCA-EMFDA features and the effects compared with MPCA based features using following values as listed below:

In SVM_1
Gamma value in redial basis kernel =0.0009456
In SVM_2
Value of parameter a in Polynomial kernel=0.0001
Value of parameter b in Polynomial kernel=0.000009
Dimension c of Polynomial function=3

The values of gamma, a, b and c are the best found experimented optimal values. The result shown in Table-4, based on the above parameters, indicate that the redial basis hyper-plane separates more precisely the training data points than the third order polynomial hyper-plane as in case of AT&T Database.

The results based on SVM classifiers, when compared with KNN classifiers, shows higher recognition accuracy. The results has been shown in Fig. 6.

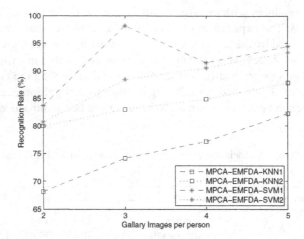

Fig. 6 Effect of all classifiers over YALE data-set with changing Gallery Images from 2 to 5

Table 4 Results along with their respective dimensions based on YALE Dataset in effect of SVM classifiers

		Sample Ratio (Training/Testing)	2/8	3/7	4/6	5/5
MPCA	SVM_1	Features	43	43	43	43
		Accuracy(%)	76.30	79.17	83.81	87.78
	SVM_2	Features	43	43	43	43
		Accuracy(%)	74.07	75.83	80.00	85.56
MPCA-EMFDA	SVM_1	Features	14	14	14	14
		Accuracy(%)	83.70	89.17	91.43	94.44
	SVM_2	Features	14	14	14	14
		Accuracy(%)	80.74	88.33	90.48	93.33

7 Conclusion

In this study KNN and SVM based classifiers, with different measurement function, have been tested over ORL and YALE datasets with and without pre-processing techniques. Based on observations it could be concluded that in case of both MP-CAEMFDA and MPCA based features the cosine based KNN and redial basis function based SVM shows their superiority over the same classification methods with different functions. It has also been observed that while increasing the feature dimension the recognition rate become steady after some dimension and decrease after some values. Finally it can be concluded that SVM based classifiers are more effective than KNN classifiers.

References

1. Zhao, W., Chellappa, R., Phillips, P.J., Rosenfeld, A.: Face recognition: A literature survey. ACM Computing Surveys (CSUR) 35(4), 399–458 (2003)
2. Turk, M., Pentland, A.: Eigenfaces for recognition. Journal of Cognitive Neuroscience 3(1), 71–86 (1991)
3. Belhumeur, P.N., Hespanha, J.P., Kriegman, D.J.: Eigenfaces vs. fisherfaces: Recognition using class specific linear projection. IEEE Transactions on Pattern Analysis and Machine Intelligence 19(7), 711–720 (1997)
4. Zhao, W., Krishnaswamy, A., Chellappa, R., Swets, D.L., Weng, J.: Discriminant analysis of principal components for face recognition. In: Face Recognition, pp. 73–85. Springer, Heidelberg (1998)
5. Singh, K.P., Kumar, M., Tripathi, C.: Face Recognition using Eigen Tensor based Linear Discriminant Analysis with SVM Classifier, M.Tech Thesis, Indian Institute of Information Technology Allahabad (2011)
6. Mika, S., Ratsch, G., Weston, J., Scholkopf, B., Mullers, K.R.: Fisher discriminant analysis with kernels. In: Proceedings of the 1999 IEEE Signal Processing Society Workshop on Neural Networks for Signal Processing IX, pp. 41–48. IEEE (August 1999)
7. Yang, J., Zhang, D., Frangi, A.F., Yang, J.Y.: Two dimensional PCA: a new approach to appearance-based face representation and recognition. IEEE Transactions on Pattern Analysis and Machine Intelligence 26(1), 131–137 (2004)
8. Yu, H., Bennamoun, M.: 1D-PCA, 2D-PCA to nD-PCA. In: 18th International Conference on Pattern Recognition, ICPR 2006, vol. 4, pp. 181–184. IEEE (August 2006)
9. Lu, H., Plataniotis, K.N., Venetsanopoulos, A.N.: Multilinear principal component analysis of tensor objects for recognition. In: 18th International Conference on Pattern Recognition, ICPR 2006, vol. 2, pp. 776–779. IEEE (August 2006)
10. Xu, D., Yan, S., Zhang, L., Zhang, H.J., Liu, Z., Shum, H.Y.: Concurrent subspaces analysis. In: IEEE Computer Society Conference on Computer Vision and Pattern Recognition, CVPR 2005, vol. 2, pp. 203–208. IEEE (June 2005)
11. Lu, H., Plataniotis, K.N., Venetsanopoulos, A.N.: MPCA: Multilinear principal component analysis of tensor objects. IEEE Transactions on Neural Networks 19(1), 18–39 (2008)
12. Yan, S., Xu, D., Yang, Q., Zhang, L., Tang, X., Zhang, H.J.: Discriminant analysis with tensor representation. In: IEEE Computer Society Conference on Computer Vision and Pattern Recognition, CVPR 2005, vol. 1, pp. 526–532. IEEE (June 2005)
13. Wang, J., Plataniotis, K.N., Venetsanopoulos, A.N.: Selecting discriminant eigenfaces for face recognition. Pattern Recognition Letters 26(10), 1470–1482 (2005)
14. Gates, K.E.: Fast and accurate face recognition using support vector machines. In: IEEE Computer Society Conference on Computer Vision and Pattern Recognition-Workshops, CVPR Workshops, p. 163. IEEE (June 2005)
15. Gold, C., Sollich, P.: Model selection for support vector machine classification. Neurocomputing 55(1), 221–249 (2003)
16. Tripathi, C., Singh, K.P.: A new method for face recognition with fewer features under illumination and expression variations. In: 2012 19th International Conference on High Performance Computing (HiPC), pp. 1–9. IEEE (December 2012)

A Syllabus-Fairness Measure for Evaluating Open-Ended Questions

Dimple V. Paul and Jyoti D. Pawar

Abstract. A modularized syllabus containing weightages assigned to different units of a subject proves very useful to both teaching as well as student community. Different criteria like Bloom's taxonomy, learning outcomes etc., have been used for evaluating the fairness of a question paper. But we have not come across any work that focuses on unit-weightages for computing the syllabus fairness. Hence in this paper we address the problem of evaluating the syllabus-fairness of open-ended questions of an examination question paper by analyzing the questions on different criteria. Text mining techniques are used to extract keywords from textual contents in the syllabus file and also in the question paper. Similarity Coefficient is used to compute the similarity between question content and syllabus content. Similarity measure is identified by computing the similarity matrix between question vectors and syllabus vectors. The similarity matrix is used as a guideline in grouping the unit-wise questions; matching its weightage against Syllabus File and evaluating the syllabus fairness of the question paper. The result of syllabus fairness evaluation can be used as a measure by the subject expert or question paper setter or question paper moderator to revise the questions of examination question paper accordingly.

1 Introduction

Examination, as the evaluation of teaching and learning methods, has fixed its important position in the practice of education [1, 2]. Examination is considered as one of the common methods to assess knowledge acceptance of the students.

Dimple V. Paul
DM's College of Arts, Science and Commerce
e-mail: dimplevp@rediffmail.com

Jyoti D. Pawar
Goa University
e-mail: jyotidpawar@gmail.com

S.M. Thampi, A. Gelbukh, and J. Mukhopadhyay (eds.), *Advances in Signal Processing and Intelligent Recognition Systems*, Advances in Intelligent Systems and Computing 264, DOI: 10.1007/978-3-319-04960-1_6, © Springer International Publishing Switzerland 2014

Based on the examination result, student's thoughts, behavior and skills can be studied. Written examination is a conventional yet a universal tool to evaluate the student's performance in educational area. Whether or not the written examination is able to assess the student's ability is very much dependent on the questions asked in the examination paper [3].

This research work tries to adopt a Similarity Coefficient based comparison of open-ended questions of a question paper against the University Prescribed Syllabus File in order to verify the effectiveness of an examination question paper for theoretical courses such as Software Engineering, Information Technology etc.,. Each unit in the syllabus file is given a weightage that correspond to the number of lecture hours to be used by the instructor to teach that unit. The weightage also indicates the importance assigned to that unit which is used by the instructor to decide on the depth to which the topics in that unit should be covered, considered by the paper setter to decide on the allocation of marks under each unit and used by the students to allocate time-schedule for each unit while preparing for an examination. Similarity measure is computed using a similarity matrix which is a two dimensional matrix representing the pair-wise similarity of question content with unit content. The computed similarity matrix is used in forming unit-wise question groups; matching its weightages against syllabus content and evaluate the syllabus fairness of the question paper. This paper is organized as follows. Section 2 discusses the literature review. Methodology adopted is explained in section 3. The problem statement and experimental results are given in section 4 and 5 respectively. Finally section 6 concludes the paper.

2   Literature Review

Current studies in educational field has resulted in generating data mining models to improve the quality of decision making process in higher education, assess student performance , classify the results of students, identify the students who are likely to drop out, identify students at risk of failure etc.,[4,5,6,7]. An interesting work in this area is carried out by mining the examination question paper based on Bloom's taxonomy [3]. In [3] the fairness of a question paper is evaluated by measuring the relevance of its questions with Bloom's Taxonomy. Bloom's taxonomy is a classification system of educational objectives based on the level of student understanding necessary for achievement or mastery. Educational researcher Benjamin Bloom and colleagues have suggested six different cognitive stages in learning such as Knowledge, Comprehension, Application, Analysis, Synthesis and Evaluation. In [8-9] Yutaka and Lui discuss the mechanisms of keyword extraction by focusing on information retrieval of text mining. Question papers can include either open-ended or closed-ended questions. The easiest type of questions is closed questions or multiple-choice questions. However, multiple-choice questions cannot determine the skills of students in writing and expressing.

At present, educators prefer to have essay questions to grade more realistically the students' skills. Open questions are considered to be the most appropriate, because they are the most natural and they produce a better degree of thought. They help to evaluate the understanding of ideas, the students' ability to organize material and develop reasoning, and to evaluate the originality of the proper thoughts. Use of open-ended question evaluation tools is good for understanding the different cognitive skills. Several methodologies have been proposed to solve the problems in automatic evaluation of open-ended and closed-ended questions. Some of them are summarized as follows-

In [10] Chang et al. made a comparative study between the different scoring methods. They also studied the different types of exams and their effect on reducing the possibility of guessing in multiple choice questions. In multiple choice question type, the evaluation by using the set of correct answers is the traditional method. But this method does not respect the order of answers [11]. There is also the evaluation by using vector concept. It is more complicated but respects the order of the answer so that the solution must be exactly similar to the template of the model answer. Reference [12] proposed a fuzzy cognitive map to determine the concepts dependences. It applies the network graphic representation. Fuzzy concepts are used to represent domain concepts' knowledge dependencies and adaptive learning system knowledge representation [13]. It also represents the concept's impact strength over the other related concepts.

In our work, we have used text mining techniques for extracting the keywords from contents of question papers and syllabus file. The extracted keywords are used for the Similarity Coefficient based comparison of open-ended questions of a question paper against the University Prescribed Syllabus File. A syllabus file provide information about how to plan for the tasks and experiences of the semester, how to evaluate and monitor students' performance, and how to allocate time and resources to areas in which more learning is needed. The particular structure that a syllabus includes varies greatly with the type of course that it details. A syllabus can serve students as a model of professional thinking and writing.

The Vector Space Model (VSM) is a popular information retrieval system implementation which facilitates the representation of a set of documents as vectors in the term space [14]. Similarity matrix consisting of a set of questions in a Question paper and the units of a syllabus file is a two dimensional matrix representing the pair-wise similarity of question content with unit content. Pair-wise similarity computation can be performed on different similarity measures. We have used Cosine Similarity Coefficient to assign a similarity score to each pair of compared question vectors and syllabus vectors. Cosine similarity is a measure of similarity between two vectors by measuring the cosine of the angle between them. The cosine of the angle is generally 1.0 for identical vectors and is in the range of 0.0 to 1.0 for non-similar or partially similar vectors. Cosine similarity remains as the most popular measure because of its simple interpretation, easy computation and document length exclusion [15][16].

Table 1 Terminology Used

Term	Meaning
Question Paper(QP)	QP includes questions with its details such as question-no, question-content and question-marks
Syllabus File(SF)	SF includes unit-wise syllabus contents organized as a set of topics and is assigned with unit-wise weightage
N	N is the total number of questions in a QP
M	M is the total number of units in the SF
n_i	n_i refers to the number of questions in which term i appears
m_j	m_i refers to the number of units in which term i appears
Question-Term-Set (question qst$_i$)	It is a set of terms extracted from each question by performing its tokenization,, stop word removal, taxonomy verb removal and stemming
Syllabus-Term-Set (unit unt$_j$)	It is a set of terms extracted from each unit by performing its tokenization,, stop word removal, taxonomy verb removal and stemming
qfreq$_{ij}$	qfreq$_{ij}$ is the frequency of question-term i in question j
sfreq$_{ij}$	freq$_{ij}$ is the frequency of syllabus-term i in unit j
qmaximum frequency(qmaxfreq$_{ij}$)	qmaxfreq$_{ij}$ is the maximum frequency of a question-term in question j
smaximum frequency(smaxfreq$_{ij}$)	smaxfreq$_{ij}$ is the maximum frequency of a syllabus-term in unit j
qterm frequency(qtf$_{ij}$)	qtfij refers to the importance of a question-term i in question j. It is calculated using the formula- $qtfij = \dfrac{qfreq}{qmaxfreq_{ij}}$ (1)
sterm frequency(stf$_{ij}$)	qtfij refers to the importance of a syllabus-term i in unit j. It is calculated using the formula- $stfij = \dfrac{sfreq}{smaxfreq_{ij}}$ (2)
qInverse Document Frequency(qidf$_i$)	qidf$_i$ refers to the discriminating power of question-term i and is calculated as - $qidf_i = \log_2 (N/n_i)$ (3)
sInverse Document Frequency(sidf$_i$)	sidf$_i$ refers to the discriminating power of syllabus-term i and is calculated as - $sidf_i = \log_2 (M/m_i)$ (4)
qtf-qidf weighting(QW$_{ij}$)	It is a weighting scheme to determine weight of a term in a question. It is calculated using the formula- $QW_{ij} = qtf_{ij} \times qidf_i$ (5)
stf-qidf weighting(SW$_{ij}$)	It is a weighting scheme to determine weight of a term in a unit. It is calculated using the formula- $SW_{ij} = stf_{ij} \times sidf_i$ (6)

3 Methodology

The procedure for finding similarity between question content and syllabus content using cosine similarity measure follows the steps as below-

- Pre-processing of Question Content and Syllabus Content
- Computing Question-Vs-Syllabus Similarity Matrix
- Displaying Syllabus Fairness of Question Paper

A brief description of the approaches used for performing each of the above steps is given below-

3.1   Pre-processing of Question Content and Syllabus Content

The five sub-steps involved in pre-processing the question content and respective syllabus content is as follows:

a) **Tokenization:** The set of questions of a question paper as well as the unit-wise contents of syllabus file are treated as collection of strings (or bag of words), which are then partitioned into a list of terms.

b) **Filtering Stop Words:** Stop words are frequently occurring, insignificant words within the question content and also in the syllabus content and are eliminated.

c) **Filtering Taxonomy Verbs:** The taxonomy verbs within the question content are identified and eliminated. Details of verbs and question examples that represent intellectual activity at each level of blooms taxonomy can be found in [17].

d) **Stemming Terms:** Stemming is a heuristic process of cutting off the ends of terms of question content as well as syllabus content for getting the correct root form of the term. There are various word stemmers [18] available for English text and the most commonly used Porter stemmer is considered.

e) **Normalization:** The idea behind normalization is to convert all terms which mean the same, but written in different forms (e.g. CPU and C.P.U) into the same form. We are using the following techniques for performing normalization-

- Lowercase the terms
- Remove special characters

3.2   Computing Question-vs-Syllabus Similarity Matrix

The similarity matrix computation is carried out by considering the matrix representation of vectors which is a natural extension of the existing VSM. Matrix representation considers the questions of a question paper as the row headers and units of the syllabus file as the column headers of the matrix. Each question is represented as a vector of question-terms and each unit of the syllabus file is considered as a vector of syllabus-terms. In the multidimensional matrix of N questions and M units say N × M matrix, each pair of question-term vector and syllabus-term vector gets compared to determine how identical they are by using cosine similarity measure. The term weight scores are calculated according to tf-idf weighting method [19]. tf-idf is the most commonly used scheme to assign weights to individual terms based on their importance in a collection of unit-wise questions as well as unit-wise syllabus contents. Each weight score is calculated as a product of tf and idf. Higher values of idf correspond to question-terms which characterize a question more distinctly than others under a unit. Detail description of tf-idf calculation is shown in Table1. The cosine similarity of question-vector and syllabus-vector say qv and sv is calculated by performing the dot product of question-vector terms and syllabus-vector terms. The calculation of cosine similarity is performed using the following formula-

Similarity $(qv1, sv1) = \text{Cos } \theta = \dfrac{qv1.sv1}{|qv1|.|sv1|}$ \hfill (7)

, where '.' denotes the dot product between vectors qv1 and sv1. |qv1| and |sv1| are the Euclidean norm of qv1 and sv1 vectors. The above formula can be expanded in the following manner.

$$\text{Cos } \theta = \frac{\sum\limits_{i=1}^{n} wi, qv1 \times wi, sv1}{\sqrt{\sum\limits_{i=1}^{n} wi * wi, qv1} \times \sqrt{\sum\limits_{i=1}^{n} wi * wi, sv1}} \tag{8}$$

Sample of a similarity matrix with computed pair-wise similarity say smx,y for n questions and m units is represented in Table 2 below. The computation of similarity of n question with m units is carried out by calculating the similarity of $n \times m$ pairs of question vectors and syllabus vectors.

Table 2 Similarity Matrix Representation

	SV1	SV2	SV3	SV4	SV5	SV6	SV7	...	SVm
QV1	sm_{11}	sm_{12}	sm_{13}	sm_{14}	sm_{15}	sm_{16}	sm_{17}	...	sm_{1m}
QV2	sm_{21}	sm_{22}	sm_{23}	sm_{24}	sm_{25}	sm_{26}	sm_{27}		sm_{2m}
QV3	sm_{31}	sm_{32}	sm_{33}	sm_{34}	sm_{35}	sm_{36}	sm_{37}		sm_{3m}
QV4	sm_{41}	sm_{42}	sm_{43}	sm_{44}	sm_{45}	sm_{46}	sm_{47}		sm_{4m}
QV5	sm_{51}	sm_{52}	sm_{53}	sm_{54}	sm_{55}	sm_{56}	sm_{57}	...	sm_{5m}
QV6	sm_{61}	sm_{62}	sm_{63}	sm_{64}	sm_{65}	sm_{66}	sm_{67}	...	sm_{6m}
QV7	sm_{71}	sm_{72}	sm_{73}	sm_{74}	sm_{75}	sm_{76}	sm_{77}	...	sm_{7m}
...	...	...	...	...		...	...	...	...
QVn									sm_{nm}

3.3 Displaying Syllabus Fairness of Question Paper

The computed similarity matrix is used to generate unit-wise question groups, calculate unit-wise question groups' weightage and compare the calculated weightage against the actual unit-weightage in the syllabus file. The syllabus fairness evaluation carried out using the similarity matrix is considered as a good measure by the subject expert or question paper setter or question paper moderator to revise the questions of a question paper. Choosing suitable threshold value for similarity computation is a difficult task and it is problem dependent. We have considered 0.75 as the threshold value for better recall of questions while generating question groups that satisfy the paper setter specified requirements.

Precision and Recall are commonly used as the metrics to evaluate the accuracy of predictions and the coverage of accurate pairs of comparisons in the information retrieval system. They are computed as –

Precision   = $\dfrac{\text{Number of relevant question-syllabus matches retrieved by the tool}}{\text{Total number of question-syllabus matches retrieved by the tool}}$

Recall   = $\dfrac{\text{Number of relevant question-syllabus matches retrieved by tool}}{\text{Total number of relevant question-syllabus match given by paper setter}}$

4   Problem Formulation

4.1   Problem Statement

Given a question paper of subject S consisting of N questions represented as $QP(S)=\{qst_1, qst_2, \ldots, qst_N\}$ and a syllabus file of S consisting of M units represented as $SF(S)= \{unt_1, unt_2, \ldots, unt_M\}$, the problem is to find unit-wise similar question groups UQG_1, $UQG_2, \ldots, UQG_k$. A question qst_i can be said to belong to unt_j if *similarity (qst_i, unt_j)*$>= \partial$ where ∂ is the user input threshold value to find the similarity.

The *similarity (qst_i, unt_j)* function could use any of the similarity measures available. We have used Cosine similarity to perform the experimental study.

The main modules of this algorithm are shown below.

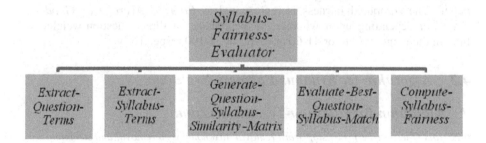

Fig. 1 Main modules of Syllabus-Fairness Evaluator

The brief details of modules are presented below–

1)   Extract-Question-Terms: Input qst_i (i=1 to N) and for each qst_i it extracts terms qt_{ij} (j=1 to N_i).

2)   Extract-Syllabus-Terms: Input unt_j (j=1 to M) and for each unt_j in the syllabus file, it extracts terms st_{jk} (k=1 to M_j).

3)   Generate-Question-Syllabus-Similarity-Matrix: Input question-terms qt_{ij} (j=1 to N_i) for all qst_i (i=1 to N) and also syllabus-terms st_{jk} (k=1 to M_j) for all unt_j (j=1 to M). For each pair of question-terms qst_i and syllabus-terms unt_j, compute similarity (qst_i, unt_j) for i=1 to N and j=1 to M using any standard similarity measuring scheme. Represent the result as a Question-Syllabus-Similarity-Matrix.

4) Evaluate-Best-Question-Syllabus-Match: For each question in the Question-Syllabus-Similarity-Matrix, it finds the highest value of similarity among the set of computed *similarity (qst_i, unt_j)>=∂,* for i=1 to N and j=1 to M. If the highest value of similarity does not get identified for a question, then the question is considered to be indirectly associated with the syllabus and is represented as indirect question else the question is considered to be directly associated with the syllabus and is represented as direct question.

5) Compute-Syllabus-Fairness: Under each unit, it performs the summation of the marks of direct questions and represents the result of summation as unit-direct-question-mark. Also it identifies whether the unit-direct-question-mark of each unit *satisfies- the- unit-weightage* of the syllabus file. The term *satisfies- the- unit-weightage* means that the unit-direct-question-mark is less-than-or-equal-to (<=) the unit-weightage. If the unit-direct-question-mark is greater-than-or-equal-to (>=) the unit-weightage, then the value of unit-direct-question-mark gets replaced with the value of unit-weightage. This replacement process is carried out to limit the value of unit-direct-question-mark to the extent to which it matches with the unit-weightage. At the next stage, it adds up the unit-direct-question-mark of all the units and represents the result of addition as direct-question-weightage. Using the direct-question-weightage, it computes the Syllabus Fairness of the question paper. The computed fairness is represented as *Poor or Average or Good or Excellent* depending upon whether the percentage of direct-question-weightage falls in the range of 0-40 or 41-60 or 61-80 or 81-100 respectively.

4.2 Algorithm for Syllabus-Fairness-Evaluator

1) Algorithm for Question-Term-Extraction

// Extract Question-Terms by stop-word removal, taxonomy verb removal and stemming
    $QT = \{ \}$
    For i=1 to N
            Extract terms from qst_i and store it in array $qt_i[ ]$
            Remove the stop-words from $qt_i[ ]$
            Remove taxonomy verbs from $qt_i[ ]$
            Extract the stem of each term in $qt_i[ ]$
            $QT = QT \cup qt_i[ ]$
    End For
    Output of Question-Term-Extraction, $QT = \{qt_1[ ], qt_2[ ], qt_3[ ], ..., qt_N[ ]\}$

2) Algorithm for Syllabus-Term-Extraction

// Extract Syllabus-Terms by stop-word removal, taxonomy verb removal and stemming
    $ST = \{ \}$

For j=1 to M
  Extract terms from unt_j and store it in array st_j[ ]
  Remove the stop-words from st_j [ ]
  Extract the stem of each term in st_j [ ]
  ST= ST U st_j [ ]
End For
Output of Syllabus-Term-Extraction, ST= {st_1[ ], st_2[ ], st_3 [ ], …, st_M[ ]}

3) *Algorithm for Syllabus-Fairness-Evaluation*

//Evaluate-Syllabus-Fairness
Input: QT= {qt_1[ ],qt_2[ ],qt_3[ ], …, qt_N[ ]} // set of questions in the question paper
*   ST= {st_1[ ], st_2[ ], st_3 [ ], …, st_M[ ]} // set of units in the syllabus file*
where

  qt_i = { qt_{i1} , qt_{i2} , qt_{i3} ,…,qt_{ip}} for p=1 to count(qst_i terms) //set of terms in question i
  st_j = { st_{j1} , st_{j2} , st_{j3} ,…,st_{jq}} for q=1 to count(unt_j terms) //set of terms in syllabus j
  N = {qst_1, qst_2,…,qst_N} // number of selected questions
  M= {unt_1, unt_2,…,unt_M} // number of selected units
  Threshold =∂   // threshold value for similarity computation

Output: k Unit_Question_Group UQG_1, UQG_2,…, UQG_k where UQG_k consist of a set of qst'_i questions of QP(S) // Form unit-wise question groups and verify its syllabus fairness

Begin
//Initialization
  cnt=0 //counter for number of question-groups
  direct_question_percentage=0 // counter for percentage of direct questions
unit_question_set= [ ]
  //Unitwise-Question-Group-Formulation
//Compare the unit-wise marks of questions of QP(S) with the corresponding unit-wise weightages in the syllabus
For i=1 to M
    cnt=cnt+1
    //Formulate unit-wise new question groups
    UQG_i= New_Unit_Question_Group (unt_i,cnt)
  For j=1 to N
    If qst_j not in unit_question_set then
      // Evaluate Best-Question-Syllabus-Match using the Similarity-Matrix
      If similarity (qst_j, unt_i) >=∂ then temp= similarity (qst_j, unt_i)
      For k=1 to M
        If similarity (qst_j, unt_k) >temp   then
          Exit for
        End if
        // Iterative stages of appending questions to each question-groups
        Add qst_j to New_Unit_Question_Group p
          unit_question_set = unit_question_set + qst_j
      End For
    End If
  End If
      End For
      End For

// Evaluate Syllabus-Fairness using marks of unit-wise questions in question-groups
For i=1 to cnt
Accept Unit_Question_Group (unt$_i$,i)
marks_unt$_i$ =sum (marks of all questions of Unit_Question_Group (unt$_i$,i))
 If marks_unt$_i$ <=syllabus-weight(unt$_i$) then
      direct_question_percetage= direct_question_percentage+ marks_unt$_i$
Else
    direct_question_percentage= direct_question_percentage+ syllabus-weight (unt$_i$)
          End If
         End For
         End

5   Implementation Details

5.1   Hardware and Software Platform Used

Implementation is done using Microsoft Visual Basic .NET as Front End Tool and SQL Server as Back End Tool on a 2 GHz processor with 1GB RAM.

5.1.1   Datasets Used

The question paper of the third year of three year bachelor's degree course of computer science (B.Sc Computer Science) for Software Engineering subject examination at Goa University contains 34 questions. The syllabus file for this subject includes 9 units. Details of experimental data used for similarity computation is as follows-

1) S= sub1=Software Engineering (SE)
2) QP(S)=qp1={quest1,quest2,...,quest10,.., quest34}
3) SF(S)={unit1,unit2,..,unit9}
4) ∂=0.75
5) A snapshot of the set of units {sub1unit1,sub1unit2,...,sub1unit9} with its extracted list of terms {approach,characterist,definit...},{category,document,engine,...},....,{black,box,case,debug,...} etc., and set of questions {qp1quest1, qp1quest10, qp1quest11..., qp1quest34} with its extracted list of terms {scope, software, term}, {benefits, code, peer, review}, {cycle, evolutionary, life, model, prototype, software, throwaway}, {alpha, testing} etc., for qp1 is displayed in Fig.2. Extraction of terms from SE subject's syllabus file and SE subject's question paper were carried out by performing four different pre-processing stages such as Tokenization, Filtering of Stop Words, Filtering Taxonomy Verbs and Normalization of Terms.

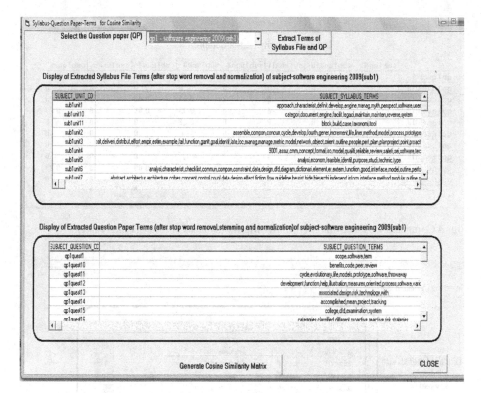

Fig. 2 Extracted list of Terms of Syllabus File and Question Paper

5.1.2 Results Obtained

Fig. 3 below shows a sample screen shot of the similarity matrix computation using cosine similarity. Each of the questions among the 34 questions in the QP of SE was compared against the 9 units in the SE syllabus file. SE-Cosine-Similarity-Matrix generated 34×9 combination of values using cosine similarity. If no similarity exist between a pair of SE question-terms and SE unit-terms, cosine returned a value of zero and in every other case, cosine returned a value in the range of 0.0-1.0.

Fig. 4 below represents the process of generation of SE-unit-question-groups. For each question in the SE-Cosine-Similarity-Matrix, the highest value of similarity among a set of pair-wise *similarity (SE- quest $_i$, SE- unit$_j$)>=75*, for i=1 to 34 and j=1 to 9 were found. When the highest value of similarity could get computed for an *SE-quest*, the question was termed as direct; else the question was represented as indirect.

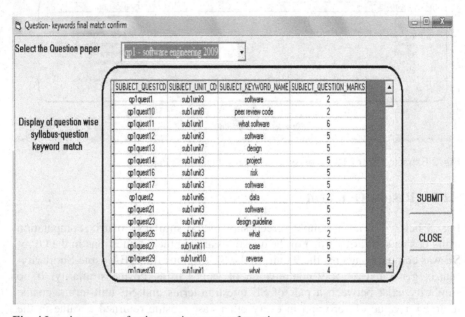

Fig. 3 Computed Cosine Similarity Matrix Matrix (Syllabus File Vs Question Paper)

Fig. 4 Iterative stages of unit-question-groups formation

Fig. 5 below shows the evaluated measure of syllabus fairness. Under each *SE-unit*, marks of direct questions were added up and were named as SE-unit-direct-question-mark. Whether or not the SE-unit-direct-question-mark of each unit could satisfy the SE-unit-weightage of SE-syllabus file was identified and the SE-unit-direct-question-mark is updated accordingly. Summation of the SE-unit-direct-question-mark of all the units was carried out to generate SE-direct-question-weightage. Using SE-direct-question-weightage, SE-Syllabus fairness was computed and was represented as "Good", as the percentage of SE-direct-question-weightage was in the range of 61-80. Performance analysis of the results indicates that cosine similarity is a good measure in grouping similar questions.

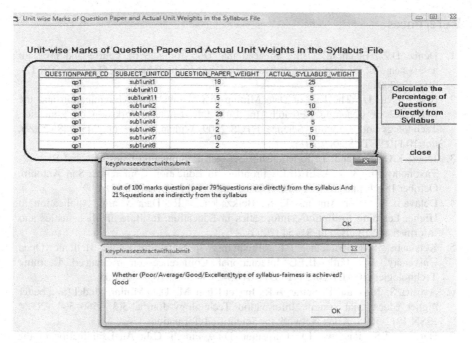

Fig. 5 Computed syllabus-fairness measure using unit-question-groups

6 Conclusion and Future Work

This paper focused on a new approach for syllabus fairness evaluation of a question paper by performing $(n \times m)$ pair-wise question-vector and syllabus-vector comparisons. Similarity matrix computation was carried out using cosine similarity which is a commonly used similarity measure for short documents. Even though cosine similarity has a main disadvantage of term independence, this is not a major concern in our work as we deal with short text documents. Results obtained indicate that cosine similarity is a good measure in formulating unit-wise question groups. The formulated question groups has been found successful in identifying the syllabus fairness of a question paper by comparing the generated unit-wise question groups' weightages against the actual unit-weightage specified in the syllabus file. The generated question groups are useful in situations where novice instructors or the question paper setter or question paper moderator needs to evaluate the syllabus fairness of a question paper and revise the questions of examination question paper accordingly. The primary objective of this study was to identify the effectiveness of statistical measures in formulating similar question groups and evaluating the fairness of a question paper. Our future work will focus on replacing the statistical approaches of similarity matrix generation by latent semantic approaches.

References

1. Deniz, D.Z., Ersan, I.: Using an academic DSS for student, course and program assessment. In: International Conference on Engineering Education, Oslo, pp. 6B8-12–6B8-17 (2001)
2. Liu, J., Wu, D.: The Use of Data Mining in Contents of the Examination and the Relevance of Teaching Research. In: First International Workshop on Education Technology and Computer Science, ETCS 2009, Wuhan, Hubei, pp. 745–748 (2009), doi:10.1109/ETCS.2009.701
3. Jones, K.O., Harland, J.: Relationship between Examination Questions and Bloom's Taxonomy. In: 39th ASEE/IEEE Frontiers in Education Conference, San Antonio, October 18-21, pp. 978–971 (2009) 978-1-4244-4714-5/09/$25.00 ©2009
4. Delavari, N., Phon-Amnuaisuk, S., Beikzadeh, M.R.: Data Mining Application in Higher Learning Institutions, Informatics in Education, Institute of Mathematics and Informatics. Vilnius 7(1), 31–54 (2008)
5. Kotsiantis, S.B., Pintelas, P.E.: Predicting students' marks in Hellenic Open University. In: Fifth IEEE International Conference on Advanced Learning Technologies (ICALT 2005) (2005) 0-7695-2338-2/05 $20.00 © 2005
6. Ayesha, S., Mustafa, T., Sattar, A.R., Inayat Khan, M.: Data Mining Model for a better higher educational system. Information Technology Journal 5(3), 560–564 (2006) ISSN 1812-5638; Asian Network for Scientific Information
7. Harding, T.S., Passow, H.J., Carpenter, D.D., Finelli, C.J.: An Examination of the relationship between academic dishonesty and professional behaviour. In: 33rd ASEE/IEEE Frontiers in Education Conference (November 2003) 0-7803-7444-4/03/$17.00 2003
8. Matsuo, Y., Ishizuka, M.: Keyword Extraction from a Single Document using Word Co-occurrence Statistical Information. Journal on Artificial Intelligence Tools (2003)
9. Lui, Y., Brent, R., Calinescu, A.: Extracting significant phrases from text. In: 21st International Conference on Advanced Information Networking and Applications Workshops, AINAW 2007, Niagara Falls, Ont. (2007), doi:10.1109/AINAW.2007.180
10. Chang, S.-H., Lin, P.-C., Lin, Z.-C.: Measures of partial knowledge and unexpected responses in multiple-choice tests. International Forum of Educational Technology & Society (IFETS), 95–109 (2007)
11. Radoslav, F., Michal, R.: Networked digital technologies, CCIS, Cairo, vol. 88, pp. 883–888. Springer (2010), doi:10.1109/ISDA.2010.5687151
12. Konstantina, C., Maria, V.: A knowledge representation approach using fuzzy cognitive maps for better navigation support in an adaptive learning system. Chrysafiadi and Virvou Springer Plus 2, 81 (2013), http://www.springerplus.com/content/2/1/81, doi:10.1186/2193-1801-2-81
13. Réka, V.: Educational ontology and knowledge testing. Electronic Journal of Knowledge Management, EJKM 5(1), 123–130 (2007)
14. Wong, K.M., Raghavan, V.V.: Vector space model of information retrieval: a reevaluation. In: Proceedings of the 7th Annual International ACM SIGIR Conference on Research and Development in Information Retrieval, SIGIR 1984, pp. 167–185. British Computer Society, Swinton (1984)

15. Tsinakos, A., Kazanidis, I.: Identification of Conflicting Questions in the PARES System. The International Review of Research in Open and Distance Learning 13(3) (June 2012)
16. Hage, H., Aimeru, E.: ICE: A system for identification of conflicts in exams. In: Proceedings of IEEE International Conference on Computer Systems and Applications, pp. 980–987 (2006)
17. Krathwohl, D.R.: A revision of Bloom's taxonomy: An overview. Theory into Practice 41(4), 212–219 (2002)
18. Paice Chris, D.: An evaluation method for stemming algorithms. In: Proceedings of the 17th Annual International ACM SIGIR Conference on Research and Development in Information Retrieval, pp. 42–50 (1994)
19. Aizawa, A.: The feature quantity: an information-theoretic perspective of tfidf-like measures. In: Proceedings of the 23rd ACM SIGIR Conference on Research and Development in Information Retrieval, pp. 104–111 (2000)

Automatic Stemming of Words for Punjabi Language

Vishal Gupta

Abstract. The major task of a stemmer is to find root words that are not in original form and are hence absent in the dictionary. The stemmer after stemming finds the word in the dictionary. If a match of the word is not found, then it may be some incorrect word or a name, otherwise the word is correct. For any language in the world, stemmer is a basic linguistic resource required to develop any type of application in Natural Language Processing (NLP) with high accuracy such as machine translation, document classification, document clustering, text question answering, topic tracking, text summarization and keywords extraction etc. This paper concentrates on complete automatic stemming of Punjabi words covering Punjabi nouns, verbs, adjectives, adverbs, pronouns and proper names. A suffix list of 18 suffixes for Punjabi nouns and proper names and a number of other suffixes for Punjabi verbs, adjectives and adverbs and different stemming rules for Punjabi nouns, verbs, adjectives, adverbs, pronouns and proper names have been generated after analysis of corpus of Punjabi. It is first time that complete Punjabi stemmer covering Punjabi nouns, verbs, adjectives, adverbs, pronouns, and proper names has been proposed and it will be useful for developing other Punjabi NLP applications with high accuracy. A portion of Punjabi stemmer of proper names and nouns has been implemented as a part of Punjabi text summarizer in MS Access as back end and ASP.NET as front end with 87.37% efficiency

Keywords: Punjabi Stemming, Punjabi Noun Stemming, Punjabi Verb Stemmer, Punjabi Names Stemmer, Punjabi Adjective Stemmer.

1 Introduction

Stemming [1] is a method that reduces morphologically similar variant of words into a single term called stems or roots without doing complete morphological

Vishal Gupta
University Institute of Engineering & Technology,
Panjab University Chandigarh, India
e-mail: vishal@pu.ac.in

S.M. Thampi, A. Gelbukh, and J. Mukhopadhyay (eds.), *Advances in Signal Processing and Intelligent Recognition Systems*, Advances in Intelligent Systems and Computing 264,
DOI: 10.1007/978-3-319-04960-1_7, © Springer International Publishing Switzerland 2014

analysis. A stemming method stems the words talking, talked, talk and talks to the word, talk. The major task of a stemmer is to find root words that are not in original form and hence absent in the dictionary. The stemmer after stemming finds the word in the dictionary. If a match is not found in dictionary, then it can be some incorrect word or a name, otherwise the word is correct. For any language in the world, stemmer is a basic linguistic resource required to develop any type of application in Natural Language Processing (NLP) with high accuracy like: machine translation, document classification, document clustering, text question answering, topic tracking, text summarization and keywords extraction etc. In information Retrieval system the Stemming is used to enhance performance. When a particular user searches the word aligning, he may want to find all documents containing words align and aligned as well.

The stemmer's development is based on language and it needs some linguistic knowledge about the language and spelling checker for that language. Most of the simple stemmer involves suffix stripping using a list of endings while some complex stemmer requires morphological knowledge of the language to find root word.

In comparison with English and other European languages very little work has been done for Indian regional languages in the area of stemming. Punjabi language has rich inflectional morphology in contrast to English. A verb in Punjabi has nearly 48 verb forms that are based on gender, number, tense, person and aspect value but a verb in English has only five different inflectional forms. For example different verb forms of 'eat' in English are eat, ate, eaten, eats and eating. Based on syntax and semantics Punjabi language completely differs from other languages of the world.

This paper concentrates on complete automatic stemming of Punjabi words covering Punjabi nouns, verbs, adjectives, adverbs, pronouns and proper names. A suffix list of 18 suffixes for Punjabi nouns and proper names and a number of other suffixes for Punjabi verbs, adjectives and adverbs and different stemming rules for Punjabi nouns, verbs, adjectives, adverbs, pronouns and proper names have been generated after analysis of Punjabi corpus. Some examples of stemmer for names/nouns in Punjabi with distinct endings are: ਕੁੜੀਆਂ kuriāṃ "girls" → ਕੁੜੀ kuri "girl" with ending ਿਆਂ iāṃ, ਲੜਕੇ larkē "boys" → ਲੜਕਾ muṇḍā "boy" with ending ੇ ē, ਫਿਰੋਜ਼ਪੁਰੋਂ phirōzpurōṃ → ਫਿਰੋਜ਼ਪੁਰ phirōzpur with ending ੋਂ ōṃ and ਰੁੱਤਾਂ ruttāṃ "seasons" → ਰੁੱਤ rutt "season" with ending ਾਂ āṃ etc. It is first time that complete Punjabi stemmer covering Punjabi nouns, verbs, adjectives, adverbs, pronouns, and proper names has been proposed and it will be useful for developing other Punjabi NLP applications with high accuracy. A portion of Punjabi stemmer for proper names and nouns has been implemented as part of Punjabi text summarizer in MS Access as back end and ASP.NET as front end with 87.37% efficiency

2   Related Work

Porter (1980) [1] proposed a rule based method for suffix stripping which is used widely for stemming of English words. The rule based removal of suffixes is suitable for less inflectional languages such as English and is useful in the area of retrieval of Information. The removal of suffixes leads to reduction in the total terms in Information Retrieval system which reduces the data complexity and is hence quiet useful. Jenkins and Smith (2005) [2] also proposed a rule based stemmer that works in steps and perform conservative stemming for both searching and indexing. It detects the word that need not be stemmed like proper nouns and stems only orthographically correct words. Mayfield and McNamee (2003) [3] suggested that N-gram can be selected as a stem is quiet useful and efficient method for certain languages. It analyzes the distribution of N-grams in the text where the value of N is selected empirically and even some high values of N such as 4 or 5 is selected. Massimo and Nicola (2003) [4] suggested a Hidden Markov based statistical method for stemming of words. It makes use of unsupervised training at the time of indexing and does not require any linguistic knowledge. The transition functions of finite-state automata of HMMs are described with the help of probability functions. Goldsmith (2001) [5] described the text in compact manner and suggested a method for morphology of language with the help of minimum description length (MDL). Creutz and lagus (2005) [6] performed morpheme segmentation using statistical maximum a posteriori (MAP) technique. The stemmer takes a corpus and gives segmentation of various word in it as an output which is similar to the linguistic morpheme segmentation

In comparison with English and other European languages very little work has been done for Indian regional languages in the area of stemming. Rao and Ramanathan (2003) [7] suggested a Hindi lightweight stemmer that performs longest match stripping of suffixes based on manually created suffix list. Islam et al. (2007) [8] proposed an approach similar to the approach used in Hindi lightweight stemmer for Bengali language that can also be used as spelling checker. It also performs longest match suffix stripping by making use of 72 endings for verbs, 22 for nouns and 8 for adjectives for Bengali language. Majumder et al. (2007) [9] proposed an approach YASS: Yet Another Suffix Stripper. It calculates four different string distances and then on the basis of these distances makes clusters of lexicon such that each cluster points to a single root word. Their stemmer did not require any prior language knowledge and suggested that stemming enhances recall of Information Retrieval Systems for regional languages. Dasgupta and Ng (2006) [10] performed unsupervised morphological analysis of Bengali wherein the words are segmented into suffixes, prefixes, stems without any knowledge of morphology of the language. Pandey and Siddiqui (2008) [11] proposed split-all method based unsupervised stemmer for Hindi. EMILLE corpus has been used for unsupervised training. The words are divided to provide n-gram (n=1, 2, 3 … k) endings, where k is the word length. And then endings and stem probability is calculated and compared. Majgaonker and

Siddiqui (2010) [12] proposed combination of stripping, rule based and statistical stripping for generation of suffix rules and developed an unsupervised stemmer for Marathi language.

3   Stemming of Punjabi Words

Punjabi stemmer finds the stem/root of the word and then checks it in the Punjabi, morph/dictionary, Punjabi noun dictionary or names dictionary for finding Punjabi nouns, verbs, adjectives, pronouns, or Punjabi names. Punjabi noun corpus has 37297 nouns. First the Punjabi Language stemmer segments the Punjabi text that is entered into words. Analysis of Punjabi corpus containing nearly 2.03 lakh distinct words with a total of 11.29 million words from famous Punjabi newspaper Ajit has been done.

3.1   Stemming for Punjabi Verbs

Thirty six (36) suffixes for Punjabi verbs were found after analyzing Punjabi news corpus. These suffixes are given in Table1.

Table 1 Suffix List for Punjabi verbs

Sr. No.	Punjabi verb suffix	Example Punjabi verb	Punjabi verb root word
1	੦ਉਦਿਆਂ	ਗਾਉਦਿਆਂ (singing) Gender: Feminine; Plural	ਗਾ (sing)
2	ਉਂਦਾ	ਪੜ੍ਹਉਂਦਾ (Teaches) Gender: Masculine; Singular	ਪੜ੍ਹ (Teach)
3	ਉਂਦੀ	ਪੜ੍ਹਉਂਦੀ (Teaches) Gender: Feminine; Singular	ਪੜ੍ਹ (Teach)
4	ਉਂਦੇ	ਪੜ੍ਹਉਂਦੇ (Teach) Gender: Masculine; Plural	ਪੜ੍ਹ (Teach)
5	ਉਣੀਆਂ	ਪੜ੍ਹਉਣੀਆਂ (Teach) Gender: Feminine; Plural	ਪੜ੍ਹ (Teach)
6	ਉਣਾ	ਪੜ੍ਹਉਣਾ (Teach) Gender: Masculine; Singular	ਪੜ੍ਹ (Teach)
7	ਉਣੀ	ਪੜ੍ਹਉਣੀ (Teach) Gender: Feminine; Singular	ਪੜ੍ਹ (Teach)
8	ਉਣੇ	ਪੜ੍ਹਉਣੇ (Teach) Gender: Masculine; Plural	ਪੜ੍ਹ (Teach)
9	੦ਦਾ	ਪੜ੍ਹਂਦਾ (Teaches) Gender: Masculine; Singular	ਪੜ੍ਹ (Teach)
10	੦ਦੀ	ਪੜ੍ਹਂਦੀ (Teaches) Gender: Feminine; Singular	ਪੜ੍ਹ (Teach)
11	੦ਦੇ	ਪੜ੍ਹਂਦੇ (Teach) Gender: Masculine; Plural	ਪੜ੍ਹਂ (Teach)
12	ਿਆ	ਭੱਜਿਆ (ran) Gender: Masculine;: Singular	ਭੱਜ (run)
13	੦ਆਂ੦	ਭੱਜੀਆਂ (ran) Gender: Feminine; Singular	ਭੱਜ (run)
14	ਇਆ	ਪੜ੍ਹਇਆ (Taught) Gender:X; Singular/Plural	ਪੜ੍ਹ (Teach)

Table 1 (*continued*)

15	ਈ	ਪੜ੍ਹਾਈ (studying) Gender: X; Singular	ਪੜ੍ਹ (study)
16	ਏ	ਪੜ੍ਹਾਏ (Taught) Gender: X; Singular/Plural	ਪੜ੍ਹਾ (Teach)
17	ਵਾਂ	ਪੜ੍ਹਾਵਾਂ (will teach) Gender: Masculine; Singular	ਪੜ੍ਹਾ (Teach)
18	ਵੀ	ਪੜ੍ਹਾਵੀਂ (will teach) Gender: Feminine; Singular	ਪੜ੍ਹਾ (Teach)
19	ਵੇਂ	ਪੜ੍ਹਾਵੇਂ (will teach) Gender:X; Plural	ਪੜ੍ਹਾ (Teach)
20	੍ਹਾ	ਪੜ੍ਹਾ (read) Gender:X; Singular	ਪੜ੍ਹ (read)
21	ਈ	ਖਾਈ (ate) Gender:Feminine; Plural	ਖਾ (eat)
22	ਏ	ਖਾਏ (eat) Gender:Masculine; Plural	ਖਾ (eat)
23	ਦਾ	ਪੜ੍ਹਦਾ (reads) Gender:Masculine; Singular	ਪੜ੍ਹ (read)
24	ਦੀ	ਪੜ੍ਹਦੀ (reads) Gender:Feminine; Singular	ਪੜ੍ਹ (read)
25	ਦੇ	ਪੜ੍ਹਦੇ (read) Gender:X; Plural	ਪੜ੍ਹ (read)
26	ੀ	ਭੱਜੀ (ran) Gender: Feminine; Singular	ਭੱਜ (run)
27	ਾ	ਭੱਜਾ (ran) Gender: Masculine; Singular	ਭੱਜ (run)
28	ੇ	ਭੱਜੇ (should run) Gender: X; Plural	ਭੱਜ (run)
29	ੇ	ਭੱਜੇ (escaped) Gender:Masculine; Plural	ਭੱਜ (escape)
30	ੇਗਾ	ਭੱਜੇਗਾ (will run) Gender: Masculine; Singular	ਭੱਜ (run)
31	ੇਗੀ	ਭੱਜੇਗੀ (will run) Gender: Feminine; Singular	ਭੱਜ (run)
32	ੇਗੇ	ਭੱਜੇਗੇ (will run) Gender: X; Plural;	ਭੱਜ (run)
33	ਣਗੇ	ਭੱਜਣਗੇ (will run) Gender: Masculine; Plural	ਭੱਜ (run)
34	ਣਗੀਆਂ	ਭੱਜਣਗੀਆਂ(will run) Gender: Feminine; Plural	ਭੱਜ (run)
35	ਏਗਾ	ਗਾਏਗਾ (will sing) Gender: Masculine; Singular	ਗਾ (sing)
36	ਏਗੀ	ਗਾਏਗੀ (will sing) Gender: Feminine; Singular	ਗਾ (sing)

Algorithm for Punjabi verb stemming

Algorithm Input:
ਪੜ੍ਹਦੇ, ਭੱਜਣਗੇ, ਗਾਏਗਾ

Algorithm Output:
ਪੜ੍ਹ, ਭੱਜ, ਗਾ

Step 1: The Punjabi word is given as input to Punjabi verb stemmer.

Step 2: If the entered word matches with a word in Punjabi dictionary then it is returned as output. Otherwise if suffix of the entered word lies in any of the verb suffixes: ਾਉਂਦਿਆਂ, ਉਂਦਾ, ਉਂਦੀ, ਉਂਦੇ, ਉਣੀਆਂ, ਉਂਾ, ਉਣੀ, ਉਣੇ, ੰਦਾ, ੰਦੀ, ੰਦੇ, ਿਆ,

ਗੀਆਂਹ, ਇਆ, ਈ, ਏ, ਵਾਂ, ਵੀਂ, ਵੇਂ, ਣਾ, ਣੀ, ਣੇ, ਦਾ, ਦੀ, ਦੇ, ਹੀ, ਹਾ, ਹੇ, ਹੇ, ਹੇਗਾ, ਹੇਗੀ, ਹੇਗੇ, ਣਗੇ, ਣਗੀਆਂ, ਏਗਾ, ਏਗੀ then delete ending from end and search the stemmed Punjabi word in Punjabi dictionary. If resultant word is found then it is returned as output.

Step 3: If the resultant Punjabi word is not found in Punjabi dictionary then verb stemmer is unable to locate the actual verb and error message is displayed.

3.2 Stemming for Punjabi Adjectives/Adverbs

Five suffixes for Punjabi adjectives/adverbs were found after analyzing the Punjabi news corpus. The suffix list is given below:

Table 2 Suffix List for Punjabi Adjectives/Adverbs

Sr. No.	Punjabi Adjective / Adverb suffix	Example Punjabi Adjective/Adverb	Punjabi Adjective/Adverb root word
1	ਆਂ	ਚੰਗੀਆਂ(good) Gender: Feminine; Plural	ਚੰਗੀ (good)
2	ਈਆਂ	ਉਚਾਈਆਂ (Heights) Gender:X; Plural	ਉਚਾਈ (Height)
3	ਿਆ	ਸੋਹਣਿਆ (Handsome)Gender: Masculine; Singular	ਸੋਹਣਾ (Handsome)
4	ਿਏ	ਸੋਹਣਿਏ (Beautiful) Gender: Feminine; Singular	ਸੋਹਣੀ (Beautiful)
5	ਿਓ	ਸੋਹਣਿਓ (Beautiful) Gender: X; Plural	ਸੋਹਣਾ (Beautiful)

Algorithm for Punjabi adjective/adverb stemming

Algorithm Input:
ਚੰਗੀਆਂ, ਸੋਹਣਿਆ, ਸੋਹਣਏ

Algorithm Output:
ਚੰਗੀ, ਸੋਹਣਾ, ਸੋਹਣੀ

Step 1: The Punjabi word is given as input to Punjabi adjective/adverb stemmer.

Step 2: If the entered Punjabi word matches with a word in Punjabi dictionary then it is returned as output. Otherwise if ending of entered word is ਆਂ or ਈਆਂ then remove the respective ending from the end and then search the stemmed Punjabi word in Punjabi dictionary. If the word is found then it is returned as output.

Step 3: If ending of entered word is ਿਆ or ਿਓ then delete the ending and add ਾ at the end and search the stemmed Punjabi word in Punjabi dictionary. If resultant word matches with a word in the dictionary then it is returned as output.

Step 4: If ending of entered word is ਿਏ or ਿਉ then delete and add ੀ at the end and search the stemmed Punjabi word in Punjabi dictionary. If resultant word is found then it is returned as output.

Step 5: If resultant word is not found then adjective/adverb stemmer is unable to locate the actual adverb/adjective and error message is displayed.

3.3 Stemming for Punjabi Pronouns

13 suffixes for Punjabi pronouns were found after analyzing the Punjabi news corpus these are as given in Table 3.

Table 3 Suffix List for Punjabi Pronouns

Sr. No.	Punjabi Adjective/Adverb suffix	Example Punjabi Adjective/Adverb	Punjabi Adjective/Adverb root word
1	ੋ	ਇਹੋ (This one) Gender: X; Singular	ਇਹ (This)
2	ੇ	ਸਾਡੇ (ours) Gender:X; Plural	ਸਾਡਾ (our)
3	ਾਂ	ਆਪਾਂ (we) Gender:X; Plural	ਆਪ (ourselves)
4	ਾਂ	ਤੁਹਾਡੀਆਂ (Your's) Gender:X; Singular	ਤੁਹਾਡੀ (Your)
5	ਦਾ	ਉਹਦਾ (His) Gender:Masculine; Singular	ਉਹ (He)
6	ਦੀ	ਉਹਦੀ (Her) Gender:Feminine; Singular	ਉਹ (She)
7	ਦੇ	ਉਹਦੇ (His/her) Gender:X ; Plural	ਉਹ (He/She)
8	ਜੀ	ਆਪਜੀ (Your's) Gender:X; Singular	ਆਪ (Your)
9	ੀ	ਇਹੀ (This only) Gender:X; Singular	ਇਹ (This)
10	ਨੀ	ਇਹਨੀ (Their) Gender:Feminine; Singular	ਇਹ (This)
11	ਨਾ	ਇਹਨਾ (Their) Gender:Masculine; Singular	ਇਹ (This)
12	ਨੇ	ਇਹਨੇ (His) Gender: Masculine; Singular	ਇਹ (This)
13	ਨਾਂ	ਇਹਨਾਂ (Theirs) Gender:X; Plural	ਇਹ (This)

Algorithm for Punjabi adjective/adverb stemming

Algorithm Input:
ਇਹਨਾ, ਆਪਣੇ, ਆਪਜੀ
Algorithm Output:
ਇਹ, ਆਪਣਾ, ਆਪ

Step 1: The Punjabi word is entered as input to Punjabi Pronoun stemmer.

Step 2: If entered word is found in dictionary then it is returned as output. Otherwise if ending of entered word is ਿ or ਿਆ or ਆਂ or ਦਾ or ਦੀ or ਦੇ or ਜੀ or ਿੀ or ਨੀ or ਨਾ or ਨੇ or ਨਾਂ then remove the ending from the end and then search the resultant stemmed Punjabi word in Punjabi dictionary. If word is found then it is returned as output.

Step 3: If ending of entered word is ਿ then remove the ending from the end and add ਾ at the end and then search the stemmed Punjabi word in Punjabi dictionary. If stemmed word is found then it is returned as output.

Step 4: If resultant word is not found in dictionary then Punjabi pronoun stemmer is unable to locate the actual pronoun and error message is displayed.

3.4 Stemming for Punjabi Proper Names and Nouns

Punjabi stemmer for names/nouns [14][15], the goal is to locate the root words and then the root words are matched with the words in dictionary for Punjabi names and nouns dictionary. 18 endings were listed for Punjabi names/nouns after analyzing the Punjabi corpus like ਆਂ āṃ, ਿਆਂ iāṃ, ਉਆਂ ūāṃ and ਿੀਆਂ īāṃ etc. and distinct rules for Punjabi name/noun stemming have been developed. Some examples of Punjabi names/nouns for distinct endings are:

ਕੁੜੀਆਂ kurīāṃ "girls" → ਕੁੜੀ kuri "girl" with ending ਿੀਆਂ īāṃ,, ਫਿਰੋਜ਼ਪੁਰੋਂ phirōzpurōṃ → ਫਿਰੋਜ਼ਪੁਰ phirōzpur with ending ਿੋਂ ōṃ, ਲੜਕੇ larkē "boys" → ਲੜਕਾ muṇḍā "boy" with ending ਿੇ ē and ਰੁੱਤਾਂ ruttāṃ "seasons" → ਰੁੱਤ rutt "season" with ending ਆਂ āṃ etc.

Algorithm for Punjabi Nouns/ Proper Names:

Algorithm Input:
ਲੜਕੇ larkē "boys" and ਰੁੱਤਾਂ ruttāṃ "seasons"
Algorithm Output:
ਲੜਕਾ muṇḍā "boy" and ਰੁੱਤ rutt "season"

Step 1: The Punjabi word is given as input to Punjabi noun/proper name stemmer.

Step 2: If entered Punjabi word matches with a word in names dictionary/ nouns dictionary then it is returned as output. Else If ending of entered word is ਆਂ āṃ (

in case of ਉਆਂ ūāṃ, ਿਆਂ iāṃ and ੀਆਂ īāṃ), ਏ ē (in case of ਿਏ īē), ਓ ō (in case of ਿਓ īō), ਆ ā (in case of ੀਆ ā, ਈਆ īā), ਵਾਂ vāṃ, ਈ ī , ਾਂ āṃ, ੀਂ īṃ, ਜ/ਜ਼/ਸ ja/z/s and ੋਂ ōṃ then remove the ending and then go to Step 5.

Step 3: If entered has ending ੋ ō, ਿਓ īō, ੇ ē, ਿਆ iā and ਿਉਂ iuṃ then remove the ending and add kunna at the end and then go to Step 5.

Step4: Else entered word is either incorrect word or incorrect word.

Step 5: Resultant Stemmed word is matched with the words of Punjabi names-dictionary/noun-morph. If it matches, then it is Punjabi-name or noun.

3.5 Overall Algorithm for Complete Stemming of Punjabi Words

Algorithm Input:
ਮੁੰਡੇ (boys), ਫੁੱਲਾਂ (flowers), ਇਹਨਾ (Theirs), ਆਪਣੇ (yours), ਪੜ੍ਹਦਾ (reads), ਭੱਜਣਗੇ (will run),
Algorithm Output:
ਮੁੰਡਾ (boy), ਫੁੱਲ (flower), ਇਹ (this), ਆਪ (your), ਪੜ੍ਹ (read), ਭੱਜ (run)

The overall algorithm for Complete stemming of Punjabi words is given below:

Step 0: The Punjabi word is given as input to Punjabi Stemmer.

Step 1: If entered word is found in dictionary/ noun morph/ names dictionary then it is returned as output. Else go to step 2.
Step 2: Apply stemmer for Proper Names and Nouns [14] [15]. If the stemmed word after stemming is located in the Punjabi names dictionary/ noun morph then it is given as output. Else go to step 3.

Step 3: Apply stemmer for Punjabi verbs. If the stemmed word after stemming is located in the Punjabi dictionary then it is given as output. Else go to step 4.

Step 4: Apply stemmer for Punjabi Adjectives/Adverbs. If the stemmed word after stemming is located in the Punjabi dictionary then it is given as output. Else go to step 5.

Step 5: Apply stemmer for Punjabi pronouns. If the stemmed word after stemming is located in the Punjabi dictionary then it is given as output. Else stemmer is unable to locate the actual root word and error message is displayed.

4   Results and Discussions

Punjabi stemming algorithm has been applied on fifty documents of news corpus of Punjabi and efficiency is found to be 87.37%. Table 4 shows efficiency and error percentage analysis of Punjabi verb stemmer, adverb/adjective stemmer, noun/proper name stemmer, pronoun stemmer and overall Punjabi stemmer.

Table 4 Accuracy and Error Percentage Analysis of Punjabi Stemmer

Punjabi Verb Stemmer	Adjective/Adverb Punjabi Stemmer	Punjabi Pronoun Stemmer	Punjabi Nouns/Proper Names	Overall Accuracy of Punjabi Stemmer
Accuracy: 84.67%	Accuracy: 85.87% Errors: 14.13%	Accuracy: 90.88%	Accuracy: 87.37%	Accuracy: 87.2%
Errors:15.33%		Errors: 9.12%	Errors: 12.63%	Errors: 12.8%

Errors are because of non existence of some of suffixes in Punjabi stemmer and also due to absence of certain root words in Punjabi dictionary. The efficiency of stemmer is words that are stemmed correctly is to the total number of words stemmed by stemmer. In the same way efficiency of each rule of stemmer is number of correct outputs of that rule is to total outputs produced by the rule.

In case of stemmer for proper names and nouns [14][15], three categories of errors are possible: 1) Syntax mistakes 2) Dictionary errors 3) Violation of stemming-rules. Syntax errors are the errors due to incorrect syntax and it occurs due to typing mistakes. Dictionary errors are the errors due to dictionary i.e. the stemmed word is not present in names/noun dictionary but it is a noun or a name actually. The typing

Errors are 0.45%, dictionary errors are 2.4% and rules violation errors are 9.78%.

Examples of rules violation errors are as given below:

The word ਹਲਕੇ halkē "light weight" is an adjective and ਬਦਲੇ badlē "in lieu of" is an adverb. The words are not matched with the words in names/nouns dictionary, though they come under ੋ ē rule which takes them as noun, which is not true.

Dictionary errors examples are given below:

The words like ਮੁਨਾਫ਼ਿਆਂ munāphaiāṃ "profits" and ਪ੍ਰਦੇਸਾਂ pradēsāṃ "foreign" are nouns but are absent in Punjabi dictionary /noun morph. The words come under ਾਂ āṃ rule and ਿਆਂ iāṃ rule of stemmer and after noun stemming their root words ਮੁਨਾਫ਼ਾ munāphā "profit" and ਪ੍ਰਦੇਸਾਂ pradēsāṃ "foreign" are also absent in dictionary /noun morph.

Syntax errors examples are as given below:

It is possible that the spellings of certain noun words are typed incorrectly, words such as ਆਕ੍ਰਤੀ ākrtī "shape" and ਚਿੜੀਆ chiḍīā "sparrow" instead of their correct spellings ਚਿੜੀਆ chiṛīā "sparrow" and ਆਕ੍ਰਿਤੀ ākritī "shape" respectively.

5   Conclusions

In this paper, I have proposed the Complete Punjabi language stemmer covering Punjabi verbs, Punjabi nouns, Punjabi proper names, Punjabi adjectives/adverbs, and Punjabi pronouns. The resources used like Punjabi noun morph, Punjabi proper names list etc. had to be done from scratch because no work had been done in the area. A detailed analysis of news corpus is done using manual and automatic tools for development of these resources. The resources for Punjabi language are developed for first time and they can be useful in a lot of Punjabi NLP applications. A part of stemmer in Punjabi covering stemmer for Punjabi proper names and nouns [14][15] is used   successfully in Text Summarization for Punjabi language [16].

References

1. Porter, M.: An Algorithm for Suffix Stripping Program 14, 130–137 (1980)
2. Jenkins, M., Smith, D.: Conservative Stemming for Search and Indexing. In: Proceedings of SIGIR 2005 (2005)
3. Mayfield, J., McNamee, P.: Single N-gram stemming. In: Proceedings of the 26th Annual International ACM SIGIR Conference on Research and Development in Information Retrieval, pp. 415–416 (2003)
4. Massimo, M., Nicola, O.: A Novel Method for Stemmer Generation based on Hidden Markov Models. In: Proceedings of the Twelfth International Conference on Information and Knowledge Management, pp. 131–138 (2003)
5. Goldsmith, J.A.: Unsupervised Learning of the Morphology of a Natural Language. Computational Linguistics 27, 153–198 (2001)
6. Creutz, M., Lagus, K.: Unsupervised Morpheme Segmentation and Morphology Induction from Text Corpora using Morfessor 1.0. Publications of Computer and Information Science, Helsinki University of Technology (2005)
7. Ramanathan, A., Rao, D.D.: A Lightweight Stemmer for Hindi. In: Proceedings of Workshop on Computational Linguistics for South-Asian Languages, EACL (2003)
8. Islam, M.Z., Uddin, M.N., Khan, M.: A Light Weight Stemmer for Bengali and its Use in Spelling Checker. In: Proceedings of. 1st Intl. Conf. on Digital Comm. and Computer Applications (DCCA 2007), Irbid, Jordan, pp. 19–23 (2007)
9. Majumder, P., Mitra, M., Parui, S.K., Kole, G., Datta, K.: YASS Yet Another Suffix Stripper. Association for Computing Machinery Transactions on Information Systems 25, 18–38 (2007)
10. Dasgupta, S., Ng, V.: Unsupervised Morphological Parsing of Bengali. Language Resources and Evaluation 40, 311–330 (2006)

11. Pandey, A.K., Siddiqui, T.J.: An Unsupervised Hindi Stemmer with Heuristic Improvements. In: Proceedings of the Second Workshop on Analytics For Noisy Unstructured Text Data, vol. 303, pp. 99–105 (2008)
12. Majgaonker, M.M., Siddiqui, T.J.: Discovering Suffixes: A Case Study for Marathi Language. Proceedings of International Journal on Computer Science and Engineering 2, 2716–2720 (2010)
13. Suba, K., Jiandani, D., Bhattacharyya, P.: Hybrid Inflectional Stemmer and Rule-based Derivational Stemmer for Gujarati. In: Proceedings of the 2nd Workshop on South and Southeast Asian Natural Language Processing (WSSANLP) IJCNLP 2011, Chiang Mai, Thailand, pp. 1–8 (2011)
14. Gupta, V., Lehal, G.S.: Punjabi Language Stemmer for Nouns and Proper Names. In: Proceedings of the 2nd Workshop on South and Southeast Asian Natural Language Processing (WSSANLP) IJCNLP 2011, Chiang Mai, Thailand, pp. 35–39 (2011)
15. Gupta, V., Lehal, G.S.: Preprocessing Phase of Punjabi Language Text Summarization. In: Singh, C., Singh Lehal, G., Sengupta, J., Sharma, D.V., Goyal, V. (eds.) ICISIL 2011. CCIS, vol. 139, pp. 250–253. Springer, Heidelberg (2011)
16. Gupta, V., Lehal, G.S.: Automatic Punjabi Text Extractive Summarization System. In: Proceedings of International Conference on Computational Linguistics COLING 2012, pp. 191–198 (2012)

A Knowledge Based Approach
for Disaggregation of Low Frequency
Consumption Data

Krishnan Srinivasarengan, Y.G. Goutam, and M. Girish Chandra

Abstract. Smart electric meters that are capable of measuring and transmitting the power consumed by a household are being installed across the world. The low frequency average consumption data generated by these smart meters are now available widely for application development. A key enabler for the applications is the disaggregation of the consumption data into its constituent components, which can be useful to a wide spectrum of target audience from end user to utilities. Disaggregation of active power data sampled once in a few seconds has been studied extensively in the literature. However, there are only limited attempts to use the once in a few minutes consumption data for disaggregation. This paper discusses some preliminary results obtained using a knowledge based approach for disaggregating the low frequency average consumption data into consumption classes. The approach utilizes the spatial and temporal characteristics extracted from the data to estimate class consumption. The initial results from this approach are promising and have inspired to develop a larger probabilistic framework, which is also described in this paper. This proposed framework will be useful in generalizing and scaling the disaggregation algorithms across data of different sampling rates.

1 Introduction

The smart electric meter installation across the world is getting accelerated through a number of government regulation and standards. Smart meters can generate two different types of data about the household consumption. A low sampling rate (about once in 15 minutes) of consumption/energy data and the other, a higher sampling rate (once in a few seconds) of active power and consumption data. The former, referred to as Advanced Metering Infrastructure (AMI) data, is accessed directly by

Krishnan Srinivasarengan · Y.G. Goutam · M. Girish Chandra
Innovation Labs, Tata Consultancy Services,
Whitefield, Bangalore, India
e-mail: {krishnan.srinivasarengan,goutam.yg,m.gchandra}@tcs.com

S.M. Thampi, A. Gelbukh, and J. Mukhopadhyay (eds.), *Advances in Signal Processing*
and Intelligent Recognition Systems, Advances in Intelligent Systems and Computing 264,
DOI: 10.1007/978-3-319-04960-1_8, © Springer International Publishing Switzerland 2014

the utility companies and used for various purposes, significant of them is the implementation of time-differential tariff. This data is also made available to the end user and to third party service providers upon request from the end user. The higher sampling rate data interface, also known as Home Area Network (HAN), requires an extra hardware in the form of a gateway, which can interface with the smart meter and access the data at the rate of once in a few seconds (typical being 10 seconds). This data can then be either processed in-home or transmitted through the home Internet connection for a remote processing. Disaggregating into the constituent appliances or appliance categories is considered one of the important analytics that can be done on smart meter data [1]. Studies have indicated that energy consumption breakdown feedback can help in improving energy savings in households [3]. Apart from energy savings, disaggregation can be useful to schedule appliances to take advantage of differential tariff, analyze appliance health through consumption analysis over the years, etc.

In this paper, we explore how consumption categorization can be performed on a 15-min sampled data. The algorithm whose results are presented, provide an illustration of the usefulness of the approach. It uses the knowledge about various consumption categories in the form of their temporal and spatial characteristics to estimate their consumption. The impact of the variation of the appliance characteristics and their impact on the accuracy on the consumption estimation is discussed. An outline of how this algorithm would fit into the proposed probabilistic framework for consumption estimation is given. In the recent times, a number of start up companies in the load disaggregation domain provide consumption categorization based on AMI data [1]. However to the best of our knowledge, the literature does not discuss the details of an approach to implement this.

In illustrating these ideas and works, the paper is organized as follows: Sec 2 touches upon a few relevant works which have tackled the problem of disaggregation. The knowledge based approach algorithm and inferences are discussed in Sec 3, which also briefly outlines the knowledge based approach algorithms. The results are discussed in the Sec 4. The Sec 5 provides a summary and the future outlook of the proposed approach.

2   Relevant Works

Non-intrusive load Monitoring has been of interest since Hart's seminal work [4]. A good majority of the works (see [10], [11] and references thereof) have focused on using a number of specialized sensors and/or multiple electrical parameters. These features will not be provided by the standard smart meters being installed across households. Works focusing on using only the active power meter data are emerging in the recent times (see for instance [8], [5], [9] and references in [10], [11]). However, these works require the instantaneous active power data to be available, and use the patterns in the data to identify appliances and compute their consumption.

---

[1] See for instance, Bidgely: www.bidgely.com

AMI systems will be ubiquitous in future and the data from these systems would be available by default. However, the literature is limited for the problem of disaggregation on low frequency AMI data. In [7], a technique based on discriminative sparse coding is used to disaggregate hourly-consumption data. The overcomplete dictionary capable of providing sparse representation and discriminative ability for different classes of appliances is learnt through the training data. The requisite sparse *activation* matrices for the classes leading to the disaggregation results are obtained through coordinate descent based approach. Even with very coarse data, the accuracy results reported appear to be useful in terms of energy consumption (47%) and classification (55%). In [2], the authors develop an approach for partial disaggregation on hourly whole-house data using the inverse modelling technique. They compare the categories of consumption and some key consumption related parameters against the external temperatures and develop models to represent these relationships.

3 Algorithm and Framework

Disaggregation of 15-min sampled data into consumption categories is a challenging task due to a number of reasons. First, the data has a number of subjective elements that depends on demographic, climatic and cultural factors, among others. For instance, the highest power consuming appliance category in North America (HVAC) is different from that of Europe (Cold Storage/Laundry)[2]. Second, due to the large sampling interval, multiple appliances' ON/OFF characteristics are captured into a single average consumption measurement, making separation difficult. The consumption classes can hence vary between different geographies. A typical set of consumption classes shall be as follows:

- Phantom Loads
- Cold Storage
- Laundry
- Evening Consumption
- Unknown High Power rating appliances

which are used in this paper. Phantom loads correspond to those consumption that are ON through out the day without any appreciable consumption changes (e.g. fire alarm, router, TV standby). Cold storage comprises of refrigerators and freezers which are used throughout, but show a variation in consumption. Evening consumption corresponds to consumption from low power appliances that show up prominently during the evenings due to collective usage. While this doesn't account for all the low power consumption appliances, it enables the end-user to understand consumption. The Laundry class accounts for both washer and dryer and the unknown

---

[2] This was learned during a project interaction

high power class combines all other high power rating appliances, including microwave, stove etc.

Before we describe the components of the knowledge based disaggregation approach, the proposed larger probabilistic framework is discussed first due to the ease of illustration. The framework was arrived at after an exhaustive set of experiments using various datasets and different types of appliance characteristics. The experiences are discussed in the Sec 4.

3.1 Proposed Framework

The framework aims to estimate the consumption of different classes and weigh them with confidence measure obtained by Maximum a Posteriori (MAP) estimate. The key blocks of the framework are shown in the Fig. 1. In a nutshell, the framework does the following: the features are extracted from the data with which an estimate of the consumption of each classes are obtained. The MAP estimate of each class consumption is then computed absorbing any apriori information available.

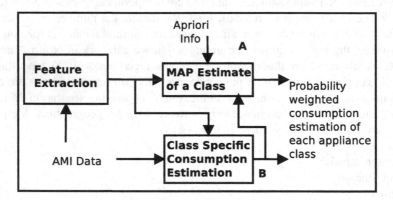

Fig. 1 Consumption Classification Framework

The Feature extraction involves processing the given data (96 samples per day). After a prolonged analysis of the data and their implications on the consumption estimation, the following categories of features were arrived at:

- Ordered Consumption extremas
- Rate of Change of Variation (Positive and negative) between local extremas
- Duration & Consumption between apparent ON & OFF
- Parameters of Probability Distribution (for learning)

A simplistic view of the MAP estimation process could be as follows. If Ω_i is the random variable representing the appliance class and f represent the feature set, then the MAP estimate is given by:

$$\Pr(\Omega_i|f) = \frac{\Pr(\Omega_i, f)}{\Pr(f)} \tag{1}$$

$$\propto \Pr(\Omega_i, f) \tag{2}$$

since the extracted featured are constants. The above representation assumes that the features have a representation which can uniquely identify classes. However, some of the features (e.g., the consumption between apparent ON & OFF) can only represent the combination of consumption classes active at that time. Hence the above representation can be rewritten as:

$$\Pr(\bar{\Omega}_i|f) \propto \Pr(\bar{\Omega}_i, f) \tag{3}$$

where, $\bar{\Omega}_i$ stands for the class combination i. Further,

$$\Pr(\bar{\Omega}_i, f) = \Pr(f|\bar{\Omega}_i) \times \Pr(\bar{\Omega}_i) \tag{4}$$

where, $\Pr(f|\bar{\Omega}_i)$ is the likelihood of the consumption class combination. One approach to compute the likelihood would be in the form of:

$$\Pr(f|\bar{\Omega}_i) = \frac{1}{\sum_j w^i_j |f^i_j - f_j|} \tag{5}$$

where, w^i_j is the weight associated with each feature, which can be variable depending upon the particular class combination i, f^i_j is the assumed feature value (either assumed or learned) for the jth feature for the class combination i and f_j is the extracted feature. The second term on the right hand side of (4) is a prior probability of the presence of the combination of consumption class. A typical prior probability could be given based on the Time of Day (T_D) and Day of the Week (D_w) correlation with the usage of each consumption class and hence that of their combination. For the sake of simplicity, if we assume that the timing correlation of each appliance class is independent of each other, we get,

$$\Pr(\bar{\Omega}) = \Pr(\Omega_1) \times \Pr(\Omega_2) \times ...\Pr(\Omega_n) \tag{6}$$

$$= \Phi_1(T_D, D_w) \times \Phi_2(T_D, D_w) \times ...\Phi_n(T_D, D_w) \tag{7}$$

The prior probability and the features corresponding to each class f^i_j are the components that would enable the framework to learn from the history of power data, contextual information based inputs (e.g. temperature based analysis as in [2]), any specific end-user feedback and so on. This MAP estimate of a class combination can be marginalized to obtain a confidence measure for individual consumption class. This could be enabled using a factor graph approach with Sum-Product or Max-Sum algorithm. Another key aspect of this MAP estimate (Block A) is the possible interaction with the consumption estimation (Block B) as indicated in the Fig. 1. This can lead to two: First, if a consumption class i is dependent on that of class j, then the MAP estimate algorithm corresponding to i can include the consumption estimate of j as a dependency factor. Second, in a scenario where the consumption measurement itself is the feature, the MAP estimate gives the confidence measure on the class combination. If the confidence measure is low, there is a scope to improve by iteratively considering different combination of the consumption classes to arrive at more accurate results.

The consumption estimation block is an essential component of the proposed framework. This block can be implemented through a variety of different techniques centered around pattern recognition, discrimination etc. In this paper, we have used a simplistic knowledge based approach to estimate the consumption of different classes. In case the accuracy is not good enough, there is always a scope, within the proposed broader framework, to improvise the results as discussed in the previous paragraph.

3.2 Knowledge Driven Consumption Estimation Algorithm

The algorithm derives its steps from the exhaustive analysis of data to identify the pertinent features. The key components of the knowledge based algorithm can be summarized as follows:

- Correlating features with consumption classes
- Estimating the boundaries of consumption class usage
- Estimating the consumption of a consumption class

These components would wary depending upon the type of consumption class being estimated. For instance, the always ON classes like phantom loads and cold storage are assumed to be present throughout the day and hence wouldn't have a defined boundary. The features required by these classes are restricted to the consumption extremas at various times of the day. For instance, the simplest way to estimate the phantom load is by considering the minima of the whole-day data. However, this could be erroneous due to overlap with cold storage appliances like refrigerator/freezer. The estimation for cold storage consumption assumes that almost no other appliances apart from the phantom loads are likely to be used between 2AM to 5AM. The accuracy of these classes could however be improved when compared over different days and appropriate inferences about the underlying characteristics could be made.

The separation of the Always ON or slow changing low power appliances and that of high power rating appliances is established through an adaptive quantization procedure. This would track the changes in power consumptions such that only low amplitude changes are accommodated. The resulting data could help in identifying the boundaries of consumption of high power appliance usage. In case of high power appliances, the Laundry analysis assumes that the washer and dryer usage would be close to each other temporally. The analysis hence looks for a pattern with a large ON time correlated to appropriate time of day. The accuracy of the high power appliances could be improved by training them over several days and through inputs from the end-user.

These inferences are captured through an algorithm and is used to disaggregate the low frequency data into its constituent consumption classes. The flow of a sample algorithmic implementation used to generate the results in this paper shall be as follows:

1. Separate the time series data into low power and high power region using an adaptive quantization procedure
2. Phantom load as a function of minimum value of the time series
3. Cold Storage as the Phantom load subtracted value of consumption during night time
4. Laundry as the high power consumption region of sufficiently long duration with appropriate time of day correlation. Any temporal distance between washer and dryer usage are accommodated through an extended search in the adjacent high power regions.
5. The high power regions negating the laundry forms the Unknown high power appliance class

4 Results and Discussion

4.1 Data Generation Approach

The data used for illustrating the algorithm is generated by averaging a higher frequency active power data. This is due to the lack of ground truth with the available 15-minutes data as well as the availability of annotated 10-seconds sampled data through a number of open data sets (e.g. REDD [6]) and from a pilot project. This helps in evaluating the algorithm's performance in a quantitative manner. The data is generated in a bottom-up way, starting from creating a base load by fusing phantom loads and evening extra consumption. Variations in these loads are brought by varying the amplitudes of these loads. To this base data, the plug data available for appliances like washer, dryer, dishwasher, refrigerator, heater are added. The temporal positions of these appliances are varied during the addition to enable a truly representative power data of a household. This hybrid, 10-second sampled data is averaged over windows of 15 minutes length to generate the data necessary for analysis. This generated data represents the first difference data of the AMI data available from a smart meter. A typical 10-second sampled generated data is shown in Fig. 2 and its corresponding 15 minutes sampled data is shown in Fig. 3. It is to be noted that the 15-minutes sampled data represents the first difference of the typical cumulative consumption data of a smart meter.

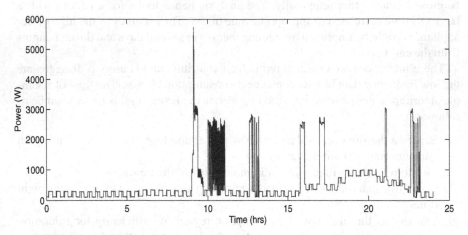

Fig. 2 An instance of generated 10-second sampled data

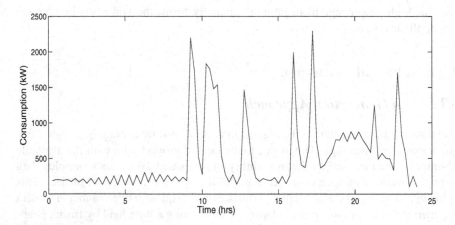

Fig. 3 An instance of generated 15-minutes sampled data

4.2 Results

The results obtained through the knowledge based approach are summarized in Table 1. The data for the three houses given in the table were generated by varying the characteristics of underlying appliances in the 10-seconds data and their time of day usage. The results were generated for several days for the three households and the statistics (minimum, maximum and average) of the accuracies of the consumption estimation are provided in the Table 1.

Table 1 Summary of Results

Consumption Class	House 1			House 2			House 3		
	min	max	avg	min	max	avg	min	max	avg
Phantom Load	0.33	0.97	0.79	0.51	0.91	0.79	0.5	0.92	0.68
Cold Storage	0.34	0.96	0.73	0.23	0.75	0.50	0.13	0.88	0.64
Evening Consumption	0.23	0.94	0.55	0.24	0.67	0.43	0.14	0.66	0.40
Laundry	0.81	0.99	0.86	0.63	0.98	0.87	0.39	0.99	0.86
Unknown High Power	0.73	0.94	0.85	0.56	0.90	0.77	0.52	0.98	0.81

4.3 Discussion

A number of assertions could be made based on the correlation between the accuracy and the characteristics of the underlying appliances/consumption classes. For instance, the accuracy of phantom loads and cold storage depends heavily on the usage of unknown high power appliances during the night. The evening consumption depends on the accuracy of the phantom loads and cold storage. The Laundry accuracy is sensitive to any significant difference in the time of day usage between washer and dryer. The Laundry accuracy and other high power rating appliances are interdependent and are affected by the temporal closeness of their operations. These inter-dependencies should be incorporated into the larger framework to improve the accuracy of the results.

5 Concluding Remarks

A probabilistic and knowledge based framework to disaggregate low frequency smart meter data into consumption classes has been discussed. The results obtained only by utilizing the knowledge based component of the framework looks promising and has opened up several important issues to be addressed in future. This includes the tight coupling between the blocks of the proposed framework. Further, the algorithm should be evaluated on large amount of diverse data with more complex user and appliance characteristics. Work is underway to develop the complete framework and also look at ways to improve the consumption estimation approach.

References

1. Armel, C., Gupta, A., Shrimali, G., Albert, A.: Is disaggregation the holy grail of energy efficiency? the case of electricity. Technical report, Precourt Energy Efficiency Center, Stanford (May 2012)
2. Birt, B.J., Newsham, G.R., Beausoleil-Morrison, I., Armstrong, M.M., Saldanha, N., Rowlands, I.H.: Disaggregating categories of electrical energy end-use from whole-house hourly data. Energy and Buildings 50, 93–102 (2012)
3. Froehlich, J.: Promoting energy efficient behaviors in the home through feedback: The role of human-computer interaction. In: Proc. HCIC Workshop, vol. 9 (2009)

4. Hart, G.: Nonintrusive appliance load monitoring. Proceedings of the IEEE 80(12), 1870–1891 (1992)
5. Kim, H.: Unsupervised disaggregation of low frequency power measurements. PhD thesis, University of Illinois (2012)
6. Kolter, J., Johnson, M.: REDD: A public data set for energy disaggregation research. In: Workshop on Data Mining Applications in Sustainability (SIGKDD), San Diego, CA (2011)
7. Kolter, J.Z., Batra, S., Ng, A.: Energy disaggregation via discriminative sparse coding. In: Advances in Neural Information Processing Systems, pp. 1153–1161 (2010)
8. Parson, O., Ghosh, S., Weal, M., Rogers, A.: Nonintrusive load monitoring using prior models of general appliance types. In: 26th AAAI Conference on Artificial Intelligence (2012)
9. Srinivasarengan, K., Goutam, Y., Chandra, M., Kadhe, S.: A framework for non intrusive load monitoring using bayesian inference. In: 2013 Seventh International Conference on Innovative Mobile and Internet Services in Ubiquitous Computing (IMIS), pp. 427–432. IEEE (2013)
10. Zeifman, M., Roth, K.: Nonintrusive appliance load monitoring: Review and outlook. IEEE Transactions on Consumer Electronics 57(1), 76–84 (2011)
11. Zoha, A., Gluhak, A., Imran, M., Rajasegarar, S.: Non-intrusive load monitoring approaches for disaggregated energy sensing: A survey. Sensors 12(12), 16838–16866 (2012)

A Group Decision Support System for Selecting a SocialCRM

Arpan Kumar Kar and Gaurav Khatwani

Abstract. As more firms adopt SocialCRM as their marketing strategy it becomes important to evaluate various SocialCRMs on the basis of certain factors. Previous SocialCRM studies have been limited to post-deployment performance evaluation using various quantitative techniques. In this paper, we propose a method to evaluate SocialCRMs which can assist an organization in selecting appropriate SocialCRM based on their needs and functionalities that they would like to perform. Firstly we evaluate criteria and finally we evaluate each SocialCRM based on each criterion using the fuzzy extension of the Analytical Hierarchy Process for group decision making.

1 Introduction

SocialCRM is defined as a tool which can be used by organizations to enable two way communications and for peer to peer management with customers. The first and foremost task for selecting social customer relationship management (CRM) for purpose of managing customer relationship activities is determining the important criteria for their evaluation keeping conscience of customers. These criteria are a part of customer management system which enables us to select CRM according to strength of each CRM. CRM consists of four dimensions namely a) Customer Identification, b) Customer Attraction, c) Customer Retention and d) Customer Development.[1]

In this research paper theory of group decision making has been utilized to address the gap of weighing different SocialCRMs on different dimensions. The consensual preferences of group with the help of fuzzy Analytical Hierarchy Process (AHP) can be estimated for prioritization and aggregation.

Arpan Kumar Kar · Gaurav Khatwani
IIM Rohtak, MDU Campus, Haryana, India, Pin 124001
e-mail: {arpan_kar,g_khatwani}@yahoo.co.in

S.M. Thampi, A. Gelbukh, and J. Mukhopadhyay (eds.), *Advances in Signal Processing*   95
and Intelligent Recognition Systems, Advances in Intelligent Systems and Computing 264,
DOI: 10.1007/978-3-319-04960-1_9, © Springer International Publishing Switzerland 2014

2 Literature Review

The review of literature has been organized into four sub-sections, focusing on the impact of social media on business, how AHP has been used for evaluating CRM solutions, AHP as a tool for group decision making and the review of decision support literature in the domain of SocialCRM.

2.1 Overview of Social Media Platforms Impact on Business

It was found that real time search engines are specialized in providing information and insight related to how data is being shared on different social media platforms and classified them based on approaches used by them to harvest information [40]. In fact there is strong correlation between holiday planning and social media influencers and holiday planners believe more on user-generated information than from travel management authorities [21]. Abrahams et. al. [1] developed vehicle defect and discovery from social media by studying conversations of various consumers of brands on social media platforms. They developed a framework which mines data from social media platforms and classifies and prioritize keyword related to vehicle defect. Further it was revealed that selective attractive gamification features in websites contribute more than 50% of the attractiveness of websites, which further contribute to their success [26]. Further attempts were made for measuring information quality by adapting AHP methodology to evaluate and prioritize selected criteria in weblogs [29]. Further model for assessing the security risks of social networking sites were developed at the enterprise by identifying criteria like confidentiality, integrity and availability [30]. Further it was investigated how social media can have an influence on companies' activities in regard to their value chain [8]. Further combining features of Web 2.0 and social networking with current CRM system a SocialCRM system as a web service was developed [33].

2.2 Overview of Decision Support Literature in SocialCRM

There has been an investigation to study the effects of variable selection and class distribution on the performance of specific logit regression and artificial neural network implementations in a CRM setting [31]. Further studies discussed CRM practice and expectations, the motives for implementing it, evaluated post-implementation experiences and investigated the CRM tools functionality in the strategic, process, communication, and business-to-customer organizational context [28]. In addition there was an examination of how sales representative can enhance their performance through their acceptance of information technology tools [4]. Further a model was developed based upon the premise that business value is enhanced through the alignment of complementary factors occurring along three dimensions, intellectual, social, and technological to assess firm's readiness for CRM [35]. A new approach in CRM based on Web 2.0 which will help healthcare providers in improving their customer support activities and assist patients with the customized personal service was introduced [5]. Further factors were identified that may influence businesses relationships and customers' adoption of SocialCRM and an enhancement to Technology Acceptance Model was proposed [6]. In fact the problem of the automatic customer segmentation was addressed by processing data collected in SocialCRM systems using Kohonen networks [14]. Further some valuable information on using a multi-agent approach for designing SocialCRM systems was proposed [36]. Further a model of CRM in e-government system that includes process specification, metrics for evaluation of the system performance and recommendations for implementation was proposed [45]. The exploration related to the role of analytical SocialCRM and examination of available tools with the required functional and technological components was accomplished [41]. A novel technical framework (GCRM) based on methods such as group detecting, group evolution tracking and group life-cycle modeling in telecom applications to manage social groups by analyzing relationships between social groups and finding potential customers in these groups in massive telecom call graphs was proposed [47]. Further web-based platforms that provide SocialCRM solution in software as a service model as well as the applications and tools that complement traditional CRM systems and possible challenges that businesses could face in adopting social media technologies in customer management processes and systems were examined [7]. Further how brands work to maintain relationships with people in social media by analyzing SocialCRM and strategies that encourage participation and involvement was examined and the brands that correspond to the different levels of Maslow's hierarchy of needs were investigated [23].

2.3 Overview of Using AHP in Evaluating CRM

The literature available on social media and fuzzy AHP is enormous but it is mainly focused on optimizing resources, decision making, identifying important features, evaluating different components in marketing and developing a model for integrated marketing activities. The use of SocialCRM by organizations is one of the major marketing activities in modern times. The literature available on CRM and fuzzy AHP is for evaluating post-deployment performance of CRM but there is hardly any study on selecting SocialCRM using group decision making that emphasizes on functionalities of an organization in marketing. This can be justified as there was an attempt to design first-rank evaluation indicators of CRM and substantiate the applicability of fuzzy comprehensive evaluation in the CRM [49]. In fact AHP has been used to prioritize the customer's satisfaction, loyalty and credit [39]. The fuzzy evaluation model of CRM performance using correlation theories of AHP by creating the index system that effects CRM performance and gives the four key dimensions that influence CRM performance was established [27]. Further studies were conducted to review and rate effective factors in customer classification that aids management in understanding customers smoothly and to have prioritized indications [43]. Further there was an attempt to classify and explain the most important aspects and factors which affect CRM readiness for business-to-business markets by suggesting the hierarchical model for CRM readiness in organizations [22]. Further a framework was proposed which helped to evaluate CRM by combining evaluation criteria for CRM on-demand systems at a functional and general level [37].

2.4 Using AHP for Group Decision Making

In our study we have used fuzzy AHP for group decision making to estimate the collective preference of group decision makers. The solution to hierarchic problems can be achieved by solving hierarchy of sub-problems iteratively for which AHP theory of measurement can be used. The priorities can be derived from continuous and discrete paired comparisons obtained from relative strength of judgments of reflecting scale. The AHP theory has many advantages in group decision making. Firstly, to estimate the consistencies of priorities of decision makers appropriate theories are available [42, 3, 2, 17, 34]. Secondly, there are theories to improve the consistency of priorities [19, 48, 12]. Thirdly, there are robust approaches to aggregate group preferences [16, 25, 20, 10, 13, 9, 18]. Additionally there are robust theories for building consensus within groups [11, 24, 17, 34, 15, 46]. However there is a lot of scope for these group decision making theories in SocialCRM selection. However none of these approaches have been explored for application in the selecting SocialCRM, which is the focus of this study.

3 The Research Gap and the Contribution

As from literature review we firstly found that how different firms utilize data available on social media platforms to manage their services and operations. Secondly, we found that how AHP theories can be used for estimating and improving consistencies of priorities. Thirdly, there is hardly any contribution towards evaluating SocialCRM. There has been extensive contribution in evaluation of traditional CRMs and day-to-day marketing. Finally we saw some quantitative techniques being used by various firms to evaluate SocialCRM based on various activities. The use of AHP in evaluating SocialCRM helps an organization to select SocialCRM based on their goals and functionalities that they would like to perform. This pre-evaluation model also helps SocialCRM vendors to develop SocialCRMs as per market demand. This model can also assist in saving cost and time related to post deployment evaluation of SocialCRM. The fuzzy extension of the AHP for group decision making helps us in evaluating five different social CRMs on four different dimensions.

4 Computational Method

The prioritization of above mentioned dimensions is achieved through an integrated approach. This process involves capturing users' linguistic judgments and correspondingly mapping to quantifiable fuzzy judgments. Further crisp priorities are derived using fuzzy linguistic judgments and AHP theory. To estimate the trade-offs for different dimensions combination of these crisp priorities and aggregated geometric mean method for prioritization is used.

Let $V = (v_1, \ldots\ldots v_n)$ be set of n users having a relative importance of λ_i such that $\lambda = (\lambda_1, \ldots\ldots\lambda_n)$ is the weight vector of the individual user who prioritize one dimension over other. Triangular functions have been used to convert individual preferences into fuzzy judgments. Comparative fuzzy judgments $N = (n_{ij})_{m \times m}$ would be coded as illustrated in Table 1.

Table 1 Scale for conversion of linguistic preferences

Definition	Fuzzy sets for the fuzzy AHP	
Equal importance	$\tilde{1}$	$\{(1,0.25),(1,0.50),(3,0.25)\}$
Moderate importance	$\tilde{3}$	$\{(1,0.25),(3,0.50),(5,0.25)\}$
Strong importance	$\tilde{5}$	$\{(3,0.25),(5,0.50),(7,0.25)\}$
Very strong importance	$\tilde{7}$	$\{(5,0.25),(7,0.50),(9,0.25)\}$
Extreme high importance	$\tilde{9}$	$\{(7,0.25),(9,0.50),(9,0.25)\}$

The simple pair wise comparison approach for fuzzy set operations has been used for fuzzy sets $\tilde{n}_i = (n_{i1}, n_{i2}, n_{i3})$ and $\tilde{n}_j = (n_{j1}, n_{j2}, n_{j3})$ as illustrated:

$$\tilde{n}_i + \tilde{n}_j = ((n_{i1} + n_{j1}),(n_{i2} + n_{j2}),(n_{i3} + n_{j3})) \tag{1}$$

$$\tilde{n}_i - \tilde{n}_j = ((n_{i1} - n_{j1}),(n_{i2} - n_{j2}),(n_{i3} - n_{j3})) \tag{2}$$

$$\tilde{n}_i \times \tilde{n}_j = ((n_{i1} \times n_{j1}),(n_{i2} \times n_{j2}),(n_{i3} \times n_{j3})) \tag{3}$$

$$\tilde{n}_i \div \tilde{n}_j = ((n_{i1} \div n_{j1}),(n_{i2} \div n_{j2}),(n_{i3} \div n_{j3})) \tag{4}$$

$$\tilde{n}_i^m = (\tilde{n}_{i1}^m, \tilde{n}_{i2}^m, \tilde{n}_{i3}^m) \text{ for } (m \in P) \tag{5}$$

$$\tilde{n}_i^{1/m} = (\tilde{n}_{i1}^{1/m}, \tilde{n}_{i2}^{1/m}, \tilde{n}_{i3}^{1/m}) \text{ for } (m \in R) \tag{6}$$

The individual priorities are obtained by solving the following system:

$$\min \sum_{i=1}^{m} \sum_{j>i}^{m} (\ln \tilde{n}_{ij} - (\ln \tilde{r}_i - \ln \tilde{r}_j)^2) \text{ s.t. } \tilde{n}_{ij} \geq 0; \tilde{n}_{ij} \times \tilde{n}_{ji} = 1; \tilde{r}_i \geq 0, \sum \tilde{r}_i = 1 \tag{7}$$

The individual priority vector [11] is obtained by $\tilde{r}_i = \dfrac{\sqrt[1/m]{\prod_{j=1}^{m} \tilde{n}_{ij}}}{\sum_{i=1}^{m} \sqrt[1/m]{\prod_{j=1}^{m} \tilde{n}_{ij}}}$ (8)

Where $\tilde{r}_i$ is the priority of decision criteria i such that $\tilde{R}_i = \{\tilde{r}_1, \tilde{r}_2 \ldots\ldots \tilde{r}_5\}$ for user i. In further steps before computing aggregation rules consistencies of these priorities needs to be evaluated. The Geometry Consistency Index (GCI) is used to estimate consistency of individual priorities [2].

$$GCI(A^{d_i}) = \frac{2}{(m-1)(m-2)} \times \sum_{j > i}^{m} (\log|\tilde{n}_{ij}| - (\log|\tilde{r}_i| - \log|\tilde{r}_j|)^2) \qquad (9)$$

$GCI(A^{d_i}) \leq \overline{GCI}$ is the criteria for consistency. For $m \geq 4$, $\overline{GCI}$ is 0.35. Collective preferences of the group for deriving the decision vector can be estimated subsequently by the aggregation of individual priorities such that the aggregate priorities (i.e. the collective priority vector) are defined as $\tilde{R}^{(c)} = \{\tilde{r}_1^{(c)}, \tilde{r}_2^{(c)} \dots \dots \tilde{r}_p^{(c)}\}$ where $\tilde{r}_i^{(c)}$ is obtained by the aggregation of priorities.

$$\tilde{r}_i^{(c)} = \frac{\Pi_1^n (r_i^{(m)})^{\lambda_i}}{\sum_1^r \Pi_1^n (r_i^{(m)})^{\lambda_i}} \qquad (10)$$

These aggregated priorities have been used for evaluating the relative importance of the evaluation criteria and the criterion specific performance of the five solutions being evaluated.

5   Case Study

A case study was conducted for identifying important criteria for selection of SocialCRM and identifying which CRM users prefer based on four different criterion discussed in beginning. As per existing literature [32, 44, 38], CRM solutions are evaluated often on four major dimensions: Customer Identification, Customer Attraction, Customer Retention and Customer Development. In this study, these dimensions have been used for evaluating the solutions for their suitability of usage. Out of the listed 9 SocialCRM solutions available in the public domain (Wikipedia), a Delphi study was conducted to identify more relevant solutions for the specific context, based on the features of these solutions. Based on the Delphi study, five SocialCRM tools were selected for the next stage of the evaluation process: Social Gateway, Kony Mobile CRM, Radian6, SocialText and Nimble.

Subsequently, four dimensions for evaluating CRM solutions were prioritized using the fuzzy extension of the AHP for group decision making. The following table highlights the individual priorities of three decision makers, as well as the aggregated priorities, for the four dimensions for evaluating SocialCRM solutions.

Table 2 Individual and aggregated priorities for the four evaluating criteria

	Customer Identification	Customer Attraction	Customer Retention	Customer Development
User 1	0.0961	0.1045	0.7164	0.0830
User 2	0.1250	0.1250	0.3750	0.3750
User 3	0.1215	0.1215	0.4799	0.2771
Aggregated Score	0.1206	0.1241	0.5372	0.2181

For the subsequent stage, the context specific and evaluation criteria specific prioritization of these five solutions were done by the group of decision makers. The following table indicates individual priorities of three different users for five different SocialCRM tools based on dimension customer identification.

Table 3 Individual and aggregated priorities based on the criteria - customer identification

	Social Gateway	Kony Mobile	Radian6	Social Text	Nimble
User 1	0.3644	0.1591	0.1415	0.2925	0.0425
User 2	0.2893	0.2323	0.2323	0.1497	0.0964
User 3	0.2468	0.1981	0.1590	0.1981	0.1981
Aggregated Score	0.3078	0.2017	0.1802	0.2134	0.0969

The following table indicates individual priorities of three different users for five different SocialCRM tools based on dimension customer attraction.

Table 4 Individual and aggregated priorities based on the criteria - customer attraction

	Social Gateway	Kony Mobile	Radian6	Social Text	Nimble
User 1	0.2522	0.0798	0.0759	0.1800	0.4122
User 2	0.1111	0.1111	0.1111	0.3333	0.3333
User 3	0.1439	0.1156	0.1156	0.2783	0.3467
Aggregated Score	0.1629	0.1032	0.1015	0.2615	0.3709

The following table indicates individual priorities of three different users for five different SocialCRM tools based on dimension customer retention.

Table 5 Individual and aggregated priorities based on the criteria - customer retention

	Social Gateway	Kony Mobile	Radian6	Social Text	Nimble
User 1	0.2084	0.0660	0.0973	0.3407	0.2876
User 2	0.4256	0.1419	0.1419	0.1139	0.1767
User 3	0.1111	0.3333	0.1111	0.1111	0.3333
Aggregated Score	0.2394	0.1632	0.1288	0.1818	0.2868

The following table indicates individual priorities of three different users for five different SocialCRM tools based on dimension customer development.

Table 6 Individual and aggregated priorities based on the criteria - customer development

	Social Gateway	Kony Mobile	Radian6	Social Text	Nimble
User 1	0.3249	0.2126	0.1890	0.1517	0.1218
User 2	0.3602	0.3252	0.0979	0.1084	0.1084
User 3	0.0979	0.2097	0.1351	0.1084	0.4489
Aggregated Score	0.2485	0.2688	0.1496	0.1337	0.1995

The following table indicates aggregated priorities of 3 users and the overall performance score of all the 5 solutions that are being evaluated.

Table 7 Aggregate priorities and overall tool score

Criteria vs tools	Criteria weights	Social Gateway	Kony Mobile	Radian6	Social Text	Nimble
Customer Identification	0.1206	0.3078	0.2017	0.1802	0.2134	0.0969
Customer Attraction	0.1241	0.1629	0.1032	0.1015	0.2615	0.3709
Customer Retention	0.5372	0.2394	0.1632	0.1288	0.1818	0.2868
Customer Development	0.2181	0.2485	0.2688	0.1496	0.1337	0.1995
Aggregate score of tools		0.2402	0.1834	0.1362	0.1850	0.2553

6 Results

Based on the analysis of case specific requirements, it was found that Nimble had the highest suitability score (0.2553) for the specific context, based on the aggregated priorities of the 3 decision makers. This was followed subsequently by Social Gateway, with a score of 0.2402, based on the aggregated priorities of the three decision makers. The other three solutions had a much lower score, i.e. Kony Mobile, Social Text and Radian 6 had a score of 0.1834, 0.1850 and 0.1362 respectively. Higher score indicates higher adherence to context specific requirements. Thus Nimble was selected as the most suitable tool for the context specific usage.

7 Conclusion

This paper proposes an approach to select SocialCRM tool based on aggregated priorities from multi-user perspective, for the same decision making context. Such an approach can be useful for the firms who are in initial phase of SocialCRM tool deployment for their marketing activities. Further this can also help firms in operating with these tools smoothly as informed decision making helps them to choose appropriate tool.

One of the limitations of the case study is that it has been conducted for specific company for a specific context hence the rankings are not generalizable. However, this can be addressed in future research whereby the judgments of a much larger sample can be evaluated to bring out an aggregated performance score for these solutions. Further, the implications of consensus achievement can also be explored in such studies.

References

1. Abrahams, A.S., et al.: Vehicle defect discovery from social media. Decision Support Systems 54, 87–97 (2012), doi:10.1016/j.dss.2012.04.005
2. Aguarón, J., Moreno-Jiménez, J.M.: The geometric consistency index: Approximated thresholds. European Journal of Operational Research 147, 137–145 (2003), doi:10.1016/S0377-2217(02)00255-2
3. Aguaron, J., et al.: Consistency stability intervals for a judgment in AHP decision support systems. European Journal of Operational Research 145, 382–393 (2003), doi:10.1016/S0377-2217(02)00544-1
4. Ahearne, M., et al.: Why sales reps should welcome information technology: Measuring the impact of CRM-based IT on sales effectiveness. International Journal of Research in Marketing 24, 336–349 (2007), doi:10.1016/j.ijresmar.2007.09.003
5. Anshari, M., Almunawar, M.N.: Evaluating CRM implementation in healthcare organization. arXiv preprint arXiv:1204.3689 (2011)
6. Askool, S., Nakata, K.: A conceptual model for acceptance of social CRM systems based on a scoping study. Ai & Society 26, 205–220 (2011), doi:10.1007/s00146-010-0311-5
7. Ayanso, A.: Social CRM: Platforms, Applications, and Tools. In: Yang, H., Liu, X. (eds.) Software Reuse in the Emerging Cloud Computing Era, vol. 20. IGI Global, Pennsylvania (2012), doi:10.4018/978-1-4666-2625-6.ch065
8. Bork, M., Behn, N.: Utilization of social media in an organizational context. Dissertation, Luleå University of Technology (2012)
9. Beynon, M.J.: A method of aggregation in DS/AHP for group decision-making with the non-equivalent importance of individuals in the group. Computers & Operations Research 32, 1881–1896 (2005), doi:10.1016/j.cor.2003.12.004
10. Bolloju, N.: Aggregation of analytic hierarchy process models based on similarities in decision makers' preferences. European Journal of Operational Research 128, 499–508 (2001), doi:10.1016/S0377-2217(99)00369-0
11. Bryson, N.: Group decision-making and the analytic hierarchy process: Exploring the consensus-relevant information content. Computers & Operations Research 23, 27–35 (1996), doi:10.1016/0305-0548(96)00002-H
12. Cao, D., et al.: Modifying inconsistent comparison matrix in analytical hierarchy process: A heuristic approach. Decision Support Systems 44, 944–953 (2008), doi:10.1016/j.dss.2007.11.002
13. Condon, E., et al.: Visualizing group decisions in the analytic hierarchy process. Computers & Operations Research 30, 1435–1445 (2003), doi:10.1016/S0305-0548(02)00185-5

14. Czyszczoń, A., Zgrzywa, A.: Automatic Customer Segmentation for Social CRM Systems. In: Kwiecień, A., Gaj, P., Stera, P. (eds.) CN 2013. CCIS, vol. 370, pp. 552–561. Springer, Heidelberg (2013), doi:10.1007/978-3-642-38865-1_55

15. Dong, Y., et al.: Consensus models for AHP group decision making under row geometric mean prioritization method. Decision Support Systems 49, 281–289 (2010), doi:10.1016/j.dss.2010.03.003

16. Dyer, R.F., Forman, E.H.: Group decision support with the analytic hierarchy process. Decision Support Systems 8, 99–124 (1992), doi:10.1016/0167-9236(92)90003-8

17. Escobar, M.T., et al.: A note on AHP group consistency for the row geometric mean prioritization procedure. European Journal of Operational Research 153, 318–322 (2004), doi:10.1016/S0377-2217(03)00154-1

18. Escobar, M.T., Moreno-Jiménez, J.M.: Aggregation of individual preference structures in AHP-group decision making. Group Decision and Negotiation 16, 287–301 (2007), doi:10.1007/s10726-006-9050-x

19. Finan, J.S., Hurley, W.J.: The analytic hierarchy process: does adjusting a pairwise comparison matrix to improve the consistency ratio help? Computers and Operations Research 24, 749–755 (1997), doi:10.1016/S0305-0548(96)00091-3

20. Forman, E., Peniwati, K.: Aggregating individual judgments and priorities with the analytic hierarchy process. European Journal of Operational Research 108, 165–169 (1998), doi:10.1016/S0377-2217(97)00244-0

21. Fotis, J., et al.: Social media impact on holiday travel planning: The case of the Russian and the FSU markets. International Journal of Online Marketing 1, 1–19 (2011), doi:10.4018/ijom.2011100101

22. Fotouhiyehpour, P.: Assessing the Readiness for implementing e-CRM in B2B Markets Using AHP Method. Dissertation, Luleå University of Technology (2008)

23. Ginman, C.: How are Brands Engaging and Building Relationships with Fans and Customers in Social Media? Dissertation, Uppsala University (2011)

24. Honert, R.C.V.: Stochastic group preference modelling in the multiplicative AHP: a model of group consensus. European Journal of Operational Research 110, 99–111 (1998), doi:10.1016/S0377-2217(97)00243-9

25. Honert, R.C.V., Lootsma, F.A.: Group preference aggregation in the multiplicative AHP: The model of the group decision process and Pareto optimality. European Journal of Operational Research 96, 363–370 (1997), doi:10.1016/0377-2217(95)00345-2

26. Hsu, S.H., et al.: Designing Attractive Gamification Features for Collaborative Storytelling Websites. Cyberpsychology, Behavior, and Social Networking 16 (2013), doi:10.1089/cyber.2012.0492

27. H.Y., et al.: Research on CRM Performance Evaluation Based on Analytic Hierarchy Process. Journal of Changchun University 11, 6 (2011), doi:CNKI:SUN:CDXB.0.2011-11-006

28. Karakostas, B., et al.: The state of CRM adoption by the financial services in the UK: an empirical investigation. Information & Management 42, 853–863 (2005), doi:10.1016/j.im.2004.08.006

29. Kargar, M.J.: Prioritization of information quality criteria in the Blog context. In: International Conference on Uncertainty Reasoning and Knowledge Engineering, vol. 1, pp. 205–208 (2011), doi:10.1109/URKE.2011.6007798

30. Kim, H.J.: Online Social Media Networking and Assessing Its Security Risks. International Journal of Security and Its Applications 6, 11–18 (2012)

31. Kim, Y.: Toward a successful CRM: variable selection, sampling, and ensemble. Decision Support Systems 41, 542–553 (2006), doi:10.1016/j.dss.2004.09.008

32. Kracklauer, A.H., et al.: Customer management as the origin of collaborative customer relationship management. Collaborative Customer Relationship Management - taking CRM to the next level, 3–6 (2004)
33. Mohan, S., et al.: Conceptual modeling of enterprise application system using social networking and web 2.0 "social CRM system". In: International Conference on Convergence and Hybrid Information Technology, pp. 237–244 (2008), doi:10.1109/ICHIT.2008.263
34. Moreno-Jiménez, J.M., et al.: The core of consistency in AHP group decision making. Group Decision and Negotiation 17, 249–265 (2008), doi:10.1007/s10726-007-9072-z
35. Ocker, R.J., Mudambi, S.: Assessing the readiness of firms for CRM: a literature review and research model. In: Proceedings of the 36th Annual Hawaii International Conference on System Sciences (2003), doi:10.1109/HICSS.2003.1174390
36. Olszak, C., Bartuś, T.: Multi-Agent Framework for Social Customer Relationship Management Systems. Issues in Informing Science and Information Technology 10 (2013)
37. Özcanli, C.: A proposed Framework for CRM On-Demand System Evaluation: Evaluation Salesforce.com CRM and Microsoft Dynamics Online, Dissertation, Royal Institute of Technology (2012)
38. Parvatiyar, A., Sheth, J.N.: Customer relationship management: Emerging practice, process, and discipline. Journal of Economic & Social Research 3, 1–34 (2001)
39. Qian, Y., Zhang, Z.: Application of AHP in CRM. Microcomputer Development 11, 14 (2005), doi:cnki:ISSN:1005-3751.0.2005-11-014
40. Rappaport, S.D.: Listening Solutions A Marketer's Guide to Software and Services. Journal of Advertising Research 50, 197–213 (2010)
41. Reinhold, O., Alt, R.: Analytical social CRM: concept and tool support. In: BLED Proceedings, pp. 226–241 (2011)
42. Saaty, T.L.: The Analytic Hierarchy Process. McGraw Hill International, New York (1980)
43. Seraj, S., Khayatmoghadam, S.: Identifying and Prioritizing Effective Factors on Classifying A Private Bank Customers by Delphi Technique and Analytical Hierarchy Process (AHP). Journal of Soft Computing and Applications, 1–12 (2013), doi:10.5899/2013/jsca-00013
44. Swift, R.S.: Accelarating customer relationships: Using CRM and relationship technologies. Prentice Hall PTR, Upper Saddle River (2001)
45. Vulić, M., et al.: CRM E-Government Services in the Cloud (2013), http://www.fos.unm.si/media/pdf/CRM_e_government_services_in_the_cloud_maj.pdf (accessed on November 10, 2013)
46. Wu, Z., Xu, J.: A consistency and consensus based decision support model for group decision making with multiplicative preference relations. Decision Support Systems 52, 757–767 (2012), doi:10.1016/j.dss.2011.11.022
47. Wu, B., et al.: Group CRM: a new telecom CRM framework from social network perspective. In: Proceedings of the 1st ACM International Workshop on Complex Networks Meet Information & Knowledge Management (2009), doi:10.1145/1651274.1651277
48. Xu, Z.S., Wei, C.P.: A consistency improving method in analytic hierarchy process. European Journal of Operational Research 116, 443–449 (1999), doi:10.1016/S0377-2217(98)00109-X
49. Liu, Z.-W.: Design & Research of CRM Evaluation Index System of the Third Party Logistics Enterprises. Journal of Guangxi University of Fi-nance and Economics 3, 27 (2009)

Evaluating E-Commerce Portals from the Perspective of the End User – A Group Decision Support Approach

Gaurav R. Kalelkar, Gaurav Kumbhare, Varun Mehta, and Arpan Kumar Kar

Abstract. The stories of Kirana shops are now becoming folklore as Digital Retail has reared its head in the Indian sub-continent. To capitalize on this growth there has been a deluge of e-retail portals. However, the demographic scenario and the needs of Indian population are changing rapidly. The current study is an attempt to understand the dynamics of the major factors that consumers look into any of these e-tailers and identify the major dimension that helps in binding the consumers with these portals. For meeting this objective, the top 5 e-retail portals have been selected, based on multiple parameters for evaluating the traffic and importance of a website. These sites are Flipkart, Ebay, SnapDeal, Jabong and Myntra. Subsequently, these websites have been evaluated using the dimensions extended from SERVQUAL. A systematic approach has been taken in evaluating these portals using the theories of Analytic Hierarchy Process for group decision making.

1 Introduction

At the turn of the 21st century, when retailing started shifting from its traditional brick-and-mortar image to the digital image, people were initially sceptical about its acceptance and its growth. However, from being the '*next big thing*' to being an '*everyday reality*', e-tailing has surely come a long way. For the consumers, it is as if they are having a shopping exercise through the "digital mall". A wide variety of products, wider reach, purchase action any place any time and low cost are some of the factors that have contributed to the speed with which e-retail has

Gaurav R. Kalelkar · Gaurav Kumbhare · Varun Mehta · Arpan Kumar Kar
Indian Institute of Management Rohtak, Haryana, India
e-mail: {gaurav.kalelkar,gauravkumbhare,mehta.vnit}@gmail.com,
        arpan_kar@yahoo.co.in

S.M. Thampi, A. Gelbukh, and J. Mukhopadhyay (eds.), *Advances in Signal Processing
and Intelligent Recognition Systems*, Advances in Intelligent Systems and Computing 264,
DOI: 10.1007/978-3-319-04960-1_10, © Springer International Publishing Switzerland 2014

grown in scope and acceptance. India is currently in the transition phase of a Digital revolution. India, with 120 million users connected to the net, ranks third in the world with reference to the number of internet users and this number is expected to grow to 330-370 million by 2015. Economically, revenue from Internet sales contributes to about 1.6% of the Gross Domestic Product (GDP) and this is projected to rise to about 2.8-3.3% of GDP by 2015. If the internet reach is widened with timely and rapid deployment of quality infrastructure India can target the digital inclusion of nearly 40% of its population. This aggressive strategy, built upon many factors like, low-cost internet usage, improved internet literacy and favourable regulatory environment, can enable India to reach out to nearly 500 million of its population.

In this context, there is a growing need to understand the dynamics of the consumers and their needs. With online market space in the country burgeoning in terms of offerings ranging from travel, movies, hotel reservations and books to the likes of matrimonial services, electronic gadgets, fashion accessories and even groceries, consumers are being drowned in a pool of portals. What do the consumers need and what they perceive as being of qualitative value to them are some questions that the e-portals need to ask themselves and answer in the form of service implementation. A non-existent switching cost to the consumer and the low-cost competitiveness amongst e-portals presumably are the main business growth catalysts. Attracting new customers, retaining them and enhancing their interaction experience with the e-portal are some of the key processes e-tailers are focussing on. The quality of service provided by the e-tailers to the consumers will help in building up the loyal customer base of any e-portal. This creates a need for the e-tailers to identify the key service attributes that consumers look for in their e-portal usage. Leveraging upon these attributes positively could build up a relationship between the e-tailer and the customer which will sustain for a longer period and in turn result in revenue benefits for the e-tailer.

2   Literature Review

2.1   Analytic Hierarchy Process

Analytic Hierarchy Process (AHP) is a multi-criteria decision-making approach developed and introduced by Saaty [35, 37]. The process has appropriate measures which will help in prioritising amongst the evaluating parameters as represented in a study [2, 9]. The pair-wise comparison methodology provides a more meaningful analysis for developing a competitive set of service attributes that will satisfy customers and assist the e-tailers in outperforming its competitors. AHP uses a fundamental scale of absolute numbers that has been proven in practice and validated by physical and decision problem experiments. The fundamental scale has been shown to be a scale that captures individual preferences with respect to quantitative and qualitative attributes just as well or better than other scales as the study suggests [35, 37]. It converts individual preferences into ratio scale weights that can be combined into a linear additive weight for each alternative. Several papers have highlighted the AHP success stories in very different fields for decision making involving a complex problem with multiple conflicting and subjective criteria as well as multiple hierarchies of decision making processes [16, 19, 23, 28, 40]. A fuzzy extension of the AHP has been used in this study to accommodate the subjectivity of the individual respondent which making the tradeoffs between the relative prioritization of the evaluation criteria and performance scores against the evaluation criteria [47].

2.2 SERVQUAL

SERVQUAL parameters have long been a scale for measuring the quality provisions of service or retail organizations. Based on this premise, we have tried to gauge the perception of people towards the above selected e-tailers against the five SERVQUAL parameters, Tangibles, Reliability, Responsiveness, Assurance and Empathy. Many research papers [29, 30, 31, 32] have suggested adapting the SERVQUAL instrument to measure service quality in relation to competition. SERVQUAL is a well-established "gap-assessment" methodology that can be used to develop service-improvement initiatives by examining the "gap" between expectations and perceptions. The adapted SERVQUAL instrument uses a non-comparative evaluation model – that is, customers visiting the e-tailer are asked to evaluate the firm against a particular parameter based on what they perceive is the value provided by the firm for the specific parameter. This perception gives an indication as to the customers' perception alongside their expectations of the service. The SERVQUAL scores thus calculated provide the foundation for highlighting the gaps which the consumers presume about any firm. Individual SERVQUAL parameters have been ranked using the AHP methodology combined with the pair-wise comparison approach.

Many researchers unanimously agree on the fact that though there are different aspects to service quality, emphasis should be on the customers' perception of the service [46, 29, 30, 31, 32, 46].Organizations can measure business excellence through quality control in services. Service quality is considered as the difference between customer expectations of service and perceived service. If expectations are greater than performance, then perceived quality is less than satisfactory and hence customer dissatisfaction occurs [22]. There is general agreement that the aforementioned constructs are important aspects of service quality, but many scholars have been sceptical about whether these dimensions are applicable when evaluating service quality in other service industries [3]. This has more explanatory power than measures that are based on the gap between expectation and performance. In addition, it has been argued that SERVQUAL focuses more on the service delivery process than on other attributes of service, such as service-encounter outcomes (i.e. technical dimensions). While there have been efforts to study service quality, there has been no general agreement on the measurement of the concept. The majority of the work to date has attempted to use the

SERVQUAL methodology in an effort to measure service quality [8, 24, 38]. In the past few decades, service quality has become a major area of attention for practitioners, managers and researchers owing to its strong impact on business performance, low costs, customer satisfaction, customer loyalty and profitability [3, 13, 15, 20, 21, 40, 41]. There has been a continued research on the definition, modelling, measurement, data collection, procedure, data analysis etc., issues of service quality, leading to development of sound base for the researcher.

2.3   E-tailers

With the advent of internet in the 1990's and its adoption in retailing, the whole canvas of shopping changed. Internet's unmatched potential for global connectivity, through its ability to 'open up new avenues for business' [33] attracted retailers to the digital world. E-tailing became the new beacon for shopaholics and brick-and-mortar was replaced by click-and-brick. Retailers who viewed e-commerce as a new front on which to compete and gain advantage over their rivals managed to get that first mover advantage [1]. The initial review of internet literature influenced considerably the scheme of things to follow. Internet's ability to provide information, facilitate two-way communication with customers, collect market research data, promote goods and services online and ultimately to support the on-line ordering of merchandise provided an extremely rich and flexible new retail. Relevant exogenous factors which moderate consumer adoption of new self-service technologies and internet shopping systems [4, 27] are "consumer traits" [4, 5], "situational factors" [43], "product characteristics" [12], "previous online shopping experiences", and "trust in online shopping" [44]. Indeed, e-shopping is now estimated to be the fastest growing arena for Internet usage and thus greater exploration is required on the major dimensions that users look forward to, in evaluating the service provided by the e-tailers.

3   Focus of the Paper

There have been heaps of studies to understand the implementation of AHP for e-tailer selection. However, speaking from the Indian perspective, there has been virtually no study to focus on implementation of AHP for e-tailer selection. Academically there was a need to address this gap. The current study is an attempt to exhibit the application of theories for consensual group decision making using the AHP along with the SERVQUAL parameters for e-tailer selection.

Firstly, the study aims at exhibiting the AHP theory for prioritization and aggregation of the e-portal preferences of a group of decision makers based on the SERVQUAL parameters. Secondly, vis-á-vis the weighted average of the individual responses for the SERVQUAL parameters, aggregated decision from the viewpoint of 101 respondents was achieved on the preferred e-portal amongst the consumers and on the major dimensions for evaluating service quality.

Thirdly, the study highlights the core competency of the leading e-tailers in terms of the service evaluation dimensions.

However it should be noted that the focus of this study is on providing group decision support for the e-tailer service evaluation problem by listening to the voice of the customer, and no attempt has been made to explore other e-portal selection issues, like the suitability of existing or new evaluation criteria for the context.

4 Computational Approach

To evaluate the service quality of e-tailers in India, a survey was conducted to rate them on five dimensions (reliability, assurance, tangibility, empathy and responsiveness); and analyze the importance of each dimension by comparing them with each other; then we evaluated each e-trailer on the five dimension by using Likert scale. After collecting the responses, the judgments and performances were mapped according to logical approach of AHP which been described subsequently.

1. Estimation of individual judgments

In this stage, the relative importance given to a pair of evaluation criteria is estimated. Let $A = (a_1 \ldots a_5)$ be the multi-dimensional consensus vector such that a_i is the aggregated priority of 'criteria i' estimated as described in the following section such that $\Sigma a_i = 1$. Let $D = (d_1 \ldots d_n)$ be the set of n decision makers having a relative importance of ψi such that $\psi = (\psi_1 \ldots \psi_n)$ is the weight vector of the decision makers and $\Sigma \psi_i = 1$.

Comparative fuzzy judgments $M = (a)_{nxn}$ would be coded from linguistic comparisons as described in Figure 2. A triangular fuzzy function has been used for coding the judgments since there is equal probability of the response of the next level as is to the response of the previous level, when a comparative judgment is made by an expert decision maker. The entropy of an individual judgment has been optimized by maximizing the Shanon function for the middle element defined as $S(\mu) = {}_\mu \ln_\mu - {}_{(1-\mu)} \ln_{(1-\mu)}$ [47].

Table 1 Mapping linguistic judgments to fuzzy judgments

Linguistic judgment		Judgment values in fuzzy set ($\tilde{k}$)
Equal importance	$\tilde{1}$	{(1,0.25), (1,0.50), (3,0.25)}
Moderate importance	$\tilde{3}$	{(1,0.25), (3,0.50), (5,0.25)}
Strong importance	$\tilde{5}$	{(3,0.25) (5,0.50) (7,0.25)}
Very strong importance	$\tilde{7}$	{(5,0.25), (7,0.50), (9,0.25)}
Extreme importance	$\tilde{9}$	{(7,0.25), (9,0.50), (9,0.25)}

After taking multiple responses from N decision makers, it would be coded as fuzzy rules to generate the judgment matrix which will be used for subsequent rounds of prioritization.

2. Collective preference of the group for delivering the decision vector can be estimated subsequently by the aggregation of individual priorities for consensus development such that the aggregate priorities (collective vector) is defined as

$$W_C = W_{C1}, W_{C2} \ldots\ldots W_{Cr} \tag{1}$$

$$W_{Ci} = (\pi_{n1} (W_{ki}))^{\psi} / \Sigma_r 1 \ \pi_{n1}(Wki)^{\psi} \tag{2}$$

Here WC is collective priority vector derived by GMM and   is weight vector or relative importance of decision maker.

3. Let 'e-tailer i' have a performance vector of E_i, a set of performance score against 5 dimension $E_i = (e_{i1}\ldots e_{i5})$. Here, e_{ij} is the score on a 5 point Likert scale of e-tailer i within a predetermined range for dimension j. The score against a particular criterion offering the highest utility would be coded as 5, and the score offering the lowest utility would be coded as 1. The intermediate scores can be computed from a linear transformation function as demonstrated:

$$S_{i,j} = \frac{e_{i,j} - e_{j(max)}}{e_{j(max)} - e_{j(min)}} \tag{3}$$

Here, e_{ij} is the absolute score of e-tailer i for criteria j, while $e_{j(max)}$ and $e_{j(min)}$ are the maximum and minimum absolute score on criteria j for all the e-tailers. The final performance score for 'e-tailer i' will be computed by the sum-product approach.

$$S_i \bullet X = (S_{i1}, \ldots S_{i7}) \bullet (x_1, \ldots, x_7) = |S_{i1} \times x_1 + \ldots S_{i7} \times x_7| \tag{4}$$

Based on this sum-product score, the e-tailers may be ranked such that a higher score would indicate a more suitable e-tailer and a lower score would indicate a less suitable e-tailer based on the priorities of a specific respondent.

5   Research Methodology

Five e-retail web-sites, namely Flipkart, Ebay, SnapDeal, Jabong and Myntra, were selected for this study. These websites were selected based on the Yahoo India Finance ranking as well as based on their Google page ranks and Alexa Traffic Rank. An extended questionnaire was implemented online to all for capturing their responses. The questionnaire was not limited to or restricted for only the GenY. Gen Y represents the demographic group of people who were born in the 1980's and early 90's. Responses were encouraged and captured from all willing to take part in the research survey. The questionnaire, means for secondary research, was designed such that participants were made to compare these portals based on the SERVQUAL parameters. The questionnaire captured what parameters were important for the user and how the users rated the five selected e-tailers on SERVQUAL parameters.

Since the questionnaire had 19 questions, a sample size exceeding 95 respondents was sufficient to generalize the outcome of the study. A purposive sampling technique was used for identifying the participants of this study. A total of 101 highly consistent responses (from 84 males and 17 females) were collected with 15.84% of responses from age group between 20-25, 81.19% from age group between 26-30 and rest from age >30. The responses were collected from people having diverse educational level such as doctorate, post graduates, graduates, higher secondary and secondary.

6   Analysis and Findings

Step 1: The responses were collected for comparing every dimension with one another and a reciprocal matrix was formulated to evaluate the final weight for each dimension which is shown in table below.

Table 2 Aggregate priority for the individual SERVQUAL parameters

Criteria	Reliability	Assurance	Tangibility	Empathy	Responsiveness
Priority	0.2068	0.1977	0.206	0.1824	0.2072

Step 2: Responses were collected by using five point Likert scale for the top five e-tailer on the service quality dimensions (Reliability, assurance, tangibility empathy and responsiveness) and geometric mean is taken for all the dimensions which is shown in the table below for different e-tailers:

Table 3 Geometric mean of the responses of participants for the individual parameters

	Reliability	Assurance	Tangibility	Empathy	Responsiveness
Flipkart	3.3708	3.6523	3.4312	3.4025	3.5985
Ebay	3.5717	3.4819	3.4755	3.1766	3.7026
Jabong	3.1523	3.2848	3.3268	3.0060	3.5605
Myntra	3.4705	3.1880	3.4264	3.3849	3.4079
Snapdeal	3.1873	3.2049	3.5335	3.3768	3.4088

Step 3: The final weight (step 1) of each dimension is then multiplied with mean of respective dimension (step 2) for all the e-tailers. Results are shown in the below table.

Table 4 Product of priority weights with geometric mean of responses

	Reliability	Assurance	Tangibles	Empathy	Responsiveness
Flipkart	0.6971	0.7221	0.7068	0.6204	0.7455
Ebay	0.7386	0.6884	0.7159	0.5793	0.7671
Jabong	0.6519	0.6494	0.6853	0.5482	0.7376
Myntra	0.7177	0.6303	0.7058	0.6172	0.7060
Snapdeal	0.6591	0.6336	0.7279	0.6158	0.7062

The results shows that consumer prefer flipkart, Ebay and Jabong because of responsiveness whereas Myntra for Relibility and Snapdeal for tangibility.

Step 4: For evaluating the best e-tailer on the five service dimension, Sum-product method is applied on the geometric mean of dimensions and final weights and thus we obtain the rating for the individual e-tailers. Flipkart scored the highest among all these e-tailers.

Table 5 Final scores for ranking the e-tailers

E-tailer	Flipkart	Ebay	Jabong	Myntra	Snapdeal
Overall Rate	3.4919	3.4892	3.2723	3.3770	3.3425

7   Conclusion

The study revealed that users viewed Flipkart, overall, as the preferred e-tailer closely followed by eBay. It was found that users felt that the employees of Flipkart, Ebay and Jabong showed more willingness in giving timely service and resolving their queries and were thus rated high in responsiveness dimension. Flipkart was rated highest on assurance which meant Flipkart and its employees were able to convey trust and confidence in their communication. Flipkart was also rated highest on empathy showing that they were more caring towards the customers' needs.

Myntra and eBay were rated highest on reliability dimension implying that they delivered what was promised on more occasions than others. Snapdeal was rated highest on the tangibility dimension and were thus most physically visible when compared to other e-tailers. The research showed that consumers weighted reliability, tangibility and responsiveness as more important dimensions than assurance and empathy when measuring service quality. Flipkart was rated the highest as it scored well on these three dimensions. E-tailers that want to gain customer loyalty would have to strive hard and respond quickly to the ever changing dynamics of the e-tailing industry.

References

1. Amit, R., Zott, C.: Value creation in e-business. Strategic Management Journal 22(6-7), 493–520 (2001)
2. Aguarón, J., Moreno-Jiménez, J.M.: The geometric consistency index: Approximated thresholds. European Journal of Operational Research 147(1), 137–145 (2003)
3. Cronin, J.J., Taylor, S.A.: Measuring service quality: a re-examination and extension. Journal of Marketing 6, 55–68 (1992)
4. Dabholkar, P.A., Bagozzi, R.P.: An attitudinal model of technology-based self-service: moderating effects of consumer traits and situational factors. Journal of the Academy of Marketing Science 30(3), 184–201 (2002)
5. Eastin, M.S., LaRose, R.: Internet self-efficacy and the psychology of the digital divide. Journal of Computer-Mediated Communication 6(1) (2000)
6. Edvardsen, B., Tomasson, B., Ovretveit, J.: Quality of Service: Making it Really Work. McGraw-Hill (1994)
7. Edvardsson, B.: Kundmissnöje och klagomålshanteringsstudier av kollektivtrafik med kritiskhändelse-metoden. Forskningsrapport 97:3, Samhällsvetenskap, Högskolan i Karlstad, Sverige (1997)
8. Edvardsson, B., Larsson, G., Setterlind, S.: Internal service quality and the psychological work environment: an empirical analysis of conceptual interrelatedness. Service Industries Journal 17(2), 252–263 (1997)
9. Escobar, M.T., Moreno-Jiménez, J.M.: Reciprocal distributions in the analytic hierarchy process. European Journal of Operational Research 123(1), 154–174 (2000)
10. European Journal of Operational Research, 147(1), 137–145 (2003)
11. Forman, E.H., Gass, S.I.: The Analytic Hierarchy Process – An Exposition. Operations Research 49(4), 469–486 (2001)
12. Grewal, D., Iyer, G.R., Levy, M.: Internet retailing: enablers, limiters and market consequences. Journal of Business Research (2002)
13. Gammie, A.: Stop at nothing in the search for quality. Human Resources 5, 35–38 (1992)
14. Gummesson, E.: Productivity, quality and relationship marketing in service operations. International Journal of Contemporary Hospitality Management 10(1), 14–15 (1998)
15. Guru, C.: Tailoring e- service quality through CRM. Managing Service Quality 13(6), 520–531 (2003)
16. Golden, B., Wasil, E., et al.: The Analytic Hierarchy Process: Applications and Studies (1989)
17. Grönroos, C.: A service quality model and its marketing implications. European Journal of Marketing 18, 36–44 (1984)
18. India Brand Equity Foundation, http://www.ibef.org/download/The-Rise-and-Rise-of-E-commerce-in-India.pdf
19. Kumar, S., Vaidya, O.S.: Analytic Hierarchy Process-An overview of applications. European Journal of Operational Research 169(1), 1–129 (2006)
20. Lasser, W.M., Manolis, C., Winsor, R.D.: Service quality perspectives and satisfaction in private banking. Journal of Services Marketing 14(3), 244–271 (2000)
21. Leonard, F.S., Sasser, W.E.: The incline of quality. Harvard Business Review 60(5), 163–171 (1982)
22. Lewis, M.: Dimensions of service quality: a study in Istanbul. Managing Service Quality 5(6), 39–43 (1990)

23. Liberatore, M., Nydick, R.: The analytic hierarchy process in medical and health care decision making: A literature review. European Journal of Operational Research 189(1), 194–207 (2008)
24. Lings, I.N., Brooks, R.F.: Implementing and measuring the effectiveness of internal marketing. Journal of Marketing Management 14, 325–351 (1998)
25. McKinsey, http://is.gd/92nHUv
26. Newman, K.: Interrogating SERVQUAL: a critical assessment of service quality measurement in a high street retail bank. International Journal of Bank Marketing 19(3), 126–139 (2001)
27. O'Cass, A., Fenech, T.: Web retailing adoption: Exploring the nature of Internet users' web retailing behavior. Journal of Retailing and Consumer Services (2002)
28. Omkarprasad, V., Sushil, K.: Analytic hierarchy process: an overview of applications. European Journal of Operational Research 169(1), 1–29 (2006)
29. Parasuraman, A., Berry, L.L., Zeithaml, V.A.: An Empirical Examination of Relationships in an Extended Service Quality Model. Marketing Science Institute (1990)
30. Parasuraman, A., Zeithaml, V., Berry, L.L.: Refinement and reassessment of the SERVQUAL scale. Journal of Retailing 67(4), 420–450 (1991)
31. Parasuraman, A., Zeithaml, V., Berry, L.L.: Research note: More on improving service quality measurement. Journal of Retailing 69(1), 140–147 (1993)
32. Parasuraman, A., Zeithaml, V., Berry, L.L.: Reassessment of expectations as a comparison standard in measuring service quality: Implications for future research. Journal of Marketing 58, 111–124 (1994)
33. Pyle, R.E.: Persistence and Change in the Protestant Establishment (1996)
34. Reynoso, J., Moore, B.: Towards the measurement of internal service quality. International Journal of Service Industry Management 6(3), 64–83 (1995)
35. Saaty, T.L.: A Scaling Method for Priorities in Hierarchical Structures. Journal of Mathematical Psychology 15, 57–68 (1977)
36. Saaty, T.L.: The Analytic Hierarchy Process. McGraw-Hill International (1980)
37. Saaty, T.L.: Fundamentals of Decision Making and Priority Theory with the AHP. RWS Publications (1994)
38. Sahney, S., Banwet, D.K., Karunes, S.: A SERVQUAL and QFD approach to total quality education: A student perspective. International Journal of Productivity and Performance (2004)
39. Shim, J.: Bibliography research on the analytic hierarchy process (AHP). Socio-Economic Planning Sciences 23(3), 161–167 (1989)
40. Silvestro, R., Cross, S.: Applying service profit chain in a retail environment. International Journal of Service Industry Management 11(3), 244–268 (2000)
41. Sureshchander, G.S., Rajendran, C., Anatharaman, R.N.: The relationship between service quality and customer satisfaction: a factor specific approach. Journal of Services Marketing 16(4), 363–379 (2002)
42. Vargas, L.: Comments on Barzilai and Lootsma Why the Multiplicative AHP is Invalid: A Practical Counterexample. Journal of Multi-Criteria Decision Analysis 6(4), 169–170 (1997)
43. Wolfinbarger, M., Gilly, M.C.: Shopping online for freedom, control, and fun. California Management Review 43(2), 34–55 (2001)

44. Yoon, S.J.: The antecedents and consequences of trust in online purchase decisions. Journal of Interactive Marketing 16(2), 47–63 (2002)
45. Zahedi, F.: A Simulation Study of Estimation Methods in the Analytic Hierarchy Process. Socio-Economic Planning Sciences 20, 347–354 (1986)
46. Zeithaml, V.A., Berry, L.L., Parasuraman, A.: The nature and determinants of customer expectations of service, pp. 91–113. Marketing Science Institute (1991)
47. Zimmerman, B. J.: Theories of self-regulated learning and academic achievement: An overview and analysis (2001)

Development of Assamese Speech Corpus and Automatic Transcription Using HTK

Himangshu Sarma, Navanath Saharia, and Utpal Sharma

Abstract. Exact pronunciation of words of a language is not found from the written form of the language. Phonetic transcription is a step towards the speech processing of a language. For a language like Assamese it is most important because it is spoken differently in different regions of the state. In this paper we report automatic transcription of Assamese speech using Hidden Markov Model Tool Kit (HTK). We obtain accuracy of 65.26% in an experiment. We transcribed recorded speech files using IPA symbols and ASCII for automatic transcription. We used 34 phones for IPA transcription and 38 for ASCII transcription.

Keywords: Automatic Transcription, Speech corpus, Assamese, HTK.

1 Introduction

The pronunciation and written forms of words are different. A transcription intended to represent each distinct speech sound with a separate symbol is known as phonetic transcription. Phonetic transcription of a language is important because written form of a language does not tell us how to pronounce it. Mainly International Phonetic Alphabet (IPA) symbols is used in phonetic transcription.

Assamese is an Eastern Indo-Aryan language spoken by about 20 million speakers[1]. It is mainly spoken in Assam of North-East of India. Assamese is a less computationally aware language. Language analysis or research about a language is a very large task. For analyzing speech of a language, one needs to work on different aspects. Firstly one needs to collect a balanced speech

Himangshu Sarma · Navanath Saharia · Utpal Sharma
Tezpur University, Napaam, Assam, India-784028
e-mail: `himangshu.tezu@gmail.com`,`{nava_tu,utpal}@tezu.ernet.in`

[1] `http://www.iitg.ac.in/rcilts/assamese.html` ; Access date: 19 July 2013

S.M. Thampi, A. Gelbukh, and J. Mukhopadhyay (eds.), *Advances in Signal Processing and Intelligent Recognition Systems*, Advances in Intelligent Systems and Computing 264,
DOI: 10.1007/978-3-319-04960-1_11, © Springer International Publishing Switzerland 2014

corpus with different native speakers of that language with different categories. Assamese is pronounced differently in different regions of Assam. So, for Assamese speech corpus it is important to collect speech data from different regions of Assam. Different categories are like read speech, extempore speech and conversation speech. Every speech can be represented phonetically by some finite set of symbols which are called phones of the language. To analyze and construct a complete speech corpus of a language it needs to be processed in different stages, such as IPA transcription[2], syllabification, prosodical feature extraction, and break-marking. Automatic transcription of a language is a big challenge till now. We find that till now it has not been possible to give 100 percent accuracy for any language.

In Section 2, we describe some previously reported work relevant to our work and phonetic characteristics of Assamese. Section 3 describes the speech corpus developed in the course of our experiments. We discuss the different issues related to transcription in Section 4. Transcription using ASCII (that is transliteration of IPA symbols to ASCII) symbols and our automatic transcription model is illustrated in Section 5. Results of automatic transcription and comparison with other data sets are discussed in Section 6. Section 7 concludes our report.

2 Background Study

Phonetic transcription is an important research issue in speech processing. Phonetic Transcription is the phonetic representation of a speech file. Phonetic representations are the use of phonetic symbols to represent a speech sound. It is used to find dynamic stress, vowel duration, precise vowel quality etc. from the written form. Usually IPA symbols are used for phonetic transcription[34]. For speech language pathologists it is most important part. IPA symbols[5] are used for new writing systems for previously unwritten language. Difference of two sounds can be found from loudness, pitch and quality of those sounds Ladefoged (1995). Every phonetic symbol represents an utterance of a sound Wells (2006). It is actually perfect pronunciation of a speech. For Assamese speech it is not so easy because same words pronounced differently by different regions of the state. A phoneme is defined as a distinctive sound unit of a language and a phone can be defined as a unit sound of a language Coxhead (2007). Every speech sound has some unique signal, but same speech sound is spoken differently by different speakers. Based on this Ladefoged and Johnstone (2011) divided phonetics in three major sub fields, which are: *Articulatory Phonetics*— how speech sounds are produced,

---

[2] http://www.langsci.ucl.ac.uk/ipa/index.html
[3] http://www.madore.org/~david/misc/linguistic/ipa
[4] http://eweb.furman.edu/~wrogers/phonemes
[5] http://www.langsci.ucl.ac.uk/ipa/IPA_chart_(C)2005.pdf

Acoustic Phonetics— how speech sounds are transmitted from speaker to listener and *Perceptual Phonetics*— how speech sounds are perceived.

There are a lot of research works done in this field till now. Chang et al (2000) describe a system of automatic transcription for American English which generated a phonetic labels of 80% concordant. Levinson et al (1989) reports text independent transcription of fluent speech database using Hidden Markov Model with 3020 trained and 180 tested sentences with an with 52% correct and 12% insertions. Approach of Liang et al (2007) uses speech recognition technique, where they try to show phonetic transcription of Chinese text into Taiwanese pronunciation. They found 12.74% transcription error rate using multiple pronunciation lexicon and 17.11% of error rate reduction was achieved using pronunciation lexicon with pronunciation variation. Giurgiu and Kabir (2012) analyzed automatic transcription and speech recognition for Romanian language using HTK Young et al (2009) and PRAAT[6]. Laurent et al (2009) describes a method to extract proper nouns a large set of phonetic variants using rule-based generation. Leung and Zue (1984) developed a system for automatic alignment of phonetic transcriptions with continuous speech. Stefan-Adrian and Doru-Petru (2009) reports a rule based transcription tool for the Romanian language, using 102 letter to sound rules they found an accuracy of 95% from a database of 4779 words and found 91.46% accuracy from the database which contains 15699 words.

For Indian languages we find the following reported work on transcription. Patil et al (2012) addresses encountered mainly issues and ambiguities during phonetic transcription of two Indian languages mainly Gujarati and Marathi languages. They shows this ambiguity in case of fricative and aspirated plosive. Sarada et al (2009) reports automatic transcription into syllable like units for different Indian languages using group delay algorithm. They found 48.7% and 45.36% accuracy for Tamil and Telugu language respectively. Nagarajan et al (2004) approach automatic transcription of a continuous speech signal without the use of manually annotated speech corpora. They found an accuracy of 43.3% and 32.9% for Tamil and Telugu language respectively. Sarma et al (2013) describes some issues and interesting observation during phonetic transcription of Assamese speech corpus.

Assamese script consist 11 vowels, 37 consonants, and 10 digits. In addition, there are 7 symbols corresponding to certain consonant sounds Sharma et al (2008). Assamese phone inventory consists of eight oral vowels, three nasalized vowels, fifteen diphthongs and twenty one consonants[7]. When, we speak Assamese, we actually mean standard pronunciation of Assamese, not the dialects.

---

[6] http://www.fon.hum.uva.nl/praat

[7] www.iitg.ernet.in/rcilts/pdf/assamese.pdf

3 Data Collection and Tools Used

To construct our Assamese Speech corpus we collect approximately seventeen hours of data from total of 27 speakers, out of them 14 male and 13 female within age group of 20 to 40 years. For a balanced speech corpus it is important to have all standard variety of the language. Assamese is spoken differently by different regions of the state. So, we collect data from speakers that belong to different regions of the state. Speech are recorded in a noise free room using SONY ICD-UX533F device with 44.1 KHz sampling rate.

3.1 Mode of Recording

For speech processing it is important to have a balanced speech corpus. So, we recorded our speech samples in different categories. Below we show those different categories:

1. Read Speech
2. Extempore Speech
3. Conversation Speech

For read speech, we collect data from two sources. AIR news and reading text. AIR news are collection of broadcasting news by All India Radio[8] collected from June 2012 to May 2013. We recorded the reading of printed news, story and articles as the second source of read speech. As the name indicates, extempore speech are evidence of normal talk on random topics. In conversation mode, speech are recorded from more than one speakers discussing together with some topics.

3.2 Tools Used

- *Wavesurfer*: Wavesurfer Sjölander and Beskow (2000) is used to listening the sound files for transcription. Each sound file and associated label file (such as pitch labelling, syllabification etc.) can be viewed using multiple time aligned sub-windows, or panes.
- *IPA Typing Tool*: For transcription we use IPA typing tool[9].

4 Transcription

IPA or International Phonetic Alphabets are the symbols which represent a sound. We transcribed approximately 7 hours of data out of 17 hours of data from our speech corpus. During transcription, we find that in Assamese labiodental, dental, retroflex, uvular, pharyngeal, trill and lateral fricative

---

[8] http://www.newsonair.com
[9] http://www.tezu.ernet.in/~nlp/ipa.htm

phone are not present. We also find that dialect of other language (For example: Hindi, English, Bengali) influence Assamese speakers, particularly in loan words. There are thirty four IPA symbols used for Assamese out of which twenty five are consonants and nine are vowels. The IPA symbols used in our transcription correspondence to Assamese phones are shown in Table 1. We calculate the frequencies of IPA symbols in Table 2 from transcribed data.

Table 1 IPA symbols for Assamese letters

	Letter	IPA	Letter	IPA	Letter	IPA
C	ক	k	ন	n	ৱ	w / bɒ
O	খ	kʰ	ত	t	শ	x
N	গ	g	থ	tʰ	ষ	x
S	ঘ	gʰ	দ	d	স	s / x
O	ঙ	ŋ	ধ	dʰ	হ	h
N	চ	s / ʃ / tʃ	ন	n	ক্ষ	kʰj
A	ছ	s / ʃ / tʃ	প	p	য়	j
N	জ	dʒ	ফ	pʰ	ড়	ɹ
T	ঝ	dʒʰ	ব	b	ঢ়	ɹh
S	এ	ɲ	ভ	bʰ	ৎ	t̪
	ট	t	ম	m	◌ঁ	ŋ
	ঠ	tʰ	য	dʒ	◌ঃ	
	ড	d	ৰ	ɹ	◌	'
	ঢ	dʰ	ল	l		
V	অ	ɔ / ɒ	উ	u	ঐ	oi
O	আ	a	ঊ	u	ও	ʊ / o
W	ই	i	ঋ	ɹi	ঔ	ou
E	ঈ	i	এ	e / ɛ		
L						
S						

- **Most and Least Frequent Phone.** From Table 2 we find that ɹ is the highest used consonant and ɒ is the most used vowel phone find during our transcription. We find that ɲ is the less used consonants and o is less used vowel in our transcription database.

4.1 Issues and Interesting Observation

We faced some issues and interesting observation related to consonants and vowels. The pronunciation rules are different in different regions of Assam. The following are some interesting observation and problems faced during the transcription.

Table 2 IPA Symbols with frequency in percentage (F) for Assamese

IPA Symbol(F)	IPA Symbol(F)	IPA Symbol(F)
a (10.1265398)	ʊˑ (0.0008391)	ñ (0.0025174)
aː (0.0025174)	ʊ (0.0730037)	ń (0.0025174)
aˑ (0.0654516)	ɒ (7.391837)	nˑ (0.0159433)
ă (0.0192998)	ɒˑ (0.006713)	p (2.4611486)
á (0.0100695)	x (2.6449166)	pʰ (0.2718759)
ă (0.0192998)	x́ (0.0008391)	p̆ (0.0016782)
e (5.6036653)	xˑ (0.0025174)	ṕ (0.0025174)
eː (0.0033565)	b (4.3198067)	s (2.0936126)
eˑ (0.0578995)	bʰ (0.6696204)	š (0.0025174)
é (0.006713)	ɓ (0.0025174)	ś (0.0025174)
ĕ (0.0293693)	ƀ (0.0041956)	sˑ (0.0402779)
ĕ (0.0016782)	d (2.1993421)	ṭ (6.3410462)
i(10.3086295)	dʰ (0.4657134)	tʰ (0.7938106)
iː (0.0100695)	đ (0.0008391)	ƚ (0.0025174)
iˑ (0.0923036)	ɗ (0.0008391)	tˑ (0.0025174)
ì (0.0008391)	g (0.0897862)	ť (0.0033565)
î (0.006713)	gʰ (0.0897862)	ŋ (0.3625013)
ĭ (0.0402779)	h (1.8368409)	ŋ̆ (0.0016782)
ï (0.0008391)	ħ (0.0008391)	g (0.7728325)
o (3.4026449)	ħ (0.0025174)	gʰ (0.0562213)
oː (0.0008391)	ħ (0.0075521)	gˑ (0.008391)
oˑ (0.0159433)	j (1.3224583)	ğ (0.0008391)
ó (0.0008391)	ĵ (0.0176216)	g̀ (0.0008391)
ŏ (0.0184607)	k (4.8988017)	ʃ (0.8114322)
ɔ (4.2040077)	kʰ (0.782902)	t (0.0318867)
ɔː (0.0008391)	ƙ (0.0008391)	ɹ (7.3112812)
ɔˑ (0.0335649)	ƙ (0.0041956)	ɹː (0.0041956)
ɔ́ (0.0033565)	ƙ (0.0100695)	ɹˑ (0.041917)
ɔ̆ (0.0436344)	l (3.1685295)	ɹ̆ (0.0268519)
u (4.4062364)	ĺ (0.0008391)	ɹ́ (0.0008391)
uː (0.0092304)	lː (0.0016782)	ɛ (0.0050347)
uˑ (0.0092304)	lˑ (0.0083912)	w (0.0092304)
ù (0.0008391)	l (0.006713)	ɲ (0.0629342)
ú (0.0008391)	m (2.9117578)	ʧ (0.3054409)
ŭ (0.0209781)	mː (0.0008391)	ʤʰ (0.0008391)
ŭ (0.0008391)	ḿ (0.0041956)	ʤˑ (0.0008391)
ú (0.0008391)	mˑ (0.006713)	ʤ̆ (0.0016782)
ŏ (0.0008391)	n (4.3273588)	ʤ (1.7957238)

4.1.1 Consonants

- w is used for ৱ if ৱ is used in middle of a word or the first letter of the word is ৱ but bɒ is used for if ৱ is the last letter of the word. e.g

ছোৱালী = ʃowali

ৱাহিদ = wahid

কেশৱ = kexɒbɒ

- ʧ is actually not used in proper Assamese pronunciation. But ʧ is used in English, Bengali and Hindi languages. In contrast to earlier times, currently many Assamese people speak English, Bengali and Hindi. So, when they speak Assamese they used ʧ. For example:

  আচ্চা = aʧa
  বাচ্চা = baʧa

- During transcription we saw that there is a ɒ occurs after every consonant. But some words i.e. ঙ, ৎ, ০ং, ০ঃ are not follow this rule. For example:

  অংক = ɒŋkɒ
  আঙুৰ = ɒŋuɹ
  উৎসৱ = uʈxɒb
  নিঃকিন = niǩkin

- We find another one interesting observation for স. When we used singly than we used x. But, in conjunct characters the pronunciation is s. e.g

  সাধাৰণ = xadʰaɹɒn
  ব্যৱস্থাত = bjbɒstʰat

- We also find that in some consonant both sequence happens. In some words ɒ occurs after the consonant and in some words ɒ does not occurs after the consonant. For example:

  ৱাইজক = ɹaiʤɒk
  ৱাজপথ = ɹaʤpɒtʰ

4.1.2   Vowels

In case of vowel we observed that ʊ and u is sometimes placed for same word which meaning is same, but placed different phone because of different region of speakers. For example:

বোলেও = bʊleo
বুলেও = buleo

4.1.3   Clustering

Vowel ɒ is inserted at the end with all consonant cluster except কান্ধ (kandʰ), বান্ধ (bandʰ). For example:

মন্তব্য=mɒntɒbjɒ
অস্তিত্ব=ɒstitbɒ

But, in some cases if a consonant cluster occurs final position then ɒ is not inserted at the last position of the word. For example:

কান্ধ=kandʰ
বান্ধ=bandʰ

5 Automatic Transcription

Automatic transcription is the process where transcriptions of speech files are automatically generated. In this paper we use HTK model for automatic transcription of Assamese speech. For transcribed files we firstly split a huge speech file into segment of approximately 10 seconds. In Figure 1 we show how it works:

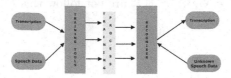

Fig. 1 Automatic Transcription Process

5.1 *Manual ASCII Transcription*

To construct our automatic transcription model, firstly we transcribed approximately 3 hours speech data. To transcribe our data, we use *thirty eight* ASCII symbols out of those twenty seven consonants, nine vowels and sil (used during transcription at the end of each segment of the speech) and sp (used in between every words). In Table 3 we show the ASCII symbols used in correspondence to Assamese phone.

5.2 *Files*

We construct some files to construct HTK model − *Configuration Files, Dictionary, Master Label Files.*

5.2.1 Configuration File

In our *HTK* model we firstly need to make a configuration file containing all the needed conversion parameters. The configuration file is a standard file which is constructed manually Young et al (2009).

5.2.2 Grammar

To construct our automatic transcription model, we need to make a grammar for ASCII symbols used in our transcription. Below we have shown the structure how we construct the grammar. Using HParse we convert below format to wdnet, which is the standard format for Hidden Markov Model.

$WORD= oa | a | i | u | ri | e | oi | o | ou | k | kh | g | gh | ng | s | j | jh | yo | t | th | d | dh | n | p | f | b | v | m | r | l | w | x | h | khy | y | rh | ta | sil;

Table 3 Representation of Assamese sound in ASCII; where L—Letters of Assamese alphabet

	Letter	Symbol	Letter	Symbol	Letter	Symbol
C	ক	k	ণ	n	ৱ	w
	খ	kh	ত	t	শ	x
C	গ	g	থ	th	ষ	x
O	ঘ	gh	দ	d	স	x
N	ঙ	ng	ধ	dh	হ	h
S	চ	s	ন	n	ক্ষ	khy
O	ছ	s	প	p	য়	y
N	জ	j	ফ	ph	ড়	r
A	ঝ	j	ব	b	ঢ়	rh
N	ঞ	yo	ভ	bh	ৎ	ta
T	ট	t	ম	m	◌ং	ng
S	ঠ	th	য	j		
	ড	d	ৰ	r		
	ঢ	dh	ল	l		
V						
O	অ	oa	উ	u	ঐ	oi
W	আ	a	ঊ	u	ও	o
E	ই ঈ	i	ঋ	ri	ঔ	ou
L	ই ঈ	i	এ	e		
S						

5.2.3 Master Label File

Master Label File(**MLF**), which includes the transcription of the speech files. We include our all segmented transcribed files into one .mlf file. MLF files have some standard format that is shown below Young et al (2009):

```
#!MLF!#
"*/filename.lab"
        sil
         w
         e
        sp
         a
         r
         e
        sil
         .
```

(starting transcription after **sil**, **sp** between words and ending transcription with **.** followed by **sil**)

5.3 Mel-Frequency Cepstral Coefficients(MFCC)

MFCC is based on known variation of the human ears critical bandwidth with frequency. MFCC has two types of filter which are spaced linearly at low frequency below 1000 Hz and logarithmic spacing above 1000Hz Hasan et al (2004).

In this process we convert all of our .wav files to .mfc(Mel Frequency Cepstrum) files. To convert .wav to .mfc we used HCopy command. For conversion of .wav to .mfc, we construct a file, where destination of .wav files corresponds to destination of .mfc files were present. In Figure 2 we show how HCopy works Young et al (2009).

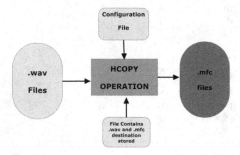

Fig. 2 HCopy Operation

5.4 Training

We trained approximately two and half hours of data contained in 527 files for our automatic transcription model. All files contained not more than 10 seconds of speech. For our training, we use HCompV − *calculate the speakers mean and variance* and HERest − *refine the parameters of existing Hidden Markov Models.*

5.5 Testing

For testing automatic transcription models approximately half an hour of data with 127 files are used. Duration of testing files are also not more than 10 seconds. To recognize, .mfc files and make a perfect automatic transcription model we use HVite tool and for saving our results we used HResults tool.

6 Result Analysis

Results of automatic speech transcription are analyzed in this Section. The results of automatic transcription are calculated by an equation, which is shown below:

$$Accuracy = \frac{H - I}{N}$$

Again,

$$H=N-S-D$$

where,
N = Total Number of phones.
I= Number of newly inserted phones.
S= Number of substitutions of phones.
D= Number of deletions of phones.

During automatic transcription, we used 527 files for training. Then, we tested over 127 files. Extempore speech are used for training and testing for the automatic transcription model. The Accuracy has found *65.65 %* with 38 phones. We find that accuracy increased depend upon phones also, if less phones used for the model than accuracy is higher. In Table 4 we show accuracy of all phone used in our transcription.

We compare the results for different sets of training and testing files. Below we describe how we compare our training and testing files.

Table 4 Phone-wise accuracy list

Symbol/ Phone	Accuracy	Symbol/ Phone	Accuracy
a	90.8	oa	60.6
b	75.5	oi	82.4
d	79.9	ou	0
dh	40	p	78.3
e	84.1	r	79.4
f	22.2	rh	0
g	75.3	ri	16.7
gh	5	s	90.1
h	61.3	sil	60.9
i	90.9	sp	49
j	75.6	t	89.1
k	75.4	ta	0
kh	41.5	th	84.1
khy	51.4	u	59.1
l	93.5	v	53.2
m	87.4	w	44.1
n	91.7	x	70.8
ng	50	y	82.4
o	62.9	yo	16.7

- First, we take the results fixing the training files and change the number of testing files. In Figure 3 shows the changes.
- In second phase, we fix the testing files and changes the number of training files. Figure 4 shows the changes.

From our results shown in Figure 4 we find that with more training data accuracy results for testing data is increasing. But, it is flattering out after some time. We hope that if we trained more data accuracy will more.

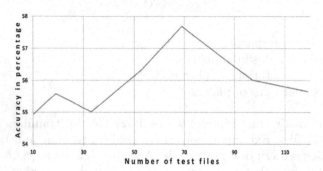

Fig. 3 Fixing training files

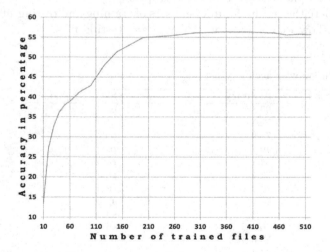

Fig. 4 Fixing testing files

7 Conclusion and Future Work

In this report we have presented the manual IPA transcription and automatic transcription of Assamese speech. Because IPA transcription is based on pronunciation, so during pronunciation we faced issues and some interesting observations of development and transcription of Assamese Speech Corpus. We could find the relative frequencies of different phone and the most frequent ones. Also we find the more frequently used word in the language. We make some interesting observations regarding the phonetic transcription of Assamese speech. During our experiment of automatic transcription model we find that for different numbers of training and testing files we compare our accuracy results of our automatic transcription model. After analyzing it has been find that accuracy increasing simultaneously with the increase of trained data. But at a stage accuracy is static for some sets of trained data. In future we shall try to design a plug-in by which we can convert ASCII to IPA from HTK models. In further work of the Assamese speech corpus we are going to collect more different categories of data from more speakers from different. Also in future we are going to work about break-marking of our speech corpus to have a more useful Assamese speech processing resource.

Acknowledgments. The work reported here has been supported by DeitY sponsored project Development of Prosodically Guided Phonetic Engine for Searching Speech Databases in Indian Languages.

References

Chang, S., Shastri, L., Greenberg, S.: Automatic phonetic transcription of spontaneous speech (american English). In: Proceedings of the INTERSPEECH, Beijing, China, pp. 330–333 (2000)

Coxhead, P.: Phones and Phonemes (2007)

Giurgiu, M., Kabir, A.: Automatic transcription and speech recognition of Romanian corpus RO-GRID. In: Proceedings of the 35th International Conference on Telecommunications and Signal Processing, Czech Republic, pp. 465–468 (2012)

Hasan, M.R., Jamil, M., Rabbani, M.G., Rahman, M.S.: Speaker identification using Mel frequency cepstral coefficients. Variations 1, 4 (2004)

Ladefoged, P.: Elements of acoustic phonetics. University of Chicago Press (1995)

Ladefoged, P., Johnstone, K.: A course in phonetics (2011), CengageBrain.com

Laurent, A., Merlin, T., Meignier, S., Esteve, Y., Deléglise, P.: Iterative filtering of phonetic transcriptions of proper nouns. In: Proceedings of the IEEE International Conference on Acoustics, Speech and Signal Processing, Taiwan, pp. 4265–4268 (2009)

Leung, H., Zue, V.: A procedure for automatic alignment of phonetic transcriptions with continuous speech. In: Proceedings of the IEEE International Conference on Acoustics, Speech, and Signal Processing, San Diego, USA, vol. 9, pp. 73–76 (1984)

Levinson, S.E., Liberman, M.Y., Ljolje, A., Miller, L.: Speaker independent phonetic transcription of fluent speech for large vocabulary speech recognition. In: Proceedings of the International Conference on Acoustics, Speech, and Signal Processing, Glasgow, Scotland, pp. 441–444 (1989)

Liang, M.S., Lyu, R.Y., Chiang, Y.C.: Phonetic transcription using speech recognition technique considering variations in pronunciation. In: Proceedings of the IEEE International Conference on Acoustics, Speech and Signal Processing, Honolulu, Hawaii, vol. 4, pp. 109–112 (2007)

Nagarajan, T., Murthy, H.A., Hemalatha, N.: Automatic segmentation and labeling of continuous speech without bootstrapping. In: Proceedings of EUSIPCO, Vienna, Austria, pp. 561–564 (2004)

Patil, H.A., Madhavi, M.C., Malde, K.D., Vachhani, B.B.: Phonetic Transcription of Fricatives and Plosives for Gujarati and Marathi Languages. In: Proceedings of International Conference on Asian Language Processing, Hanoi, Vietnam, pp. 177–180 (2012)

Sarada, G.L., Lakshmi, A., Murthy, H.A., Nagarajan, T.: Automatic transcription of continuous speech into syllable-like units for Indian languages. Sadhana 34(2), 221–233 (2009)

Sarma, H., Saharia, N., Sharma, U., Sinha, S.K., Malakar, M.J.: Development and transcription of Assamese speech corpus. In: Proceedings of National seminar cum Conference on Recent threads and Techniques in Computer Sciences, Bodoland University, India (2013)

Sharma, U., Kalita, J.K., Das, R.K.: Acquisition of morphology of an Indic language from text corpus. ACM Transactions on Asian Language Information Processing 7(3), 9:1–9:33 (2008)

Sjölander, K., Beskow, J.: Wavesurfer-an open source speech tool. In: Proceedings of ICSLP, Beijing, China, vol. 4, pp. 464–467 (2000)

Stefan-Adrian, T., Doru-Petru, M.: Rule-based automatic phonetic transcription for the Romanian language. In: Proceedings of the Computation World: Future Computing, Service Computation, Cognitive, Adaptive, Content, Patterns, Athens, pp. 682–686 (2009)

Wells, J.C.: Phonetic transcription and analysis. Encyclopedia of Language and Linguistics, pp. 386–396. Elsevier, Amsterdam (2006)

Young, S., Evermann, G., Gales, M., Hain, T., Kershaw, D., Liu, X., Moore, G., Odell, J., Ollason, D., Povey, D., et al.: The HTK book, version, 3.4th edn. Cambridge University Engineering Department (2009)

Hybrid Approach for Punjabi Question Answering System

Poonam Gupta and Vishal Gupta

Abstract. In this paper a hybrid algorithm for Punjabi Question Answering system has been implemented. A hybrid system that works on various kinds of question types using the concepts of pattern matching as well as mathematical expression for developing a scoring system that can help differentiate best answer among available set of multiple answers found by the algorithm and is also domain specific like sports. The proposed system is designed and built in such a way that it increases the accuracy of question answering system in terms of recall and precision and is working for factoid questions and answers text in Punjabi. The system constructs a novel mathematical scoring system to identify most accurate probable answer out of the multiple answer patterns.The answers are extracted for various types of Punjabi questions. The experimental results are evaluated on the basis of Precision, Recall, F-score and Mean Reciprocal Rank (MRR). The average value of precision, recall, f-score and Mean Reciprocal Rank is 85.66%, 65.28%, 74.06%, 0.43 (normalised value) respectively. MRR values are Optimal. These values are act as discrimination factor values between one relevant answer to the other relevant answer.

Keywords: Natural language processing, Text Mining, Punjabi Question answering system, Information Extraction, Information retrieval.

1   Introduction

Question Answering is an application area of natural language processing which is concerned with how to find the best answer to a question. Question Answering utilized multiple text mining techniques like question categorization to categorize the questions into various types (who, where, what, when etc.) and information

Poonam Gupta · Vishal Gupta
UIET, Panjab University Chandigarh
e-mail: poonam_123z@yahoo.co.in, vishal@pu.ac.in

S.M. Thampi, A. Gelbukh, and J. Mukhopadhyay (eds.), *Advances in Signal Processing*   133
and Intelligent Recognition Systems, Advances in Intelligent Systems and Computing 264,
DOI: 10.1007/978-3-319-04960-1_12, © Springer International Publishing Switzerland 2014

extraction (IE) for the extraction of the entities like people, events etc [1]. In today's context, the major components of Question answering system (QAS): question classification, information retrieval, and answer extraction. Question classification play chief role in QA system for the categorization of the question based upon its type. Information retrieval method involves the identification & the extraction of the applicable answer post by their intelligent question answering system [4]. As the users tried to navigate the amount of information that is available online by means of search engines like GOOGLE, YAHOO etc., the need for the automatic question answering becomes more urgent. Since search engines only return the ranked list of documents or the links, but they do not returning an exact answer to the question.[2]. Question answering is multidisciplinary. It consists of various technologies like information technology, artificial intelligence, natural language processing, and cognitive science. There are two types of information sources in question answering: Structured data sources like relational databases and Unstructured data sources like audio, video, written text, speech [5][6]. The Internet today has to face the difficulty of dealing with multi linguism [3]. All the work in Question answering system is done for various other languages but as per knowledge, no work is done for Punjabi Language.

2   Related Work

Case based reasoning question answering system (CBR) i.e. the new questions are answered by retrieving suitable and existing historical cases & had the advantages that it can easily access to knowledge & maintain the knowledge database[7]. A supervised multi-stream approach that consists of various features like (i) the compatibility between question types and the answer types, (ii) the redundancy of answers across various streams (iii) the overlap and the non-overlap information between the question and the answer pair and its related text [8]. Answer ranking framework which was a combination of answer relevance & similarity features. Relevant answers can be finding out by knowing the relationship between answer and the answer type or between answer and the question keyword [9]. If the user is not satisfied with the answers then the fuzzy relational operator is defined to find the recommended questions [10].

3   Problem Formulation and Objectives of the Study

There is an urgent need to develop the system which does not rely basically on pattern matching of strings to find the answers of questions which are most relevant and accurate for running systems that can be used for mass implementation. The effectiveness of scoring system is another issue which needs to be evaluated for finding most accurate answer or the solution to the question for which the answer is to be searched on, then timings are critical factor for evaluating performance as quick and correct answer is the basic need requirement of the QA systems. Therefore, we propose a hybrid system that works on various kinds of Punjabi question types using the concepts of pattern matching as well as mathematical expression for developing a scoring system that can help differentiate best answer among available set of multiple answers found by the algorithm and is also domain specific like sports.

3.1   Objectives of the Study

1. Develop a representative dataset of sports domain Punjabi input text. 2. Develop algorithm to identify the factoid types of question from text. 3. Develop a scoring framework to calculate score of questions and answers for developing Q-A similarity matrix. 4. Using Q-A similarity matrix and pattern matching identify more accurate answer for questions. 5. Evaluate the performance of system in terms of Recall, Precision, F-score and Mean Reciprocal Rank (MRR).

4   Algorithm for Punjabi Question Answering System

The proposed algorithm takes Punjabi question and a paragraph text as input from which answers are to be extracted. The algorithm proceeds by segmenting the Punjabi input question into Words. For each word in the question follow following steps:

Step 1 Query Processing:- In this step it will classify the question based upon its categories: ਕੀ(what), ਕਦੋਂ (when), ਕਿੱਥੇ/ਕਿੱਥੋਂ/ਕਿਹੜੇ/ਕਿਹੜਾ/ਕਿਹੜੀ (where/which), ਕੋਣ//ਕਿਸ/ਕਿਸੇ (who) & ਕਿਉਂ (why) then system checks the corresponding rules.

Step 2 If the input question consists of ਕੋਣ/ਕਿਸ/ਕਿਸੇ (WHO/WHOM/WHOSE) then go to the procedure for ਕੋਣ (WHO/WHOM/WHOSE). elseIf the input question consists of ਕੀ (WHAT) then go to the procedure for ਕੀ(WHAT). elseIf the input question consists of ਕਦੋਂ (WHEN) then go to the procedure for ਕਦੋਂ (WHEN). elseIf the input question consists of ਕਿੱਥੇ/ਕਿੱਥੋਂ/ਕਿਹੜੇ/ਕਿਹੜਾ/ਕਿਹੜੀ (WHERE/WHICH)then go to the procedure for ਕਿੱਥੇ/ਕਿੱਥੋਂ/ਕਿਹੜੇ/ਕਿਹੜਾ/ਕਿਹੜੀ(WHERE/WHICH).elseIf the input question consists of ਕਿਉਂ(WHY) then go to the procedure for ਕਿਉਂ (WHY).Else Algorithm cannot find the answer.

4.1 Procedure for ਕੋਣ ਕੋਣ/ ਕਿਸ/ ਕਿਸੇ (Who/Whom/Whose)

This procedure is used for finding the named entities from input paragraph. The algorithm for ਕੋਣ ਕੋਣ/ਕਿਸ/ਕਿਸੇ (Who/Whom/Whose) has been given below:

Step 1 Extract the substring before ਕੋਣ ਕੋਣ/ਕਿਸ/ਕਿਸੇ (WHO/WHOM/WHOSE) from the given question and say it is X1.

Step 2 Extract the substring after ਕੋਣ ਕੋਣ/ਕਿਸ/ਕਿਸੇ (WHO/WHOM/WHOSE) upto the end of the question and say it is X2.

Step 3 Initialize Answer = null.

Step 4 If X1 ≠ null then depending upon X1 set the answer {

If X1 contains the name of the person// Name is identified by searching current word in the Punjabi Dictionary if it is not in the Punjabi Dictionary & found in the Punjabi name list (it consists of 200 names) then it is a name//{Search the paragraph by matching the strings in X1 & X2 and append the matching details to the end of X1. Say it is X1'. Append X2 after X1' & store it in the Answer. }} (End of Step 4 loop)

Step 5 else If X1 contains the words ਤੁਸੀ (You), ਤੁਹਾਨੂੰ (You), ਤੂੰ (You)then{Replace

the word as follows: ਤੁਸੀ (You) to ਮੈਂ (I), ਤੁਹਾਨੂੰ (You) to ਮੈਨੂੰ (I), ਤੂੰ (You) to

ਮੈਂ (I)}Search the name of person, thing or concept in the paragraph based upon X1. //Name can be identified by searching in the Punjabi dictionary if it is not in the dictionary and it is found in the Punjabi name list(it consists of 200 names) then it is considered as a name// and append it after X1. Say it is X1'.

Change the last word of X2 according to the words in X1. {

If ਤੁਸੀ (You) then last word is ਹੋ (is) changes to ਹਾਂ (is).

If ਤੁਹਾਨੂੰ (You) then last word is same as in substring X2.

If ਤੂੰ (You) then last word is ਹੈ (is) changes to ਹਾਂ (is) and store it in X2'. }

Append X2' to the end of X1' and store it in Answer. } (end of step 5 loop)

The procedure has got two rules out of which one must be satisfied to generate final result. If step 4 of procedure satisfies the sentence then the question is of the form "ਤੁਸੀ ਕੋਣ ਹੋ? Who are you?" And the result generated contain pattern similar to the sentence "ਮੈਂ ਕਮਲਜੀਤ ਹਾਂ। I am Kamaljit". Similarly if step 4 satisfies the sentence then the question is of one of the two forms, it can be "ਤੁਹਾਨੂੰ ਕੋਣ ਜਾਣਦਾ ਹੈ? Who knows you?" And the result generated contain pattern similar to the sentence "ਮੈਨੂੰ ਮਨਦੀਪ ਜਾਣਦਾ ਹੈ। Mandeep knows me." The second form is "ਤੂੰ ਕੋਣ

ਹੈ? Who are you?" And the result generated contain pattern similar to the sentence "ਮੈਂ ਲਵਲੀਨ ਹਾਂ । "I am Loveleen."

4.2 Procedure for ਕੀ (WHAT)

This procedure is used for finding the reason for ਕੀ(WHAT) type of questions from the input paragraph. The algorithm for ਕੀ(WHAT) has been given below:

Step 1 Extract the substring before ਕੀ (WHAT) from the given question and say it is X1.

Step 2 Extract the substring after ਕੀ (WHAT) upto the end of question and say it is X2.

Step 3 Initialize Answer = null.

Step 4 If X1= null then { Store ਹਾਂ/ ਨਹੀ (Yes/No) in Answer by matching X2 with the paragraph} (end of step 4) else

Step 5 If X1≠ null & contains the words ਤੁਸੀ (You) , ਤੁਹਾਡਾ (Your), ਤੈਨੂੰ (You),ਤੂੰ (You) then{ Replace the words as follows ਤੁਸੀ (You) to ਅਸੀ (We) , ਤੁਹਾਡਾ (Your) to

ਮੇਰਾ (My), ਤੈਨੂੰ (You) to ਮੈਨੂੰ (I), ਤੂੰ (You) to ਮੈਂ (I).}

Search the name of person, thing or concept in the paragraph based upon X1
// name can be identified by searching in the Punjabi Noun dictionary if it
is not in the dictionary then it is considered as a name// Append it after
X1. Say it is X1'.

Change the last word of X2 according to X1. {

If ਤੁਸੀ (You) then last word ਹੋ (is) changes to ਹਾਂ (is).

If ਤੁਹਾਡਾ (your) then last word is same as in substring X2.

If ਤੈਨੂੰ (you) then last word is same as in substring X2.

If ਤੂੰ (you) then last word is ਹੈ (is) changes to ਹਾਂ (is).

Store it in X2'.}Append X2'to X1' and store it in answer.}(end of step 5)

This procedure has got two rules out of which one of the rule must satisfy to generate a final result. The Rule_1 is if the substring before ਕੀ (WHAT) is null then it stores ਹਾਂ/ ਨਹੀ (Yes/No) in Answer. The Rule_2 is if the substring before ਕੀ (WHAT) is not null then the question is of the form "ਤੁਸੀ ਕੀ ਖੇਡ ਰਹੇ ਹੋ? What are you playing?" And the result generated contain the pattern "ਅਸੀ ਫੁੱਟਬਾਲ ਖੇਡ ਰਹੇ ਹਾਂ। We are playing Football." It can be "ਤੁਹਾਡਾ ਕੀ ਨਾਂ ਹੈ? What is your name?" And the result generated contain the pattern "ਮੇਰਾ ਨਾਂ ਚਰਨਦੀਪ ਹੈ। My name is Charandeep."

4.3   Procedure for ਕਦੋਂ*(WHEN)*

This procedure is used for finding the time and the date expression for a particular question from the input paragraph.

The algorithm for ਕਦੋਂ (WHEN) has been given below:

Step1 Extract the substring before ਕਦੋਂ (WHEN) from the given question and say it is X1

Step 2 Extract the substring after ਕਦੋਂ (WHEN) upto the end of question and say it is X2.

Step 3 Initialize Answer = null

Step 4 If X1= null and X2 containing the words ਤੁਹਾਨੂੰ (you), ਤੁਸੀ (you), ਤੁਹਾਡੀ (your),ਮੈਂ (I) then{   Replace the word ਕਦੋਂ (WHEN) with the time and date expression and it can be identified if it is in the following formats:

> DD/MM/YY like 02/04/97
> DD/MM/YYYY like 02/04/1997
> YY/MM/DD like 97/04/02
> YYYY/MM/DD like 1997/04/02
> MM/DD/YY like 05/26/97
> MM/DD/YYYY like 05/26/1997
> DD.MM.YY like 02.04.97
> DD.MM.YYYY like 02.04.1997
> DD-MM-YY like 02-04-97
> DD-MM-YYYY like 02-04-1997
> String, Numeric like May 26
> Numeric, String like 26 May
> Numeric : Numeric like 6:45 and
> append ਨੂੰ and then store it in X1'.

> Change the words in X2{
> ਤੁਹਾਨੂੰ (you) to ਮੈਨੂੰ (I), ਤੁਸੀ (You) to ਅਸੀ (We), ਤੁਹਾਡੀ (Your) to ਮੇਰੀ (My), ਮੈਂ (I) to ਤੂੰ (You). Store it in X2.}

> Append X2' to the end of X1' and store it in Answer. }(end of step 4) else

Step 5 If X1≠ null and is containing the word ਤੁਹਾਡੀ (Your) {

> Then Replace the word ਤੁਹਾਡੀ (Your) to ਮੇਰੀ. Store it in X1'.

> Replace the word ਕਦੋਂ (WHEN) with the time and date expression and it can be identified if it is in the following formats:
> DD/MM/YY like 02/04/97
> DD/MM/YYYY like 02/04/1997
> YY/MM/DD like 97/04/02

> YYYY/MM/DD like 1997/04/02
> MM/DD/YY like 05/26/97
> MM/DD/YYYY like 05/26/1997
> DD.MM.YY like 02.04.97
> DD.MM.YYYY like 02.04.1997
> DD-MM-YY like 02-04-97
> DD-MM-YYYY like 02-04-1997
> String, Numeric like May 26
> Numeric, String like 26 May
> Numeric : Numeric like 6:45 and append ਨੂੰ and then store it in X1'.

> Append X2 to the end of X1'& store it in Answer. }(end of step 5 loop)

This procedure has one rule that must be satisfied to generate a final result. The Rule is if the substring before ਕਦੋਂ (WHEN) is null then replace the word ਕਦੋਂ (WHEN) with the time and date expression. The question is of the form "ਕਦੋਂ ਤੁਸੀ ਇੱਥੇ ਤੋਂ ਜਾਵਾਗੇ? When did you go from here?" And the result generated contains the pattern "ਕਲ ਨੂੰ ਅਸੀ ਇੱਥੇ ਤੋਂ ਜਾਵਾਗੇ। Tomorrow we will leave from here." The question can be of the form "ਕਦੋਂ ਤੁਹਾਨੂੰ ਇੱਥੇ ਪੰਜ ਸਾਲ ਹੋ ਜਾਵਗੇ? When you complete your five years here?" And the result generated contain the pattern "02-04-2012, ਨੂੰ ਮੇਨੂੰ ਇੱਥੇ ਪੰਜ ਸਾਲ ਹੋ ਜਾਵਗੇ। My five years will be completed on 02-04-2012." The question can be of the form "ਕਦੋਂ ਮੈਂ ਤੁਹਾਡੀ ਸਹਾਇਤਾ ਕਰ ਸਕਦਾ ਹਾਂ? When will I help you? And the result generated contains the pattern "04/12/2011 ਨੂੰ ਤੋਂ ਮੇਰੀ ਸਹਾਇਤਾ ਕਰ ਸਕਦਾ ਹਾਂ। You can help me on 04/12/2011."

4.4   Procedure for ਕਿੱਥੇ ਕਿੱਥੋਂ ਕਿੱਥੋਂ ਕਿਹੜੇ ਕਿਹੜਾ ਕਿਹੜੀ (WHERE/WHICH)

This procedure is used for finding the location name for a particular question. The Algorithm for ਕਿੱਥੇ/ਕਿੱਥੋਂ/ਕਿਹੜੇ/ਕਿਹੜਾ/ਕਿਹੜੀ (WHERE/WHICH) has been given below:

Step 1 Extract the substring before from the given question and say it is X1.
Step 2 Extract the substring after ਕਿੱਥੇ/ਕਿੱਥੋਂ/ਕਿਹੜੇ/ਕਿਹੜਾ/ਕਿਹੜੀ (WHERE/WHICH) upto the end of question and say it is X2.
Step 3 Initialize Answer = null.

Step 4 If X1≠ null and X2≠ null and X1 is containing the words ਤੁਸੀ (You), ਤੁਹਾਨੂੰ (You) then {   Replace the words as follows

ਤੁਸੀ (You) to ਮੈਂ (I), ਤੁਹਾਨੂੰ (You) to ਮੈਨੂੰ (I) and store it in X1'.

Replace the word ਕਿੱਥੇ (WHERE) with the matching locations like ਅੰਦਰ

(In),ਤੇ (At), ਨੇੜੇ (Near) from the paragraph or name of the location can also be identified by searching it in the Punjabi dictionary if it is not in the dictionary then it is considered as a location name and store it in X1'.
Change the last word of X2 according to X1 {

  If ਤੁਸੀ (You) then last word is ਹੋ (is) changes to ਹਾਂ (is).

  If ਤੁਹਾਨੂੰ (You) then last word is same as in substring X2 & store it in X2'. }
    }(end of step 4 loop) else
Step 5 If X1 is not containing the words ਤੁਸੀ (You), ਤੁਹਾਨੂੰ (You) then {

  Keep X1 and X2 as such and store it in X1' and X2' respectively.
  Just Replace ਕਿੱਥੇ (WHERE) with the matching locations like ਅੰਦਰ (In), ਤੇ

  (At), ਨੇੜੇ (Near) from the paragraph or name of the location can also be

  identified by searching it in the Punjabi dictionary if it is not in the dictionary then it is considered as a location name and store it in X1'.
  Append X2' to the end of X1' and store it in answer.} (end of step 5 loop)

This procedure has one rule that must be satisfied to generate a final result. The Rule is if the substring before and after ਕਿੱਥੇ/ਕਿੱਥੋਂ/ਕਿਹੜੇ/ਕਿਹੜਾ/ਕਿਹੜੀ (WHERE) is not null then replace the word ਕਿੱਥੇ/ਕਿੱਥੋਂ/ਕਿਹੜੇ/ਕਿਹੜਾ/ਕਿਹੜੀ (WHERE) with the matching locations like ਅੰਦਰ (In), ਤੇ (At), ਨੇੜੇ (Near). The question is of the form "ਤੁਸੀ ਕਿੱਥੇ ਜਾ ਰਹੇ ਹੋ? Where are you going?"& the result generated contain the pattern "ਅਸੀ ਅੰਬਾਲਾ ਜਾ ਰਹੇ ਹਾਂ। We are going to Ambala. It can be of the form "ਉਹ ਤੁਹਾਨੂੰ ਕਿੱਥੇ ਮਿਲੇਗਾ ? Where did you meet him?"& the result generated contain the pattern "ਉਹ ਮੈਨੂੰ ਬੱਸ ਸਟੈਂਡ ਤੇ ਨੇੜੇ ਮਿਲੇਗਾ । I will meet him near the bus stand."

4.5   Procedure for ਕਿਉ (WHY)

This procedure is used for finding the reason for ਕਿਉ (WHY) type of questions from the input paragraph. The algorithm for ਕਿਉ (WHY) has been given below:
Step 1 Extract the substring ਕਿਉ (WHY) before from the
given question & say it is X1.

Step 2 Extract the substring after **ਕਿਉ** (WHY) upto the end of question & say it is X2.

Step 3 Initialize Answer = null.

Step 4 If X1 ≠ null and X2≠ null then{   Keep X1 as such.

    Replace the word **ਕਿਉ** (WHY) with some reason by matching the strings in

    X1 and X2 with the paragraph. Store it after X1.Keep X2 as such.

    Append X2 to the end of X1 and store it in the Answer. } (end of step 4) else

Step 5   If X1 ≠ null , X2≠ null and X1 is containing the words **ਤੁਹਾਡੇ** (your), **ਤੁਸੀ**

    (you) {Replace the words as follows {

    ਤੁਹਾਡੇ (your) to **ਮੇਰੇ** (my) , **ਤੁਸੀ** (you) to **ਅਸੀ** (we) }

    Replace the word **ਕਿਉ** (WHY) with some reason by matching the strings

    in X1 and X2 with the paragraph. Store it in X1. Keep X2 as such.

    Append X2 to the end of X1 & store it in the Answer. } (end of step 5)

This procedure has one rule that must be satisfied to generate a final result. The Rule is if the substring before and after **ਕਿਉ** (WHY) is not null then replace the word **ਕਿਉ** (WHY) with some reason by matching the substrings of the question. The question is of the form "**ਬੱਚੇ ਜਮਾਤ ਵਿੱਚ ਖੁਸ਼ ਕਿਉ ਲਗ ਰਹੇ ਸਨ**? Why the students are happy in the classroom?" and the result generated contain the pattern "**ਬੱਚੇ ਜਮਾਤ ਵਿੱਚ ਖੁਸ਼ ਕਮਲ ਦੇ ਜਨਮਿਦਨ ਤੇ ਲਗ ਰਹੇ ਸਨ।** Students are happy in the classroom because of the Kamal's birthday.

4.6   Procedure for ਕਿਨਾਂ ਕਿੰਨੇ/ ਕਿੰਨੀ/ ਕਿੰਨੀਆਂ (HOW MANY)

Step 1 Let Q be the question .

Step 2 Let S1 be extracted question string of question sentence before question word.

Step 3 Let S2 be extracted question string of question sentence after question word.

Step 4 Let 'answer' be string representing final answer found.

Step 5 Let N_m be the set of {Whole numbers, Real numbers and the numbers represented in Punjabi.}

Step 6 for each chunk in paragraph

    Find match from N_m

      If found match

        Find index, position

        Extract the string based upon threshold.

        Find score for each answer.

        Sort the answers in ascending order.

        Select the answer with highest score.

      end

     end

    end

5   Proposed Hybrid Algorithm for Punjabi Question Answering

A hybrid system that works on various kinds of question types ਕੀ (what), ਕਦੋਂ (when), ਕਿੱਥੇ/ਕਿੱਥੋਂ/ਕਿਹੜੇ/ਕਿਹੜਾ/ਕਿਹੜੀ/ਕਿਹੜੀਆਂ (where/ which), ਕੌਣ/ਕਿਸ/ਕਿਸੇ (who/whom/whose) and ਕਿਉਂ (why) using the concepts of pattern matching as well as mathematical expression for developing a scoring system that can help differentiate best answer among available set of multiple answers found by the algorithm. The hybrid algorithm works in combination with mathematical expressions in tandem with the steps discussed in section 4.

Step 1 The paragraph and question text which is sports based is input to the system.

Step 2 The questions are preprocessed i.e. we can remove the stop words & the questions are classified based upon the category of the question i.e. ਕੀ (what), ਕਦੋਂ (when), ਕਿੱਥੇ/ਕਿੱਥੋਂ/ਕਿਹੜੇ/ਕਿਹੜਾ/ਕਿਹੜੀ (where / which), ਕੌਣ ਕੌਣ/ਕਿਸ/ਕਿਸੇ (who/whom/whose) & ਕਿਉਂ (why).

Step 3 The questions and the paragraph are tokenized in context of verbs, nouns, adjectives and adverbs.

Step 4 The question score is calculated using $W = W_1 + W_2$;   (1)

   Where W= total score

   W_1= Frequency table dataset of artifacts (nouns, verbs, adjectives, adverbs, pronouns) with respect to the dictionary.

   W_2= Relative Frequency of terms common in both the sets of questions and paragraph text chunks.

Step 5 The multiple answers are extracted from the paragraph based upon the threshold by calling thealgorithms ਕੀ (what), ਕਦੋਂ (when), ਕਿੱਥੇ/ਕਿੱਥੋਂ/ਕਿਹੜੇ/ਕਿਹੜਾ/ਕਿਹੜੀ (where), ਕੌਣ/ਕਿਸ/ਕਿਸੇ (who/whom/whose) and ਕਿਉਂ (why) based upon the corresponding question type discussed in section 4.

Step 6 The answer score for each multiple answers is calculated using AW{ AW_1, AW_2} where AW_1= It is the frequency table of artifacts (nouns, verbs, adverbs, adjectives, pronouns) found in the answer tokenized dataset with respect to dictionary.

   AW_2= It is the relative frequency of terms common in question token and answer token.

Step 7 Calculate the smallest distance between question and the answer using

   Similar score=0; dissimilar score;

   For each value in AW_{set}

   Find similar value in QW_{set}

   If { AW value equals QW value}

   Similar score= similar score+1

   else

   Dissimilar score= Dissimilar score+1

End
End
Distance= Dissimilar score / similar score + dissimilar score
If the distance is less than (0.99) then find the index of paragraph chunk
with respect to the "Pattern Matched" string selected on the basics of the score.
End
Step 8 Select the answer that has score closest to zero.

Algorithm Input Example (Domain based):

ਭਾਰਤ ਕ੍ਰਿਕਟ ਦੀ ਖੇਡ ਦਾ ਹੁਣ ਸ਼ਹਿਨਸ਼ਾਹ ਹੈ। ਟਵੰਟੀ-20 ਵਿਸ਼ਵ ਕੱਪ ਅਤੇ ਟੈਸਟ ਕ੍ਰਿਕਟ ਵਿੱਚ

ਪਹਿਲਾ ਸਥਾਨ ਮੱਲ੍ਹਣ ਤੋਂ ਬਾਅਦ ਭਾਰਤ ਨੇ ਇਕ ਰੋਜ਼ਾ ਕ੍ਰਿਕਟ ਦਾ ਵਿਸ਼ਵ ਕੱਪ ਜਿੱਤ ਕੇ ਕ੍ਰਿਕਟ ਦੇ

ਸਾਰੇ ਰੂਪਾਂ ਵਿੱਚ ਆਪਣੀ ਸਰਦਾਰੀ ਕਾਇਮ ਕਰ ਲਈ ਹੈ। ਆਪਣੇ ਹੀ ਦਰਸ਼ਕਾਂ ਮੁਹਰੇ ਦੇਸ਼ ਦੀ

ਪਹਿਲੀ ਨਾਗਰਿਕ ਰਾਸ਼ਟਰਪਤੀ ਪ੍ਰਤਿਭਾ ਦੇਵੀ ਸਿੰਘ ਪਾਟਿਲ ਦੀ ਮੌਜੂਦਗੀ ਵਿੱਚ ਭਾਰਤੀ ਟੀਮ ਨੇ

ਵਾਨਖੇੜੇ ਸਟੇਡੀਅਮ ਵਿਖੇ ਇਤਿਹਾਸ ਨੂੰ ਮੋੜਾ ਦਿੰਦਿਆਂ ਵਿਸ਼ਵ ਕੱਪ ਦੀ ਟਰਾਫੀ ਚੁੰਮੀ। ਮੁੰਬਈ ਦੇ

26/11 ਅਤਿਵਾਦੀ ਹਮਲੇ ਵੇਲੇ ਪੂਰਾ ਦੇਸ਼ ਸੁੰਨ ਹੋ ਗਿਆ ਸੀ । ਉਸ ਮਾਇਆਨਗਰੀ ਵਿੱਚ ਮਿਲੀ ਜਿੱਤ

ਨੇ ਹੁਣ ਪੂਰੇ ਦੇਸ਼ ਨੂੰ ਜਿੱਤ ਦੇ ਨਸ਼ੇ ਨਾਲ ਹਿਲਾ ਕੇ ਰੱਖ ਦਿੱਤਾ। ਆਸਟਰੇਲੀਆ ਤੇ ਵੈਸਟ ਇੰਡੀਜ਼ ਤੋਂ

ਬਾਅਦ ਭਾਰਤ ਦੁਨੀਆਂ ਦਾ ਤੀਜਾ ਅਤੇ ਏਸ਼ੀਆ ਦਾ ਪਹਿਲਾ ਦੇਸ਼ ਬਣ ਗਿਆ ਜਿਸ ਨੇ ਦੋ ਵਾਰ ਵਿਸ਼ਵ

ਕੱਪ ਜਿੱਤਿਆ ਹੈ। ਇਸ ਤੋਂ ਪਹਿਲਾਂ 1983 ਵਿੱਚ ਲਾਰਡਜ਼ ਵਿਖੇ ਕਪਿਲ ਦੇ 'ਦੇਵਾਂ' ਨੇ ਭਾਰਤ ਨੂੰ ਵਿਸ਼ਵ

ਚੈਂਪੀਅਨ ਬਣਾਇਆ ਸੀ। ਇਹ ਵੀ ਇਤਫਾਕ ਹੈ ਕਿ ਭਾਰਤ ਨੇ ਉਸ ਵੇਲੇ ਵਿਸ਼ਵ ਕ੍ਰਿਕਟ ਦੀ

ਬਾਦਸ਼ਾਹਤ ਦੇ ਮਾਲਕ ਅਤੇ ਪਿਛਲੇ ਦੋ ਵਿਸ਼ਵ ਕੱਪ ਦੇ ਵਿਜੇਤਾ ਵੈਸਟ ਇੰਡੀਜ਼ ਨੂੰ ਹਰਾ ਕੇ ਵਿਸ਼ਵ ਕੱਪ

ਜਿੱਤਿਆ ਸੀ ਅਤੇ ਇਸ ਵਾਰ ਵੀ ਭਾਰਤ ਨੇ ਵਿਸ਼ਵ ਕ੍ਰਿਕਟ ਦੇ ਸਰਤਾਜ ਅਤੇ ਪਿਛਲੇ ਤਿੰਨ ਵਾਰ ਦੇ

ਵਿਸ਼ਵ ਚੈਂਪੀਅਨ ਆਸਟਰੇਲੀਆ ਨੂੰ ਕੁਆਰਟਰ ਫਾਈਨਲ ਵਿੱਚ ਹਰਾ ਕੇ ਵਿਸ਼ਵ ਕੱਪ ਜਿੱਤਿਆ ।

Now, India is the emperor of cricket. After occupying twenty-20 and test cricket, India won one-day cricket in World Cup. In front of home countries, first lady president Pratibha Devi Singh Patil, India made history by winning the World Cup at Wankhede stadium. Mumbai terror attack on 26/11 anesthetized the whole country. India's victory filled the whole country with joy and happiness. India is the World's third & Asia's first country after Australia & West Indies who has won World Cup twice. Earlier in 1983 at Lord's, Kapil's team made India World's champion. Coincidentally, India made history by defeating World's champion West Indies and this time, India won the World Cup by defeating the emperor of cricket and three time World's champion team Australia in quarter final.

Questions Asked from the Above Paragraph as Below:

1. ਮੁੰਬਈ ਤੇ ਅਤਿਵਾਦੀ ਹਮਲਾ ਕਦੋਂ ਹੋਇਆ? When Mumbai Terror attack happens?

2. ਭਾਰਤ ਨੇ ਪਹਿਲਾ ਵਿਸ਼ਵ ਕੱਪ ਕਦੋਂ ਜਿੱਤਿਆ? When did India win first world cup?

3. ਕ੍ਰਿਕਟ ਦੀ ਖੇਡ ਦਾ ਹੁਣ ਸ਼ਹਿਨਸ਼ਾਹ ਕੌਣ ਹੈ ? Now who is the emperor of cricket?

Algorithm Output:

1. ਮੁੰਬਈ ਦੇ 26/11 ਅਤਿਵਾਦੀ ਹਮਲੇ ਵੇਲੇ ਪੂਰਾ ਦੇਸ਼ ਸੁੰਨ ਹੋ ਗਿਆ ਸੀ । Mumbai terror attack on 26/11 anesthetized the whole country.

2. ਇਸ ਤੋਂ ਪਹਿਲਾਂ 1983 ਵਿੱਚ ਲਾਰਡਜ਼ ਵਿਖੇ ਕਪਿਲ ਦੇ ਦੇਵਾਂ ਨੇ ਭਾਰਤ ਨੂੰ ਵਿਸ਼ਵ ਚੈਂਪੀਅਨ ਬਣਾਇਆ ਸੀ। Earlier in 1983 at Lord's, Kapil's team made India World's champion.

3. ਭਾਰਤ ਕ੍ਰਿਕਟ ਦੀ ਖੇਡ ਦਾ ਹੁਣ ਸ਼ਹਿਨਸ਼ਾਹ ਹੈ। Now, India is the emperor of cricket.

6 Experimental Results and Discussion

This section discusses the experimental evaluation of our hybrid algorithm on our test data sets. The experiments are conducted on 40 Punjabi documents of sports domain, which consists of 400 sentences, 10000 words and 200 questions on which the system is evaluated.

6.1 Results on the Basis of Precision, Recall, F-score

$$\text{Precision} = \frac{No.of\ relevant\ retrieved\ answers\ given\ by\ the\ system}{Total\ no.of\ retrieved\ answers\ given\ by\ the\ system} \tag{2}$$

$$\text{Recall} = \frac{No.of\ relevant\ retrieved\ answers\ given\ by\ the\ system}{Total\ no.of\ relevant\ answers\ that\ should\ have\ been\ retrieved} \tag{3}$$

F-score is calculated using the values of precision and recall.

$$\text{F-Score} = 2 * \frac{Precision*Recall}{Precision+Recall} \tag{4}$$

Table 1 Average Precision, Recall and F-score of each question type (in %)

S.No.	Question Type	Average Precision	Average Recall	Average F-Score
1.	What (ਕੀ)	88.88	66.66	76.18
2.	When (ਕਦੋਂ)	86.66	68.42	76.46
3.	Why (ਕਿਉਂ)	85.00	62.96	72.33
4.	Who (ਕੋਣ/ਕਿਸ/ਕਿਸੇ)	84.44	61.29	71.02
5.	Where (ਕਿੱਥੇ/ਕਿੱਥੋਂ/ਕਿਹੜੇ/ਕਿਹੜਾ/ਕਿਹੜੀ)	83.33	67.07	74.32

The graph given in Fig.1 shows the values of precision for each type of factoid question type, which shows how the system finds the best, possible relevant answer from possible dataset of answers it has predicted, the system basically searches for information which is best suited for answer based on gains. It can make by finding more common artifacts between the question tokens and the answer tokens, the more common verbs, nouns, adjectives, adverbs, pronouns in both token sets, have more is the possibility, the results would be more precise.

The graph given in Fig.2 shows the percentage of recall for each question type. It may happen the actual answer of possible answers may be either more or less as compared to what the question answering system finds. How many possible answers are possible depends on the evaluator and the degree for ground truth.. Therefore, we have also calculated the recall value of the system and results show fair amount of recall percentage due to obvious high value of precision.

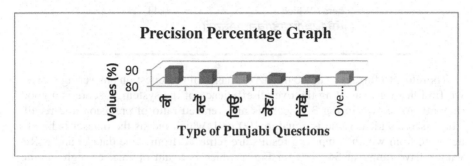

Fig. 1 Average Precision of each Question Type (%)

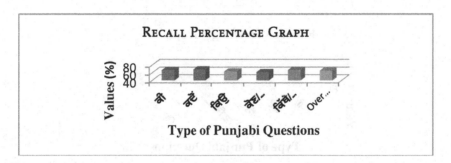

Fig. 2 Average Recall of each Question Type (%)

6.2   *Results on the Basis Mean Reciprocal Rank (MRR)*

MRR is used for the computation of the overall performance of the system. MRR

is calculated as: $MRR = \dfrac{\sum_{k=1}^{|Q|} \frac{1}{r_k}}{|Q|}$ (5)

Where Q is the question collection & r_k is the highest rank of the correct answers & its value is 0 if no correct answer is returned. The Table2 shows the average values of MRR for each question type.

Table 2 Average Mean Reciprocal Rank (MRR) of each Question Type (in %)

S.NO.	Question Type	Average MRR
1.	What(ਕੀ)	0.49
2.	When (ਕਦੋਂ)	0.44
3.	Why (ਕਿਉ)	0.41
4.	Who/Whose/Whom (ਕੋਣ/ਕਿਸ/ਕਿਸੇ)	0.39
5.	Where / Which (ਕਿੱਥੇ/ਕਿੱਥੋਂ/ਕਿਹੜੇ/ਕਿਹੜਾ/ਕਿਹੜੀ)	0.43

Therefore to further evaluate the performance of our question answering system we find that for evaluating the overall efficiency of the system, F-score is a good method. As shown in Fig 3 F-score is an averaged ratio of precision and recall. The F-score's ideal value is equal to 100 or 1, which means the dataset is highly accurate from which competing results are retrieved from same dataset and MRR values have range of 34- 38% in Fig 4 which is again optimal enough for the system to find dissimilarity as well as similarity of the systems.

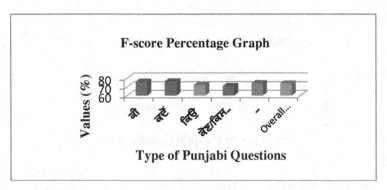

Fig. 3 Average F-score of each Question Type (%)

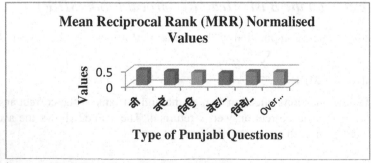

Fig. 4 Average MRR of each question Type (in %)

7   Error Analysis

Overall F-score of Punjabi Question Answering System is 74.06%. So the overall errors of the system are 25.94% which includes: What (ਕੀ) = 2.26%, Who (ਕੋਣ/ਕਿਸ/ਕਿਸੇ) = 7.94% , Why (ਕਿਉ) = 6.77%, Where (ਕਿੱਥੇ/ਕਿੱਥੋਂ/ਕਿਹੜੇ/ਕਿਹੜਾ/ਕਿਹੜੀ/ ਕਿਹੜੀਆਂ) = 5.52%, When (ਕਦੋਂ) = 2.45%.The F-score value for What (ਕੀ) is 76.18%, 23.32% errors are due to the questions which are searching for name keys entries from Name dictionary which is not meant for this category. As the question type for named entity is Who (ਕੋਣ/ਕਿਸ/ਕਿਸੇ). The examples of these types of questions are:

1.   ਰਣਦੀਪ ਸਿੰਘ ਦੇ ਪਿਤਾ ਦਾ ਨਾਮ ਕੀ ਸੀ? What is the name of Randip Singh's father?

Since in this type of question, the question type is What (ਕੀ), but these questions are searching for the named entities. The F-score value for When (ਕਦੋਂ) is 76.46%, 23.54% errors are due to if the Punjabi input text consists of multiple input sentences that are based on time and date expressions and there may be an error because of the wrong scoring of these answers. The F-score value for Who (ਕੋਣ/ਕਿਸ/ਕਿਸੇ) is 71.02%, 28.98% errors are due to if Punjabi input text consists of multiple name sentences. The example of this type of error is like "ਸੋਨੂ ਕੋਣ ਸੀ? Who is Sonu?" The candidate answers to this question consist of each answer of the same named entity. Sometimes there may be names that are not available in the domain specific name list. Sometimes it can recognize the names but are not able to recognize the question patterns. Sometimes it can recognize the names but are not able to recognize the question patterns. The F-score value for Why (ਕਿਉ) is 72.33%, 27.67% errors are due to if there is less discrimination between similar and dissimilar Punjabi input text sentences.

The F-score value for Where (ਕਿੱਥੇ/ਕਿੱਥੋਂ/ਕਿਹੜੇ/ਕਿਹੜਾ/ਕਿਹੜੀ/ ਕਿਹੜੀਆਂ) is 74.32%, 25.68% errors are due to the wrong calculation of scores of answers due to very large number of candidate answers of locations with respect to the question. These errors occured when the Punjabi input document contains multiple sentences that consists of location names. The system collects the sentence that are more relevant in terms of pattern matching and also based on multiple locations.

Sometimes the errors occured due to the question types ਕਿਨਾਂ/ ਕਿੰਨੇ/ ਕਿੰਨੀ/ ਕਿੰਨੀਆਂ (HOW MANY), ਕਿਵੇਂ (HOW) since we have not used these question types like:1. ਭਾਰਤ ਨੇ ਸ੍ਰੀਲੰਕਾ ਨੂੰ ਕਿੰਨੇ ਵਿਕਟਾਂ ਨਾਲ ਹਰਾਇਆ? By how many wickets India beat Sri Lanka? 2. ਅਨਜਾਣ ਲੋਕਾਂ ਨੂੰ ਕੱਬਡੀ ਬਾਰੇ ਕਿਵੇਂ ਪਤਾ ਚੱਲਿਆ? How ignorant people know about Kabbadi ? Sometimes the errors occurred if there is more number of stop words in the question; the system is not able to identify the candidate answer of that particular question. Screenshot of this system is given in Fig5.

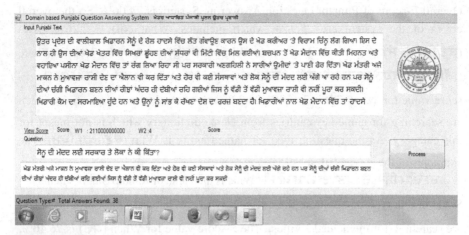

Fig. 5 Screenshot of Punjabi Question Answering System

8 Conclusion and Future Scope

The work illustrated here in context of Punjabi Question Answering System is first of its kind. In this paper we have discussed a hybrid algorithm for the implementation of Punjabi Question Answering System. In the current research work, we have used the concept making a hybrid that works in pattern matching (regular expressions) and new proposed answer finding scoring system, which has yielded for better Recall and Precision values. The focus of the system has been on various kinds of questions of types ਕੀ (what), ਕਦੋਂ (when), ਕਿੱਥੇ/ਕਿੱਥੋਂ/ਕਿਹੜੇ/

ਕਿਹੜਾ/ਕਿਹੜੀ (where/which), ਕੌਣ ਕੋਣ/ਕਿਸ/ਕਿਸੇ (who/whom/whose) and ਕਿਉਂ (why).

For future work, more types of questions can be added for question classification. Question answering system can be developed for different languages, for multiple domains and for multiple documents. Its accuracy can be improved by making the scoring system more accurate.

References

1. Gupta, V., Lehal, G.S.: A Survey of Text Mining Techniques and Applications. J. Emerging Technologies in Web Excellence 1 (2009)
2. Hirschman, L., Gaizauskas, R.: Natural language question answering: The view from here. Natural Language Engineering 7(4), 275–300
3. Sahu, S., Vasnik, N., Roy, D.: Proshanttor: A Hindi Question Answering System. J. Computer Science & Information Technology (IJCSIT) 4, 149–158 (2012)

4. Ramprasath, M., Hariharan, S.: A Survey on Question Answering System. J. International Journal of Research and Reviews in Information Sciences (IJRRIS) 2, 171–178 (2012)
5. Kolomiyets, O., Moens, M.-F.: A survey on question answering technology from an information retrieval perspective. J. Info Sciences 181, 5412–5434 (2011)
6. Frank, A., Krieger, H.-U., Xu, F., Uszkoreit, H., Crysmann, B., Jörg, B., Schäfer, U.: Question Answering from structured knowledge sources. J. of Applied Logic 5, 20–48 (2006)
7. Zhenqiu, L.: Design of Automatic Question Answering base on CBR. J. Procedia Engineering 29, 981–985 (2011)
8. Tallez-Valero, A., Montes-y-Gomez, M., Villasenor-Pineda, L., Padilla, A.P.: Learning to select the correct answer in multi-stream question answering. J. Information Processing and Management 47, 856–869 (2011)
9. Ko, J., Si, L., Nyberg, E.: Combining evidence with a probabilistic framework for answer ranking and answer merging in question answering. J. Information Processing and Management 46, 541–554 (2010)
10. Ahn, C.-M., Lee, J.-H., Choi, B., Park, S.: Question Answering System with Recommendation using Fuzzy Relational Product Operator. In: iiWAS 2010 Proceedings, pp. 853–856. ACM (2010)

Design of an Adaptive Calibration Technique Using Support Vector Machine for LVDT

K.V. Santhosh and Binoy Krishna Roy

Abstract. Design of an adaptive calibration technique for displacement measurement using Linear Variable Differential Transformer (LVDT) is proposed in this paper. The objectives of this proposed work are (i) to extend linearity range of LVDT to full scale of input range, (ii) to make system capable of measuring displacement accurately with variations in dimensions of primary and secondary coil, number of primary and secondary windings, and excitation frequency. Output of LVDT is differential AC voltage across secondary coils. It is converted to DC voltage by using a suitable data conversion circuit. Support Vector Machine (SVM) model is added in cascade to data conversion unit replacing the conventional calibration circuit to achieve desired objectives. The system once designed is subjected to test data with variations in physical parameters of LVDT and excitation frequency for particular displacement. Results show the proposed technique has achieved its set objectives. Designed displacement measurement technique using proposed adaptive calibration technique yields a root mean percentage of error of 0.0078, with linearization over the range 0 to 100 mm.

1 Introduction

Displacement is one of the identified three quantities that are used as base units for any measurement. Displacement measurement has a critical need in many processes. Many sensors are used for this purpose like potentiometer, capacitance picks, LVDT etc; LVDT finds a very wide application because of its high sensitivity and ruggedness. However, the problems of offset, non-linear response characteristics,

K.V. Santhosh
Manipal Institute of Technology, Manipal
e-mail: kv.santhu@gmail.com

Binoy Krishna Roy
National Institute of Technoloy Silchar
e-mail: bkr_nits@yahoo.co.in

S.M. Thampi, A. Gelbukh, and J. Mukhopadhyay (eds.), *Advances in Signal Processing and Intelligent Recognition Systems*, Advances in Intelligent Systems and Computing 264, DOI: 10.1007/978-3-319-04960-1_13, © Springer International Publishing Switzerland 2014

dependence of output on physical parameters of LVDT have restricted use of LVDT. Several techniques have been suggested in literature to overcome the effects of non-linear response characteristics of LVDT, but these techniques have restricted the linearity range to a certain range of input scale. Further, the process of calibration needs to be repeated every time LVDT is replaced with another having different physical parameters like number of primary and secondary winding, dimensions of primary and secondary winding, and/or when excitation frequency is changed.

Literature review suggests in [1], linearization of LVDT output is reported using digital signal processing algorithms. Linearization of LVDT using least mean square polynomial equations is reported in [2], [7]. In [3], [9], artificial neural network block is trained to produce linear output for LVDT. A Look up table implemented on a field programmable gate array for linearization of LVDT is reported in [4]. In [5], linearization of LVDT is discussed using analog circuits. Polynomial equations are used for linearization of displacement sensor in [6]. Radial basis function neural network is used for extension of linear range of LVDT is reported in [8]. In [10], [11], LVDT calibration is performed using discrete and analog circuits. LVDT data acquisition is designed using PCI-1716; calibration of sensor is done using digital Fourier transforms in [12].Most of the reported works discussed linearization of measurement technique for a certain input range. Though a few works have been reported on extension of linearization range but not over the full scale of input range.

Secondly, calibration needs to be repeated there exists a condition of replacing the sensor in the measurement system, with sensor of different physical parameters. Effect of LVDT parameters like number of primary, secondary coil windings, and dimension of primary and secondary winding is discussed in [13]. Effect of frequency and dimension of LVDT is discussed in [14].

To overcome the limitation of the reported survey papers a technique is proposed in this paper using support vector machine. Support Vector machine is trained to produce linear output for variation in input displacement over the full scale of input, and make the output adaptive for variations in physical parameters of LVDT like number of primary and secondary winding, dimension of primary and secondary coils, and excitation frequency.

The paper is organised as follows: after introduction in Section-1, a brief description on LVDT model is given in Section-2. The output of the LVDT is AC voltage; summary on data conversion i.e. AC to DC converter is presented in Section-3. Section-4 deals with the problem statement followed by proposed solution in Section-5. Finally, result and conclusion is given in Section-6.

2   Displacement Measurement

2.1   LVDT Sensor

LVDT is used to measure linear displacement. LVDT operates on the principle of a transformer. As shown in Fig.1, LVDT consists of a coil assembly and a core. The coil assembly is typically mounted to a stationary form, while the core is secured to the object whose position is being measured. The coil assembly consists of three coils of wire wound on the hollow form. A core of permeable material can slide freely through the center of the form. The centre coil is the primary, which is excited by an AC source as shown. Magnetic flux produced by the primary is coupled to the two secondary coils placed on both sides of primary coil, inducing an AC voltage in each secondary coil [17], [15], [16], [18].

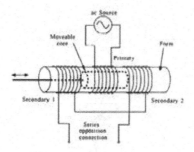

Fig. 1 Schematic diagram of LVDT

$$v_1 = \frac{1.24}{10^5} \frac{fI_p n_p n_s}{\ln(\frac{r_o}{r_i})} \frac{2L_1 + b}{mL_a} x_1^2 \tag{1}$$

$$v_2 = \frac{1.24}{10^5} \frac{fI_p n_p n_s}{\ln(\frac{r_o}{r_i})} \frac{2L_2 + b}{mL_a} x_2^2 \tag{2}$$

where
I_p - primary current for the excitation V_p
x_1 - distance penetrated by the armature towards the secondary coil 1
x_2 - distance penetrated by the armature towards the secondary coil 2
n_p - number of primary windings
n_s - number of secondary winding
f - frequency of excitation in Hz

Taking $L_a = 3b$, the differential voltage v is thus given by $v_1 - v_2$

$$v = \frac{fI_p n_p n_s bx(3.3 * 10^{-5})}{m \ln \frac{r_o}{r_i}} (1 - \frac{x^2}{2b^2}) \tag{3}$$

where x is the displacement of core given by $(x_1 - x_2)/2$

$$I_p = \frac{V_p}{\sqrt{(R_p^2 + (6.28fL_p)^2)}} \tag{4}$$

where
L_p-Primary inductance
R_p-Primary resistance

2.2   *Data Conversion Unit*

The data conversion circuit consists of a rectifier circuit followed by instrumentation amplifier. The rectifier circuit converts the AC voltage of secondary coil to DC output. Instrumentation amplifier is used to find difference between two secondary coils as shown in Fig.2.

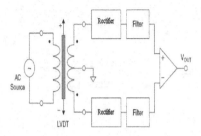

Fig. 2 DCC of LVDT

3   Problem Statement

In this section, characteristics of LVDT are simulated to understand difficulties associated with available measurement technique. For this purpose, simulation is carried out with three different ratios between outer and inner diameter of coil. These are (ro/ri) = 2, 4, and 6. Three different values of ratio of length of primary coil and secondary coil (b/m) are considered. These are b/m = 0.25, 0.5, and 0.75. Three different primary winding turns are taken. These are n_p = 100, 200, and 300. Three different secondary winding turns are taken. These are n_s = 100, 200, and 300. Three different excitation frequencies are considered. These are f = 2.5 KHz, 5 KHz, and 7.5 KHz are used to find the outputs of LVDT with respect to various values of input displacement, considering a particular set of physical parameter of LVDT, and excitation frequency. These output voltages from LVDT are passed through data conversion circuit. The output DC voltages of data conversion circuit for full range of displacement are noted. These output voltages from data conversion circuit are separately plotted in the following figures.

Fig.3 to Fig.7 show the variations of voltages with change in input displacement considering different values of f, ro/ri, b/m, n_s, and n_p. It has been observed from the above graphs that the relation between input displacement and output voltage has a non-linear relation. Datasheet of LVDT suggests that 10% to 80% of input range is used in practice for linearity constraint. The output voltage also varies with the change in ro/ri, b/m, n_s, n_p, and excitation frequency. These are the reasons for which various calibration techniques have been designed. These conventional techniques have drawbacks that its time consuming, and need to be calibrated every time whenever an LVDT is changed in the system, and/or there is a change in their excitation frequency as seen from Fig.3 to Fig.7. Further, the use is restricted only

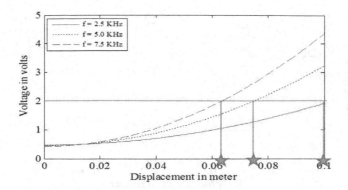

Fig. 3 Affect of frequency on displacement measurement with LVDT

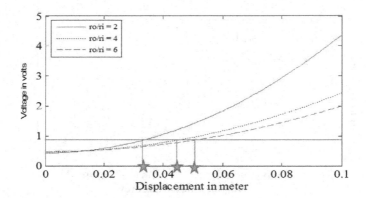

Fig. 4 Affect of r_o/r_i on displacement measurement with LVDT

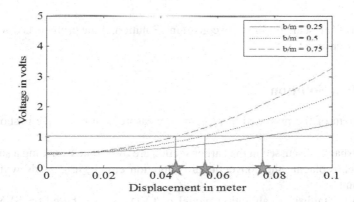

Fig. 5 Affect of b/m on displacement measurement with LVDT

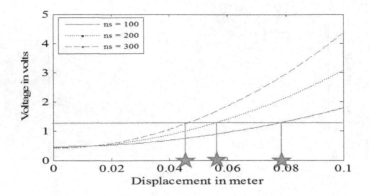

Fig. 6 Affect of n_s on displacement measurement with LVDT

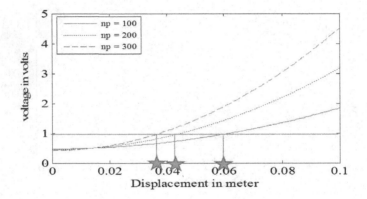

Fig. 7 Affect of n_p displacement measurement with LVDT

to a portion of full scale for linearity constraint. Solution of the problem is discussed in the next section.

4   Problem Solution

Block diagram of the proposed displacement measurement technique is shown in Fig. 8

The drawbacks discussed in the earlier section are overcome by adding a suitable SVM model, replacing the conventional calibration circuit, in cascade with data converter unit.

When we establish a calibration model of LVDT sensor based on SVM, we should solve a regression problem in deed. Support Vector Machines provide a framework for the regression problem and it can be applied to regression analysis [20], [19]. The calibration principle based on SVM makes use of the input parameters mapped to high-dimensional space by nonlinear transformation function,

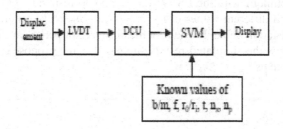

Fig. 8 Block diagram of proposed displacement measurement with LVDT

thus regression analysis can be performed in the high-dimensional space, and finally the Input/ Output function can be obtained [21], [22], [23]. The regression function based on LSSVM is denoted as in eq.5

$$f(x) = w.g(x) + b \tag{5}$$

Where

$w.g(x)$ is the inner product of w and $g(x)$

w is the vector in hign-dimensional space computed by training.

b is the bias.

$g(x)$ is the input function of pf SVM model

Considering a set of training data for displacement measurement using LVDT with input and output: x_{ik}, y_{ik}, where x_{ik} is the output of data conversion output for variations in displacement, for different ratio of primary and secondary windings, dimension of primary and secondary windings, and excitation frequency. y_{ik} R is target parameter of the proposedmeasurement technique having a linear relation with displacement, and independent of ratio of primary and secondary windings, dimension of primary and secondary windings, and excitation frequency.

w and b are computed by using the optimization problem given by

$$min_{w,z}[\frac{1}{Z} \parallel w \parallel^2 + C \sum_{{}^n i=1} (Z + Z^2)] \tag{6}$$

By using the relaxation variable Z, $Z^* \geq 0$,

5   Results and Conclusion

Once training is over, the proposed system is subjected to various test inputs corresponding to different displacements at a particular ro/ri, b/m, n_s, n_p, and excitation frequency, all within the specified range. For testing purposes, the range of displacement is considered from 0 to 100 mm, the range of ro/ri is 2 to 6, the range of b/m is 0.25 to 0.75, the range of ns is 100 to 300, the range of n_p is 100 to 300, and range of excitation frequency is 2.5 to 7.5 kHz. The outputs of the proposed displacement measurement technique with the LSSVM are noted corresponding to various input displacement with different values of ro/ri, b/m, n_s, n_p, and excitation frequency. The results are listed in Table-1. The root mean square of percentage error is found to be 0.0078. Thus the simulated results suggest that the proposed technique has fulfilled its objectives.

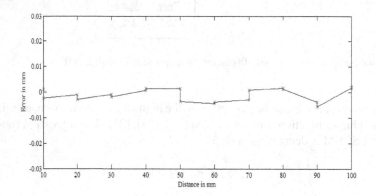

Fig. 9 Percentage Error Vs actual displacement graph

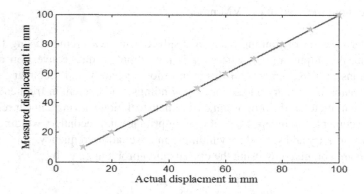

Fig. 10 Measured Vs actual displacement graph

Available reported works displacement measurement are not adaptive of variations in physical parameters of LVDT, supply frequency. Hence, repeated calibration is required for any change in physical parameters of LVDT, and excitation frequency. Some time the calibration circuit may itself need to be replaced. Such step is time consuming and tedious procedure. Further, some reported works have not utilized the full scale of measurement. In comparison to these, the proposed displacement measurement technique achieves linear input-output characteristics for full scale input range and makes the output adaptive of variations in physical parameters of LVDT, and excitation frequency. All these have been achieved by using LSSVM.

Table 1 Simulation results of proposed intelligent displacement measurement with LVDT

AD mm	r_o/r_i	b/m	n_s	n_p	f kHz	MD mm	% error
10	2	0.25	100	300	7.5	9.9989	0.011
10	3	0.30	120	290	2.9	10.0023	-0.023
20	4	0.35	140	280	5.2	20.0011	-0.006
20	5	0.40	160	270	3.6	20.0030	-0.015
30	6	0.45	180	260	6.4	30.0009	-0.003
30	6	0.50	200	250	7.0	30.0018	-0.006
40	5	0.55	220	240	2.6	39.9992	0.002
40	4	0.60	240	230	6.2	39.9985	0.004
50	3	0.65	260	220	3.8	49.9986	0.003
50	2	0.70	280	210	7.4	50.0037	-0.007
60	2	0.75	300	200	4.0	60.0044	-0.007
60	3	0.75	300	190	5.9	60.0038	-0.006
70	4	0.70	280	180	2.7	69.9991	0.001
70	5	0.65	260	170	3.1	70.0029	-0.004
80	6	0.60	240	160	7.2	79.9983	0.002
80	6	0.55	220	150	6.7	79.9988	0.001
90	5	0.50	200	140	2.8	90.0038	-0.004
90	4	0.45	180	130	6.5	90.0057	-0.006
AD - Actual displacement in mm							
MD - Measured displacement in mm							

References

1. Ford, R.M., Weissbach, R.S., Loker, D.R.: A novel DSP-based LVDT signal conditioner. IEEE Transactions on Instrument and Measurement 55(3), 768–774 (2001)
2. Flammini, A., Marioli, D., Sisinni, E., Taroni, A.: Least Mean Square Method for LVDT Signals Processing. In: Proc. Instrumentation and Measurement Technology Conference, Corno, Italy (2004)
3. Mishra, S.K., Panda, G.: A Novel Method for Designing LVDT and its Comparison with Conventional Design. In: Proc. Sensors Applications Symposium, Houston, USA (2006)
4. Martins, R.C., Ramos, H.G., Proence, P.: A FPGA-based General Purpose Multi-Sensor Data Acquisition System with Nonlinear Sensor Characteristics and Environment Compensation. In: Proc. International Conference on Instrumentation and Measurement Technology Conference, Sorrento, Italy (2006)
5. Drumea, A., Vasile, A., Comes, M., Blejan, M.: System on Chip Signal Conditioner for LVDT Sensors. In: Proc. 1st Electronics System Integration Technology Conference, Dresden, Germany (2006)
6. Tudic, V.: Uncertainty Evaluation of Slope Coefficient of High Precision Displacement Sensor. In: Proc. 48th International Symposium, Zadar, Croatia (2006)
7. Flammini, A., Marioli, D., Sisinni, E., Taroni, A.: Least Mean Square Method for LVDT Signal Processing. IEEE Transactions on Instrumentation and Measurement 56(6), 2294–2300 (2007)

8. Wang, Z., Duan, Z.: The Research of LVDT Nonlinearity Data Compensation based on RBF Neural Network. In: Proc. 7th World Congress on Intelligent Control and Automation, Chonqging, China (2008)
9. Mishra, S.K., Panda, G., Das, D.P.: A Novel Method for Extending the Linearity Range of Linear Variable Differential Transformer Using Artificial Neural Network. IEEE Transaction of Instrumentation and Measurement 59(4), 947–953 (2010)
10. Wang, L., Wang, X., Sun, Y.: Intelligent Acquisition Module for Differential Transformer Position Sensor. In: Proc. International Conference on Intelligent System Design and Engineering Application, Hunan, China (2010)
11. Spiezia, G., Losito, R., Martino, M., Masi, A., Pierno, A.: Automatic Test Bench for Measurement of Magnetic Interface on LVDTs. IEEE Transactions on Instrumentation and Measurement 60(5), 1802–1810 (2011)
12. Fan, J., Jia, S., Lu, W., Wang, Z., Li, X., Sheng, J.: Application of LVDT Sensor Data Acquisition System Based on PCI-1716. In: Proc. International Conference on Computer Science and Automation Engineering, Shanghai, China (2011)
13. Yun, D., Ham, S., Park, J., Yun, S.: Analysis and design of LVDT. In: International Conference on Ubiquitous Robots and Ambient Intelligence, Incheon, South Korea (2011)
14. Macione, J.: Design and analysis of a novel mechanical loading machine for dynamic in vivo axial loading. Journal on Review of Scientific Instruments 83(2), 25113–25114 (2012)
15. Neubert, H.K.P.: Instrument Transducers: An Introduction to Their Performance and Design. In: Handbook of Experimental Pharmacology, 2nd edn. Oxford University Press, UK (2003)
16. Liptak, B.G.: Instrument Engineers Handbook-Process Measurement and Analysis, 4th edn. CRC Press, UK (2003)
17. Murty, D.V.S.: Transducers and Instrumentation. PHI Publication, India (2003)
18. Regitien, P.P.L.: Inductive and Magnetic Sensors, pp. 125–159. Sensor for Mechatronics (2012)
19. Chih-Chung, C., Lin, C.-J.: LIBSVM (2001) A library for support vector machines available at NTU, http://www.csie.ntu.edu.tw/~cjlin/libsvm
20. Xuegong, Z.: Statical learning theory and support vector machines. Acta Automatica Sinica 26(1), 32–42 (2000)
21. Suykens, J.A.K., Vandewalle, J.: Least Squares Support Vector Machine Classifiers. Kluwer Academic Publisher, Netherlands (1999)
22. Schokopf, B., Smola, A.: Learning with Kernels: Support Vector Machines, Regularization, Optimization and Beyond. MIT Press, USA (2002)
23. Gestel, T.V., Suykens, J.A.K., Baesens, B., Viaene, S., Vanthienen, J., Dedene, G., De Moor, B., Vandewallel, J.: Benchmarking least squares support vector machine classifiers. Journal on Machine Learning 54(1), 5–32 (2004)

Speech Processing for Hindi Dialect Recognition

Shweta Sinha, Aruna Jain, and Shyam S. Agrawal

Abstract. In this paper, the authors have used 2-layer feed forward neural network for Hindi dialect recognition. A Dialect is a pattern of pronunciation of a language used by a community of native speakers belonging to the same geographical region. In this work, speech features have been explored to recognize four major dialects of Hindi. The dialects under consideration are Khariboli (spoken in West Uttar Pradesh, Delhi and some parts of Uttarakhand and Himachal Pradesh), Bhojpuri (spoken by population of East Uttar Pradesh, Bihar and Jharkhand), Haryanvi (spoken in Haryana, parts of Delhi, Uttar Pradesh and Uttarakhand) and Bagheli (spoken in Central India). Speech corpus for this work is collected from 15 speakers (including both male and female) from each dialect. The syllables of CVC structure is used as processing unit. Spectral features (MFCC) and prosodic features (duration and pitch contour) are extracted from speech for discriminating the dialects. Performance of the system is observed with spectral features and prosodic features as input. Results show that the system performs best when all the spectral and prosodic features are combined together to form input feature set during network training. The dialect recognition system shows a recognition score of 79% with these input features.

Keywords: Hindi Dialects, spectral features, prosodic features, Feed forward neural networks.

1 Introduction

State of the art speech recognition systems are concentrating on variances in speech due to the dialect or accent of the spoken language. The dialect of a given

Shweta Sinha · Aruna Jain
Birla Institute of Technology, Mesra, Ranchi, India
e-mail: meshweta_7@rediffmail.com, arunajain@bitmesra.ac.in

Shyam S. Agrawal
KIIT College of Engineering, Gurgaon, India
e-mail: ss_agrawal@hotmail.com

S.M. Thampi, A. Gelbukh, and J. Mukhopadhyay (eds.), *Advances in Signal Processing and Intelligent Recognition Systems*, Advances in Intelligent Systems and Computing 264,
DOI: 10.1007/978-3-319-04960-1_14, © Springer International Publishing Switzerland 2014

language is a pattern of pronunciation and/or vocabulary of a language used by the community of native speakers belonging to the same geographical region. Studies show that recognition of speaker's dialect prior to automatic speech recognition(ASR) helps in improving performance of ASR systems by adapting to appropriate acoustic and language model [1].

Different approaches namely, phonotactic and spectral have been used for studying dialectal variations [2]. The phonotactic approach based studies concentrate on phone sequence distribution. In this approach, linguistic patterns such as vowel inventory, diphthong formation, tense marking, tones etc works as a base for these studies [3, 4]. Where as, in the spectral modeling approach discrimination of spectral features of speech due to dialects are highlighted [5, 6]. Dialect specific information in speech signal is contained at the entire segmental, sub segmental and supra segmental level. By observing the shapes of vocal tract during the production of a sound, information specific to spoken dialect can be obtained. This shape can be characterized by the spectral envelope and values can be used at segmental level to represent the information. The shape of glottal pulse can be studied to extract dialect specific information at the segmental level where as, syllable duration and pitch contour can be useful in studying dialectal influences at supra segmental level.

A few works have been done for studying dialectal variations of any language. Most of them are based on phonological approach of dialectal study [3, 7, 8]. Available resources show that automatic dialect identification studies were carried out for languages of western and eastern countries [4, 10], but for Indian languages little has been done in this direction. Literature highlights few accent based studies done for Indian languages using spectral features of speech [5], speech database used for this work consists of utterances from native and non-native speakers of the language.

Hindi is the most common language of India with several dialectal variations. It has become necessary to discriminate dialects for building a speaker independent robust speech recognition system. Hindi language has at least 50 different dialects with speaker population varying from thousands in one dialect to millions in another. The standard dialect arose through a considerable borrowing of vocabulary from another dialects and languages [3]. This work is based on four major dialects of Hindi. Khariboli, Bhojpuri, Haryanvi and Bagheli dialect is chosen for this study. The dialect Khariboli is spoken in western part of Uttar Pradesh, Delhi and some parts of Uttarakhand and Himachal Pradesh. The dialect Bhojpuri is spoken in eastern Uttar Pradesh, Bihar and Jharkhand. Haryanvi is mainly spoken in Haryana, but parts of population of Delhi, Uttar Pradesh and Uttarakhand also uses this dialect. Another dialect under consideration is Bagheli which is spoken in central India, mainly Madhya Pradesh and parts of Chhattisgarh.

In this work authors have explored both spectral and prosodic features of speech for recognizing the spoken dialects. Mel frequency cepstral coefficients (MFCC) is used as spectral feature and prosodic features are captured in terms of duration and pitch contour values. To execute the work, syllables from initial and final position of spoken utterances have been extracted. Feed forward neural network (FFNN) is used to capture the dialect specific information from the proposed spectral and prosodic features. The rest of the paper is organized as; section 2 explains the database created for the execution of this work, section 3 presents the techniques and feature sets used for the development of dialect recognition system, section 4 discusses results and observations with different feature sets and section 5 presents conclusion of the work.

2 Database

Survey of existing literatures show that no standard database exists for Hindi language and its dialects. A new database from speakers of four dialects; Khariboli (KB), Bhojpuri (BP), Haryanvi (HR) and Bagheli (BG) is created to fulfill the requirement of this task. Recording is done in a sound proof room at 16 KHz. Since, our interest lies in the study of dialectal influences on pronunciation of words hence all the speakers were asked to utter same clauses. Final selection of the speakers for this work was based on perception test conducted for all the recorded utterances. It is observed that few speakers who moved out of their native place at an early age or who did not speak their native language regularly were minimally influenced by their dialect. Recording of only those speakers were selected for whom perception test showed a confidence score greater than 95%. Finally, a database of 60 speakers, 15 from each of the four dialects is obtained. Out of these 15 speakers from each dialect, 10 are male speakers and 5 are female speakers. The system was trained with 7 male and 3 female speakers from each of the dialect and the rest of the data was used for testing the system performance. All the selected speakers are of the age group 21 years to 50 years. 400 utterances were collected from each speaker. In total, the database consists of 24000 utterances.

3 Development of Hindi Dialect Recognition System

Two key concepts of artificial intelligence are automatic knowledge acquisition and adaptation, one way in which these concepts can be implemented is through collection of neurons. Neurons are information processing fundamental unit of neural networks. For this work, feed forward neural network (FFNN) is used. Studies show that syllables are the most basic units for analysis of speech in Indian languages. They implicitly capture shape of vocal tract and co-articulation and represent the production and perception of speech in Indian languages [9]. Research [7, 8] shows that duration of syllables and phonemes are greatly influenced by dialects and their influence is most significant at the initial and final position. Syllables of the form consonants-vowel-consonants (CVC) are captured from the initial and final position of the spoken utterances for this study. The syllables used for the development of the model are categorized in ten groups based on Hindi vowel category i.e. /a/, /aa/, /i/, /ii/, /u/, /uu/, /e/, /ei/, /o/, /ou/. The Spectral and prosodic features of these syllables are fed as input to the neurons of FFNN.

3.1   Feed Forward Neural Networks

A feed forward neural network is expected to model behavior of certain phenomena by establishing functional relationship between input and output vectors of the network and subsequently be able to classify different aspects of those behaviors. The general structure of multilayer feed forward neural network is shown in figure 1 where w's represents associated weights and b's represents bias to the neurons.

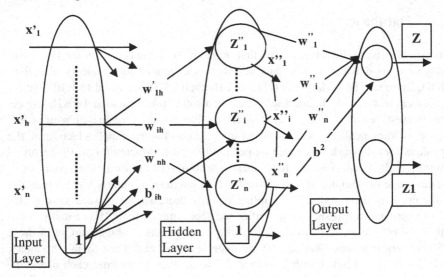

Fig. 1 2-layer feed forward neural network

Two layer FFNN is defined to be capable of handling any complex pattern classification problem [10]. For this work 2-layer FFNN is implemented. The input layer known as 0th layer is fed by the features extracted from the utterances. The first layer is the hidden layer. Activation function at this layer is sigmoid function which is known to be stable in the range 0-1. The learning rate of the system is fixed at 0.01, as large learning rate will cause the system to oscillate and too fewer rates will cause the system to take very long time to converge. The network is trained with scaled conjugate back propagation with momentum. This momentum helps to dynamically treat all weights in similar manner and to avoid a situation of no learning. Number of epochs needed for training depends upon the behavior of training error. It is observed that 80 epoch is sufficient for training existing dataset.

3.2   Feature Set

For dialect recognition FFNN model is created using spectral and prosodic features. For the spectral features MFCCs are extracted. These coefficients are obtained by reducing the frequency information of speech signal into values that emulate separate critical bands in the basilar membrane of the ear. The Speech signals were divided into short frames of 25ms with a frame shift of 10ms. To obtain the information about amount of energy at each frequency band Discrete Fourier Transform (DFT) of windowed signal is obtained. A bank of 26 filters is applied to collect energy from each frequency band. Ten of these filters are spaced linearly below 1000Hz and remaining are spread logarithmically above 1000Hz. Twelve MFCCs, one zero coefficient and one energy coefficient totaling to fourteen coefficients are extracted from each frame.

For the prosodic features, syllable duration and pitch contour is extracted from the speech signal. The syllables are extracted based on position in words. Syllables from initial and final position only are taken into consideration. The duration of syllables is measured in milliseconds. These values are normalized and stored as input features for the network. For representing durational information two feature vectors are used. One represents the normalized duration of syllables and second records its positional value; 1 or 0 for initial and final position respectively. Pitch is referred as speaker characteristics. Studies show that it is also influenced by dialectal variations [5]. Same speaker when utters in two different dialect his pitch varies. So pitch is an important parameter to represent the dialectal characteristics of speech signal. The sequence of fundamental frequency values constitute pitch contour. In this work pitch contours are extracted from the given utterance using autocorrelation method. The size of syllables varies based upon the dialect as well as the position. FFNN requires fixed size of input in each iteration so the number of pitch component has to be fixed. Based upon inspection of the utterances under consideration fifteen pitch values are finalized for this work. Depending upon length of input utterances frame size is modified to keep the number of parameters fixed for each frame. Finally, fifteen features representing pitch contour values are selected.

The recorded utterances under consideration belong to both male and female speakers. An input unit is used to represent the voice as belonging to male or female; 1 or 0 respectively. In totality thirty two features are used for representing the overall information about the input utterances.

3.3   Implementation

2-layer (one hidden and one output) FFNN is implemented for dialect recognition system. Thirty two features extracted from the spoken utterances are fed as input to thirty two neurons of the input layer. The four dialects under consideration for recognition are represented at the output layer by four neurons. The choice of number of neurons at hidden layer is arbitrary. It was observed that the best result of the system performance is obtained by keeping the number of hidden layer neurons fixed at forty five. Back propagation algorithm is used for supervised training of the network. This algorithm works iteratively in two phases for reducing the error by updating synaptic weights of connecting synapse. Figure 2 represents the implementation details of dialect recognition system. Features extracted from the speech frames are fed as input to the network. W and B represent weight and Bias respectively. Value of these variables are adjusted during the training of the network Target set represents the input during supervised training of the network. Depending upon learning of the network with assigned parameters classification performance of the system is evaluated.

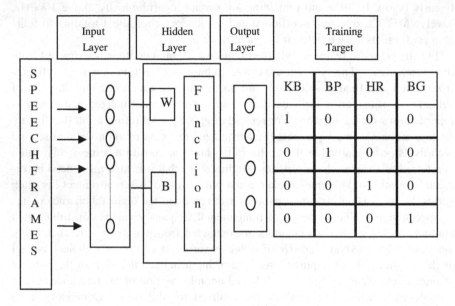

Fig. 2 Implementation of dialect recognition system

4   Results and Observations

As the data representing each of the four dialect were sufficient for carrying this experiment, we have developed a single FFNN that accepts spectral features, pitch contour values, gender information, syllable duration and its position in spoken utterance as input. The system is trained with 70% (10 speakers) of the available data from each dialect and the rest 30% (5 speakers) are used for testing the system. Based upon input features the trained network classifies input utterances as one of the dialects: Khariboli (KB), Bhojpuri (BP), Haryanvi (HR) and Bagheli (BG). To study the influence of spectral and prosodic features on dialect recognition system the experiment is executed in two phases.

In the first phase, the system is trained with 15 features only, where 14 are spectral features (12 MFCCs, 0th coefficient, and frame energy) and one feature represents gender identity of the input utterance. Values to the rest of the 17 neurons at the input layer are all set to 0. Performance of the system based on spectral features is presented in Table 1. The average recognition score is observed to be

66%. The entries in the diagonal of the table represent correct classification based on input and target of the system. Rest of the cells of the table represents misclassification of the input utterance of one dialect to that of other dialects.

Table 1 Performance of dialect recognition system developed using spectral features

Recognition Performance(%)				
Dialects	KB	BP	HR	BG
KB	62	4	22	12
BP	12	71	8	9
HR	20	11	61	8
BG	6	14	9	71

The entries in the table represents percentage of recognition (KB: Khariboli, BP: Bhojpuri, HR:Haryanvi, BG: Bagheli)

In the second phase of execution prosodic features are included along with the spectral features of the syllables taken from the spoken utterance. Initially, only durational measure along with positional information of the input syllables is included in the training of the system. Number of features used in this training set is 17 (12 MFCCs, 0th coefficient, frame energy, one feature representing gender identity of the input utterance, duration of input syllable and its position as initial or final) . Input to rest of the neurons is all set to zero. Average performance of the system improved significantly and recognition score is observed to be 74%. Table 2 represents the recognition performance of the system with a set of 17 features. Diagonals represent correct classification and other cells show misclassification of the input.

Table 2 Performance of dialect recognition system developed using spectral features and duration of input as prosodic feature

Recognition Performance (%)				
Dialects	KB	BP	HR	BG
KB	69	8	14	9
BP	11	78	0	11
HR	14	4	76	6
BG	6	10	10	74

The entries in the table represents percentage of recognition (KB: Khariboli, BP: Bhojpuri, HR: Haryanvi, BG: Bagheli)

With the aim to further improve the system performance pitch contour values are also included in the training of the system. These values are extracted as sequence of fundamental frequencies of each frame. The system is trained with all 32 features and its performance is measured. Results show that average performance of the system further improved and recognition score is observed to be 79%. Table 3 shows the recognition score of dialect recognition system when all 32 features are taken into consideration for training of the network.

Table 3 Performance of dialect recognition system developed using spectral features and prosodic features

Dialects	Recognition performance (%)			
	KB	BP	HR	BG
KB	77	5	11	7
BP	7	83	0	10
HR	2	3	84	11
BG	14	9	5	72

The entries in the table represents percentage of recognition (KB: Khariboli, BP: Bhojpuri, HR: Haryanvi, BG: Bagheli)

5   Conclusions

In this paper a dialect recognition system has been proposed. Four major dialects of Hindi, Khariboli (spoken in western part of Uttar Pradesh, Delhi and some parts of Uttarakhand and Himachal Pradesh) , Bhojpuri (spoken by population of eastern Uttar Pradesh, Bihar and Jharkhand), Haryanvi (spoken in Haryana, parts of Delhi, Uttar Pradesh and Uttarakhand) and Bagheli (spoken in Madhya Pradesh and parts of Chhattisgarh) have been considered for this study. Speech corpus for this work was created based on few fixed utterances spoken by all speakers and final selection was done on the basis of confidence score obtained in the perception test of the recorded utterances. Spectral (MFCC) and prosodic features (duration and pitch contour) extracted from input speech were explored. 2-layer Feed forward neural network was used to capture dialect specific information from the input features. Average performance of the system with only spectral features is found to be 66%, performance of the system improved significantly to 74% by including position based durational feature. This performance boost shows that time taken by speakers of different dialect for same utterance differ considerably and duration of syllable is a major factor in differentiating the Hindi dialects. Performance of the system further improved to 79% when pitch contour was also included in the prosodic feature set. This boost in recognition performance can be attributed to complementary nature of spectral and prosodic features. Study of glottal closure and other excitation source features can be explored and their inclusion in the existing feature set may further improve the system performance.

References

[1] Liu, M., Xu, B., Hunng, T., Deng, Y., Li, C.: Mandarin accent adaptation based on context independent/context dependent pronunciation modeling. In: Proceedings of The Acoustic, Speech and Signal Processing, ICASSP 2000, Washington DC, USA, pp. 1025–1028 (2000)

[2] Behravan, H.: Dialect and accent recognition. Dissertation, University of Eastern Finland (2012)

[3] Mishra, D., Bali, K.: A comparative phonological study of the dialects of Hindi. In: Proceedings of ICPhS XVII, Hong Kong, pp. 17–21 (2011)

[4] Zue, W., Hazen, T.J.: Automatic language identification using segment based approach. In: Proceedings of Eurospeech, pp. 1303–1306 (1993)

[5] Rao, K.S., Nandy, S., Koolagudi, S.G.: Identification of Hindi dialect using speech. In: Proceedings of WMSCI 2010- the 14th World Multi-Conference on Systemics, Cybernetics and Informatics, Orlando, Florida, USA (2010)

[6] Mehrabani, M., Boril, H., Hansen, J.H.L.: Dialect distance assessment method based on comparision of pitch pattern statistical models. In: Proceedings of ICASSP, Dallas, USA, pp. 5158–5161 (2010)

[7] Lee, H., Seong, C.J.: Experimental phonetic study of the syllable duration of Korean with respect to the positional effect. In: Proceedings of the Fourth International Conference on Spoken Language Processing (ICSLP), Philadelphia, pp. 1193–1196 (1996)

[8] Sinha, S., Agrawal, S.S., Jain, A.: Dialectal influences on acoustic duration of Hindi phonemes. In: Proceedings of Oriental-COCOSDA-2013, Gurgaon, India (2013)

[9] Aggarwal, R.K., Dave, M.: Integration of multiple acoustic and language models for improved Hindi speech recognition system. International Journal of Speech Technology (IJST) 15, 165–180 (2012)

[10] Haykin, S.: Neural Networks: A comprehensive foundation. Pearson Education Asia, Inc., New Delhi (2002)

Classifying Sonar Signals Using an Incremental Data Stream Mining Methodology with Conflict Analysis

Simon Fong, Suash Deb, and Sabu Thampi

Abstract. Sonar signals recognition is an important task in detecting the presence of some significant objects under the sea. In military sonar signals are used in lieu of visuals to navigate underwater and/or locating enemy submarines in proximity. Specifically, classification in data mining is useful in sonar signal recognition in distinguishing the type of surface from which the sonar waves are bounced. Classification algorithms in traditional data mining approach offer fair accuracy by training a classification model with the full dataset, in batches. It is well known that sonar signals are continuous and they are collected in streaming manner. Although the earlier classification algorithms are effective for traditional batch training, it may not be practical for incremental classifier learning. Because sonar signal data streams can amount to infinity, the data pre-processing time must be kept to a minimum to fulfill the need for high speed. This paper introduces an alternative data mining strategy suitable for the progressive purging of noisy data via fast conflict analysis from the training dataset without the need to learn from the whole dataset at one time. Simulation experiments are conducted and superior results are observed in supporting the efficacy of the methodology.

1 Introduction

In general sonar which stands for Sound Navigation And Ranging is a sound propagation technology used in underwater navigation, communication and/or

Simon Fong
University of Macau, Taipa, Macau SAR
e-mail: ccfong@umac.mo

Suash Deb
Cambridge Institute of Technology, Ranchi, India
e-mail: suashdeb@gmail.com

Sabu Thampi
Indian Institute of Information Technology & Management, Kerala, India
e-mail: sabu.thampi@iiitmk.ac.in

S.M. Thampi, A. Gelbukh, and J. Mukhopadhyay (eds.), *Advances in Signal Processing and Intelligent Recognition Systems*, Advances in Intelligent Systems and Computing 264,
DOI: 10.1007/978-3-319-04960-1_15, © Springer International Publishing Switzerland 2014

detection of submarine objects. The relevant techniques have been recently reviewed in [1]. In particular it was highlighted that detection/classification task of sonar signals is one of the most challenging topics in the field.

Choosing the right classification model for sonar signals recognition is an important matter in detecting the presence objects of interest under the sea. In military sonar systems are used instead of visuals to navigate underwater and/or detecting the presence of enemy submarines in proximity. Whichever combatant vessels may gain an upper-hand in a battle, if they possess the ability to detect their enemy fleet from farther the distance away and with higher detection accuracy [2]. However sonar signals that propagate underwater especially in long haul are prone to noise and interferences. In particular, classification techniques in data mining have been employed widely in sonar signal recognition in distinguishing the surface of the target object from which the sonar waves are echoed [3, 4, 5].

Classification algorithms in traditional data mining approach may be able to achieve substantial accuracy by inducing a classification model using the whole dataset. The induction however is usually done in batches, which implies certain decline in accuracy between the models updates may be expected [6]. Moreover the update time may become increasingly long as the whole bulk of dataset gets larger when fresh data accumulates. Just like any data stream, it is known that sonar signals are incessant and they are sensed in continual manner. Even though the batch-mode classification algorithms produce an accurately trained model, it may not be suitable in streaming scenarios such as sonar sensing. Since sonar signal data streams can potentially sum to infinity, it is crucial to keep the data processing time very short for real-time sensing and scouting.

In this paper we present an alternative data stream mining methodology designed for the incrementally purging of noisy data using fast conflict analysis from the stream-based training dataset. It is called Incremental Data Stream Mining Methodology with Conflict Analysis or iDSM-CA (in acronym). The methodology has an advantage of learning a classification model from the stream data incrementally. Simulation experiments are carried out to illustrate the efficacy of the proposed methodology, especially in overcoming the task of removing noisy data from the sonar data while they are streaming.

The rest of the paper is organized as follows. Section 2 highlights some popular computational techniques for the detection and removal of noise from training datasets. Section 3 describes our new data stream mining methodology, and the "conflict analysis" mechanism used for removing misclassified instances. Section 4 presents a sonar classification experiment for validating our methodology. Section 5 concludes the paper.

2   Related Work

Researchers have attempted different techniques for detecting and removing noisy data, which are generally referred to as random chaos in the training dataset. Basically these techniques identify data instances that confuse the training model and diminish the classification accuracy. In general they look for data irregularities, and how they do affect classification performance. Most of these techniques can fit under these three categories: statistics-based, similarity-based, and classification-based methods.

2.1 Statistics-Based Noise Detection Methods

Outliers, or data with extraordinary values, are interpreted as noise in this kind of method. Detection techniques proposed in the literature range from finding extreme values beyond a certain number of standard deviations to complex normality tests. Comprehensive surveys of outlier detection methods used to identify noise in pre-processing can be found in [7] and [8]. In [9], the authors adopted a special outlier detection approach in which the behavior projected by the dataset is checked. If a point is sparse in a lower low-dimensional projection, the data it represents are deemed abnormal and are removed. Brute force, or at best, some form of heuristics, is used to determine the projections.

A similar method outlined by [10] builds a height-balanced tree containing clustering features on non-leaf nodes and leaf nodes. Leaf nodes with a low density are then considered outliers and are filtered out.

2.2 Similarity-Based Noise Detection Methods

This group of methods generally requires a reference by which data are compared to measure how similar or dissimilar they are to the reference.

In [11], the researchers first divided data into many subsets before searching for the subset that would cause the greatest reduction in dissimilarity within the training dataset if removed. The dissimilarity function can be any function returning a low value between similar elements and a high value between dissimilar elements, such as variance. However, the authors remarked that it is difficult to find a universal dissimilarity function. Xiong et al. [12] proposed the Hcleaner technique applied through a hyper-clique-based data cleaner. Every pair of objects in a hyper-clique pattern has a high level of similarity related to the strength of the relationship between two instances. The Hcleaner filters out instances excluded from any hyper-clique pattern as noise. Another team of researchers [13] applied a k-NN algorithm, which essentially compares test data with neighboring data to determine whether they are outliers by reference to their neighbors. By using their nearest neighbors as references, different data are treated as incorrectly classified instances and removed. The authors studied patterns of behavior among data to formulate Wilson's editing approach, a set of rules that automatically select the data to be purged.

2.3 Classification-Based Noise Detection Methods

Classification-based methods are those that rely on one or more preliminary classifiers built as references for deciding which data instances are incorrectly classified and should be removed.

In [14], the authors used an n-fold cross-validation approach to identify mislabeled instances. In this technique, the dataset is partitioned into n subsets. For each of the n subsets, m classifiers are trained on the instances in the other n-1 subsets and the instances in the excluded subset are classified. Each classifier tags an instance as misclassified if it is classified incorrectly. Majority voting or a consensus approach can be used in the filtering process. Another team of researchers [15] presented a robust decision tree method for the removal of outliers. In this method, a pruning tree is built on the training dataset and is used to classify the training data. Instances the pruned tree classifies incorrectly are removed from the training dataset. These processes are repeated until the pruned tree correctly classifies all instances in the training dataset. In the study reported in [16], the researchers innovatively used a genetic algorithm (GA) to create a set of suspicious noisy instances and select a prototype to identify the set of actual noisy instances. The fitness function of the GA is a generic classifier built in advance, and the GA uses it to search heuristically for misclassified instances.

3   Our Proposed Methodology

The above-mentioned techniques were designed for data pre-processing in batch mode, which requires a full set of data to determine which instances are to be deleted. The unique data pre-processing and model learning approach proposed here is different from all those outlined in Section 2.

Traditionally this pre-processing method has been seen as a standalone step which takes place before model learning starts. The dataset is fully scanned at least for once to determine which instances should be removed because they would cause misclassification at a later stage. The filtered training set is then inputted into the learning process expecting that it will facilitate noise-free learning.

In contrast, iDSM-CA is embedded in the incremental learning process, and all of the steps—noise-detection, misclassified data removal and learning—occur within the same timeframe. In this dual approach, pre-processing and training are followed by testing work as the data stream flows in. As an illustration, Fig. 1 shows a window of size W rolling along the data stream. Within the window, the data are first subject to conflict analysis (for noise detection), then to misclassified data removal and training (model building). After the model is duly trained, incoming instances are tested. Since this approach allows for intermediate performance results to be obtained, the average performance level can also be calculated at the end of the process based on the overall performance results.

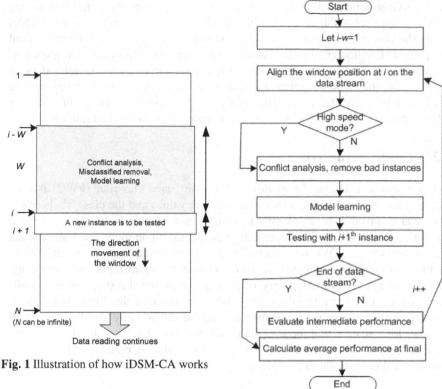

Fig. 1 Illustration of how iDSM-CA works

Fig. 2 Workflow of the incremental learning method

3.1   Workflow of the Pre-processing & Incremental Learning Model

The full operational workflow of the iDSM-CA is shown in Fig. 2. Both pre-processing and training occur within the same window, which slides along the data stream from the beginning and is unlikely to require all available data. This is called *anytime* method in data mining, which means the model is ready to use without waiting for all the training data (for testing) at any time. Whenever new data come in, the window progressively covers the new instances, fades out the old (outdated) instances. When the analysis kicks in again, the model is updated incrementally in real time.

By this approach, there is no need to assume the dataset is static and bounded, and the advantages of removing misclassified instances are gained. Each time the window moves on to fresh data, the training dataset framed in the window W is enhanced and the model incrementally learns from the inclusion of fresh data within W. Another benefit of the proposed approach is that the statistics retained by the rolling window W can be cumulative. By accumulating statistics on the contradiction analysis undertaken within each frame of the window as it rolls forward, the characteristics of the data are subtly captured from a long-run global perspective. Contradiction analysis can be improved by employing such global information, and it can possibly become more accurate in recognizing noisy data. In other words, the noise detection function becomes more experienced (by tapping into cumulative statistics) and refined in picking up noise. Noise is, of course, a relative concept, the identification of which requires an established reference.

3.2   Conflict Analysis

For contradiction analysis, a modified Pair-wise based classifier (PWC) is used that is based on the dependencies of the attribute values and the class labels. PWC is similar to instance-based classifier or lazy classifier which only gets activated for testing an instance and incrementally trains at most one round a classifier when the instance arrives. PWC has several incentives over other methods in addition to its fast processing speed which is a prerequisite for lightweight pre-processing. The advantages include simplicity in merely computing the supports and confidence values for estimating which target label one instance should be classified into, no persistent tree structure or trained model needs to be retained except small registers for statistics, and the samples (reference) required for noise detection can scale flexibly to any amount ($\leq W$). One example that is based on [17] about a weighted PWC is shown in Fig. 3.

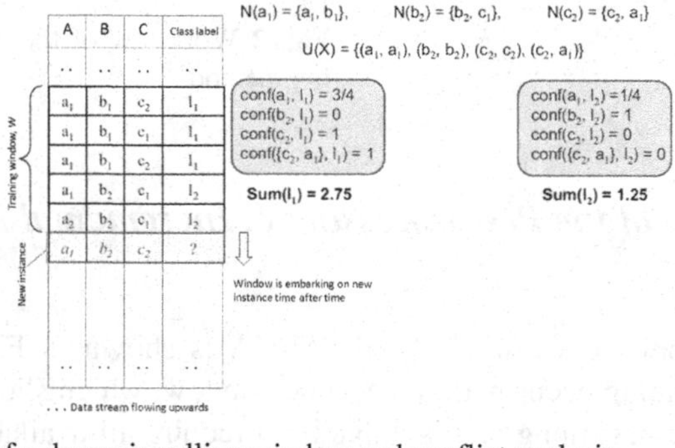

Fig. 3 Illustration of a dynamic rolling window and conflict analysis

In each round of ith iteration of incremental model update, where the current position is i, the sliding window contains W potential training samples over three attributes (A, B, C) and one target class. X is the new instance at the position $i+1$ which is just ahead of the preceding end of the window, and it has a vector of values $\{a_1, b_2, c_2\}$ as an example. Assume $k=W/2$ which rounds up to 2, the neighborhood sets for each attribute values of X are shown in the upper-right area of the figure. For example, $N(a_1)=\{a_1, b_1\}$ because $conf(a_1, a_1)=1$ and $conf(a_1, b_1)=0.75$ are the highest two for a_1. The resulting $U(X)$ set is found below. For instance, associated with a_1 in $U(X)$ is only a_1 itself, forming the pair (a_1, a_1). It does not belong in X and it shall be excluded from $U(x)$ although $b_1 \in N(a_1)$. The same applies for c_1 with respect to $N(b_2)$, whereas both c_2 and a_1, which belong in $N(c_2)$ are included in $U(X)$ associated with c_1. For each member of $U(X)$, PWC examines the confidence values against the two target classes l_1 and l_2. For instance, for (a_1, l_1) we calculate a $\mathrm{conf}(a_1, l_1) = \frac{support(\{a_1, l_1\})}{support(a_1)} = 3/4 = 0.75$, which checks a first-order dependency between a_1 and l_1. In contrast, for pair (c_2, a_1) we examine a second-order dependency by calculating $conf(\{c_2, a_1\}, l_1) = \frac{support(\{a_1, c_2, l_1\})}{support(a_1, c_2)} = 2/2 = 1$. Taking the sum of confidence values for each class, we obtain $Sum(l_1) = 2.75$ and $Sum(l_2) = 1.25$, therefore the new instance should belong to class l_1. In this way, conflict is determined by checking whether the calculated membership of the class matches the class label for the new instance. If the new instance is in agreement with the PWC calculation, no conflict is assumed and the window proceeds forward by incrementing one row, leaving out the last row and including the new instance in the training set. If the class label of the new instance contradicts to the result of the calculated class label, the new instance is deemed as a conflict and be purged.

One modification we made in our process is the version of neighbour sets. The current neighbor sets store only the most updated confidence values of the pairs within the current window frame, which are called Local Sets. The information in the Local Sets gets replaced (re-computed) by the new results every time when the window moves to a new position with inclusion of a new instance. In our design a similar buffer called Global Sets are used that do not replace but accumulate the newly computed confidence values corresponding to each pair in the window frame. Of course the conflict analysis can be implemented by similar algorithms. It can be seen that by using PWC the required information and calculation are kept as minimal as possible, which ensures fast operation.

4   Experiment

The objective of the experiment is to investigate the efficacy of the proposed iDSM-CA strategy. In particular we want to see how iDSM-CA works in classifying sonar signal data in data stream mining environment.

A total of seven algorithms were put under test of sonar classification using iDSM-CA. For traditional batch-based learning, representative algorithms include, Neural Network (NN) and Support Vector Machine (SVM). Instance based classifiers which learn incrementally as fresh data stream in, include Decision Table (DT), K-nearest Neighbours Classifier (IBK), K*- An Instance-based Learner Using an Entropic Distance Measure (KStar), and Locally Weighted Learning (LWL). An incremental version of algorithm that is modified from traditional Naive Bayesian called Updateable Naïve Bayesian (NBup) is included as well for intellectual curiosity.

The experiment is conducted in a Java-based open source platform called Weka which is a popular software tool for machine learning experiments from University of Waikato. All the fore-mentioned algorithms are available as either standard or plug-in functions on Weka which have been well documented in the Weka repository of documentation files (which is available for public download at http://www.cs.waikato.ac.nz/ml/weka). Hence their details are not repeated here. The hardware used is Lenovo Laptop with Intel Pentium Dual-Core T3200 2GHz processor, 8Gb RAM and 64-bits Windows 7.

The test dataset used is called "Connectionist Bench (Sonar, Mines vs. Rocks) Data Set", abbreviated as Sonar, which is popularly used for testing classification algorithms. The dataset can be obtained from UC Irvine Machine Learning Repository (http://archive.ics.uci.edu/ml/datasets). The pioneer experiment in using this dataset is by Gorman and Sejnowski where sonar signals are classified by using different settings of a neural network [18]. The same task is applied here except we use a data stream mining model called iDSM-CA in learning a generalized model incrementally to distinguish between sonar signals that are bounced off the surface of a metal cylinder and those of a coarsely cylindrical rock.

This sonar dataset consists of two types of patterns: 111 of them that are empirically acquired by emitting sonar signals and let them bounced off a metal cylinder at distinctive angles and under different conditions. The other 97 patterns are signals bounced off from rocks under similar conditions. The sonar signal transmitted is mainly a frequency-modulated acoustic chirp in increasing frequency. A wide variety of aspect angles at which the signals are transmitted, cover between 180 degrees for the rock and 90 degrees for the metal cylinder. Each pattern is made up of a vector of 60 decimal numbers [0, 1] as attributes or features of the bounced signal. Each attribute represents the amount of energy within a certain frequency band, collected over a fixed period of time. The target class that describes each record is binary, with the word "Rock" if the obstacle is a piece of rock or "Metal" if the object that reflected the signal is a metal cylinder. The attributes in the dataset or sorted in an ascending order of angles although the exact values of the angles are not encoded.

In our experiment we use this data set to test the learning speed, accuracy of prediction by the classification model and receiver operating characteristic (ROC) indices. Learning speed is measured by the average time-taken to build a classification model per model update or rebuilt in the case of batch learning. Accuracy is defined as the amount of correctly classified instances over the all the instances. ROC serves as a unified degree of accuracy ranging from 0 to 1, by exhibiting the limits of a test's ability (which is the power of discrimination in classification) to separate between alternative states of target objects over the full spectrum of operating conditions.

To start the experiment, the dataset is first subject to a collection of seven algorithms in the calibration stage for testing out the optimal size of W. In practice, only a relatively small sample shall be used, and calibration could repeat periodically or whenever the performance of the incremental learning drops, for fine-tuning the window size. In our experiment, the window size is set increasingly from the sizes of 49, 80 and 117 to 155. For simple convention, the various window sizes are labeled as 25%, 50%, 75% and 100% in relative to the full dataset.

The performance results of the seven classification algorithms that are run under the iDSM-CA are shown in the following charts, in terms of classification accuracy (c.f. Fig. 4), model induction time in seconds (c.f. Fig. 5) and ROC index (c.f. Fig. 6) respectively. Accuracy, in percentage is the direct indication of the accuracy of the classifier in discriminating the target objects, rock or metal. Model induction time implies the suitability of the applied algorithms in data stream mining environment. Ideally, the algorithms should take almost no time in updating/refreshing the model on the fly. When window size $W=0$, it means there is no conflict analysis to be done hence no instance is to be removed from the dataset.

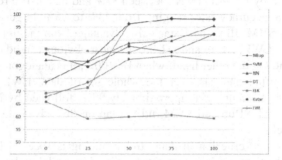

Fig. 4 Accuracy of sonar classification by various algorithms under iDSM-CA

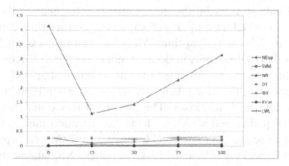

Fig. 5 Model induction time of sonar classification by various algorithms under iDSM-CA

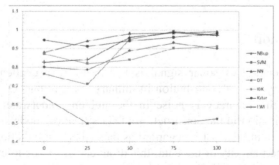

Fig. 6 ROC of sonar classification by various algorithms under iDSM-CA

The accuracies of the classification models differ by algorithms and by %W size. As we observe in Fig. 4, the traditional classification algorithms generally perform worse than the incremental group of algorithms, except NN. NN in general can achieve high accuracies over various W%. It is capable of outperforming the incremental algorithms besides LWL and IBK. NN's accuracy exceeds that of LWL in small W, but the accuracy declines when W becomes large. Under the scenario of stream mining, it is desirable to keep W minimum. IBK has the relatively highest accuracy in small window, followed by LWL, Kstar, NBup and DT. When W=0, it means there is no pre-processing for noise removal, IBK survives the best, followed by Kstar and NN (between 87% and 82%). The rest of the algorithms perform at accuracy rates ranging from 73% to 66%. In general, all the algorithms except SVM, IBK and Kstar gain advantage from the noise removal via conflict analysis. LWL, NBup and DT are the most obvious candidate in taking up the improvement. On the other hand, when W=100%, it is analogous to conducting a throughout conflict analysis over the whole set of data. LWL and DT both achieve a very high accuracy at approximately 98%, followed by NN, Kstar and IBK. It is clear that NBup and SVM did not tap on this advantage of full noise-removal very well.

Another important criterion in data stream mining is speed, which is mainly contributed by the time consumption required in each model update. In incremental learning, the frequency of model updates is assumed fixed; i.e., each time a new instance of data streams in, the model refreshes for once in the inclusion of the new instance in model training. Fig. 5 shows the average time consumption for each model update for each algorithm. All the algorithms, except NN, take approximately less than 0.4 seconds in doing a model update. NN requires the longest time when the data are not processed for noise-removal at W=0. The curve of NN in Fig. 4 shows that the time taken gets very short when W=0, and it graduate consumes more time as W increases. The last performance indictor, ROC index, implies about the quality and stability of the classification model. Overall every algorithm is able to produce a high ROC level, except SVM which performs poorly in this aspect especially when iDSM-CA applies. NN attains the best ROC, Kstar and LWL are also good candidates of incremental learning algorithms with respect to using conflict analysis.

5   Conclusion

Accurate classification of sonar signal is known to be a challenging problem though it has a significant contribution in military applications. One major factor in deteriorating the accuracy is noise in the ambient underwater environment. Noise causes confusion in the construction of classification models. Noisy data or instance in the training dataset is regarded as those that contradicting ones that do not agree with the majority of the data; this disagreement leads to erroneous rules in classification models and disrupts homogenous meta-knowledge or statistical patterns by distorting the training patterns. Other authors refer to noise as outliers, misclassified instances or misfits, all of which are data types the removal of which will improve the accuracy of the classification model. Though this research topic has been studied for over two decades, techniques previously proposed for removing such noise assume batch operations requiring the full dataset to be used in noise detection.

In this paper we propose a novel pre-processing strategy called iDSM-CA, that stands for incremental data stream mining with conflict analysis. The main advantage of the iDSM-CA lies in its lightweight window sliding mechanism designed for mining moving data streams. The iDSM-CA model is extremely simple to use in comparison with other more complex techniques such as those outlined in Section 2. Our experiment validates its benefits in terms of its very high speed and its efficacy in providing a noise-free streamlined training dataset for incremental learning, using empirical sonar data in distinguishing metal or rock objects.

References

1. Peyvandi, H., Farrokhrooz, M., Roufarshbaf, H., Park, S.-J.: SONAR Systems and Underwater Signal Processing: Classic and Modern Approaches. In: Kolev, N.Z. (ed.) Sonar Systems, pp. 173–206. InTech (2011)
2. China reveals its ability to nuke the US: Government boasts about new submarine fleet capable of launching warheads at cities across the nation. In: Daily Mail Online (2013), Available via Mail Online,
   http://www.dailymail.co.uk/news/article-2484334/China-boasts-new-submarine-fleet-capable-launching-nuclear-warheads-cities-United-States.html (cited November 11, 2013)
3. Akbarally, H., Kleeman, L.: A Sonar Sensor for Accurate 3D Target Localization and Classification. In: Proceedings of IEEE International Conference on Robotics and Automation, pp. 3003–3008 (1995)
4. Heale, A., Kleeman, L.: Fast target classification using sonar. In: Proceedings of 2001 IEEE/RSJ International Conference on Intelligent Robots and Systems, vol. 3, pp. 1446–1451 (2001)
5. Balleri, A.: Biologically inspired radar and sonar target classification. Doctoral thesis, University College London (2010)
6. Fong, S., Yang, H., Mohammed, S., Fiaidhi, J.: Stream-based Biomedical Classification Algorithms for Analyzing Biosignals. Journal of Information Processing Systems, Korea Information Processing Society 7(4), 717–732 (2011)
7. Zhu, X., Wu, X.: Class Noise vs. Attribute Noise: A Quantitative Study. Artificial Intelligence Review 22(3), 177–210 (2004)
8. Hodge, V., Austin, J.: A Survey of Outlier Detection Methodologies. Artificial Intelligence Review 22(2), 85–126 (2004)
9. Aggarwal, C., Yu, P.: Outlier Detection for High Dimensional Data. In: Proceedings of the ACM SIGMOD International Conference on Management of Data (2001)
10. Zhang, T., Ramakrishnan, R., Livny, M.: BIRCH: An Efficient Data Clustering Method for Very Large Databases. In: Procceedings of the Conference of Management of Data (ACM SIGMOD 1996), pp. 103–114 (1996)

11. Arning, A., Agrawal, R., Raghavan, P.: A Linear Method for Deviation Detection in Large Databases. In: Proceedings of 1996 Int. Conf. Data Mining and Knowledge Discovery (KDD 1996), Portland, OR, pp. 164–169 (1996)
12. Xiong, H., Pandey, G., Steinbach, M., Kumar, V.: Enhancing Data Analysis with Noise Removal. IEEE Transactions on Knowledge and Data Engineering (TKDE) 18(3), 304–319 (2006)
13. Brighton, H., Mellish, C.: Advances in Instance Selection for Instance-Based Learning Algorithms. Data Mining and Knowledge Discovery 6(2), 153–172 (2002)
14. Brodley, C.E., Friedl, M.A.: Identifying and Eliminating Mislabeled Training Instances. In: Proceedings of the 30th National Conference on Artificial Intelligence, pp. 799–805. AAI Press, Portland (1996)
15. John, G.H.: Robust Decision Tree: Removing Outliers from Databases. In: Proceedings of the First Int'l Conf. Knowledge Discovery and Data Mining, pp. 174–179 (1995)
16. Byeon, B., Rasheed, K., Doshi, P.: Enhancing the Quality of Noisy Training Data Using a Genetic Algorithm and Prototype Selection. In: Proceedings of the 2008 International Conference on Artificial Intelligence, Las Vegas, Nevada, USA, July 14-17, pp. 821–827 (2008)
17. Nanopoulos, A., Papadopoulos, A.N., Manolopoulos, Y., Welzer-Druzovec, T.: Robust Classification Based on Correlations Between Attributes. International Journal of Data Warehousing and Mining (IJDWM) 3(3), 1–17 (2007) ISSN: 1548-3924
18. Gorman, R.P., Sejnowski, T.J.: Analysis of Hidden Units in a Layered Network Trained to Classify Sonar Targets. Neural Networks 1, 75–89 (1988)

Tracking Performance of the Blind Adaptive LMS Algorithm for Fading CDMA Systems

Zahid Ali and Ahmad Ali Khan

Abstract. Reliable and accurate time delay estimation is an important signal processing problem and is critical to a diverse set of applications. Multi-user receivers in asynchronous Code Division Multiple Access (CDMA) systems require the knowledge of several parameters including delay estimates between users. In this paper, we address this problem by proposing a novel approach based on blind least mean squares (LMS) based early-late delay tracker. Analytical expressions have been derived and simulation results of the proposed delay tracker are compared with the classical delay locked loop (DLL) approach in a multipath fading scenario. These results show that the proposed delay tracker provides very good tracking performance in challenging cases of multipath delays.

Keywords: multiple access interference (MAI), delay locked loop, code acquisition, tracking, least mean square algorithm.

1   Introduction

CDMA time delay estimation has received much attention in current literature and many different approaches have been proposed. Joint estimation techniques for single and multiuser case have been also addressed in the literature [1] and as well as multipath scenario in [2]. Filtering methods based on Kalman filter and particle filter have also been investigated [3, 4]. Other techniques based on super resolution methods [5] and expectation maximization have also been researched [6].

The synchronization task can be divided into initial acquisition of relevant delays and subsequent tracking of acquired delays. Acquisition is used to coarsely align the received signal with the locally generated PN code to within one chip duration and then tracking is initiated to minimize the delay offset to maintain

Zahid Ali · Ahmad Ali Khan
KFUPM, Saudi Arabia
e-mail: zawali@ud.edu.sa, s201074140@kfump.edu.sa

S.M. Thampi, A. Gelbukh, and J. Mukhopadhyay (eds.), *Advances in Signal Processing*   183
and Intelligent Recognition Systems, Advances in Intelligent Systems and Computing 264,
DOI: 10.1007/978-3-319-04960-1_16, © Springer International Publishing Switzerland 2014

synchronization between the signals. Code tracking based on delay-locked loop (DLL) and the tau-dither loop (TDL) [7], have extensively been used. Other variations of the basic DLL and TDL have also been proposed.

The DLL is suitable for code tracking under additive white Gaussian noise (AWGN) channels, but it suffers severe performance degradation in the presence of MAI and multipath fading. The discriminator characteristic or S-curve of the DLL is distorted and randomly biased by the time-varying multipath and MAI. This results in a tracking bias harming the tracking capability of the loop and thus degrading receiver performance.

In this paper we present a modified DLL in a multiuser environment that employs an early and late channel with LMS-type algorithm for delay update.

2   The System Model

We consider here an asynchronous DS-CDMA system with BPSK modulation. The transmitted k^{th} user's data signal $s_k(t)$ in a DS/CDMA channel can be modeled by an equivalent complex baseband representation as

$$s_k(t) = \sqrt{2P_k} b_k(i) c_k(t - iT_b) \cos \omega_c t \tag{1}$$

where ω_c is the phase of the carrier, P_k is the power and $b_k(i)$ is the ith information bit transmitted by the kth user given by

$$b_k(t) = \sum_{i=-\infty}^{\infty} b_k(i) p_{T_b}(t - iT_c)$$

and $c_k(t)$ is the spreading waveform of the kth user,

$$c_k(t) = \sum_{m=0}^{N_c} c_k(m) p_{T_c}(t - mT_c)$$

where p_{T_c} is a pulse waveform of length T_c, and N_c is the number of chips in one spreading code period given as $N_c = {T_b}/{T_c}$. The spreading sequence is binary, i.e. $c_k(m) \in \{-1,1\}$, τ_k is the transmission delay of the kth user. It is assumed that τ_k is independent and uniformly distributed over $[0, T_b]$. The combined transmitted signal due to all K users in the channel is thus given by

$$s(t) = \sum_{k=0}^{K} s_k(t - \tau_k) h_k(t - \tau_k) \tag{2}$$

where $h_k(t)$ is the channel response associated with the kth user and is given by

$$h_k(t) = a_k(t)e^{j\phi_k(t)}\delta(t)$$

Where $a_k(t)$ is the amplitude response of the channel and $\phi_k(t)$ is the phase response associated with it. The received signal is given by

$$r(t) = s(t) + n(t) \tag{3}$$

$$= \sum_{k=0}^{K} s_k(t - \tau_k)h_k(t - \tau_k) + n(t)$$

$$= \sum_{k=0}^{K} \sqrt{2P_k}b_k(i_k)h_k(t - \tau_k)c_k(t - iT_b - \tau_k)\cos(\omega_c t(t - \tau_k)) + n(t)$$

where $i_k = \left\lfloor \dfrac{(t - \tau_k)}{T_b} \right\rfloor$ and $n(t)$ is complex base band additive white Gaussian

noise with zero mean and pass band two-sided power spectral density $N_o/2$.

3   Early Late Delay Tracking Algorithm

In this work we propose an algorithm based on early late delay tracking by introducing a block for LMS for update of the delay τ_k as shown in figure 2. The two channels, an early and a late channel, are used for the purpose of delay adjustment. Each channel has a bank of matched filter (MF) and a multiuser interference estimation block shown in figure 1.

The top channel is called the early channel as the relative delay to the MF bank is "earlier than" the estimated delay $\hat{\tau}$. Similarly, the other channel is called the late channel as the relative delay to the MF bank is "delayed than" the estimated delay $\hat{\tau}$.

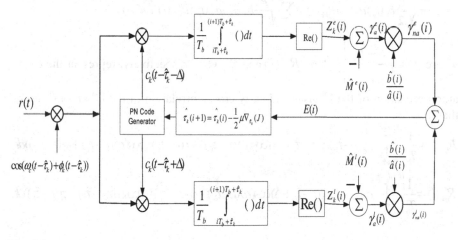

Fig. 1 Proposed Early Late delay tracking structure

Figure 2 shows how estimated delay $\hat{\tau}$ can be used to generate $\hat{M}^{e}(i)$ and $\hat{M}^{l}(i)$, where $\hat{M}^{e}(i)$ and $\hat{M}^{l}(i)$ are the early and late estimated interference from K-1 users, respectively.

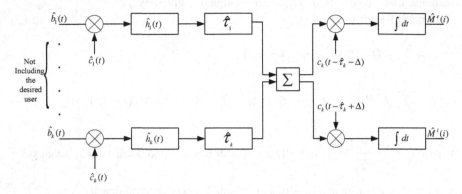

Fig. 2 Proposed $\hat{M}^{e}(i)$ and $\hat{M}^{l}(i)$ estimation block

Let $Z_{k}^{e}(i)$ and $Z_{k}^{l}(i)$, respectively, represent the output of early and late matched filters as shown in figure 2. Thus, $Z_{k}^{e}(i)$ for the kth user can be obtained by taking real part of the matched filter operation using k^{th} user's spreading waveform as follows:

$$Z_{k}^{e}(i) = Re\left\{ \frac{1}{T_{b}} \int_{iT_{b}+\hat{\tau}_{k}}^{(i+1)T_{b}+\hat{\tau}_{k}} r(t)c_{k}(t-\hat{\tau}_{k}+\Delta)\cos(\omega_{c}(t-\hat{\tau}_{k})+\phi_{k}(t-\hat{\tau}_{k}))\,dt\right\} \qquad (4)$$

$$= \sqrt{\frac{P_{k}}{2}}b_{k}(i_{k})a_{k}(i_{k})R_{kk}^{e}(i) + \sum_{\substack{m=0 \\ m \neq k}}^{k}\sqrt{\frac{P_{m}}{2}}b_{m}(i_{m})a_{m}(i_{m})R_{mk}^{e}(i) + n^{e}(i)$$

where $i_{k} = \left\lfloor \dfrac{t-T_{b}}{\tau_{k}} \right\rfloor$. Here $R_{kk}^{e}(i)$ and $R_{km}^{e}(i)$, respectively, represent the early auto correlation of the kth user and early cross-correlation of the kth and m^{th} user, that is,

$$R_{kk}^{j}(i) = \frac{1}{T_{b}} \int_{iT_{b}+\hat{\tau}_{k}}^{(i+1)T_{b}+\hat{\tau}_{k}} c_{k}(t-\tau_{k})c_{k}(t-\hat{\tau}_{k}+\Delta)\cos(\omega_{c}(t-\tau_{k})+\phi_{k}(t-\tau_{k}))\cos(\omega_{c}(t-\hat{\tau}_{k})+\phi_{k}(t-\hat{\tau}_{k}))dt$$

$$R_{mk}^{e}(i) = \frac{1}{T_{b}} \int_{iT_{b}+\hat{\tau}_{k}}^{(i+1)T_{b}+\hat{\tau}_{k}} c_{k}(t-\tau_{m})c_{k}(t-\hat{\tau}_{k}+\Delta)\cos(\omega_{c}(t-\tau_{m})+\phi_{m}(t-\tau_{m}))\times\cos(\omega_{c}(t-\hat{\tau}_{k})+\phi_{k}(t-\hat{\tau}_{k}))dt$$

Similarly, for the late channel, $Z_k^l(i)$, $R_{kk}^l(i)$ and $R_{km}^l(i)$ can be obtained as follows:

$$Z_k^l(i) = Re\left\{\frac{1}{T_b}\int_{iT_b+\hat{\tau}_k}^{(i+1)T_b+\hat{\tau}_k} r(t)c_k(t-\hat{\tau}_k-\Delta)\cos(\omega_c(t-\hat{\tau}_k)+\phi_k(t-\hat{\tau}_k))dt\right\} \quad (5)$$

$$= \sqrt{\frac{P_k}{2}}b_k(i_k)a_k(i_k)R_{kk}^l(i) + \sum_{\substack{m=0\\m\neq k}}^{k}\sqrt{\frac{P_m}{2}}b_m(i_m)a_m(i_m)R_{mk}^l(i) + n^l(i)$$

$$R_{kk}^l(i) = \frac{1}{T_b}\int_{iT_b+\hat{\tau}_k}^{(i+1)T_b+\hat{\tau}_k} c_k(t-\tau_k)c_k(t-\hat{\tau}_k-\Delta)\cos(\omega_c(t-\tau_k)+\phi_k(t-\tau_k))\cos(\omega_c(t-\hat{\tau}_k)+\phi_k(t-\hat{\tau}_k))dt$$

$$R_{mk}^l(i) = \frac{1}{T_b}\int_{iT_b+\hat{\tau}_k}^{(i+1)T_b+\hat{\tau}_k} c_k(t-\tau_m)c_k(t-\hat{\tau}_k-\Delta)\cos(\omega_c(t-\tau_m)+\phi_m(t-\tau_m))\times\cos(\omega_c(t-\hat{\tau}_k)+\phi_k(t-\hat{\tau}_k))dt$$

According to the figure 2, the adjusted matched filter output for the early channel, $\gamma_a^e(i)$, may be expressed as

$$\gamma_a^e(i) = Z_k^e(i) - \hat{M}^e(i) \quad (6)$$

$$\approx z^e(i) + \sqrt{\frac{P_k}{2}}b_k(i_k)a_k(i_k)R_{kk}^e(i)$$

Similarly, for the late channel, we have

$$\gamma_a^l(i) = Z_k^l(i) - \hat{M}^l(i) \quad (7)$$

$$\approx z^l(i) + \sqrt{\frac{P_k}{2}}b_k(i_k)a_k(i_k)R_{kk}^l(i)$$

Let $\hat{a}^e(i)$ and $\hat{a}^l(i)$ be the estimated complex amplitudes for the early and late channels, with $\hat{b}^e(i)$ and $\hat{b}^l(i)$ be the estimated symbols, then normalized adjusted output is

$$\gamma_{na}^e(i) = \frac{\gamma_a^e(i)}{\hat{a}^e(i)}$$

and

$$\gamma_{na}^l(i) = \frac{\gamma_a^l(i)}{\hat{a}^l(i)}$$

The error signal between early and late estimate of the desired symbol is given by

$$E(i) = \gamma_{na}^e(i)\hat{b}^e(i) - \gamma_{na}^l(i)\hat{b}^i(i) \tag{8}$$

$$= \sqrt{\frac{P_k}{2}}b_k(i)\hat{b}^e(i)\frac{a_k(i)}{\hat{a}_k(i)}R_{kk}^e(i) - \sqrt{\frac{P_k}{2}}b_k(i)\hat{b}^l(i)\frac{a_k(i)}{\hat{a}_k(i)}R_{kk}^l(i) + \tilde{n}(i)$$

if the amplitude and data bits are estimated close enough for each channel, then

$$\frac{a^e(i)}{\hat{a}^e(i)} \approx 1, \qquad \frac{a^l(i)}{\hat{a}^l(i)} \approx 1$$

$$b^e(i)\hat{b}^e(i) \approx 1, \qquad b^l(i)\hat{b}^l(i) \approx 1$$

so that

$$E(i) = \gamma_{na}^e(i) - \gamma_{na}^l(i)$$

$$\approx \sqrt{\frac{P}{2}}R_{kk}^e(i) - \sqrt{\frac{P}{2}}R_{kk}^l(i) + \tilde{n}(i)$$

where

$$\tilde{n}(i) = n^e(i) - n^l(i)$$

In the next section we derive expression for the delay update based on LMS algorithm.

4   Derivation of Blind Nonlinear LMS

This algorithm is termed as blind LMS algorithm as it does not have desired output available to calculate the update error and it has non-linearity because the delay is inside the cosine function. The LMS algorithm is most commonly used as an adaptive algorithm because of its simplicity and a reasonable performance. According to the well-known steepest descent approach [8], the LMS-type algorithm for the update of the delay estimate can be set up as follows [11]

$$\hat{\tau}_k(i+1) = \hat{\tau}_k(i) - \frac{1}{2}\mu\nabla_{\hat{\tau}_k}(J) \tag{9}$$

where μ is defined as the step size and J represents the cost function to be minimized which is chosen as the square of the error signal $E(i)$ and is given by

$$J = E[E^2(i)] \tag{10}$$

$$\nabla_{\hat{\tau}_k} J = \frac{\partial}{\partial \hat{\tau}_k} J$$

$$= 2E(i)\frac{\partial}{\partial \hat{\tau}_k} E(i)$$

$$= 2E(i)\left[R_{kk}^{e\,'}(i) - R_{kk}^{l\,'}(i) \right] \tag{11}$$

Now, for the above differentiations, we have used the following relation [9]

$$\frac{d}{da}\int_{\psi(a)}^{\varphi(a)} f(x,a)dx = f(\varphi(a),a)\frac{d\varphi(a)}{da} - f(\psi(a),a)\frac{d\psi(a)}{da} + \int_{\psi(a)}^{\varphi(a)} \frac{d}{da}f(x,a)dx$$

Moreover, in the differentiation of cosine terms we have used the approach of [10]

$$R_{kk}^{e\,'}(i) = \frac{\partial}{\partial \hat{\tau}_k} R_{kk}^{e}(i)$$

$$= \frac{\partial}{\partial \hat{\tau}_k}\left\{ \frac{1}{T_b}\int_{iT_b+\hat{\tau}_k}^{(i+1)T_b+\hat{\tau}_k} c_k(t-\tau_k)c_k(t-\hat{\tau}_k+\Delta)\cos(\omega_c(t-\tau_k)+\phi_k(t-\tau_k))\cos(\omega_c(t-\hat{\tau}_k)+\phi_k(t-\hat{\tau}_k))dt \right\}$$

$$= \frac{1}{T_b}\left\{ \begin{array}{l} c_k((i+1)T_b+\tau_k-\hat{\tau}_k)c_k((i+1)T_b+\Delta)\cos\left(\omega_c((i+1)T_b+\tau_k-\hat{\tau}_k)+\phi_k((i+1)T_b+\tau_k-\hat{\tau}_k)\right)\cos\left(\omega_c((i+1)T_b)+\phi_k((i+1)T_b)\right) \\[6pt] -c_k(iT_b+\tau_k-\hat{\tau}_k)c_k(iT_b+\Delta)\cos\left(\omega_c(iT_b+\tau_k-\hat{\tau}_k)+\phi_k(iT_b+\tau_k-\hat{\tau}_k)\right)\cos\left(\omega_c(iT_b)+\phi_k(iT_b)\right) \\[6pt] +\int_{iT_b+\hat{\tau}_k}^{(i+1)T_b+\hat{\tau}_k} c_k(t-\tau_k)\cos\left(\omega_c(t-\tau_k)+\phi_k(t-\tau_k)\right)\frac{\partial}{\partial \hat{\tau}_k}\left[c_k(t-\hat{\tau}_k+\Delta)\cos\left(\omega_c(t-\hat{\tau}_k)+\phi_k(t-\hat{\tau}_k)\right)\right]dt \end{array} \right\}$$

In order to evaluate the derivative $\dfrac{\partial}{\partial \hat{\tau}_k} c_k(t - \hat{\tau}_k - \Delta)$ we employ the methodology

of [10] to arrive at

$$\frac{\partial}{\partial \hat{\tau}_k} c_k(t - \hat{\tau}_k + \Delta) = \text{sign}\left(c_k\left(t - \left[\hat{\tau}_k\right] + \Delta \right) - c_k\left(t - \left[\hat{\tau}_k\right] + \Delta \right) \right)$$

As a result, the derivative of $R_{kk}^e(i)$ is found to be

$$
R_{kk}^e(i) = \frac{1}{T_b} \left\{
\begin{array}{l}
c_k((i+1)T_b + \hat{\tau}_k - \tau_k)c_k((i+1)T_b + \Delta)\cos\left(\omega_c((i+1)T_b + \hat{\tau}_k - \tau_k) + \phi_k((i+1)T_b + \hat{\tau}_k - \tau_k)\right)\cos\left(\omega_c((i+1)T_b) + \phi_k((i+1)T_b)\right) \\[2mm]
-c_k(iT_b + \hat{\tau}_k - \tau_k)c_k(iT_b + \Delta)\cos\left(\omega_c(iT_b + \hat{\tau}_k - \tau_k) + \phi_k(iT_b + \hat{\tau}_k - \tau_k)\right)\cos\left(\omega_c(iT_b) + \phi_k(iT_b)\right) \\[2mm]
+ \displaystyle\int_{iT_b + \hat{\tau}_k}^{(i+1)T_b + \hat{\tau}_k} c_k(t - \tau_k)\cos\left(\omega_c(t - \tau_k) + \phi_k(t - \tau_k)\right)
\left[
\begin{array}{l}
\mathrm{sign}\left(c_k\left[t - \left\lceil \hat{\tau}_k \right\rceil + \Delta\right] - c_k\left[t - \left\lfloor \hat{\tau}_k \right\rfloor + \Delta\right]\right)\cos\left(\omega_c(t - \hat{\tau}_k) + \phi_k(t - \hat{\tau}_k)\right) \\[2mm]
-c_k\left(t - \hat{\tau}_k + \Delta\right)\sin\left(\omega_c(t - \hat{\tau}_k) + \phi_k(t - \hat{\tau}_k)\right)\left(-\omega_c + \dfrac{\phi_k(t - (\hat{\tau}_k + \Delta \hat{\tau}_k)) - \phi_k(t - \hat{\tau}_k)}{\Delta \hat{\tau}_k}\right)
\end{array}
\right] dt
\end{array}
\right\}
$$

where $\Delta\hat{\tau}_k = \hat{\tau}_k - \hat{\tau}_{k-1}$

Similarly,

$$
R_{kk}'(i) = \frac{1}{T_b} \left\{
\begin{array}{l}
c_k((i+1)T_b + \hat{\tau}_k - \tau_k)c_k((i+1)T_b - \Delta)\cos\left(\omega_c((i+1)T_b + \hat{\tau}_k - \tau_k) + \phi_k((i+1)T_b + \hat{\tau}_k - \tau_k)\right)\cos\left(\omega_c((i+1)T_b) + \phi_k((i+1)T_b)\right) \\[2mm]
-c_k(iT_b + \hat{\tau}_k - \tau_k)c_k(iT_b - \Delta)\cos\left(\omega_c(iT_b + \hat{\tau}_k - \tau_k) + \phi_k(iT_b + \hat{\tau}_k - \tau_k)\right)\cos\left(\omega_c(iT_b) + \phi_k(iT_b)\right) \\[2mm]
+ \displaystyle\int_{iT_b + \hat{\tau}_k}^{(i+1)T_b + \hat{\tau}_k} c_k(t - \tau_k)\cos\left(\omega_c(t - \tau_k) + \phi_k(t - \tau_k)\right)
\left[
\begin{array}{l}
\mathrm{sign}\left(c_k\left[t - \left\lceil \hat{\tau}_k \right\rceil - \Delta\right] - c_k\left[t - \left\lfloor \hat{\tau}_k \right\rfloor - \Delta\right]\right)\cos\left(\omega_c(t - \hat{\tau}_k) + \phi_k(t - \hat{\tau}_k)\right) \\[2mm]
-c_k\left(t - \hat{\tau}_k - \Delta\right)\sin\left(\omega_c(t - \hat{\tau}_k) + \phi_k(t - \hat{\tau}_k)\right)\left(-\omega_c + \dfrac{\phi_k(t - (\hat{\tau}_k + \Delta \hat{\tau}_k)) - \phi_k(t - \hat{\tau}_k)}{\Delta \hat{\tau}_k}\right)
\end{array}
\right] dt
\end{array}
\right\}
$$

5   Simulation Results

For the purpose of simulation we have considered a typical asynchronous DS-CDMA reverse link with a Rayleigh fading channel for four-users-two path scenario. PN code synchronization follows a procedure of combined tracking/reacquisition tracking after a previous initial acquisition to within half a chip. The value of the update is chosen to be $\mu = 0.0125$ and the first arriving timing epoch has been detected for the estimation purposes.

First we consider the accuracy of DLL-based TOA estimation as shown in Figure 3 in the form of the histogram of the residual timing error at the mobile

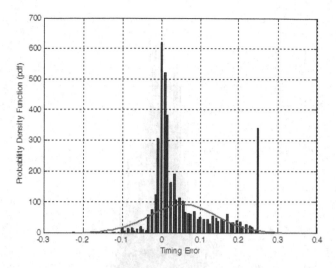

Fig. 3 Histograms for PDF's of DLL timing error (equal power users)

serving base station. It can be seen that the DLL timing error is affected with timing error distributed over $\pm T_c/2$ which is the same as initially assumed after the acquisition stage.

If we compare DLL estimate with the proposed delay tracker we immediately see improved results as shown in Figures 4 where the timing error axis has been zoomed in for error histogram. The timing error for both equal and unequal power

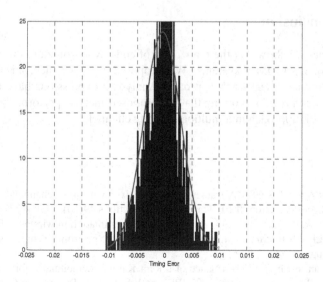

Fig. 4a Histogram of the timing error for LMS based delay tracker for equal power users

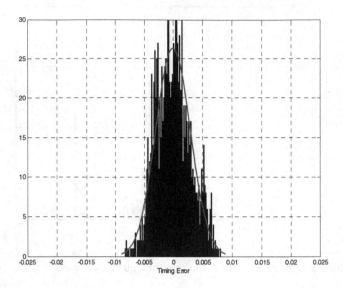

Fig. 4b Histogram of the timing error for LMS based delay tracker for unequal power users

users shows significant improvement converging to zero. It is also clear that the variance of timing error for the proposed structure is also less than the classical DLL structure.

6   Conclusion

In this paper we show that the realization of code synchronization is a challenging problem in DS-CDMA systems. We proposed a new synchronization method using a blind nonlinear LMS approach. Analytical expressions have been derived with simulation results showing that the proposed method performs better than the classical DLL approach in a multipath fading channel.

References

1. Iltis, R.A.: A DS-CDMA tracking mode receiver with joint channel/delay estimation and MMSE detection. IEEE Trans. Communications 49(10), 1770–1779 (2001)
2. Burnic, A., et al.: Synchronization and channel estimation in wireless CDMA systems. In: IEEE 9th International Symposium on Spread Spectrum Techniques and Applications, pp. 481–487 (2006)
3. Flanagan, B., et al.: Performance of a joint Kalman demodulator for multiuser detection. In: Proceedings of the 2002 IEEE 56th Vehicular Technology Conference, VTC 2002-Fall, vol. 3. IEEE (2002)

4. Ghirmai, T., et al.: Joint symbol detection and timing estimation using particle filtering. In: Proceedings of the 2003 IEEE International Conference on Acoustics, Speech, and Signal Processing, ICASSP 2003, vol. 4. IEEE (2003)
5. Ge, F.-X., et al.: Super-resolution time delay estimation in multipath environments. IEEE Transactions on Circuits and Systems I, 1977–1986 (2007)
6. Masmoudi, A., Billili, F., Affes, S.: Time Delays Estimation from DS-CDMA Multipath Transmissions Using Expectation Maximization. In: 2012 IEEE Vehicular Technology Conference (VTC Fall). IEEE (2012)
7. Glisic, S.G.: Adaptive WCDMA Theory and Practice. John Wiley & Sons (2003)
8. Yuan, Y.X.: A new step size for the steepest descent method. Journal of Computational Mathematics 24(2), 149–156 (2006)
9. Grandshteyn, I.S., Ryzhik, I.M.: Table of Integral, Series, and Products, 7th edn. Academic Press, Elsevier (2007)
10. Lim, T.J., Rasmussen, L.K.: Adaptive symbol and parameter estimation in asynchronous multiuser CDMA detectors. IEEE Transactions on Communications 45(2), 213–220 (1997)
11. Ali, Z., Memon, Q.A.: Time Delay Tracking for Multiuser Synchronization in CDMA Networks. Journal of Networks 8(9), 1929–1935 (2013)

Investigations on Indium Tin Oxide Based Optically Transparent Terahertz E-shaped Patch Antenna

S. Anand*, Mayur Sudesh Darak, and D. Sriram Kumar

Abstract. An optically transparent microstrip patch antenna is designed and its radiation characteristics are analyzed in 706 - 794 GHz band. Terahertz communication systems offer advantages such as broad bandwidth, low transmit power, secured wireless communication and compactness. It has applications in various fields like hidden object detection, imaging and sensing. In the proposed antenna, transparent conducting indium tin oxide thin film is used as a radiating patch and a ground plane material. The entire antenna structure is optically transparent in the visible spectrum region. The proposed antenna is simulated using Ansoft – HFSS, a finite element method (FEM) based electromagnetic solver.

Keywords: Transparent antenna, Terahertz, Transparent conducting materials, Patch antenna, Indium tin oxide, Terahertz antenna.

1 Introduction

Nanotechnology enables the system miniaturization and component development for future wireless communication systems [1-3]. Thin film deposition and nano lithography techniques made fabrication of MEMS and NEMS devices feasible [4, 5]. Today, advanced wireless sensors, actuators, tunable RF devices, etc. are realized at the nano scale, which has not only reduced the system size but has increased the overall efficiency and accuracy of the system [6, 7]. Future wireless communication systems require higher data rate transmission capability and low

S. Anand · Mayur Sudesh Darak · D. Sriram Kumar
Department of Electronics and Communication Engineering,
National Institute of Technology, Tiruchirappalli 620015, India.
email: {anand.s.krishna,darak.mayur}@gmail.com, srk@nitt.edu

* Corresponding author.

S.M. Thampi, A. Gelbukh, and J. Mukhopadhyay (eds.), *Advances in Signal Processing and Intelligent Recognition Systems*, Advances in Intelligent Systems and Computing 264,
DOI: 10.1007/978-3-319-04960-1_17, © Springer International Publishing Switzerland 2014

transmission power with reduced size [8]. The aforementioned requirements are met by usage of communication systems in terahertz spectrum, which offers secured communication with free spectrum availability [9]. Antennas for terahertz communication systems are being developed and studied for their application to various fields viz. astronomy, spectroscopy, imaging and sensing. Its military applications include screening of explosives and bio-hazards, detecting concealed objects and water contents [10, 11]. In the terahertz spectrum, the microstrip patch antenna is best suited due to its advantages such as low profile, light weight, low power requirement and conformity to planar and non-planar surfaces [12]. But transparent microstrip patch antenna has got some limitations like impedance bandwidth (< 5%), gain (< 2dB) and poor radiation efficiency [13].

The transparent conducting thin films are optically transparent and electrically conductive layers. They are widely used for the development of liquid crystal displays (LCD) [14], photovoltaic cells [15], light emitting diodes [16], optical displays [17] and transparent antennas [18]. Generally used transparent conducting materials for photo conductive devices are indium tin oxide (ITO), silver coated polyester film (AgHT), titanium-doped indium Oxide (TIO), gallium zinc oxide (GZO) and antimony tin oxide (ATO) [19, 20]. The challenge encountered in designing the transparent patch antenna is to optimize the electrical conductivity of the transparent conductive thin film while retaining its optical transmittance [21].

The work presented in this paper discusses design of an ITO based optically transparent microstrip patch antenna on a 20 μm thin substrate of polyimide in terahertz band (706 – 794 GHz). Following sections, present the optically transparent microstrip patch antenna design. Its impedance bandwidth, gain, directivity and radiation efficiency are analyzed in the desired band using the simulated results.

2 Transparent Patch Antenna

Microstrip antenna consists of a radiating patch, substrate and a ground plane. As shown in Fig. 1. The E-shaped patch of transparent conducting TIO is placed on a 20 μm thick substrate of transparent polyimide. Polyimide has a relative permittivity of 3.5. Transparent TIO ground plane is on the other side of the substrate. Thickness of the patch is selected such that it is 0.001 times the free space wavelength [22]. By tuning the arms of the E-shaped patch, antenna is made to resonate at 750 GHz. A microstrip line is used for impedance transition from coaxial cable to the patch of the antenna. This enables microstrip line to feed the power from coaxial cable to antenna patch with minimum mismatch loss. Antenna performance such as its bandwidth, radiation efficiency improves for substrates with lower dielectric constants and increasing thickness [23]. But this results in increased size of the antenna. Hence there is a tradeoff between performance parameters of antenna such as bandwidth, radiation efficiency and compactness of the antenna. The effect of fringing of electric field lines occurs at the edges of the patch as well as at the boundaries of microstrip line [24]. The majority of antenna radiation is from fields fringed at the edges of the patch. Most of the field lines near the boundaries of microstrip line are confined in the substrate. However, parts of lines reside in air. Thus, the effective dielectric constant to account for the combined effect of air and substrate media must be calculated [25]. The dimensions of the antenna structure are given in Table 1.

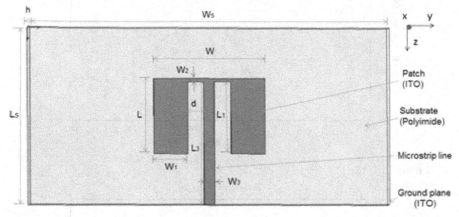

Fig. 1 ITO based optically transparent antenna

Table 1 Design parameters and dimensions of the optically transparent antenna

Parameters	Dimensions
E-shaped patch length ($L \times L_1$)	88.98 μm x 83.9 μm
E-shaped patch width ($W \times W_1 \times W_2$)	133.2 μm x 41.75 μm x 17. 9 μm
Patch thickness (t)	0.4 μm
Spacing (d)	5.08 μm
Substrate length and width ($L_s \times W_s$)	208.98 μm x 433.2 μm
Substrate thickness (h)	20 μm
Microstrip line length and width ($L_3 \times W_3$)	143.9 μm x 13.9 μm

3   Results and Discussion

The proposed transparent antenna structure is simulated using electromagnetic solver, Ansoft – HFSS. The return loss (S_{11}) and radiation characteristics are

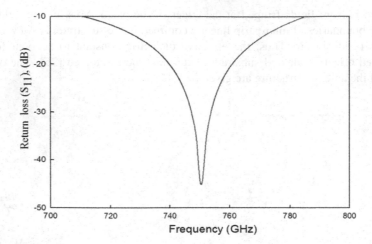

Fig. 2 Return loss (S_{11})

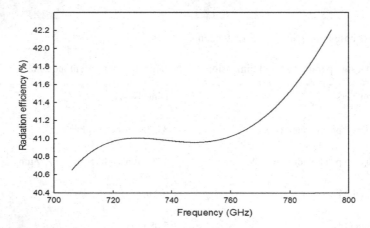

Fig. 3 Radiation efficiency (%)

investigated in 706 – 794 GHz band. As shown in Fig. 2, at resonance frequency of 750 GHz, the return loss of -40.19dB is observed. An impedance bandwidth of 11.73% is achieved with respect to centre frequency / resonance frequency in 706 – 794 GHz.

From Fig. 3, the radiation efficiency of more than 40% is achieved in desired band. At resonance frequency, it is observed that the radiation efficiency is achieved to be 40.96 %. The peak radiation efficiency of 42.2% is observed at 794 GHz. The radiation efficiency increases steeply after resonance frequency.

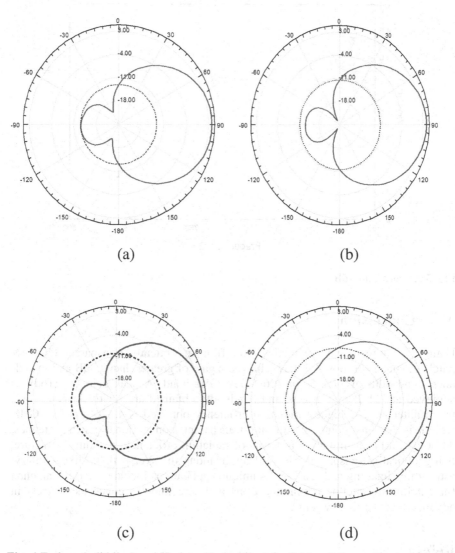

(a)   (b)

(c)   (d)

Fig. 4 E-plane (solid line) and H-plane (dashed line) far field radiation patterns at (a) 710 GHz, (b) 730 GHz, (c) 750 GHz and (d) 780 GHz.

The far field E and H plane radiation patterns for the designed antenna are shown in Fig. 4 (a) to (d). The antenna has minimum back lobe radiation of -13.38dB at 730 GHz.The plot of antenna gain versus frequency in the desired band is shown in Fig. 5. Peak gain of 4.16 dB is observed at 711 GHz. Over the entire band of interest, the gain greater than 2.8 dB is obtained.

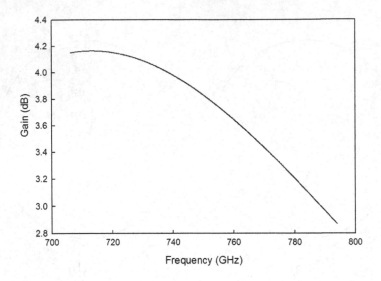

Fig. 5 Antenna Gain (dB)

4   Conclusion

Using ITO transparent conducting thin film an optically transparent terahertz patch antenna is proposed. It has achieved a gain of greater than 2.8dB and impedance bandwidth of 11.73% in 706 – 794 GHz band. A peak gain of 4.16dB is achieved at 711GHz. The gain and bandwidth obtained are relatively good for a patch antenna. The highest radiation efficiency obtained is 42.20% at 794 GHz, which is less as compared to non-transparent conventional patch antennas. Although, it has a limitation in terms of radiation efficiency, the entire structure, being optically transparent can find applications in diverse fields like military, astronomy, imaging and sensing. Its unique applications include solar cell antenna for satellite to satellite communications and screening of concealed objects in advanced recognition systems.

References

1. Dragoman, M., Muller, A.A., Dragoman, D., Coccetti, F., Plana, R.: Terahertz antenna based on graphene. Journal of Applied Physics 107, 104313–104313 (2010)
2. Llatser, I., Christian, K., Albert, C.-A., Josep, M.J., Eduard, A., Dmitry, N.C.: Graphene-based nano-patch antenna for terahertz radiation. Photonics and Nanostructures-Fundamentals and Applications 10, 353–358 (2012)
3. Jornet, J.M., Ian, F.A.: Graphene-based nano-antennas for electromagnetic nanocommunications in the terahertz band. In: 2010 Proceedings of the Fourth European Conference on Antennas and Propagation (EuCAP), pp. 1–5. IEEE (2010)

4. Tamagnone, M., Juan, S.G.-D., Juan, R.M., Julien, P.-C.: Reconfigurable terahertz plasmonic antenna concept using a graphene stack. Applied Physics Letters 101, 214102 (2012)
5. Huang, Y., Lin-Sheng, W., Min, T., Junfa, M.: Design of a beam reconfigurable THz antenna with graphene-based switchable high-impedance surface. IEEE Transactions on Nanotechnology 11, 836–842 (2012)
6. Kingsley, N., Dimitrios, E.A., Manos, T., John, P.: RF MEMS sequentially reconfigurable sierpinski antenna on a flexible organic substrate with novel DC-biasing technique. Journal of Microelectromechanical Systems 16, 1185–1192 (2007)
7. Yashchyshyn, Y.: Reconfigurable antennas by RF switches technology. In: 2009 5th International Conference on MEMSTECH 2009, pp. 155–157. IEEE (2009)
8. Tonouchi, M.: Cutting-edge terahertz technology. Nature Photonics 1, 97–105 (2007)
9. Sharma, A., Singh, G.: Rectangular microstirp patch antenna design at THz frequency for short distance wireless communication systems. Journal of Infrared, Millimeter, and Terahertz Waves 30, 1–7 (2009)
10. Kemp, M.C., Taday, P.F., Bryan, E.C., Cluff, J.A., Anthony, J.F., William, R.T.: Security applications of terahertz technology. In: International Society for Optics and Photonics, AeroSense 2003, pp. 44–52 (2003)
11. Galoda, S., Singh, G.: Fighting terrorism with terahertz. IEEE Potentials 26, 24–29 (2007)
12. Luk, K.M., Mak, C.L., Chow, Y.L., Lee, K.F.: Broadband microstrip patch antenna. Electronics Letters 34, 1442–1443 (1998)
13. Song, H.J., Tsung, Y.H., Daniel, F.S., Hui, P.H., James, S., Eray, Y.: A method for improving the efficiency of transparent film antennas. IEEE Antennas and Wireless Propagation Letters 7, 753–756 (2008)
14. Oh, B.-Y., Jeong, M.-C., Moon, T.-H., Lee, W., Myoung, J.-M., Hwang, J.-Y., Seo, D.-S.: Transparent conductive Al-doped ZnO films for liquid crystal displays. Journal of Applied Physics 99, 124505–124505 (2006)
15. Pasquier, A.D., Husnu, E.U., Alokik, K., Steve, M., Manish, C.: Conducting and transparent single-wall carbon nanotube electrodes for polymer-fullerene solar cells. Applied Physics Letters 87, 203511–203511 (2005)
16. Gu, G., Bulovic, V., Burrows, P.E., Forrest, S.R., Thompson, M.E.: Transparent organic light emitting devices. Applied Physics Letters 68, 2606–2608 (1996)
17. De, S., Thomas, M.H., Philip, E.L., Evelyn, M.D., Peter, N.N., Werner, J.B., John, J.B., Jonathan, N.C.: Silver nanowire networks as flexible, transparent, conducting films: extremely high DC to optical conductivity ratios. ACS Nano 3, 1767–1774 (2009)
18. Katsounaros, A., Yang, H., Collings, N., Crossland, W.A.: Optically transparent ultra-wideband antenna. Electronics Letters 45, 722–723 (2009)
19. Minami, T.: Transparent conducting oxide semiconductors for transparent electrodes. Semiconductor Science and Technology 20, S35 (2005)
20. Granqvist, C.G., Hultåker, A.: Transparent and conducting ITO films: new developments and applications. Thin Solid Films 411, 1–5 (2002)

21. Saberin, J.R., Cynthia, F.: Challenges with Optically Transparent Patch Antennas. IEEE Antennas and Propagation Magazine 54, 10–16 (2012)
22. Balanis, C.A.: Antenna theory: analysis and design. John Wiley & Sons (2012)
23. Jha, K.R., Singh, G.: Effect of low dielectric permittivity on microstrip antenna at terahertz frequency. Optik-International Journal for Light and Electron Optics 124, 5777–5780 (2013)
24. Pozar, D.M.: Microstrip antennas. Proceedings of the IEEE 80, 79–91 (1992)
25. Carver, K., Mink, J.: Microstrip antenna technology. IEEE Transactions on Antennas and Propagation 29, 2–24 (1981)

Efficient Hardware Implementation of 1024 Point Radix-4 FFT

Senthilkumar Ranganathan, Ravikumar Krishnan, and H.S. Sriharsha

Abstract. Since FFT algorithm is extremely demanding task and has several applications in the areas of signal processing and communication systems, it must be precisely designed to induce an efficient implementation of the parameters involving area and performance. To fulfill this requirement an optimized architecture is demonstrated in this paper for computing 1024-point, Radix-4 FFT using FPGA and is majorly compared with Xilinx LogiCore^{TM} FFT IP and found that proposed design is more efficient and effective in terms of area and performance. A novel architecture referred to as 2-D Vector Rotation and Complex Math Processor has been proposed in this paper. This single structure rotation helps in effectively carrying out the complex multiplications. The algorithm implements multiplexor hardware for computing the complex multipliers, thus consuming the minimal hardware resources. The entire RTL design is described using Verilog HDL and simulated using Xilinx ISim$^{[TM]}$. This experimental result is tested on Spartan-6 XC6SLX150T. The result shows 557 LUT's, 837 Flip Flops, 3 DSP Slices, Maximum Frequency of 215 MHz. This is about 52% improvement in resource usage and 5% upgrade in the performance.

Keywords: FPGA, FFT, 2-D Vector Rotation, Complex Math Processor.

Senthilkumar Ranganathan
Institution of Engineers (India)
e-mail: r.senthilkumar.in@ieee.org

Ravikumar Krishnan
KCG college of technology, Chennai
e-mail: r.ravikumar.be@gmail.com

H.S. Sriharsha
Asmaitha Wireless Technologies Pvt. Ltd
e-mail: sriharsha.suresh@gmail.com

S.M. Thampi, A. Gelbukh, and J. Mukhopadhyay (eds.), *Advances in Signal Processing*   203
and Intelligent Recognition Systems, Advances in Intelligent Systems and Computing 264,
DOI: 10.1007/978-3-319-04960-1_18, © Springer International Publishing Switzerland 2014

1   Introduction

DFT is a very complicated task as it involves huge complex hardware [1]. Fast Fourier Transform is a tool to travel between the time domain and frequency domain that has wide application in the area of Communication, Astronomy, Earth science and in Optics [2]. So, there's a need for low cost digital signal processing device which can compute FFT without negotiating on performance and flexibility ratio. Of the many interests in the Universal research, the prime is efficient utilization of area without compromising performance. There is an endless demand for FPGA as a result of its simplicity in implementation and low cost compared to ASIC. FFT can be effectively enforced in FPGA because it is a low volume application and cost efficient solution. In recent years, Fast Fourier transform of interest are Split Radix algorithms [3], Twisted Radix algorithms [4], Parallel FFT [5], Mixed-Radix algorithms [6], Vector-Radix [7]. This paper proposes a design of 1024 sampling point, DIF Radix-4 FFT [12, 13] considering area and power as the main factors. This design does not imply to any FFT architecture based on pipeline design [13, 14, 15, 16, 17], instead a unique architecture for computing the complex multiplications and additions has been administered using Complex Math processor and Rotational structures that is mentioned in the further sections. The design is compared with the prevailing Xilinx LogiCORE^{TM} Fast Fourier Transform IP and found that prescribed design is more effectual in terms of resource utilization and performance for the Spartan-6 for a speed grade of -2 thus making more appropriate in terms of arithmetic value and power. The rest of the work is organized as follows, section 2 describes the general architecture of Radix-4 FFT algorithm, section 3 explains the customised architecture and related optimization techniques. 4, describes the implementation and verification with FPGA board and elaborated comparison of results achieved with the Xilinx FFT IP core. The conclusion is given in section 5.

2   Radix-4 FFT Algorithm

The discrete Fourier Transform of a complex time series is given by,

$$X(k) = \sum_{n=0}^{N-1} x(n) W_N^{kn} \quad \text{where } k = 0,1,2...,N\text{-}1$$

(1)

According to the Divide and Conquer rule [8], the complex series will be decimated in time and frequency. In Radix-4 implementation, the problem is decimated by four time units with period of (N/4) after the twiddle factor generation [9]. The figure 1 shows the architecture of fundamental Radix-4 Butterfly unit. The four recurrent results of every butterfly unit is expressed as,

$$y(0) = (x(0) + x(1) + x(2) + x(3)) \tag{2}$$

$$y(1) = (x(0) - jx(1) - x(3) + jx(4)) * W_N^k \tag{3}$$

$$y(2) = (x(0) - x(1) + x(2) - x(3)) * W_N^{2k} \tag{4}$$

$$y(3) = (x(0) + jx(1) - x(3) - jx(4)) * W_N^{3k} \tag{5}$$

The detailed description of Radix-4 FFT is described in the paper [10].

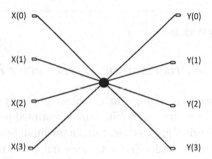

Fig. 1 Radix 4 Butterfly Diagram with 4 samples

3   Proposed Radix 4 Butterfly Architecture

The block diagram of four input sequence, Radix-4 Butterfly unit is shown in figure 2. This architecture contains the description of all the computations concerned in single Butterfly unit. Two complex Math Processors (CMPs) are used to perform the real and complex arithmetic's. Depending on the selector inputs, the output of CMP is obtained. The same selection lines are used for decisive addition or subtraction in the add/sub unit. The twiddle factors are multiplied with the output of complex adder by using the standard complex multiplier. The table describing the selection techniques is as shown below,

Table 1 Selection Technique performed by CMP

Sel[1:0]	CMP1	CMP2	ADD / SUB
00	x(0) + x(1)	x(2) + x(3)	c1 + c2
01	x(0) −jx(1)	x(2) −jx(3)	c1 − c2
10	x(0)-x(1)	x(2)-x(3)	c1 + c2
11	x(0)+jx(1)	x(2)+jx(3)	c1 - c2

Whenever the select line is 00, twiddle factor is considered as unity. Depending on the least significant bit of select inputs addition or subtraction is performed.

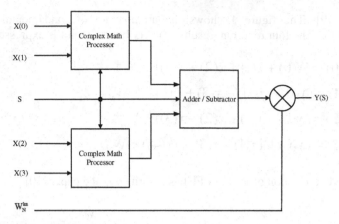

Fig. 2 Proposed Butterfly Architecture

3.1   Complex Math Processor and 2 – D Vector Rotation

This unit describes the methodology to calculate one single output of Radix-4 complex butterfly unit. A common architecture is outlined to compute all Radix-4 butterfly outputs considering just one twiddle factor input at a time. The processor simply uses the multiplexor, adder and subtractor hardware resources as shown in figure 3. The multiplication of imaginary term 'j' is achieved by the 2-D Vector Rotation technique. The vector rotation is performed for equation (3) and (5). The rotation is achieved using the multiplexor. The 2D quadrant is swapped with respect to (+j) and (-j) corresponding to +90 degree and -90 degree rotation. This sign change technique is incorporated using adder / subtractor circuit.

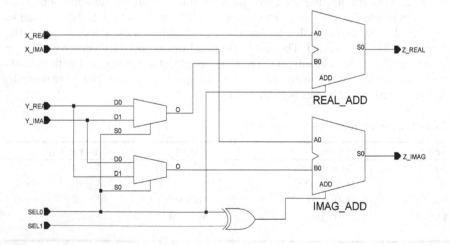

Fig. 3 Complex Math Processor Architecture

3.2 1024 Point DIF Radix-4 FFT Processor

The 1024 Radix-4 point FFT uses two Block RAMs for accessing input and output data. These two Block RAMs acts as a Ping-Pong Buffers. A multiplexor at the input side selects the source of data and stores it in Block RAM 1. After processing the sequence, Block RAM 2 holds the output data. Address Generator unit (ADG) is responsible for generating the address in every stage based on indexing. The twiddle factors are stored serially in the ROM and accordingly select lines are chosen. The outputs are generated sequentially from the Block RAM 2. The figure 4 shows the complete FFT Processor design.

Initially, at the input stage of the computation, Block RAM 1 is used for reading and Block RAM 2 will be idle. After computing the first stage of Butterfly unit, Block RAM 2 is used for writing the output sequences. The cycle of operation of two Block RAMs switches in every FFT stage as showed in the below table 2.

Table 2 Block RAM usage in every stage of computation

Stages	Block RAM 1	Block RAM 2
Input	Write	-
Stage 0	Read	Write
Stage 1	Write	Read
Stage 2	Read	Write
Stage 3	Write	Read
Stage 4	Read	Write
Output	-	Read

4 Implementation Results

The submitted novel design involving the Complex Math Processor and 2-D Vector Rotation is described in Verilog HDL and has been functionally verified. The timing simulations are carried out in Xilinx ISim™ as shown in figure 6. The entire RTL has been successfully synthesized using Spartan -6 XC6SLX150T FPGA.

The presented design is compared with Xilinx LogiCORE™ IP Fast Fourier Transform (v8.0) [11] and with [12] which is based on pointer FIFO embedded with gray code counters and found a major difference in performance and resource usage. The following table presents the architectural implementations when compared to the Xilinx LogiCore™ FFT IP synthesized for specific Spartan-6 architecture with speed grade of -2. The slice count, block RAM count, and XtremeDSP slice count are listed. The maximum clock frequency is also listed with the transform latency.

Implementation resources, as reported by Xilinx® ISE 14.5 synthesis tool, are summarized in table 3 for the proposed FFT Processor. The performance of the FFT core can be estimated by the clock cycle number, required to compute a full 1024-point input signal. There are 5 Butterfly stages containing 256 individual complex Butterflies. 1039 Clock cycle are required for the computation of each stage and then be schematized as shown in Figure 5. FFT processor runs at 215 MHz with the latency of 24.16µS as indicated in below table.

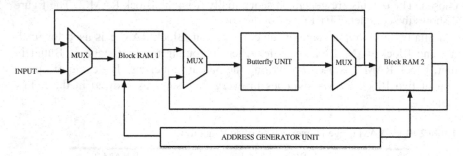

Fig. 4 Radix -4 FFT Processor

Table 3 Implementation Results of Xilinx LogiCORE IP and proposed design

Parameter	Xilinx LogiCORE FFT IP	Our Design
Channel	1	1
Point Size	1k	1k
Implementation	R4	R4
Configurable Point Size	NO	1024
Input Data Width	16	16
Phase factor Width	16	16
Scaling type	S	S
Rounding Mode	C	C
Output Ordering	N	N
Memory Type	B	B
Xilinx Part	XC6SLX150T	XC6SLX150T
LUT/FF Paris	1750	837
LUTs	1438	557
FFs	1608	837
9k Block RAMs	10	6
XtremeDSP slices	9	3
Max clock frequency	206	215
Latency (clock cycles)	3453	5195
Latency(Micro Sec)	18.27	24.16

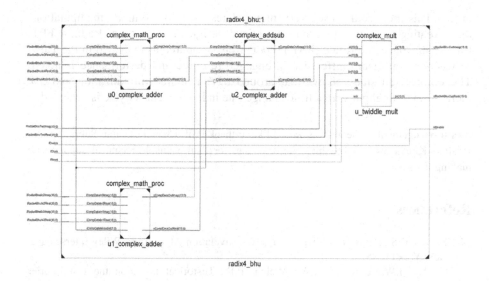

Fig. 5 RTL Design of 1024 Point DIF FFT Processor

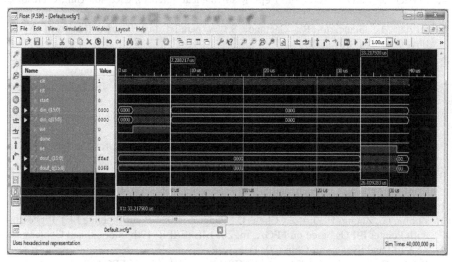

Fig. 6 Xilinx ISim[TM] Simulation Result of FFT Processor

5   Conclusion

The cumulative work of this paper is to formulate an FFT algorithm that would consume less area and achieve the greater frequency of operation than the Xilinx LogiCore[TM] FFT IP. This paper illustrates the custom designed Complex Math Processor and 2 – D Vector Rotation approach for multiplying the imaginary terms. This architecture uses very minimal resource for complex manipulations thus creating the low power and optimized design of 1024 point Radix-4 FFT. Finally the Place and Route (PAR) report were compared as indicated in the Table 3, where a significant difference in engaged resource and performance is noted. The comparison study shows that around 52% resource minimization and 5% increase in throughput with a little change rate in latency clock cycle is achieved.

Acknowledgements. The authors would like to thank Prof. Chandrasekaran Subramaniam, Jaraline Kirubavathy, Subhashini Vaidyanathan for their valuable technical support and making it success.

References

1. Burrus, C.S., Perks, T.W.: DFUFFT and Convolution Algorithms. Wiley Interscience, New York (1985)
2. Cooley, J.W., Lewis, P.A.W., Welch, P.D.: Historical notes on the fast Fourier transform. Proc. IEEE 55, 1675–1677 (1967)
3. Shu, C.J., Peng, X.: Dept. of Electron. & Eng. Tsinghua Univ., Beijing
4. Mateer, T.: Ph.D. dissertation, Dept. Mathematical Science., Clemson Uni., Clemson., SC
5. Silvia, M., Giancarlo, R., Gaetano, S.: A parallel fast Fourier transform. Mod. Phy. C 10, 781–805 (1999)
6. Grioryan, A.M., Againn, S.S.: Split mangeable efficient algorithm for Fourier and Hadamard transform. IEEE Trans. Signal Processs. 48(1), 172–183 (2000)
7. Chan, I.C., Ho, K.L.: Split vector-radix fast Fouriet transform. IEEE Trasn. Signal Process. 40, 2029–2040 (1992)
8. Cooley, J.W., Tukey, W.: An Algorithm for the Machine Calculation of Complex Fourier Series. Math. of Computations 19, 297–301 (1965)
9. Knight, W.R., Kaiser, R.: A Simple Fixed-Point Error Bound for the Fast Fourier Transform. IEEE Trans. Acoustics, Speech and Signal Proc. 27(6), 615–620 (1979)
10. Saidi, A.: Decimation-in-Time-Frequency FFT Algorithm. In: Proc. IEEE International Conf. on Acoustics, Speech, and Signal Processing, vol. 3, pp. 453–456 (1994)
11. Xilinx® Logic core™, Fast Fourier Transform V8.0, Xilinx® (2012)
12. Zhong, G., Zheng, H., Jin, Z., Chen, D., Pang, Z.: 1024-Point Pipeline FFT Processor with Pointer FIFOs based on FPGA. In: 2011 IEEE/IFIP 19th International Conference on VLSI and System-on-Chip (2011)
13. He, H., Guo, H.: The Realization of FFT Algorithm based on FPGA Co-processor. In: Second International Symposium on Intelligent Information Technology Application, vol. 3, pp. 239–243 (December 2008)
14. Oh, J.-Y., Lim, M.-S.: Area and power efficient pipeline FFT algorithm. In: IEEE Workshop on Signal Processing Systems Design and Implementation, November 2-4, pp. 520–525 (2005)

15. Wang, H.-Y., Wu, J.-J., Chiu, C.-W., Lai, Y.-H.: A Modified Pipeline FFT Architecture. In: 2010 International Conference on Electrical and Control Engineering (ICECE), pp. 4611–4614 (June 2010)
16. Sukhsawas, S., Benkrid, K.: A high-level implementation of a high performance pipeline FFT on Virtex-E FPGAs. In: Proceedings of the IEEE Computer society Annual Symposium on VLSI, February 19-20, pp. 229–232 (2004)
17. He, S., Torkelson, M.: Design and implementation of a 1024-point pipeline FFT processor. In: Proceedings of the IEEE 1998 Custom Integrated Circuits Conference, May 11-14, pp. 131–134 (1998)

Gaussian Approximation Using Integer Sequences

Arulalan Rajan, Ashok Rao, R. Vittal Rao, and H.S. Jamadagni

Abstract. The need for generating samples that approximate statistical distributions within reasonable error limits and with less computational cost, necessitates the search for alternatives. In this work, we focus on the approximation of Gaussian distribution using the convolution of integer sequences. The results show that we can approximate Gaussian profile within 1% error. Though Bessel function based discrete kernels have been proposed earlier, they involve computations on real numbers and hence increasing the computational complexity. However, the integer sequence based Gaussian approximation, discussed in this paper, offer a low cost alternative to the one using Bessel functions.

1 Introduction

Lindberg, in his work, [1], presents a family of kernels, which are the discrete analogue of the Gaussian family of kernels. The discrete Gaussian kernel in [1] uses modified Bessel function of integer order. It is well known that Bessel function evaluation is a computationally demanding process. This necessitates the need for

Arulalan Rajan
Dept. of Electronics & Communication Engg,
National Institute of Technology Karnataka, Surathkal
e-mail: perarulalan@gmail.com

Ashok Rao
Consultant
e-mail: ashokrao.mys@gmail.com

R. Vittal Rao · H.S. Jamadagni
Dept. of Electronics System Engg,
Indian Institue of Science, Bangalore
e-mail: {rvrao,hsjam}@cedt.iisc.ernet.in

S.M. Thampi, A. Gelbukh, and J. Mukhopadhyay (eds.), *Advances in Signal Processing*   213
and Intelligent Recognition Systems, Advances in Intelligent Systems and Computing 264,
DOI: 10.1007/978-3-319-04960-1_19, © Springer International Publishing Switzerland 2014

looking at computationally less-demanding alternatives to approximate Gaussian profiles. Interestingly, while exploring the use of integer sequences [2] for generating window functions for digital signal processing applications [3], [4], we found that the convolution of symmetrised integer sequences resulted in a Gaussian like profile. This made it worth to explore the degree, to which, a single convolution of two symmetrised integer sequences, approximate a Gaussian profile. It is well known that the computations on integers and integer sequences are less demanding in terms of power and complexity. The aim of this paper is to throw light on the use of non-decreasing integer sequences to approximate discrete Gaussian profile. In this regard, we provide two techniques, based on convolution of integer sequences, to approximate Gaussian distribution, as listed below:

- Symmetrising the integer sequences, followed by their linear convolution.
- Linear convolution of non decreasing integer sequences and symmetrising the resulting sequence about its maximum.

These techniques result in mean squared error, between the estimated probability density function and the obtained density function, of the order of 10^{-8} or equivalently about 1% error. We use some of the sequences listed in the Online Encyclopedia of Integer Sequences [2]. The following notations are used throughout this paper:

- N : Sequence Length
- $x_i[n], x_j[n]$: Sequences used in linear convolution, of length L and M respectively
- $y[n]$: Sequence resulting from the linear convolution of two integer sequences, given by

$$y[n] = \sum_{k=0}^{L-1} x_i[k]x_j[n-k] \quad n = 0, 1, 2, \ldots, L+M-1 \quad (1)$$

- X_{sc}^1, X_{cs}^2: Random variable that can assume values from the set $\{1, 2, 3, \ldots, N\}$, such that, the probability

$$P(X = n) = \frac{y[n]}{\sum_{k=1}^{N} y[k]} \quad \forall n = 1, 2, \ldots, N \quad (2)$$

In this paper, we restrict ourselves to integers, for the well known reason that, the computations on integers are much less complex than those on fixed point or floating point numbers. However, the comparison of complexity issues with regard to non-integer but fixed point methods with integers is beyond the scope of this paper.

---

[1] $(.)_{sc}$: Non decreasing input sequences are first symmetrized and then convolved
[2] $(.)_{cs}$: Non decreasing input sequences are first convolved and then symmetrized

2 Approximating Discrete Gaussian by Symmetrising the Integer Sequences, Followed by Convolution

2.1 Convolution of Symmetrized Arithmetic Progression Sequence

Consider, the arithmetic progression sequence $p[n] = a + nd$, where a, d, n are all integers. This sequence is labeled as A000027 in [2]. For a finite length, N, of the sequence, we symmetrize the sequence at $N/2$ or $(N+1)/2$, depending on whether the length is even or odd. Let us denote this symmetrized sequence by $x_s[n]$. This sequence is then convolved with itself to obtain $y[n]$. Let X_N be the random variable that can assume one of the values in the set $\{1, 2, 3, \ldots, 2N-1\}$. The probability that X_N can take a specific value, n, is given by

$$P(X_{sc} = n) = \frac{y[n]}{\sum\limits_{k=1}^{2N-1} y[k]}, \qquad \text{for all } n = 1, 2, 3, \ldots, 2N\text{-}1 \qquad (3)$$

The plot given in fig.1 illustrates that one can indeed approximate Gaussian distribution by the sequence obtained by convolution of symmetrized arithmetic progression sequence, with a mean squared error of the order of 10^{-8}. The obtained density function, in fig.1 is same as the convolution profile. For the profile obtained, we compute the mean, μ, and variance, σ^2. For this mean and variance, we then fit a Gaussian distribution, using the MATLAB inbuilt function $normpdf$. This corresponds to the estimated density function mentioned in fig.1.

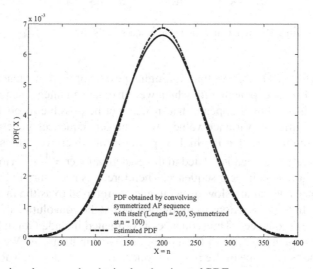

Fig. 1 Comparison between the obtained and estimated PDF

2.2   Convolution of n^2 and Higher Powers of n

We now look at integer sequences generated by higher powers of n, say n^2, n^3 etc.
Fig. 2 gives the profiles obtained by convolving symmetrized integer sequences generated by various powers of n. The distributed square error, is given in fig. 3. We find
that, the sequence n^2, after symmetrization and convolution, approximates discrete
Gaussian with an error of the order of 10^{-9} or even less. The sequence, n^3, does a
poor approximation of discrete Gaussian. Thus, it appears that there is an optimal
value of the exponent, k, between 2 and 3. However, for $2 < k < 3$, the sequence n^k
ceases to be an integer sequence.

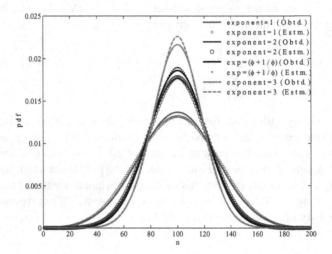

Fig. 2 Approximating Discrete Gaussian by sequences of the form n^k

Also, from fig. 4, we observe that, for higher powers of n, the variance saturates.
Moreover, the lowest exponent with which we can obtain an integer sequence in the
interval $(2,3)$ is 2. Thus it appears that, it may not be possible to obtain discrete
Gaussian with different variance values, with integer sequences generated by n^k,
for $k > 2$. We conjecture that the highest power k for which we can use n^k to approximate discrete Gaussian is related to the golden ratio, $\varphi = \frac{1+\sqrt{5}}{2}$. However, this
sequence will not be an integer sequence. Therefore, it is necessary to look at other
integer sequences which are slow growing and are referred to as the metafibonacci
sequences [5]. In the subsections to follow, we look at the convolution of other symmetrized integer sequences. The sequences investigated include Golomb sequence
[2], A005229 sequence [2], apart from those discussed in the paper. We present the
simulation results for some of the integer sequences in the following subsections.

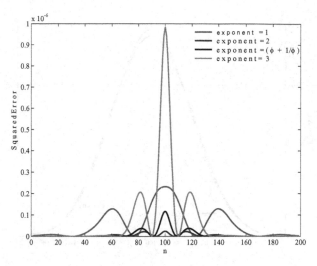

Fig. 3 Error Distribution in Approximating Discrete Gaussian by sequences of the form n^k

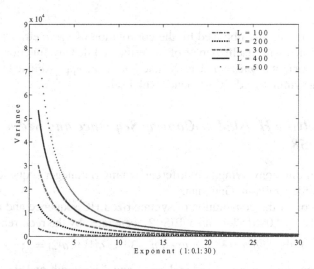

Fig. 4 Variance Saturates for higher values of k

2.3 Convolution of Symmetrized Hofstadter-Conway (HC) Sequence

The next sequence that we look at is the Hofstadter-Conway sequence [2], generated by the recurrence relation,

$$a[n] = a[a[n-1]] + a[n - a[n-1]] \qquad a[1] = a[2] = 1; \qquad (4)$$

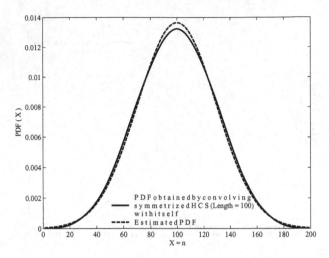

Fig. 5 Obtained PDF vs Estimated PDF

Fig.5 gives the profile obtained by the convolution of symmetrized Hofstadter Conway sequence with itself. A plot of the estimated density function is given in fig.5. This shows that a Gaussian density function can be approximated very closely by convolving symmetrized HC sequence with itself.

2.4 Convolving Hofstadter-Conway Sequence and Sequence A006158

We also found that convolving two different symmetric integer sequences also resulted in a profile similar to Gaussian.

Now, we look at the convolution of symmetrized HC sequence and another sequence labelled as A006158, in the OEIS [2], generated by the recurrence relation,

$$a[n] = a[a[n-3]] + a[n - a[n-3]] \quad a[1] = a[2] = 1; \quad (5)$$

Both the sequences were considered to be equal in length and the length was taken to be 100. From fig.5, we find that Gaussian distribution with a desired variance can be obtained by varying the length of the sequences to be convolved. Interestingly, from fig.6, we can infer that, to approximate a Gaussian distribution with a desired variance, using integer sequences, and with minimal mean squared error, there are two options, namely,

- varying the lengths, N_1 and N_2, of the two symmetric sequences
- choice of the sequences.

The plots indicate that one can closely approximate a Gaussian distribution of a specific variance by convolving symmetrized integer sequences. Fig.7 compares the

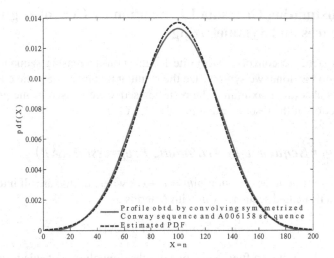

Fig. 6 Obtained PDF vs Estimated PDF

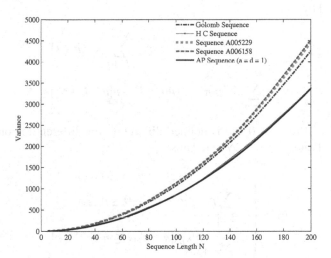

Fig. 7 Variance versus Sequence Length for Different Sequences

manner in which the variance, of distributions generated by convolving symmetrized integer sequences, depends on the length of the sequence.

From fig. 7 it is clear that we can obtain the length of the sequence required for any particular variance of the Gaussian PDF. However, this automatically fixes the mean. At present, it appears that only one of the two parameters - variance or mean, can be realised by the choice of the lengths of the sequences.

3 Approximating Gaussian Distribution by Convolving Integer Sequences and Symmetrizing

In this approach, we convolve two finite length, non-decreasing sequences. Once the convolution is done, we symmetrize the result at that N^*, where the convolution result has its absolute maximum. The resulting sequence is used as the probability density function of the discrete random variable, X.

3.1 Integer Sequence in Arithmetic Progression (AP)

Consider an arithmetic progression $p[n] = a + nd$, where a, d, n are all integers. We then define a truncated sequence $x[n]$ with N terms as

$$x[n] = p[n](u[n] - u[n-N]) \tag{6}$$

where $u[n]$ is the unit step function. Consider the convolution of this truncated sequence,(of length N) with itself. This results in a sequence, $y[n]$, of length $2N-1$ and is defined by

$$y[n] = \sum_{k=0}^{N-1} x[k]x[n-k] \tag{7}$$

$$y[n] = \sum_{k=0}^{N-1} p[k]p[n-k](u[n-k] - u[n-k-N]) \tag{8}$$

Clearly $y[n] = 0$ for $n < 0$ and is defined differently for different regions namely $0 \leq n \leq N-1$ and $N \leq n \leq 2N-1$. Thus

$$y[n] = \begin{cases} 0 & n < 0 \\ \sum_{k=0}^{n} p[k]p[n-k] & 0 \leq n \leq N-1 \\ \sum_{k=n-N+1}^{N-1} p[k]p[n-k] & N \leq n \leq 2N-1 \end{cases} \tag{9}$$

Evaluating the summations we get

$$y[n] = \begin{cases} 0 & n < 0 \\ a^2(n+1) + ad(n(n+1)) + d^2\left[\dfrac{n(n^2-1)}{6}\right] & 0 \leq n \leq N-1 \\ a^2(2N-n-1) + ad[n(2N-n-1)] + d^2 t_1 & N \leq n \leq 2N-1 \end{cases} \tag{10}$$

where

$$t_1 = \left[\frac{N(N-1)(3n-2N+1)-(n-N)(n-N+1)(n+2N-1)}{6} \right] \qquad (11)$$

The closed form of convolution of arithmetic progression is given in eq.10. The convolution plot is given in fig.8

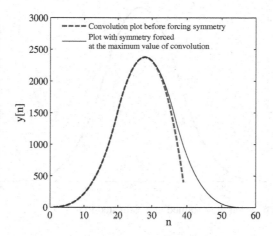

Fig. 8 Convolution of AP Sequences and Reflection about the index of maximum value

From fig.8, we find that, to get a Gaussian-like profile, we need to truncate the convolved sequence at the point of its absolute maximum and employ symmetry. To do so, it is necessary to obtain the point at which the absolute maximum occurs. Let us denote this point as N^*, the point at which, $y[n]$, defined by eq.10, attains its absolute maximum. It can be shown that the maxima is obtained at

$$N^* \approx \frac{(2N-1)}{\sqrt{2}} - \frac{(2-\sqrt{2})a}{d} \qquad (12)$$

Thus we get, from eq.12, the point at which the result of convolution of an AP sequence with itself has its maximum value. At this N^* we symmetrize the result of convolution. This results in a sequence, $y_1[n]$, which closely approximates the Gaussian curve. The length of $y_1[n]$ is $2N^*$. In this case, where we convolve first and then employ symmetry about the point of absolute maximum, the values that the random number X can assume is from the set $\{1, 2, 3, \ldots, N^*, \ldots, 2N^*\}$. Hence, the probability that the random variable X can take is defined by

$$P(X_{cs} = n) = \frac{y_1[n]}{\displaystyle\sum_{k=1}^{2N^*} y_1[k]} \qquad (13)$$

where $y_1[n]$ is the sequence obtained by the convolution of the AP sequence and employing symmetry at N^*. The following were the values taken for the AP sequence: $a = 1; d = 1; N = 100$.

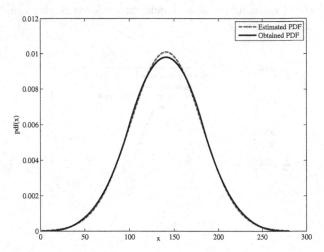

Fig. 9 Comparison between the obtained and estimated PDF

Fig.9 compares the obtained density function with the estimated one. Further investigations show that convolution of various other metafibonacci sequences indeed result in very close approximation of a Gaussian profile.

3.2 Validation Using Fourier Fit

To validate that the two techniques, presented in the previous sections, a four term Fourier fit was done. This involved the following steps:

- **Step 1:** The integer sequences were symmetrized and then convolved or vice-versa.
- **Step 2:** The mean and variance of the resulting sequence were obtained.
- **Step 3:** For that mean and variance, a discrete Gaussian distribution was obtained. The mean squared error was found out.
- **Step 4:** Assuming that it is a continuous distribution, a four term Fourier fit was obtained at random points. The number of points were the same as the length of the sequence resulting from step 1. However, the sample points were randomly chosen.
- **Step 5:** For this distribution, the mean and variance were obtained.
- **Step 6:** With this as the mean and variance, a Gaussian PDF was estimated.
- **Step 7:** The mean square error was obtained as the absolute difference between the profile obtained in step 4 and step 6.

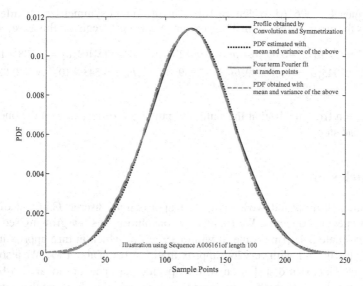

Fig. 10 Fourier Fit

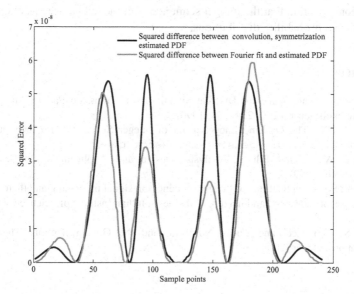

Fig. 11 Squared Error Plot

The mean square error was found to be of the order of 10^{-8}. This clearly shows that the sequence obtained by convolution of integer sequences or the sequence obtained by convolution of integer sequences and symmetrization, indeed, approximated discrete Gaussian distribution with 1% error. Fig.10 illustrates the same, using sequence A006161, of length 100, convolved and symmetrized, while fig.11 compares the error distribution. The Fourier series coefficients, in this case, are:

$$a_0 = 0.004277; a_1 = -0.005424; a_2 = 0.001288; a_3 = -0.0002101; a_4 = 9.588 \times 10^{-5};$$
$$b_1 = -0.0006025; b_2 = 0.000289; b_3 = -7.239 \times 10^{-5}; b_4 = 4.543 \times 10^{-5}; w = 0.02688;$$

It can be seen from fig.10 that the sampling points are different from the ones used in step 2 and step 3.

4 Conclusion

In this paper, we proposed two techniques to approximate discrete Gaussian distribution with integer sequences. We found that convolution of slow-growing sequences can approximate a 4 term cosine fit within 1% error, which in turn approximates a Gaussian distribution with very low approximation error of about 1%. We also found that discrete Gaussian distribution with a specific variance and mean can be controlled by varying the length of the integer sequences or by choosing the sequences or both. Also, as operations on integers are known to be less demanding in terms of computations, we find that the integer sequences can indeed be used as alternatives to approximate probability distributions.

References

1. Lindberg, T.: Scale-Space for Discrete Signals. IEEE Trans. on Pattern Analysis and Machine Intelligence 12(3), 234–254 (1990)
2. Sloane, N.J.A.: The On-Line Encyclopedia of Integer Sequences, Ed.2008 published electronically at, http://oeis.org/SequenceNumber
3. Antoniou, A.: Digital Filters: Analysis, Design and Applications, 2nd edn. Tata McGraw-Hill (1999)
4. Rajan, A.: Some applications of Integer sequences in signal processing and their implications on performance and architecture, PhD thesis, Indian Institute of Science (September 2011)
5. Vajda, S.: Fibonacci and Lucas Numbers, and the Golden Section: Theory and Applications, Dover (2008)

Inbuilt Multiband Microstrip Antenna for Portable Devices

Aditya Desai, Deepak C. Karia, and Madhuri Bhujbal

Abstract. With the recent introduction of the long term evolution (LTE) operation of mobile broadband services, the mobile devices like Smartphones, laptop computers and tablet computers are expected to meet the required bandwidths allocated for LTE/GSM/UMTS, GPS and WWAN applications. The proposed printed diamond-shaped-patch (DSP) antenna has the capability to provide multiband operation. To create this antenna, several narrow strips acting as resonance paths can be integrated with the DSP antenna. By changing the length of the added resonant strips in the notched region, the center frequency of the multi resonances below the UWB frequency can be fine tuned. This antenna is designed to merge multiple operating bands of at least 850MHz, 1.5, 2.4, 3.1 – 10.6 GHz with a return loss better than 6 dB thereby avoiding usage of multiple antennas.Proposed antenna is planner and will be printed on both the sides of FR4 substrate which adds compactness to antenna that can be embedded easily in any portable devices such as the Smartphones, laptop computers and tablet computers.

Keywords: Small antenna, multi-band antenna, wireless communication frequencies, LTE/GSM/UMTS, Ultrawideband (UWB) antenna, GPS, WWAN.

1   Introduction

Communication systems require multiple antenna to support several allocated wireless frequency bands. Therefore the design of a multi-band antenna which also covers the UWB range without deteriorating the UWB performance is of interest. Also, a major issue in communication systems is to reduce the antenna size while providing good performance over all the multiple bands. Planar monopole antennas, due to their attractive features such as low cost, simple structure, ease of fabrication, wide bandwidth, and omnidirectional radiation pattern have received great attention for UWB systems [1], [2].

Aditya Desai · Deepak C. Karia · Madhuri Bhujbal
Sardar Patel Institute of Technology, Mumbai University, Mumbai, India
e-mail: {aditya_d15,bhujbalmadhuri}@yahoo.com,
    deepakckaria@gmail.com

S.M. Thampi, A. Gelbukh, and J. Mukhopadhyay (eds.), *Advances in Signal Processing*   225
and Intelligent Recognition Systems, Advances in Intelligent Systems and Computing 264,
DOI: 10.1007/978-3-319-04960-1_20, © Springer International Publishing Switzerland 2014

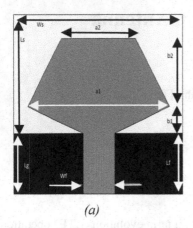 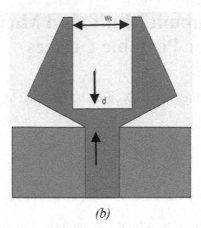

(a)   (b)

Fig. 1 (a) DSP UWB antenna. (b) DSP antenna with notched region

Two techniques are reported in the literature to create multiband printed monopole antennas. In the first, a patch antenna is designed to cover the desired wide bandwidth. The lower limit of the frequency band increases the size of the antenna to be large. To create the multi-band behavior, notches are then introduced into the antenna. In the second technique, a small-size antenna can be designed to cover the highest frequency band and by adding extra resonant elements to the main body, lower frequency bands can be created. In this paper the corner of a UWB monopole antenna, an L-shaped resonant element is added to integrate the lower frequency band with the existing UWB antenna [3].

In this communication, a base diamond-shaped-patch (DSP) antenna is considered to cover the entire UWB range. By inserting a notched region in the middle part of the DSP antenna and adding quarter-wavelength strips, a multi-band antenna is achieved. The novelty of the present method is that Parallel Resonant (PR) strip are applied to base antenna without affecting the UWB behavior and get a wide lower band at about 900 MHz, this is because the PR strip can be directly connected to the antenna's radiating patch, hence making the antenna easy to implement. The antenna covers WWAN or WLAN operation, and most of the band in LTE/GSM/UMTS operation. The size of the proposed multi-band antenna substrate is kept the same as the base antenna [4] -[6]. The simulation is carried out via the HFSS software package [7].

2   Proposed Antenna Design

The structure of a planar radiator that is used as the base for the eventual multi-band antenna, operating over an entire UWB frequency range, is shown in Fig. 1 (a). This base UWB antenna can be referred to as the DSP antenna. The antenna uses FR4 substrate with a dimension of 16 x 22 x 1 mm, with a dielectric constant of 4.4 and a loss tangent of 0.02. The width W_f of the Microstrip fed line is fixed at 1.86 mm to achieve 50 ohm characteristic impedance, and is connected to the DSP via a line of length L_f. A simple rectangular conducting ground plane of width W_s and length L_g is placed on the other side of the substrate.

In designing the multi-band antenna, the DSP antenna is considered to cover the UWB range. As shown in Fig. 1 (a), there are four parameters a1, a2, b1 and b2 that could affect the performance of the UWB antenna. To obtain a compact size antenna that covers the UWB range, parametric studies for these four parameters are carried out and results of reflection coefficient are presented [1].

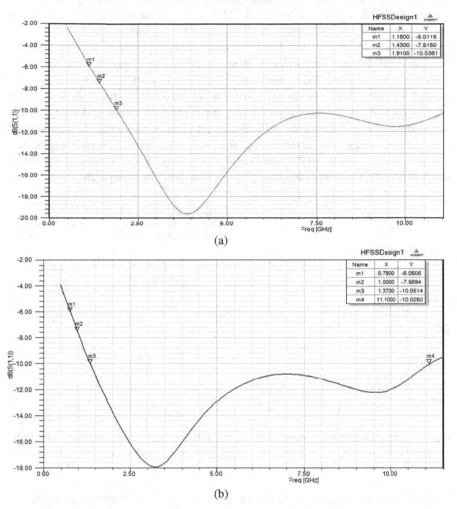

(a)

(b)

Fig. 2 (a) Simulated reflection coefficient of the simple DSP antenna (without the notched region) for a1 = 14 mm, a2 = 5 mm, b1 = 2 mm and b2 = 8 mm. (b) Simulated reflection coefficient of the simple DSP antenna with the notched region for d = 1 mm and W_c = 4 mm.

From Fig. 2 (a), it is clear that the DSP antenna of Fig. 1 (a) should be broader along the lower side (towards ground plane) to improve the reflection coefficient characteristic over the desired band. Furthermore, the value of b1 and b2 changes the lower cutoff frequency of the DSP antenna. An appropriate value of a1 can improve the reflection coefficient at higher frequencies and provide wider impedance matching over the UWB range.

The current distribution over this DSP antenna is mostly concentrated over the outside edges of the patch with negligible current in the center region. Thus, a rectangular section along the axis of the DSP antenna can be removed as shown in Fig. 1 (b), without affecting the overall antenna impedance and radiation characteristics leading to the UWB antenna. This adjustment also ensures the asymmetrical design of the antenna. Results of Fig. 2 (b) show that the parameter does not have a significant effect on the reflection coefficient. But it is large enough so that when a rectangular section is removed, as shown in Fig. 1 (b), several resonant strips can be placed within the notched region to design a multi-band antenna. Fig. 3 shows the reflection coefficient of the base DSP antenna and two different sizes of notched region. It is seen that narrow notched region has almost no effect on the reflection coefficient results. Also, increasing the notch size has no effect on cutoff frequency of the base DSP antenna patch.

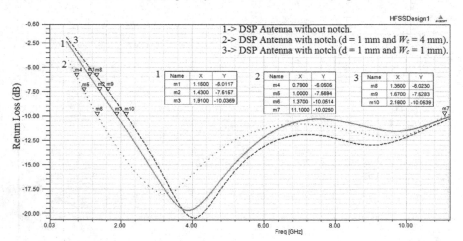

Fig. 3 Simulated reflection coefficient of the DSP antenna before and after inserting notched region for various widths of the cut out section W_c. (a1 = 14 mm, a2 = 5 mm, b1 = 2 mm, b2 = 8 mm, L_f = 8mm and Lg = 8 mm).

Center feed method is used to design small size dual and triple band antennas as it is desired to have other frequency bands at specific wireless communication frequencies below the UWB [2]. A notched region is introduced in a UWB base antenna, resulting in reducing the effect of ground plane and RF cabling loss at lower frequencies. After the creation of the notched region, the resonant length of the created path is $\lambda/2$. As shown in the previous section, one can insert a notched region into the base patch antenna without affecting the characteristics of the antenna. To obtain multi-band behavior, additional resonant strips can be placed in the notched region. These strips are efficiently excited, if placed along the direction of the current flow and near the main Microstrip feed line. This technique of creating a multi-band antenna can be referred to as the center feed method.

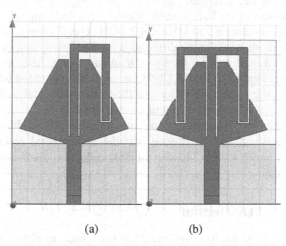

(a)   (b)

Fig. 4 Multi-band DSP antenna configurations for (a) dual-band, (b) triple band applications

Fig. 4 (a) shows a dual band antenna structure and Fig. 4 (b) shows a triple band antenna structure using this center feed method. By using the center feed method, a segment of the added strips can be placed within the notched region of the base patch while the remaining sections are brought back towards the patch. Even though strips are added to the base antenna, the overall substrate dimension has not changed, leading to small-size multi-band antenna.

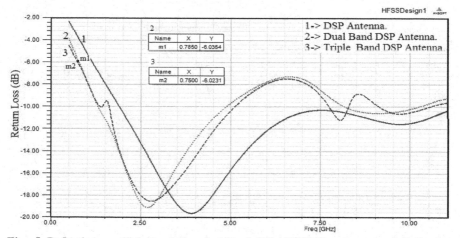

Fig. 5 Reflection coefficient of the dual-band WLAN/UWB antenna and Reflection coefficient of the tri-band GSM/WLAN/UWB antenna

In this section, the design of a dual- and triple-band antenna is presented. The structures of the proposed antennas are shown in Fig. 4. In all these structures, the width of the slot lines, between the added strips and the base patch, is set at 0.2 mm and has negligible effect on the antenna performance. Fig. 5 shows the reflection coefficient of a dual-band antenna with a single strip.

The results shown in Fig. 5 are suitable for WLAN/UWB applications. To obtain resonant frequencies between 750 MHz and 2.4 GHz (i.e. To integrate GPS, GSM or WLAN with UWB antenna) for the dual and triple-band structure, the length of the strips are to be set accordingly for the lower frequency band. Also, the dimension of the DSP antenna is kept fixed for the UWB range. As the operation frequency of the antenna increases, the guided wavelength of the structure decreases, resulting in the feed line length being several times that of the wavelengths.

The dimensions of the two strips are considered to have triple-band GSM/WLAN/UWB antenna. It is seen that the triple- band antenna has good resonances at 1.4 and 2.4 GHz and also suitable reflection coefficient below 6 dB over the UWB range.

3   Result and Discussion

The HFSS full-wave simulation software has been used to obtain the simulation data [7]. The studies are developed through optimizing the parameters step by step. Fig. 5 displays the simulated and measured return loss, from 750 to 1005 MHz, from 1670 to 3000 MHz and 3.1 to 10.6 GHz with VSWR < 3, the bandwidth of the desired lower and upper bands are observed, respectively [4].

The radiation patterns of the base DSP, dual- and triple-band antennas are investigated. It is seen that the multi-band antennas provide omnidirectional radiation patterns in the H-plane (x-z plane) and stable patterns in the form of figure-eight in the E-plane (y-z plane).

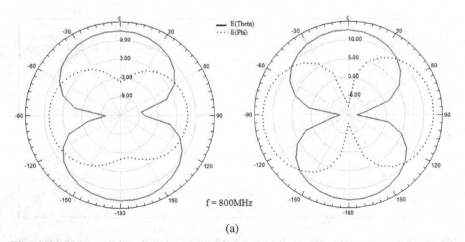

(a)

Fig. 6 E - Plane radiation pattern of the triple-band antenna at various frequencies of (a) 800MHz, (b) 1.3GHz, (c) 1.8 GHz, (d) 2.4 GHz, (e) 3 GHz, (f) 5 GHz, and (g) 8 GHz.

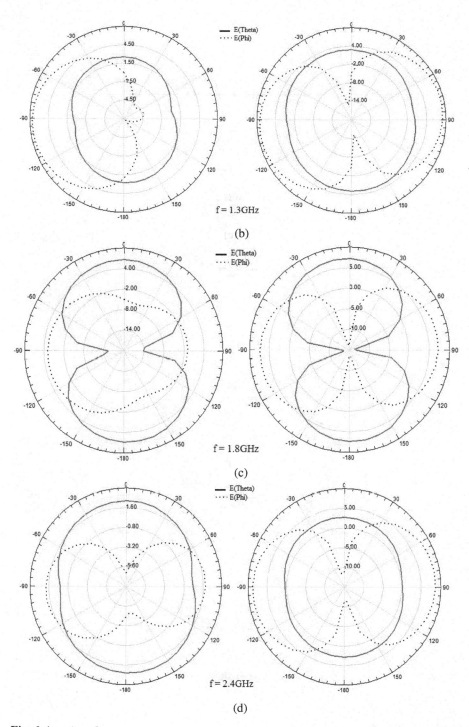

f = 1.3GHz

(b)

f = 1.8GHz

(c)

f = 2.4GHz

(d)

Fig. 6. (*continued*)

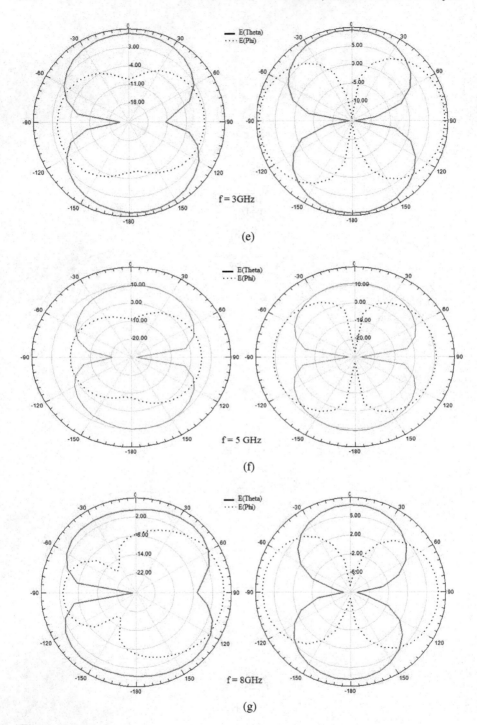

f = 3GHz

(e)

f = 5 GHz

(f)

f = 8GHz

(g)

Fig. 6. (*continued*)

These results are similar to those of ordinary dipole antennas. Also, the measured and simulated E - plane radiation patterns of the proposed triple-band antenna of Fig. 6, at the desired frequencies of 0.8, 1.3, 1.8, 2.4, 3, 5 and 8 GHz are shown in Fig. 6. It is seen that the proposed antenna has stable radiation pattern characteristics over the triple-band frequencies.

The three-dimensional (3-D) total-power radiation patterns are plotted in Fig. 7 (a). The full 3-D pattern (the plane where the display ground is located) for six representative frequencies are shown. At lower frequencies (800, 1300 and 1700 MHz in the bands), near-omnidirectional radiation is observed. While at higher frequencies (2400, 3000, 5000 and 8000 MHz in the bands), more variations are seen in the radiation patterns. This behavior is similar to that observed for the internal WWAN handset antennas [3]. The radiation patterns are not normalized to make it easy to read the gain of both components.

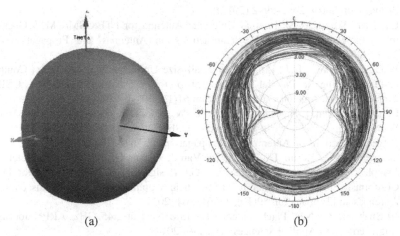

(a)   (b)

Fig. 7 (a) Simulated 3D Radiation Pattern of the proposed antenna. (b) Simulated 2D Radiation Pattern of the proposed antenna

To analyze the effect of the resonant strip, Fig. 5 shows the comparison of the simulated return loss of the proposed antenna and the reference antenna [1]. From the results shown in Fig. 4 (a), it is seen that major bands for LTE/2300/2500, GSM850/900/UMTS, GSM/1800/1900/UMTS, and UWB (3.1-10.6 GHz) are supported by the proposed antenna.

4   Conclusion

A novel diamond shaped Microstrip patch antenna with a small-size, planar structure has been proposed. The technique of using a Parallel resonant strip, which generally does not increase the antenna size, has been shown to be promising in generating a resonant mode to enhance the bandwidth of the antenna's lower band. By changing the length of the added resonant strips in the notched region, the center frequency of the multi resonances below the UWB frequency can be finely tuned. The antenna can be easily printed on both sides of a thin FR4 substrate of size 16 X 22 X 1 mm only and support frequency bands from 0.85, 1.5, 2.4 and 3.1–10.6 GHz that covers the GPS, GSM, UMTS, LTE, WLAN and the UWB, respectively. The dual and triple-band antennas show stable omnidirectional radiation patterns over entire considered frequency bands. With the obtained results, the antenna is promising for practical portable devices, like tablet computer, cell phones, and various handheld GPS and other devices.

References

1. Foudazi, A., Hassani, H.R., Nezhad, S.M.A.: Small UWB Planar Monopole Antenna With Added GPS/GSM/ WLAN Bands. IEEE Transactions on Antennas and Propagation 60(6), 2987–2992 (2012)
2. Lu, J.-H., Wang, Y.-S.: Internal Uniplanar Antenna for LTE/GSM/UMTS Operation in a Tablet Computer. IEEE Transactions on Antennas and Propagation 61(5), 2356–2364 (2013)
3. Wong, K.-L., Wu, T.-J., Lin, P.-W.: Small-Size Uniplanar WWAN Tablet Computer Antenna Using a Parallel-Resonant Strip for Bandwidth Enhancement. IEEE Transactions on Antennas and Propagation 61(1), 492–496 (2013)
4. Ban, Y.-L., Chen, J.-H., Sun, S.-C., Li, J.L.-W., Guo, J.-H.: Printed Monopole Antenna With a Long Parasitic Strip for Wireless USB Dongle LTE/GSM/UMTS Operation. IEEE Transactions on Antennas and Propagation 11, 767–770 (2012)
5. Scarpello, M.L., Kurup, D., Rogier, H., Van de Ginste, D., Axisa, F., Vanfleteren, J., Joseph, W., Martens, L., Vermeeren, G.: Design of an Implantable Slot Dipole Conformal Flexible Antenna for Biomedical Applications. IEEE Transactions on Antennas and Propagation 59(10), 3556–3564 (2011)
6. Al-Shaheen, A.: New Patch Antenna for ISM band at 2.45 GHz. ARPN Journal of Engineering and Applied Sciences 7(1), 1–9 (2012)
7. Ansoft Corporation HFSS, http://www.ansoft.com/products/hf/hfss

Mean-Variance Blind Noise Estimation for CT Images

Alex Pappachen James and A.P. Kavitha

Abstract. Noise estimation is a precursor to de-noising techniques to improve the signal and visual quality of medical images. We present a noise estimation algorithm using the local image statistics of the CT images at voxel level. The algorithm calculates the local mean variance distribution and detects the minimised error rates for identifying the tolerance range of voxel to artificial noises. The reliability of the method is experimentally verified using Gaussian noise and Speckle noise on CT scan images.

1 Introduction

In practical cases of medical image processing, the image available will be corrupted by noise. This situation is termed as the blind condition. De-noising an image will improve the performance and accuracy of further signal processing steps which follows it such as image segmentation or recognition. But many of the successful noise reduction algorithms require that the amount of noise in the image is known a-priori. As these algorithms are realized assuming the a-priori knowhow of noise level, the algorithms do not operate using any fixed parameters rather they are adapted to the amount of noise level present in the image. So in order to cater these adaptive algorithms, noise estimation becomes mandatory and forms the pre-cursor to image de-noising.

Most of the denoising methods based on filtering set the exact value of the distribution of noise as one of the essential filter parameter [3, 1]. Therefore, the statistical properties of the original image or that of the corrupting noise signal such as, the variance value or spectral distribution of power, etc must be first calculated

Alex Pappachen James
Nazarbayev University and Enview R&D Labs
e-mail: apj@ieee.org

A.P. Kavitha
University of Kerala

S.M. Thampi, A. Gelbukh, and J. Mukhopadhyay (eds.), *Advances in Signal Processing and Intelligent Recognition Systems*, Advances in Intelligent Systems and Computing 264, DOI: 10.1007/978-3-319-04960-1_21, © Springer International Publishing Switzerland 2014

solely from the noisy image signal before applying any type of image de-noising algorithms.

A variety of methods like use of wavelets, anisotropic diffusion and bilateral filtering provide highly promising and accurate denoising results [2, 13, 10, 9, 4]. But for the algorithms to work properly the basic assumption taken is that the level of noise signal is constant for all values of brightness in the image and this level is known and forms one of the input to the de-noising method. Noise is estimated either from a single image or more than one image [6, 15, 14, 7]. Both has got advantages as well as disadvantages. Multiple-image noise estimation is an over-constrained problem, whereas estimation from a single image is an under-constrained problem and as a remedial measure several other assumptions need to be considered for the noise.

While analyzing and processing the medical image data captured using computed tomography the most important factor to be considered is the estimation of variance of noise signal present in the image. All known algorithms which are used for noise removal, segmentation, clustering, image restoration and registration, etc depend on the amount of noise variance. Also, many applications finalize their results on statistical assumptions about the inherent noise characteristics, such as voxel-based morphometry that employ statistical analysis techniques. Finally, the knowledge of the level of noise or the value of the noise variance is useful in assessing the quality of the imaging system itself such as to check the performance of the preamplifier or noise characteristics of the receiver coil. The noise variance estimation algorithms [12, 11, 5, 8] which were developed over the last two decades uses one or more several common steps.

1) Separation of the signal from the noise

a) Pre-classification of homogeneous areas: Since the variance of intensity exactly equals the noise variance over these areas of the image they are the best suited for noise variance estimation.

b) Image filtering: A high-pass filter is used to convolve the processed image or alternatively the response from a low-pass filter and the difference of the processed image is computed. An edge detector is used to recognize the edges of the object which are present along with the noise in the filtered result. After that the recognized image edges are removed. Once this procedure is over what remains is only the noise, which can be directly estimated for its variance.

c) Wavelet transform: Using wavelet transform and assuming that the noise is resulted as the wavelet coefficients at the finest decomposition level gives the simplest way of estimating noise in an image. But sometimes this procedure ends up with significant over-estimates. The reason is that the image structures also affect the wavelet coefficients.

2)Analysis of the local variance estimate distribution: The references give elaborative description of the methods based on the analysis of the local variance estimate distribution. A perfect separation of noise and the signal is not possible. Therefore outliers are present in the distribution of variance estimates computed for image blocks (local). Afterwards reliable statistical methods are applied, to compute the variance of the noise present, which are insensitive to outliers.

2 Proposed Noise Variance Estimator

The proposed method is also an estimator in the spatial domain or a block-based method. The input images used are assumed to be noise-free signals. The images used are 48-slice volume images of the oral cavity of human origin having pixel resolution 512×512 pixels. These images are selected to eliminate the effect of artifacts which affect the image and also the associated noise since this part of the human body is mainly free of moving organs.

In computed tomography (CT), an X-ray tube is rotated around the patients body and many two-dimensional radiographs are acquired. Using the Radon transform the three-dimensional image can be constructed by a dedicated computer from the 2D projections obtained. Thus CT forms a 3D version of conventional X-ray radiography. CT also offers very high contrast between bone and soft tissue, but the contrast is low among different soft tissues. Therefore to artificially increase the contrast among layers of tissues and thereby enhancing the image quality, a chemical solution which is opaque to the X-rays known as contrast agent is injected into the body of the patient. The manufacturers of modern CT devices claim that the X-ray dose is sufficiently low in their machines and such radiation exposure is not a major health risk. Since CT is based on multiple radiographs and therefore the hazards and deleterious effects of ionizing electromagnetic radiation should always be taken into account. CT imaging is a modality of choice when it comes to capture images of the thoracic cavity as the image can be acquired within a single breath hold. Further as CT images are volume images, the estimator works on voxels rather than pixels.

Algorithm 1 shows the summary of steps in the proposed noise estimation technique. After acquisition of the image, the mean and standard deviation of the volume image along the third dimension is evaluated. This procedure gives a 2-D representation of the entire tissue characteristics in the part of the human body which is captured as a 3-D image. The tissue characteristic is represented by a statistical model. After this a synthetic image is constructed pixel by pixel using the value of mean and standard deviation of each pixel where an arbitrary value w is multiplied with the standard deviation of each pixel and then added with the mean of that pixel. The parameter w is a random variable whose value exist within a specific range.

The next step is to evaluate the difference between the original image intensity value of the pixel and the constructed image value of the pixel and the error signal is generated. The error signal is generated for the same pixel of each of the 48 volume images and then averaged over all the 48 images. Then the value of w is varied and the process is repeated for all the values within the specified range for w for that single pixel. The estimator then identifies the value of w for which the error signal is minimum. This value is denoted as w_{opt} for that pixel. The same step is repeated for all the pixels in the image and the distribution of w_{opt} over the entire image is plotted. The last step is the calculation of the mean value of w_{opt} over the entire image. This mean value of w_{opt} is a direct measure of the approximate level of noise in the image.

The experimental evaluation of the estimator is done by synthetic noise generation. The same noise free images are converted into noisy images by adding synthetic un-correlated Gaussian noise and Speckle noise to these images. Both the Gaussian noise and Speckle noise is generated with zero mean and varying values of variance. The same computational procedure as discussed above is done with the noisy images and mean value of w_{opt} is determined for each value of noise variance. The results show a one-to-one relationship between the value of applied noise variance and the mean w_{opt}. Thus the results obtained proves that the determination of mean w_{opt} gives a direct readout of the noise variance present in the image. The results are similar for both Gaussian noise and Speckle noise. This mathematical process when extended to all the 512 x 512 pixels, accounts for very large time for the computation. An optimization of the code developed and the use of parallel processing can considerably reduce this time while working with multi-core processors.

Data: A set of ordered image file names, $F = \{f1, f2, f3, ..., fk\}$ with a cardinality k. The width of the input image is m and height is n. A row vector w of weights having values from -3 to +3 in steps of 0.01.

Result: The mean image, M created from all input images. The standard deviation image, SD created from all input images. Graph showing change of mean value of W_{opt} vs noise variance σ.

for $i \leftarrow 1$ **to** k *in steps of 1* **do**
  | $I_i \leftarrow$ Read image f_i.
end
for $\sigma \leftarrow 0.2 \times 10^{-9}$ **to** 0.2×10^0 *in steps of 1* **do**
  | **for** $z \leftarrow 1$ **to** k *in steps of 1* **do**
  | | $I_Z = I_Z +$Gaussian noise with variance (* adding uncorrelated Gaussian noise of same with each image *)
  | **end**
end
for $x \leftarrow 1$ **to** m *in steps of 1* **do**
  | **for** $y \leftarrow 1$ **to** n *in steps of 1* **do**
  | | $$M(x,y) \leftarrow \frac{\sum_{z=1}^{k} I_Z(x,y)}{k}$$
  | | $$SD(x,y) \leftarrow \sqrt{\frac{1}{k-1}(I_Z(x,y) - M(x,y))^2}$$
  | **end**
end
for $x \leftarrow 1$ **to** m *in steps of 1* **do**
  | **for** $y \leftarrow 1$ **to** n *in steps of 1* **do**
  | | $diff_{reg} \leftarrow$A zero matrix with 1×601 elements.
  | | **for** $z \leftarrow 1$ **to** k *in steps of 1* **do**
  | | | $vir_{img} \leftarrow M(x,y) + w \times SD(x,y)$ (* multiplication is elementwise and vir_{img} is a row vector*)
  | | | $error \leftarrow |vir_{img} - I_Z(x,y)|$ (*elementwise subtraction*)
  | | | $diff_{reg} \leftarrow diff_{reg} + error$ (*elementwise addition*)
  | | | $diff_{min} \leftarrow min(diff_{reg})/k$ (*elementwise addition*)
  | | | $w_{opt} \leftarrow$ The value of w corresponding to the minimum value in $diff_{reg}$.
  | | | $diff_{min}(x,y) \leftarrow diff_{min}$
  | | | $W_{opt}(x,y) \leftarrow w_{opt}$
  | | **end**
  | | $$meanw_{opt}(\sigma) \leftarrow \frac{\sum_{x=1}^{m} \sum_{y=1}^{n} W_{opt}(x,y)}{m \times n}$$
  | **end**
  | Plot graph $meanw_{opt}(\sigma)$ vs σ.
end

Algorithm 1. Proposed noise estimation algorithm

2.1 Experiments and Results

The proposed algorithm is realized as two modules. The primary module performs image acquisition using instructions from data acquisition tool box, specifies the characteristics and type of synthetic noise generated, convert noise-free images into noisy samples, evaluate the mean value of w_{opt} for the specific value of noise variance and plot the results in graphical form. The secondary module is addressed as a sub-routine and called into the primary module to perform all computations. The secondary module evaluates the statistical characteristics of the volume images, creates the virtual image using the statistical entities and random parameter w, evaluate the error signal and the value of w_{opt}.

The availability of noise-free CT scan images is the primary requirement for performing the experiment. Not only the images should be noise-free, but also the images can be obtained from medical institutions only after proper documentation and consent of all concerned. For performing the experiments we obtained two different set of images of two patients from a leading medical institution which are 48 slice images of the oral cavity taken from the same CT scan machine at two different days. The reliability of the scanner was verified by a medical practitioner and radiographer at the institution. Both the set of images are shown in Fig. 1 and Fig. 2. As stated earlier the images of the oral cavity and forehead are more or less free of artefact and similar associated noise.

The CT scan images collected were of the DICOM format. So the images can be viewed only using DICOM-Viewer software. In MATLAB 2009b version, the Data Acquisition toolbox has got specific in-built functions to read the DICOM file directly into MATLABs workspace.

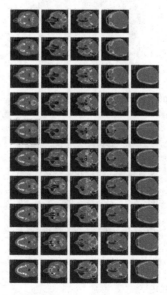

Fig. 1 48-slice CT scan images - Sample 1

After the acquisition of image the mean and standard deviation of each pixel along the third dimension, i.e., the mean and standard deviation of the same pixel for all the 48 image slices is evaluated and plotted. The obtained result for the first sample is given in the Fig 3 and Fig 4.

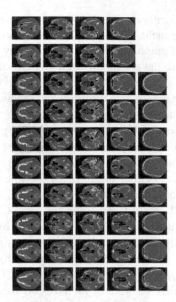

Fig. 2 48-slice CT scan images - Sample 2

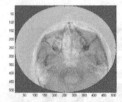

Fig. 3 The mean value of all the 48 images of sample-1

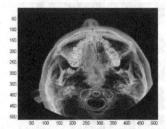

Fig. 4 The standard deviation of all the 48 images of sample - 1

The next step in the experiment is to generate a virtual image with the obtained mean and standard deviation such that the synthetic image is the sum of mean and the product of standard deviation and a random variable w. The range of values of w is optimized after several trial and error operations. Initially it was specified from 0 to 6 with increments in steps of 0.01. Thus w is represented as a matrix of 601 values.

After experimenting with different regions having varying texture in the image the range of values of w is allowed to vary from -1 to +1 and in increments of 0.01 has reduced the sample values to 201 instead of 601 which has given little advantage in total computation time. Once the virtual image is constructed for a single pixel in the image for all the possible values of w, the error signal (*error*) is generated for each image by evaluating the difference between the pixel value of the virtual image and the pixel value of the original image for each of the images in the group. Then the simple average of error is calculated over all the images. Among this average value of error, the minimum value of error ($diff_{min}$) is detected and the value of w corresponding to the value of $diff_{min}$ is denoted as w_{opt} for that pixel in the volume image. The following graph shows the variation of this error signal ($diff_{reg}$) with the values of w for a single pixel. Also it is possible to see the minimum value of $diff_{reg}$ and the corresponding w_{opt} from the graph shown in Fig. 5.

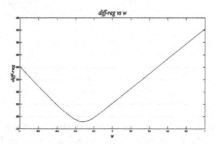

Fig. 5 Plot of $diff_{reg}$ versus w showing $diff_{min}$ occurrence at w_{opt}

The above computation procedure is then extended to all the 512 x 512 pixels of the image. This is a time consuming computation process. The next stage was the generation of synthetic noise signal. For evaluating the reliability and performance of the estimator, noisy images need to be used. To achieve this, the noise-free images available need to be corrupted with artificial noise signals. The evaluation of the estimator is done for two different types of noise namely the Gaussian noise and the Speckle noise. The reason for the selection of this type of noise is that Gaussian being the most common type of noise affecting any image signal and which is independent of the image signal and also the total noise distribution due to many different types of noise affecting an image can be well approximated as Gaussian. On the other hand Speckle noise is an image-dependant signal. So the evaluation of the estimator using these two types of noise can give a generalized as well as comprehensive performance evaluation for the most common types of noise affecting CT scan images.

The Gaussian noise and the Speckle noise signals are generated separately and the performance of the estimator is evaluated separately for individual type of noise acting alone. The common property for both the synthetic noise signals was that the mean of the signal is zero and the values of variance can be varied in such a way that the noise power varies from very small values to 20% of the peak image signal power. Also both the noise signals used are uncorrelated noise signals affecting each image in the image matrix independently. The following figures show typical examples of noisy images used in the experiment and the distribution of w_{opt} for those values of noise variance.

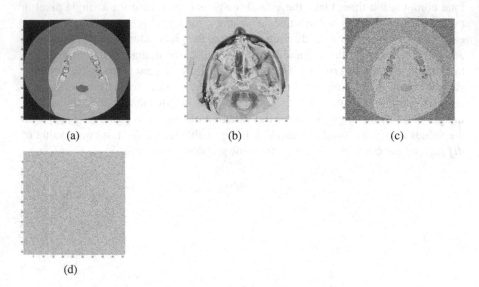

      (a)   (b)   (c)

(d)

Fig. 6 (a) showing the image with noise variance = 0. (b) shows the distribution of w_{opt} for the image in (a). (c) shows the image with Gaussian noise having variance = 0.02. (d) shows the distribution of w_{opt} for the image in (c).

3 Conclusion

We presented a noise estimation technique for detecting the noise levels in CT images using the variability changes of the images at local regions in the image. The error based learning approach indicates a local minimum on the parameter w that is best indicative of the noise level in the images. Since proposed approach is local region based approach it can be applied to detect backgrounds and also to segment out the images. The method is useful to be applied on images detected from any given CT device, and the parameter estimated each time when using a new device. The advantage of the method is its relative simplicity and general applicability of its use in multi-modal images. The disadvantage is the need to retrain the weights when the imaging device is calibrated with progress in time. The future application of the method involve the noise estimated of imaging regions with variations in real-time situations.

References

1. Achim, A., Bezerianos, A., Tsakalides, P.: Novel bayesian multiscale method for speckle removal in medical ultrasound images. IEEE Transactions on Medical Imaging 20(8), 772–783 (2001)

2. Brailean, J.C., Kleihorst, R.P., Efstratiadis, S., Katsaggelos, A.K., Lagendijk, R.L.: Noise reduction filters for dynamic image sequences: a review. Proceedings of the IEEE 83(9), 1272–1292 (1995)
3. Catté, F., Lions, P.L., Morel, J.M., Coll, T.: Image selective smoothing and edge detection by nonlinear diffusion. SIAM Journal on Numerical analysis 29(1), 182–193 (1992)
4. Gerig, G., Kubler, O., Kikinis, R., Jolesz, F.A.: Nonlinear anisotropic filtering of mri data. IEEE Transactions on Medical Imaging 11(2), 221–232 (1992)
5. Immerkaer, J.: Fast noise variance estimation. Computer Vision and Image Understanding 64(2), 300–302 (1996)
6. Long, C., Brown, E., Triantafyllou, C., Aharon, I., Wald, L., Solo, V.: Nonstationary noise estimation in functional mri. NeuroImage 28(4), 890–903 (2005)
7. McCann, A., Workman, A., McGrath, C.: A quick and robust method for measurement of signal-to-noise ratio in mri. Physics in Medicine and Biology 58(11), 3775 (2013)
8. Meer, P., Jolion, J.M., Rosenfeld, A.: A fast parallel algorithm for blind estimation of noise variance. IEEE Transactions on Pattern Analysis and Machine Intelligence 12(2), 216–223 (1990)
9. Nowak, R.D.: Wavelet-based rician noise removal for magnetic resonance imaging. IEEE Transactions on Image Processing 8(10), 1408–1419 (1999)
10. Pizurica, A., Philips, W., Lemahieu, I., Acheroy, M.: A versatile wavelet domain noise filtration technique for medical imaging. IEEE Transactions on Medical Imaging 22(3), 323–331 (2003)
11. Rank, K., Lendl, M., Unbehauen, R.: Estimation of image noise variance. In: IEE Proceedings of the Vision, Image and Signal Processing, vol. 146, pp. 80–84. IET (1999)
12. Sendur, L., Selesnick, I.W.: Bivariate shrinkage with local variance estimation. IEEE Signal Processing Letters 9(12), 438–441 (2002)
13. Sijbers, J., Den Dekker, A., Van Audekerke, J., Verhoye, M., Van Dyck, D.: Estimation of the noise in magnitude mr images. Magnetic Resonance Imaging 16(1), 87–90 (1998)
14. Veraart, J., Sijbers, J., Sunaert, S., Leemans, A., Jeurissen, B.: Weighted linear least squares estimation of diffusion mri parameters: strengths, limitations, and pitfalls. NeuroImage (2013)
15. Zhu, X., Gur, Y., Wang, W., Fletcher, P.T.: Model selection and estimation of multi-compartment models in diffusion mri with a rician noise model. In: Gee, J.C., Joshi, S., Pohl, K.M., Wells, W.M., Zöllei, L. (eds.) IPMI 2013. LNCS, vol. 7917, pp. 644–655. Springer, Heidelberg (2013)

Real-Time Hard and Soft Shadow Compensation with Adaptive Patch Gradient Pairs

Muthukumar Subramanyam, Krishnan Nallaperumal, Ravi Subban,
Pasupathi Perumalsamy, Shashikala Durairaj, S. Gayathri Devi, and S. Selva Kumar

Abstract. This research emphasizes an approach toward real-time shadow compensation for dark/thick/hard and shallow/thin/soft shadows of captured scenes. While humans are very good at estimating objects size, position, color, environmental changes, movements irrespective of occlusions and noise and hence, are able to smoothly visualize the scene. But, computing machines often lack the ability to sense their environment, in a manner comparable to humans. This discrepancy prevents the automation of certain real-time jobs and the shadows make it more cumbersome. Therefore, enhancement of shadow detected region patches with suitable compensation might change object detection and scene visualization more plausible. The authors examine the patch in shadow and non-shadow regions and make the best similar patch pair. These pair characteristics are used to reconstruct both soft and hard shadow regions. However, the hard shadows do not have scene information below the shadow area, that is filled with adaptive gradient patch in-painting technique using close neighboring information. This proposed hybrid framework shows improvement in the overall image quality in terms of both qualitative and qualitative evaluations.

Keywords: Hard shadow compensation, shadow in-painting, shadow extraction, image reconstruction, shadow restitution, De-shadowing, Image Enhancement.

Muthukumar Subramanyam
Dept of CSE, NIT, Puducherry, India
e-mail: sm.cite.msu@gmail.com

Krishnan Nallaperumal · Pasupathi Perumalsamy ·
Shashikala Durairaj · S. Gayathri Devi · S. Selva Kumar
CITE, MS University, Tirunelveli, India
e-mail: {Krishann1751968,pp.cit.msu,shashikalait85,
     gayathri.s7,smartselva2}@gmail.com

Ravi Subban
Dept of CSE, Pondicherry University, Pondicherry, India
e-mail: sravicite@gmail.com

S.M. Thampi, A. Gelbukh, and J. Mukhopadhyay (eds.), *Advances in Signal Processing*   245
and Intelligent Recognition Systems, Advances in Intelligent Systems and Computing 264,
DOI: 10.1007/978-3-319-04960-1_22, © Springer International Publishing Switzerland 2014

1   Introduction

In computer vision unexpected shadows always make object detection difficult, since shadow causes problems, such as occlusion, object merging and shape distortion [1]. Therefore, shadow extraction removal is important for object detection [2]. However, shadow field estimation is a challenging task, because it is difficult to tell if the change of a pixel value is caused by the texture or the shadow. Though several approaches to shadow removal have been proposed in recent years, still it is a nightmare. Image $I(x, y)$ formed by reflectance $R(x, y)$ and Illumination $L(x, y)$ along with the spectral color components(k) .

A classical approach to shadow removal in a single image is to identify the shadow edges, zero the derivatives of those pixels and then integrate to obtain a shadow-free image [3]. Alternatively, shadow regions can be removed by adding a constant factor in the log domain to the intensities enclosed within the shadow edge. These approaches produce good results when the shadow edges are sharp and the shadow occurs on a flat non-textured surface. However, poor results are obtained when shadows are on curved and textured surfaces. This is due to the fact that both textural information and surface gradient information existing at the shadow boundary are removed [4].

(a)   (b)   (c)   (d)

Fig. 1 (a) Umbra/penumbra (b), (c) Hard, soft (d) Static/dynamic Shadows

Our approach finds scale factors (uniform and non-uniform), which are used to compensate the effects of shadows. This method finds the scale factor of a shadow region by sampling pixels along the shadow edge, and estimating the value that minimizes the difference between pixels out-side the shadow region and those inside it [5]. However, it still does not account for scenes containing curved surfaces with shadows, since the minimization of differences between the pixels on either side of the shadow boundary assumes that the shadowed surface is geometrically flat in the penumbra region [6]. The authors use adaptive patch similarity based gradient in-Painting to compensate the missing information at hard shadow boundaries [7].

2   Proposed Methodology

Many researchers suggested different techniques for compensating shadow regions. However, they can be put into two groups: invariant color property based algorithms [8, 9] and the algorithms that assume some prior knowledge of the scene, such as geometry, color ratios [10] or object properties within the scene. This research suggests a novel reconstruction stage for shadow compensation [11] once they have been detected by any kind of method. Any shadow detection algorithm can be used for shadow region identification [6], since the proposed method is not confined to images with certain illumination and /or environmental conditions and specific applications. This research uses hybrid methodology to detect shadows from the captured scenes.

In general, when the geometry, intensity surface of the scene, is not constrained, the problem of finding the correct scale factor along with the surface in the penumbra region is ill-posed [12]. A priori assumption is introduced that the intensity surface in the penumbra region may locally act as a deformable thin plate in the shadow-free image [13]. In practice, these require that the first and second order directional derivatives of penumbra pixels in the shadow-free surface are continuous. Although this seems a strong requirement for all natural images that needs to be examined [3]. The scale factor is a value c that minimizes the following energy functional of the reconstructed deformable thin plate F:

$$E(F) = E_f(F) + E_s(F) \tag{1}$$

Such that $E_f(F)$ and $E_s(F)$ are the data fitting error and the smoothness measure, respectively, given below:

$$E_f(F) = \iint_\Omega \omega(x,y)|F(x,y) - [g(x,y) + C(x,y)]|^2 dxdy \tag{2}$$

$$E_s(F) = \iint_\Omega \left[\left(\frac{\partial^2 F}{\partial x^2}\right)^2 + 2\left(\frac{\partial^2 F}{\partial x \partial y}\right)^2 + \left(\frac{\partial^2 F}{\partial y^2}\right)^2\right] dxdy \tag{3}$$

and

$$\omega(x,y) = \begin{cases} 0 & (x,y) \ is \ a \ penumbra \ point \\ 1 & otherwise \end{cases} \tag{4}$$

And since $C(x,y)$ is assumed constant in the shadow region, $C(x,y) = c \ if f(x,y)$ is a point inside the shadow region but not the penumbra and equals 0 elsewhere [3].

Consider an image I of size $N \times M$ and its gradient magnitude field $\|\nabla I\|$. The gradient magnitude distribution image P^I is calculated as follows:

$$P^I_{xy} = Pr(\|\nabla I(x,y)\|) \tag{5}$$

where, $Pr(\|\nabla I(x,y)\|)$ is the probability of the gradient magnitude at pixel (x,y) in the scene. Adaptively thresholds t1 and t2 are chosen based on the closest similarity patch (if found) else the helinger distance neighbourhood pixels with similar energy, but differencing in edge probabilities energy, is penalized in the current patch. If a hard shadow with abrupt changing shadow boundary is detected. Then adaptive similar patch gradient information near the shadow boundary region is used for in-painting [14]. The approach to stitch the derivatives of the input patches, is given as follows:

1. Compute the derivatives of the similar patch pairs
$\partial I_1/\partial x$, $\partial I_1/\partial y$, $\partial I_2/\partial x$, $\partial I_2/\partial y$ and associate the prominent features of interest in the shadow region.

2. Stitch the derivative images to form a fields $F = (F_x, F_y)$. F_x is obtained by stitching $\frac{\partial I_1}{\partial x}$ and $\frac{\partial I_2}{\partial x}$ and F_y is obtained by stitching $\partial I_1/\partial y$, $\partial I_2/\partial y$.

3. Find the patch gradients are closest to F. This is equivalent to minimizing $d_p(\nabla I, F, \pi, U)$, where π, entire image area and U is an uniform image.

The algorithm for shadow compensation algorithm is as follows:

i.   Using hybrid techniques (Geometry, physical, texture and photometric properties) identify the shadow and non-shadow regions.

ii. Classify the shadow regions based on edge information as soft shadow or hard shadow by using fuzzy support vector machines based on crisp shadow edge or progressive shadow edge feature [15].

iii. If hard shadows edge boundary is found compensate it with adaptive similar gradient patch pair from non-shadow region (named as adaptive gradient patch pair in-painting) [14].

iv. Else if soft shadow edge boundary is found, it is compensated with global and local non-uniform scale factors wherever it is applicable.

v.   Apply post-processing image enhancement techniques to the entire scene.

3   Experimental Results

Several real scenario shadow images were tested and their resulting shadow-free images produced by our algorithm are in good quality. The obtained shadow-free images are able to recover the effects, from hard and soft shadows [16]. The textural information is preserved by this approach and produces the plausible visual scene as output. The approach is tested under benchmark datasets [17] and various shadow-illness criteria. Additionally, the shadow region in the curved scenes appears lighter in the exposed area and darker in the unexposed area [8] and to derive the correct scale factor on curved surfaces, gradient approach is used.

Fig. 2 Hard and Soft Shadow

Table 1 Comparative performance analysis in terms of percentage of accuracy

Data Set	Our Approach (%)	Avg. Accuracy (in%) by our approach
PETSV6	95.41	
Football	93.18	
CVC	98.12	
PETS 2000	94.14	
HERMS	92.14	
ATON	93.48	
GITECH	93.52	96.94
IMRA	95.12	
AVSS	98.75	
SOR	94.66	
LSVN	95.59	
NEMESIS	96.73	

Table 2 Accuracy Assessment Report

Data Set	TP	FP	FN	Error	POD
Cvc	37403	25087	10529	874	96.00
Pets	35938	23721	11313	1418	94.15
Hallway	45950	17428	21383	4832	96.64
Campus	87520	27731	59917	1133	92.56
Caviar	31299	12369	17300	2841	91.43

Note: TP-True Positive, FP-False Positive, FN-False Negative, SF-Splitting Factor, MF-Merge Factor, POD- Percentage of Object Detection, QP- Quality Percentage

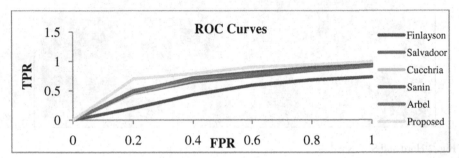

Fig. 3 ROC Curves for Various Authors Proposals

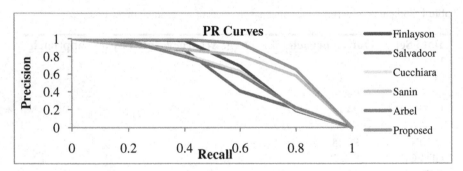

Fig. 4 PR Curves for Various Authors Proposals

4   Conclusion

In this paper, we presented a de-shadowing method that can handle images containing shadows, cast on different kind of surfaces. In the proposed method, shadow is removed by determining a shadow scale factor for each shadow pixel, and then adding these scale factors in the log domain. It also demonstrate that simply calculating the non-uniform scale factor for shadow regions, from the pixels adjacent to shadow edges can result in a high-quality shadow-free image, in which the textural information of the original image is preserved. The approach performs well, even for images containing shadows with wide penumbra and dense hard shadows. The overall accuracy (96.94 %) and quality of scene is sufficiently high in comparison with the methodologies proposed by peer researchers.

References

1. Xu, L.-Q., Landabaso, J.-L., Lei, B.: Segmentation and tracking of multiple moving objects for intelligent video analysis. BT Technology Journal 22(3), 140–150 (2004)
2. Wang, J.-M., et al.: Shadow detection and removal for traffic images. In: 2004 IEEE International Conference on Networking, Sensing and Control, vol. 1. IEEE (2004)
3. Arbel, E., Hel-Or, H.: Texture-preserving shadow removal in color images containing curved surfaces. In: IEEE Conference on Computer Vision and Pattern Recognition, CVPR 2007. IEEE (2007)
4. Bajcsy, R., Lieberman, L.: Texture gradient as a depth cue. Computer Graphics and Image Processing 5(1), 52–67 (1976)
5. Rosin, P.L., Ellis, T.J.: Image difference threshold strategies and shadow detect Graphics and Image Processing 5(1), 52–67 (1976)
6. Jiang, C., Ward, M.O.: Shadow identification. In: Proceedings of the 1992 IEEE Computer Society Conference on Computer Vision and Pattern Recognition, CVPR 1992. IEEE (1992)
7. Fredembach, C., Finlayson, G.: Hamiltonian path based shadow removal. In: Proceedings of 16th British Machine Vision Conference (BMVC), pp. 970–980 (2005)
8. Finlayson, G.D., Hordley, S.D., Drew, M.S.: Removing shadows from images. In: Heyden, A., Sparr, G., Nielsen, M., Johansen, P. (eds.) ECCV 2002, Part IV. LNCS, vol. 2353, pp. 823–836. Springer, Heidelberg (2002)
9. Salvador, E., Cavallaro, A., Ebrahimi, T.: Cast shadow segmentation using invariant color features. Comput. Vis. Image Underst. 95(2), 238–259 (2004)
10. Levine, M.D., Bhattacharyya, J.: Removing shadows. Pattern Recogn. Lett. 26(3), 251–265 (2005)
11. Ollis, M., Stentz, A.: Vision-based perception for an automated harvester. In: Proceedings of the 1997 IEEE/RSJ International Conference on Intelligent Robots and Systems, IROS 1997, vol. 3. IEEE (1997)
12. Sato, I., Sato, Y., Ikeuchi, K.: Acquiring a radiance distribution to superimpose virtual objects onto a real scene. IEEE Transactions on Visualization and Computer Graphics 5(1), 1–12 (1999)
13. McFeely, R., et al.: Removal of non-uniform complex and compound shadows from textured surfaces using adaptive directional smoothing and the thin plate model. Image Processing, IET 5(3), 233–248 (2011)
14. Subban, R., Muthukumar, S., Pasupathi, P.: Image Restoration based on Scene Adaptive Patch In-painting for Tampered Natural Scenes. In: Thampi, S.M., Abraham, A., Pal, S.K., Rodriguez, J.M.C. (eds.) Recent Advances in Intelligent Informatics. AISC, vol. 235, pp. 65–72. Springer, Heidelberg (2014)

15. Muthukumar, S., Subban, R., Krishnan, N., Pasupathi, P.: Real time insignificant shadow extraction from natural sceneries. In: Thampi, S.M., Abraham, A., Pal, S.K., Rodriguez, J.M.C. (eds.) Recent Advances in Intelligent Informatics. AISC, vol. 235, pp. 391–399. Springer, Heidelberg (2014)
16. Hasenfratz, J.-M., et al.: A Survey of Real‐time Soft Shadows Algorithms. Computer Graphics Forum 22(4) (2003)
17. Zhu, G., et al.: Detecting video events based on action recognition in complex Scenes using spatio-temporal descriptor. In: Proceedings of the 17th ACM International Conference on Multimedia. ACM (2009)

Classification and Retrieval of Focal and Diffuse Liver from Ultrasound Images Using Machine Learning Techniques

Ramamoorthy Suganya, R. Kirubakaran, and S. Rajaram

Abstract Medical Diagnosis has been gaining importance in everyday life. The diseases and their symptoms are highly varying and there is always a need for a continuous update of knowledge needed for the doctors. This forces lots of challenges as the diagnostic tools need to visualize organs and soft tissues and further classify them for diagnosis. One such application of diagnostic ultrasound is liver imaging. The existing approaches for classification & retrieval system have the following issues: speckle noise, semantic gap, computational time, dimensionality reduction and accuracy of retrieved images from large dataset. This paper proposes a new method for the classification & retrieval of liver diseases from ultrasound image dataset. The proposed work concentrates on diagnosing both focal and diffuse liver diseases from ultrasound images. The contribution of this paper relies on the following areas. Speckle reduction by Modified Laplacian Pyramid Nonlinear Diffusion (MLPND), Mutual Information (MI) based image registration, Image texture analysis by Haralick's features, Image Classification & retrieval by machine learning algorithms. The dataset used in each phase of the work are authenticated dataset provided by doctors. The results at each phase have been evaluated with doctors in the relevant field.

The CNR value for MLPND has improved 95% compared to existing speckle reduction methods. The MI based registration with optimization techniques to reduce the computation time & monitor the growth of the liver diseases. The results retrieved from different machine learning techniques indicate that the proposed methods improve the image quality and overcome the fuzzy nature of dataset.

Keywords: Speckle reduction, Mutual information, Haralick's features, machine learning algorithms, ultrasound liver.

Ramamoorthy Suganya · R. Kirubakaran
Dept. of Computer Science and Engineering, Thiagarajar College of Engg, Madurai, India
e-mail: rsuganya@tce.edu, kirubakarancse@gmail.com

S. Rajaram
Dept of Electronics and Communication Engineering Thiagarajar
College of Engg, Madurai, India
e-mail: rajaram_siva@tce.edu

S.M. Thampi, A. Gelbukh, and J. Mukhopadhyay (eds.), *Advances in Signal Processing and Intelligent Recognition Systems*, Advances in Intelligent Systems and Computing 264,
DOI: 10.1007/978-3-319-04960-1_23, © Springer International Publishing Switzerland 2014

1   Introduction

A rapid development in the field of medical and healthcare sectors are focused on the diagnosis, prevention and treatment of illness directly related to every citizen's quality of life. Image based medical diagnosis is one of the important research area in this sector. A large number of diverse radiological and pathological images in digital format are generated by hospitals and medical centers with sophisticated image acquisition devices. In the radiology department of the University Hospital of Geneva alone, the number of images produced per day in 2002 was 12,000, and it is still rising. NLF (National Liver Foundation), a voluntary and non-profit organization reported that liver diseases are one of the most common diseases in India, causing lakhs of deaths every year (Nimer Assy et al 2009). Image classification and retrieval has played a vital role in the field of medical domain.

Ultrasound imaging has several advantages over other medical imaging modalities. It is safe, relatively low cost, and allows real-time imaging. A major problem for handling the ultrasound liver images is the presence of various granular structures such as speckles. The analysis of ultrasonic images containing speckles will lead to the problem in texture classification. Most of the existing approaches use artificial neural network classifiers. But the accuracy of retrieving similar images consists of classifying more irrelevant images. The process of using machine learning is generally flexible and therefore a lot of decisions are taken in ad-hoc manner. The proposed approach consists of four phases: Image pre-processing, Image Registration, Feature extraction and classification & retrieval which are shown in Fig 1.

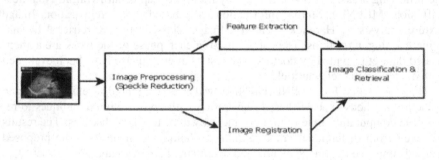

Fig. 1 Systematic flow diagram of the proposed work

The existing approaches for classification & retrieval system have the following issues: speckle noise, semantic gap, computation time, dimensionality reduction and accuracy of retrieved images from large dataset. The main objective of this paper is to suggest appropriate methods at each phase in classification & retrieval mechanisms that will eliminate speckles present in the ultrasound images and reduce the "semantic gap". This improves the accuracy of relevant retrieved images from database at a low computational time. The paper is organized as follows: Section 2 and Section 3 presents the Modified Laplacian Pyramid Nonlinear Diffusion (MLPND) for Speckle reduction and Mutual Information (MI) based

Image Registration. Section 4 proposed a Haralick's Feature Extraction. Section 5 gives Image Classification & Retrieval by Support Vector Machine with Relevance Feedback, Hybrid Kohonen Self Organizing Map (HKSOM) and Fuzzy classifiers. Section 6 discusses experimental results and Section 7 draws the conclusions and provides the future research directions.

2   Modified Laplacian Pyramid Nonlinear Diffusion (MLPND) for Speckle Reduction

The speckle refers to the variation in the intensity of pixels due to several factors. Speckle degrades the quality of ultrasound images and thereby reducing the ability of a human observer to discriminate intricate details of diagnostic examination. The presence of speckle in a medical image would hide the necessary pathological information leading to improper results in diagnosis. Several research works have been reported in the literature for analyzing the speckle removal in ultrasound images. The main reason for using Diffusion based spatial filter is to suppress the noise and preserve fine details of edges in ultrasound liver images. Yu et al (2002) and Abd-Elmoniem et al (2002) introduced the concept of diffusion based spatial filter. The above couple of methods have common limitation in retaining subtle features like cyst, lesions in ultrasound images. Zhang et al (2007) presented a Laplacian pyramid-based nonlinear diffusion (LPND) method for log compressed data.

The proposed work in this paper concentrates on improving the LPND model considering the multiscale analysis in anisotropic diffusion for improving noise suppression, edge preservation and retaining subtle features. MLPND speckle reduction method consists of two stages: Laplacian pyramid decomposition and pyramid reconstruction. Zhang et al (2007) used diffusivity function in Laplacian pyramid nonlinear diffusion method. The proposed work uses a modified diffusivity function $C_2(\| \nabla I \|)$ as

$$C_2(\| \nabla I \|) = exp\left[1 - \left(\| \nabla I \|^2/(2\lambda + 1)\right)^2\right]$$

and gradient threshold λ of the aforesaid work. The choice of the gradient threshold plays an important role in determining the parts of an image that will be blurred or enhanced in the diffusion process. The value of gradient threshold is estimated using the robust median absolute deviation (MAD) estimator. The despeckled image is then passed to image registration or feature extraction depending up on the user needs.

3   Mutual Information Based Image Registration

Medical Image Registration is the process of transforming different sets of medical image data into one coordinate system to monitor the growth of the liver diseases. Peng Wen (2008) recommended a medical image registration based on points, contour and curves. The limitation of the above approach is due to selection of correlation based registration leading to poor results. But in ultrasound image classification & retrieval system, Mutual Information (MI) based image registration plays an important role in monitoring the growth of the liver diseases that motivates this work.

The fundamental technique for mono-modal image registration involves the following steps: (i) pre-processing by proposed S-Mean filter (Speckle Reduction by Anisotropic Diffusion (SRAD) and median) on both the target image and reference image (ii) calculating MI with the help of entropy values. MI is a statistical measure that based on the marginal and joint image intensity distributions P(I), P(T(J) and P(I, T(J)), I and J being the images, T(J) the image J transformed by T and can be written mathematically as follows:

$$MI_T(I,J) = \sum_{I,J} P\left(I, T(J) \log_2 \frac{P(I,T(J))}{P(I)P(T(J))}\right)$$

(iii) Applying a rigid body transformation to relate the target image and (iv) Applying two optimization technique namely DIRECT and Nelder-Mead method in order to bring the computation time to utmost accuracy. The computation time of mono modal liver image registration is reduced by DIRECT method compared to Nelder-Mead method.

4   Haralick's Feature Extraction

Medical images are usually fused, subject to high inconsistency and composed of different minor structures. So there is necessity for feature extraction and classification of images for easy retrieval. Among visual features, texture is widely used for content-based access to medical images. Since the liver images belong to ultrasound modality, gray scale features are sufficient for extracting more relevant features. The Gray Level Co-occurrence Matrices (GLCM) is found effective in texture analysis method for medical images. Haralick et al (1973) proposed a general procedure for GLCM for extracting textural properties in medical images. Lee et al (2003) had chosen fractal feature vector based on wavelet transform for classification of ultrasonic liver tissue.

The despeckled image is then passed into feature extraction phase. Selecting a minimal set of features for classification helps in dimensionality reduction and improves the retrieval speed of the system. The main objective of this work is to select the most discriminating texture features (parameters) for classification and retrieval. The Twelve Haralick's textural features are Contrast, Energy, Correlation, Entropy, Homogeneity, Maximum probability, Dissimilarity, Angular Second Momentum (ASM), Mean, Variance, Cluster prominence, and Cluster shade. Gray Level Co-occurrence Matrices (GLCM) represents the frequency of all possible pairs of adjacent gray level values in the entire image. The above twelve

Haralick's features are applied in two steps: (i) In the first step, the whole image is considered for the calculations (ii) In the second step, only the selected Pathology Bearing Region (PBR) is considered for calculations. The main advantage of GLCM features is its construction of co-occurrence matrices that characterizes the spatial interrelationships of the gray tones in an ultrasound image. Out of twelve features, five predominant features namely Contrast, Cluster prominence, Auto correlation, Cluster Shade and Angular second moment are selected by Correlation Feature Selection (CFS) for classification. The selected features are fed into image classification & retrieval.

5   Image Classification and Retrieval

Most of the existing approaches use artificial neural network classifiers. But the accuracy of retrieving similar images consists of classifying more irrelevant images. The objective of this research paper is to develop an efficient classification & retrieval system for both focal and diffused liver by three machine learning techniques namely Support Vector Machine with Relevance Feedback, Hybrid Kohonen SOM and Fuzzy classifiers. The research concentrates on improving the learning capabilities of the model considering the fuzzy nature of the dataset.

5.1   Support Vector Machine (SVM) with Relevance Feedback (RF)

The proposed work on SVM-RF is used to address the main issue- 'semantic gap' in image classification & retrieval. The objective of SVM-RF method is to classify whether ultrasound liver image is normal or diseased one. An input query image is fed into the proposed SVM-RF system. After image registration, feature extraction and feature selection, it is compared with liver images in the database pool and the resultant images will be retrieved. The results obtained with SVM are further improved by applying the relevance feedback technique. In this research work linear kernel function for SVM based classification is utilized. Linear kernel function for binary classification of medical images K $(K-1)/2$ models are constructed, where k is the number of categories. SVM classifies both positive (normal liver) and negative (diseased liver) samples based upon the input query. The future iteration will be preceded by the physician to select appropriate images from the retrieved images by applying relevance feedback technique. The system will refine the retrieval results based on the feedback and present the new list of images to the user. Methods for performing relevance feedback using the visual features as well as the associated keywords in unified frameworks have been performed by reducing the semantic gap. For the large dataset with different stages of liver diseases, the classification is switched to unsupervised learning algorithm.

5.2   Hybrid Kohonen SOM

For a large dataset, unsupervised learning algorithm called Hybrid Kohonen SOM (HKSOM) uses a neighborhood function to preserve the topological properties of the input space. In the proposed HKSOM method, the network architecture consists of an input layer followed by a single competitive layer. Input layer consists of the six neurons where six Haralick's feature values are extracted from the images. Each unit in the input layer is connected to single neuron in the competitive layer. The merits of HKSOM are as follows: Accessing input patterns and clustering them into groups of similar patterns by modified Best Matching Unit (BMU) $W (t+1) = W (t) + \theta(t) L(t) (V(t) - W(t))/2$ & (ii) Handling higher dimensional medical data more efficiently. The key findings from this work attains that classification of liver diseases like hepatoma and hemangioma are clearly distinguished with the help of SOM, but some samples of the liver cyst and fatty is grouped into one cluster which denotes fuzzy nature of texture characteristics in liver dataset.

5.3   Fuzzy Classifier

As the ultrasound liver dataset namely liver cyst and fatty are fuzzy in nature, the above machine learning method- SVM-RF and HKSOM cannot provide accurate results. Hence this part of the work is modelled by using Fuzzy classifier. Juan et al (2011) proposed Fuzzy Object Relational Database Management System for flexible retrieval of medical images. Our proposed method consists of the following steps: (i) Pre-process the query image by MLPND with and calculate entropy value for texture analysis (ii) Fuzzify the values by triangular membership function and identify its range for liver cyst, fatty and normal. (iii) Apply fuzzy classifier to compare the query value with the database and retrieve its result. In order to handle with the imprecision and overlapping bounds, fuzzy classifier is implemented, that reduces the misclassification error and improves 100% of retrieval accuracy.

6   Experimental Results:

The experiment is carried out with 2D ultrasound liver image dataset collected from GEM Hospital, Coimbatore. Each of these images is a clinical raw image taken for diagnosis or treatment purposes. The results in this research work have been evaluated with doctors in the relevant field. This dataset is formed from 300 cases; 60 cysts, 45 cavernous hemangioma, 66 normal livers and 41 hepatoma, 40 fatty liver and 48 Cirrhosis liver images. The performance measure of MLPND for speckle reduction is evaluated based on the contrast to noise ratio (CNR) is shown in Table 1. From the Table 1, it is inferred that the image clarity is not affected but improved using MLPND in comparison to existing speckle reduction methods. Fig 2 shows the resultant image of applying MLPND on input liver image

Table 1 Comparison of CNR values for liver images using Modified LPND with existing Non-linear Diffusion Filters

Liver Diseases / Filters	Normal	Cyst	Hemangioma	Hepatoma	Fatty	Cirrhosis
Noisy	4.45	5.34	5.90	4.40	6.12	6.23
ND	9.45	11.87	17.10	9.18	9.20	10.92
SRAD	9.41	11.36	18.57	14.18	15.23	16.01
LPND	9.46	10.89	19.01	14.63	15.61	16.94
Modified LPND	10.84	12.34	19.22	15.29	17.52	18.11

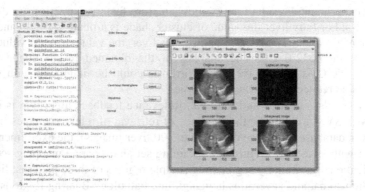

Fig. 2 Resultant image of applying MLPND on input liver image

During classification and retrieval, if the patient want to know about his history from the database which is already stored in ultrasound modality, his new ultrasound image is undergo the process of MI based image registration technique. The comparison of performance measure for MI & Correlation Coefficients (CC) similarity measures are shown in Table 2. Fig 3 shows that the rigid body transformation on MI -Mono-modal Ultrasound liver images. The performance of the two optimization techniques namely DIRECT and Nelder-Mead method has been compared & inferred that the computation time for DIRECT method is 0.719sec that produce optimal solution compared to Nealder- Mead method that computes 0.980sec for 50 iterations. DIRECT method is very well suited for clinical applications.

Table 2. Comparison of Mono-modal liver image registration by correlation coefficient and MI based method

Liver images	Rotation (degree) Rigid Body Transformation		Before Registration	CC	MI
	Target Image (US)	Reference Image (US)		After Registration	
Cyst	2	0	0.4423	0.4473	0.5658
Hepatoma	2	0	0.5012	0.5639	1.9598
Hemangioma	2	0	0.3367	0.4342	1.4821
Fatty liver	2	0	0.5174	0.5174	1.5003
Cirrhosis	2	0	0.4198	0.4667	1.5322

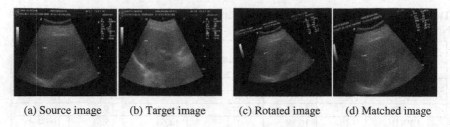

(a) Source image   (b) Target image   (c) Rotated image   (d) Matched image

Fig. 3 Rigid body transformation on MI -Mono-modal Ultrasound liver images

The experimental results for Haralick's texture feature extraction are discussed as follows: For fatty liver the value of cluster shade is around 75.5±0.6 and cluster prominence is about 1.75E+03. For cirrhosis liver, the value of ASM value should lay between 0.632-0.652. The analysis of the obtained results suggested that diseases like Cyst, Fatty Liver, and Cirrhosis could be diagnosed with minimal features and reduced overall retrieval time.

In image classification and retrieval, Relevance feedback algorithm is incorporated with SVM to improve the accuracy of retrieval of relevant images from 150 images in database. For large dataset (i.e. 300 liver images), the experiment is carried out with HKSOM. It consists of training set and separate mapping set for the classification algorithm by segmenting sample regions of three patterns: liver cirrhosis, hepatoma, and cavernous hemangioma. Pathology Bearing Regions (PBRs) of size 23×23 pixels for each kind of patterns is sampled. The output from Kohonen SOM layer is passed to Multi-Layer Perceptron (MLP) to classify the abnormal liver images into Cirrhosis, Hepatoma and Hemangioma, by means of Kohonen map. By using HKSOM, classification of ultrasound liver images is achieved 96% correctly. However this result can vary with large number of dataset containing different types of speckle noise. HKSOM uses Modified BMU to preserves the topological properties of the liver diseases like cirrhosis, hepatoma and hemangioma. But some samples of the liver cyst and fatty are grouped into one cluster which denotes fuzzy nature of texture characteristics in liver dataset. The Fuzzy classifier is implemented to overcome this issue. Triangular membership function is used to classify images like cyst and fatty which are fuzzy in nature. The experimental dataset used in the fuzzy classifier consists of a set of ultrasound liver images that include: Normal Images-40, Cyst-20, and fatty liver-25. The entropy range of normal, cyst and fatty liver were found to be ranging from 4.3-5.8, 4.1-6.2, and 3.3-5.4 respectively. From the entropy values it is concluded that for an image that has the entropy value of 4.1, it cannot be decided whether it has a cyst or fatty deposits. Therefore, to handle with the imprecision and overlapping bounds, fuzzy based classifier is implemented, that reduces the misclassification error. Results illustrate that the system developed in this paper significantly improves 100% the overall retrieval quality compared to the previous existing systems which is shown in Table 3.

Table 3. Comparison of Retrieval accuracy rate with different machine learning techniques

Machine learning techniques	Retrieval Accuracy rate
SVM with Relevance Feedback	72.1%
Hybrid Kohonen SOM (HKSOM)	96%
Fuzzy classifier	100%

7 Conclusion

This paper in overall has suggested an efficient framework for performing classification & retrieval of ultrasound liver images. Appropriate techniques have been selected at each phase namely – Pre-processing for Speckle reduction, Image registration, Feature extraction, Image classification and retrieval. The proposed work has the advantage of combining image registration for growth monitoring of liver diseases and retrieval of fuzzy nature of liver dataset. The result has been evaluated at each phase by Doctors in the relevant field. This work in overall proposes a more automated decision making system. More focus in this research has been in classification approaches for automated decision making system. Retrieval is part of the classification mechanism in each phase of the research, without concentrating on content based retrieval algorithms. The future enhancement of this paper is to develop a complete web interface system integrating the challenges in handling medical image modalities could be presented.

References

1. Abd-Elmoniem, K.Z., Youssef, A.M., Kadah, Y.M.: Real-time speckle reduction and coherence enhancement in ultrasound imaging via nonlinear anisotropic diffusion. IEEE Transaction on Biomedical Engineering 49(9), 997–1014 (2002)
2. Zhang, F., Yoo, Y.M., Koh, L.M., Kim, Y.: Nonlinear diffusion in laplacian pyramid domain for ultrasonic speckle reduction. IEEE Transaction on Medical Imaging 26(2) (2007)
3. Haralick, R.M., Shanmugam, K., Dinstein, I.: Textural features for image classification. IEEE Transaction on Systems, Man and Cybernatics 3(6), 610–621 (1973)
4. Medina, J.M., Castillo, S.J., Barranco, C.D., Campana, J.R.: On the use of a fuzzy object relational database for flexible retrieval of medical images. IEEE Transaction on Fuzzy Systems 20(4), 786–803 (2012)
5. Lee, W.L., Chen, Y.C., Hsieh, K.S.: Ultrasonic liver tissues classification by fractal feature vector based on M-band wavelet transform. IEEE Transaction on Medical Imaging 22(3), 382–392 (2003)
6. Assy, N., Nasser, G., Djibre, A., Beniashvili, Z., Elias, S., Zidan, J.: Characteristics of common solid liver lesions and recommendations for diagnostic workup. World Journal of Gastroenterology 15(26), 3217–3227 (2009)
7. Wen, P.: Medical image registration based-on points, contour and curves. In: International Conference on Biomedical Engineering and Informatics, pp. 132–136 (2008)
8. Yu, Y.J., Acton, S.T.: Speckle reducing anisotropic diffusion. IEEE Transaction on Image Processing 11(11), 1260–1270 (2002)

Tampered Image Reconstruction with Global Scene Adaptive In-Painting

Ravi Subban, Muthukumar Subramanyam, Pasupathi Perumalsamy, R. Seejamol,
S. Gayathri Devi, and S. Selva Kumar

Abstract. The objective of in-painting is to reconstruct the mislaid region of an image. This paper presents a new in-painting algorithm from the goodwill of Exemplar-based Greedy algorithms, which consist of two phases: making a decision of filling-in order and selection of good exemplars for the damaged regions. The proposed method overcomes these tribulations with the protection of edges, textures and also with lesser propagation error. This scheme upgrades the filling-in order that is based on the combination of priority terms, to encourage the early synthesis of linear structures. The subsequent contribution helps sinking the error propagation to an improved detection of outliers from the candidate patches. The proposed methodology is well suited in terms of both natural and artificial images with plausible output. This scheme dramatically outperforms earlier works in terms of both perceptual quality and computational efficiency.

Keywords: Curvature Driven Delusion, Exemplar based Approach, Texture Synthesis, and Structure Synthesis.

1   Introduction

The process of repairing the damaged regions or modifying the regions of an image into a non-detectable form is known as digital image in-painting [31]. Digital image in-painting has a variety of applications such as restoration of damaged old

Ravi Subban
Dept of CSE, Pondicherry University, Pondicherry, India
e-mail: sravicite@gmail.com

Muthukumar Subramanian
Dept of CSE, NIT, Pondicherry, India
e-mail: su.muthukumar@gmail.com

Pasupathi Perumalsamy · Seejamol · S. Gayathri Devi · S. Selvakumar
CITE, MS University, Tirunelveli, India
e-mail: pp.cit.msu@gmail.com

S.M. Thampi, A. Gelbukh, and J. Mukhopadhyay (eds.), *Advances in Signal Processing*   263
and Intelligent Recognition Systems, Advances in Intelligent Systems and Computing 264,
DOI: 10.1007/978-3-319-04960-1_24, © Springer International Publishing Switzerland 2014

Fig. 1 Example Applications a) Texture Removal b) Scratch Removal c) Object Removal

printing materials and old photographs, error recovery of images and videos, multimedia editing and replacing large regions in an image or video for privacy protection ,creation of special effects to images and for instance specific object removal [35]. The main theme of the image in-painting is to alter the scratched regions of an image or video, so that the in-painted region is smooth to a neutral observer [18]. Therefore, image in-painting restores the lost information from the images, so that the image after in-painting looks natural [40]. There are many causes for the information loss in images, for instance, transmission loss, accidental damages or corruption of art works/photographs due to aging effects or damages caused by occlusion. Image information can be classified into structure, texture and color [34]. Zhou and Zheng proposed an adaptive size patch based in-painting algorithm using gradient angle histogram. It solved the optimization problem by determining the optimal size of each patch [19]. Liu et.al derived an adaptive in-painting method using local similarity and gradient magnitude [17].

Fig. 2 (a) Tampered Image (b) In-painted Image (c) In-painting Problem (d) In-painting Methodology

2   Literature Review

There are many research proposals available for image in-painting or region filling problem. The prominent works are discussed below. An effective and simple algorithm of texture synthesis by non parametric sampling was proposed by Efors and Leung. It is based on Markov Random Field and texture synthesis. But, this algorithm is very slow because the filling-in is being done pixel by pixel [3]. Jing Xu et.al implemented the 8 neighborhood Fast sweeping Method removes the text and hidden errors on video [22]. This method is fast and feasible only for normal scenes. The method of nearest point and fast sweeping method is proposed by Tsai, which works with the distance between pixels [8]. A Curvature Driven

Delusion (CDD) is proposed by Chan and Shen use the Euler elastic encapsulate for both CCD in-painting and transportation in-painting [7]. Another algorithm for texture synthesis is introduced by Criminisi et.al, for filling the edges based on priority [12]. Fast Texture synthesis method using tree structured vector quantization algorithm is created by Wei and Levoy. Though it is efficient, in specific scenes and also produces high quality fineable textures [4, 5]. But, the Partial Differential Equation (PDE) needs lot of iterations before reaching the convergence and it is computationally expensive. Jiying-Wu developed a cross isophotes, exemplar based in-painting algorithm based on the analysis of anisotropic diffusion [14, 36]. Wong projected a non-local means approach for exemplar inpainting algorithm, inferred by a non local mean set of candidate patches [20]. Roth and Black proposed a framework for learning image priors based on prior models [13,15]. The diffusion technique used for denoising approaches , is also applied to repair the damaged images. Bertalmio et al. developed a method which combines the advantages of partial equations and texture synthesis, decomposed into structure and texture component [6, 11]. Chen and Williams extended the idea to 3D images by calculating a linear warped between corresponding 3D points of two scenes and interpolated for views. The author also derived an in-painting model by considering the image as an element of the space of bounded variation (BV) images, endowed with the Total Variation (TV) norm. The solution of the in-painting problem comes from the minimization of an appropriate function over the region. The curvature and the connectivity principle according to the human eye principle, helps to reconstruct the broken edges. Their research tries to deal with both holes and visibility ordering [2, 33]. Zhaolin-Lu proposed Exemplar based image completion algorithm, in which the size of image patch can be decided by the gradient domain. The filling priority is decided by the geometrical structuring. The known region instead of a single (best) match patches used as a feature of the image, especially in the curvature and the direction of the isophotes[23]. It introduces a better patch-matching scheme, which incorporates the curvature and color of the image. The determined source template is copied into the destination template and then the information of the destination template is updated [26]. Shen, et al, proposed a patch based in-painting scheme, which distinguishes the damaged areas and non damaged areas based on structures and textures of frame work and also redefine confidence and illumination variation[21]. It is robust to the direction of the full-front. The method spreads the isophotes. It combines at the boundaries/edges of the in-painting region, which is proposed by Bertalmio et al. This method preserves the arrival angle and smoothens the inside region. It is used in the same on text of copying the natural process to present a very simple algorithm[25]. It efficiently synthesizes a wide variety of textures. The input consists of an example texture patch and a random noise image with size specified by the user [5, 28]. Masnou and Morel implemented their method to the early related work under the word "disocclusion" rather than in-painting was done [10].

Another in-painting model is proposed by Mumford and Shah, which takes extra care of the edges on the functional to be minimized [1]. Esedoglu and Shen made extension for curvature model that was proposed by using the Euler's elastic [9]. Muthukumar et al. proposed algorithms for removing objects and in-painting damaged regions with patch based methodology. It is used in both one and two dimensional images. It improves the efficiency of filling, by Poisson method. To improve the computational cost successive elimination algorithm is employed to obtain global optimal solution for two dimensions and also proposed a methodology of adaptive patch based image in-painting for tampered natural scenes [27, 29]. They also analyze various techniques of in-painting [37]. Exemplar based algorithms have been proposed with adaptive patch size, to perk up the excellence of filling outcome with fast processing. This scheme reduces the computation time and produces better performance than other methods used in the literature [24].

3   Methodology

In existing Exemplar-based methods, it is assumed that texture patterns in the source region are distinguishable with appropriate patch size. They are not appropriate to fill the part of a target regions' structure and texture information. If the size of each patch is too large, structure may be incorrectly reconstructed. On the contrary, if the patch is too small, it is time consuming to synthesize a large region that has similar texture patterns [41]. The scheme is based on the fact that natural images may contain redundant information. Greedy in-painting using exemplars, also known to use patches, that consist of two major steps: select the patch to be filled and promulgate the texture and structure. The prior step chooses patches with linear structures, by giving them higher priority order. The latter has related to the selection of the k-most similar patches from which the information is copied. These two steps are iterated until the holes in the image are fully restored [40]. The global in-painting optimization problem with a message passing to make the solution sequential, a scheduling is proposed that requires the computation of many distances at the initialization process. More precisely, for each patch located at the boundary between the known and the unknown region, the distance between this patch and other known patches have to be computed [32]. Let Ω stands for the region to be in-painted, and for its boundary. Intuitively, the technique proposed will prolong the isophote lines arriving at $\delta\Omega$, while maintaining the angle of "arrival." We precede drawing from $\delta\Omega$ inward in this way, while curving the prolongation lines progressively to prevent them from crossing each other.

3.1   The Proposed In-Painting Algorithm

The robust in-painting algorithm based on region segmentation, can in-paint damaged regions, as well as remove the objects. A segmentation map provides local texture similarly and dominant structure region. It adaptively chooses weighting parameter values of the robust priority function for each segment with boundary information of a segmented image map.

The foremost method constructs a segmentation map M using an input image I, where a target region T is manually selected. Then, it determines factor values of the robust priority function, in which factor values are chosen for the robust priority function. To choose suitable factor values to reflect properties of every segment, the Difference of Gaussians (DoG) is used. In the robust exemplar in-painting step, the proposed method computes the priority of target patches using robust priority function and searches the best matching source patch using Criminisi algorithm. To improve the efficiency of the proposed algorithm, we use adaptive patch size selection globally and search region reduction locally.

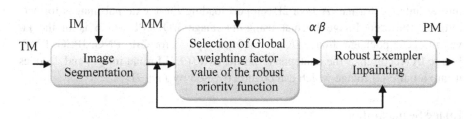

Fig. 3 Block diagram of proposed method

Detecting Global Structure

A window based pattern matching, is used to detect global structures in the target area, instead of a pixel-wise gradient. A pattern selected as a form of rectangular window. A set of projection kernels applied on the regions near border line of target area. The projection results are combined with the priority function. This approach reserves restoration of global structures with higher priority and has noise immunization caused by local gradients. It reduces the rate of zero convergence of the confidence term

Propagating Texture and Structure

Once the patch with maximum Ψp priority has been selected, the next step is to synthesize its missing region $\Psi p \cap \Omega$. This is achieved by searching K similar patches Ψqi, where $\Psi qi \, \varepsilon \, \Omega$ and i=1,2,3,4.....,k to the patch to be filled Ψp. Then the unknown region of Ψp, is filled with equivalent extracted from the K similar patches Ψqi, eg using a linear combination or simply use the information from the best match [16].

Robust Exemplar in-Painting

The exemplar based approach consists of two basic steps, in the first step priority assignment and the second step consists of the selection of the best matching patch [38]. The exemplar based approach samples the best matching patches from the known region, whose similarity is measured by certain metrics, and pastes into the target patches in the missing region[. Exemplar based in-painting iteratively synthesizes the unknown region, i.e., target region, by the most similar patch in the source region [39].

Convert to Buffered Image

The regions that use these images, first convert them from that external format into an internal format. A Java 2D support loading these external images formats into its Buffered Image Format using its image I/O API which is in the javax.image.io package. Image I/O has built-in support for GIF, PNG, JPEG, BMP. Image I/O recognizes the contents of the file as a JPEG format image, and decodes it into a buffered Image which can be directly used by Java 2D

Image Segmentation

The local segmentation approaches may produce better separations for complicated or detailed regions. Here, a local window (small area of image) is used over a segmentation region to perceive a need for more segmentation iterations. In this way, the image is segmented hierarchically until each object region becomes smooth. As a result, the segmentation of the main structures is kept, and the detailed structures are detected. The most important advantage of this approach is that the segmentation result does not depend on the initial conditions and the method is relatively fast.

Detecting Global Structure

A window based pattern matching is used to detect global structures in the target area, instead of a pixel-wise gradient. A pattern can be given as a form of rectangular window. A set of projection kernels is applied on the regions near border line of target area. The projection results are combined with the priority function. This approach reserves restoration of global structures with higher priority and has noise immunization caused by local gradients. It reduces the rate of zero convergence of the confidence term.

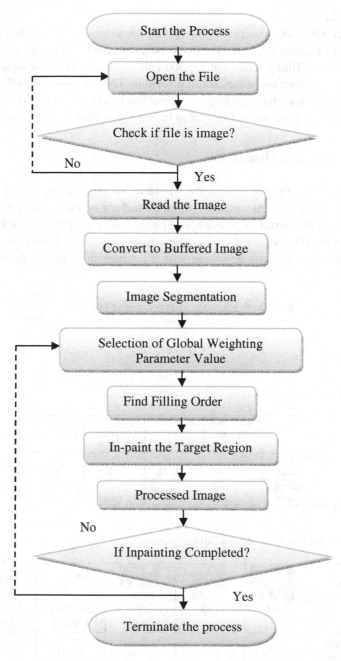

Fig. 4 Flow chart for proposed method

Filling Order

The most popular idea hitherto is propagating linear structures and then textures, in such a way that the formations of major peninsulas in the fill front are avoided. According to the filling order, the method fills structures in the missing regions using spatial information of neighboring regions. This method is an efficient approach for reconstructing large target regions. The filling order definition, is better encouraging the propagation of structures before the synthesis of textures

4   Results and Discussion

The proposed priority that utilizes the similarity of neighboring source patches, assigns priority to the structures in a more robustly way, that preserves the edge continuity and the texture consistency. The proposed algorithm achieves more pleasant results that align with the human visual perception. It takes an average of 15sec to fill in the missing region with 1025 missing pixels in the degraded image

$$MSE = \frac{1}{IJ}\sum_{p=1}^{I}\sum_{q=1}^{J}[a(p,q) - a'(p,q)]^2 \tag{1}$$

$$PSNR = 10log\frac{255^2}{MSE} \tag{2}$$

Table 1 Comparison of previous and proposed method (min)

	Patch on Corner	Patch on Edge	Patch on Regular Texture	Patch on Flat region
Previous Method	3.719	0.518	3.530	0.330
Proposed Method	0.080	0.064	0.011	0.002

Table 2 Qualitative analysis Metrics (dB)

Images					
PSNR/ SSIM	30.336 / 0.9977	33.136 / 0.996	34.398 / 0.9995	33.331 / 0.9998	33.806 / 0.997
Computation Time	2:16	1 : 12	0 : 26	0 : 31	2:47

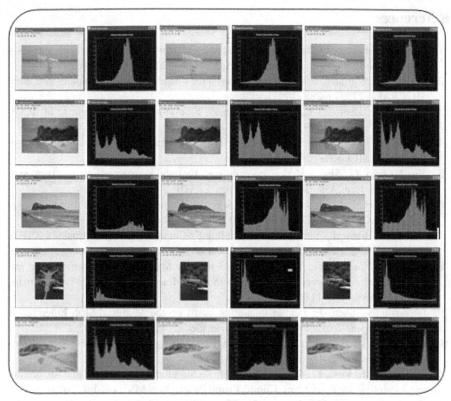

Fig. 5 a) Input b) Histogram of a c) previous output d) hist of c e)Proposed method f) Hist of e

5 Conclusion

This proposed method connects the curves in a smooth manner and gives more visually plausible filling result. It determines the suitable patch size and selects candidate source regions for reducing unnatural artifacts. The proposed method produces good results for in-painting the multiple and large missing region in complex backgrounds also. This algorithm fills the target region on an average 2 seconds, on a Pentium IV, 2.52GHz, 1GB RAM whereas similar methods, with the nearest in quality requires approximately 22 seconds. This algorithm can in-paint both structure and textures of images. It is efficient in terms of reduced computational cost and the time required for in-painting tampered regions. The experiments show that our proposed scheme is capable of effectively maintaining the texture consistency and the edge continuity for a good visual quality. Moreover, the method is robust to the high-frequency components such as artifacts in the priority assignment.

References

1. Mumford, Shah, J.: Optimal approximations by piecewise smooth functions and associated variation problems. Comm. Pure Appl. Math. 42(5), 577–685 (1989)
2. Chen, S.E., Williams, L.: View interpolation for image synthesis. Computer Graphics, SIGGRAPH 27, 279–288 (1993)
3. Efors, A.A., Leung, T.K.: Texture synthesis by non-parametric sampling. In: ICCV (2), pp. 1033–1038 (1999)
4. Levoy, W.: Fast Texture Synthesis using Tree-structured Vector Quantization. In: Proceedings of SIGGRAPH (2000)
5. Xu, Y., Guo, G., Shum, H.Y.: Chaos mosaic: Fast and memory efficient texture synthesis. Tech. Rep., Microsoft Research (April 2000)
6. Bertalmio, et al.: Image in-painting. In: Siggraph, Computer Graphics Proceedings, pp. 417–424. ACM Press/ACM SIGGRAPH (2000)
7. Chan, T.F., Shen, J.: Non Texture inpainting by Curvature-Driven Diffusions (CDD). Jounal of Vis. Comm. Image Rep. 4(12), 436–449 (2001)
8. Tsai, et al.: Curve evolution implementation of the Mumford Shah functional for image segmentation, denoising, interpolation and imagination. IEEE Trans. Image Process. 10(8), 1169–1186 (2001)
9. Esedoglu, S., Shen, J.: Digital inpainting based on the Mumford-Shah-Euler image model. European Journal of Applied Mathematics 13(4), 353–370 (2002)
10. Masnou, S.: Disocclusion: A variational approach using level lines. IEEE Transactions on Image Processing 11, 68–76 (2002)
11. Bertalmio, M., Vese, L., Sapiro, G., Osher, S.: Simultaneous Structure and texture image inpainting. In: Proc. Conf. Comp. Vision Pattern Rec. Madison, WI (2003)
12. Criminisi, A., Perez, P., Toyama, K.: Region filling and object removal by exemplar based image in-painting. IEEE Trans. on Image Processing 13, 1200–1212 (2004)
13. Roth, S., Black, M.J.: Fields of experts: A framework for learning image priors. In: Proc. IEEE Computer Society Conf. Computer Vision and Pattern Recognition, pp. 860–867 (2005)
14. Wu, J., et al.: Object Removal By Cross Isophotes Exemplar-based In-painting. In: Proceedings of the 18th International Conference on Pattern Recognition, ACM Digital Library Proceeding, ICPR 2006, vol. 3, pp. 810–813 (2006)
15. Roth, S., Black, M.J.: Steerable random fields. In: Proc. IEEE Com-puter Society Conf. Computer Vision and Pattern Recognition, pp. 1–8 (2007)
16. Hong-Bin, Z., Jia-Wen, W.: Image Inpainting by Integrating Structure and Texture Features. Journal of Beijing University of Technology 33(8), 864–869 (2007)
17. Liu, D., et al.: Image Compression With Edge-Based Inpainting. IEEE Transactions on Circuits And Systems For Video Technology 17(10), 1273 (2007)
18. Li, X., Zheng, Y.: Patch based video processing: a variation Bayesian approach. IEEE Transaction on circuits and Systems for Video Technology 19(10), 2476–2491 (2007)
19. Zhou, Zheng.: Gradient based image completion by solving the Poisson equation. Computers & Graphic Science Direct 31(1), 119–126 (2007)

20. Wong, A., Orchard, J.J.: A nonlocal-means approach to exemplar based inpainting. In: IEEE Int. Conf. Image Processing (2008)
21. Shen, B., Hu, W., Zhang, Y., et al.: Image inpainting via sparse representation. In: IEEE International Conference on Acoustics, Speech and Signal Processing, ICASSP, pp. 697–700 (2009)
22. Xu, J., et al.: An Image Inpainting Technique Based on 8-Neighborhood Fast Sweeping Method. In: Published in Proceeding CMC 2009 Proceedings of the WRI International Conference on Communications and Mobile Computing, vol. 3, pp. 626–630. IEEE Computer Society, Washington, DC (2009)
23. Barnes, C., Shechtman, E., Finkelstein, A., Goldman, D.B.: Patch match: A randomized correspondence algorithm for structural image editing. ACM Trans. Graph. 28(3), 24:1–24:11 (2009)
24. Muthukumar, S., Krishnan, N., Pasupathi, P., Deepa, S.: Analysis of Image Inpainting Techniques with Exemplar, Poisson, Successive Elimination and 8 Pixel Neighborhood Methods. International Journal of Computer Applications 9(11), 15–18 (2010)
25. Bugeau, Bertalmio, M., Caselles, V., Sapiro, G.: A comprehensive framework for image inpainting. IEEE Trans. Image Process. 19(10), 2634–2645 (2010)
26. Lu, Z., et al.: A Novel Hybrid Image Inpainting Model. presented at the IEEE International Conference on Genetic and Evolutionary Computing (2010)
27. Xu, Z., Sun, J.: Image inpainting by patch propagation using patch sparsity. IEEE Trans. Image Process. 19(5), 1153–1165 (2010)
28. Zhong, Z., Wang: Image inpainting-based edge enhancement using the eikonal equation (2011) 978-1-4577-0539-7/ IEEE
29. Jian-Bin, Y.: Image in-painting using complex 2-D dual-tree wavelet transform. Appl. Math. J. Chinese University 26(1), 70–76 (2011)
30. Zontak, M., Irani, M.: Interfinal statistics of a single natural image. In: IEEE Conference on Computer Vision and Pattern Recognition (CVPR) (2011)
31. Li, S., Zhao, M.: Image in-painting with salient structure completion and Texture propagation, 0167-8655/ Elsevier, pattern Recognition (2011)
32. Turkan, M.: Novel texture synthesis methods and their application to image prediction and image inpainting. Ph.D. thesis, Univ. Rennes 1 (2011)
33. Mart Inez-Noriega, R.: Exemplar-Based Image In-painting: Fast Priority and Coherent Nearest Neighbor Search. In: IEEE International Workshop on Machine Learning for Signal Processing, pp. 23–26 (2012)
34. Mahajan, K.S., Vaidya, M.B.: Image in Painting Techniques: A survey. IOSR Journal of Computer Engineering (IOSRJCE) 5(4), 45–49 (2012) ISSN: 2278 - 0661, ISBN: 2278 – 8727
35. Subban, R., Pasupathi, P., Muthukumar, S.: Image Restoration Based on Scene Adaptive Patch In-painting for Tampered Natural Scenes. Recent Advances in Intelligent Informatics, Advances in Intelligent Systems and Computing 235 (2013), doi:10.1007/978-3-319-01778-5-7, @Springer International Publishing Switzerland
36. Baek-Sop Kim, S., Park, J.: Exemplar Based Image Inpainting on a Projection Framework. International Journal of Software Engineering and Its Applications 7(3) (May 2013)

37. Subban, R., Pasupathi, P., Muthukumar, S., Krishnan, N.: Image Inpainting Techniques – A Survey and Analysis. In: International Conference on IIT, 978-1- 4673-6203-0© Conference on United Arab Emirates University, Dubai IEEE (2013)

38. Das, S., Reeba, R.: IJSER. Robust Exemplar based Object Removal in Video 1(2) (2013), ISSN 2347-3878

39. Sangeetha, K., Sengottuvelan, P., Balamurugan, E.: Performance analysis of exemplar based image inpainting algorithms for natural scene image completion. In: International Conference on Intelligent Systems and Control (ISCO), pp. 276–279. IEEE (2013)

40. Doria, D.: A Greedy Patch-based Image Inpainting Framework. Posted in Scientific Visualization, ITK

41. Ashikhmin, M.: Synthesizing natural textures. In: ACM Symposium on Interactive 3D

Gray Level Image Enhancement Using Cuckoo Search Algorithm

Soham Ghosh, Sourya Roy, Utkarsh Kumar, and Arijit Mallick

Abstract. In this work we have assessed the capability of a new optimization algorithm – the Cuckoo Search algorithm in tuning the image enhancement functions for peak performance. The assessment has been conducted in comparison to two of the old optimization algorithm aided enhancement, namely, Genetic Algorithms and Particle Swarm Optimization and previous enhancement techniques Histogram Equalization and Linear Contrast Stretch techniques. Results have been assimilated in this paper and conclusions have been drawn keeping the fitness of image and number of edgels in enhanced image as the benchmark. The results have illustrated the capability of Cuckoo search algorithm in optimizing the enhancement functions.

Keywords: Cuckoo Search Optimization, Edgels, Enhancement, Image fitness, Levy flight, Metaheuristic Algorithm.

1 Introduction

Images are one of the most important modes for transmission and presentation of information. Various areas such as biomedical science, fault detection, machine learning, etc. mainly use images as input [7, 8, 12, 13]. It is therefore necessary that we be able to produce and transmit images which are uncorrupted. Also, it is important to ensure that the information required for a particular process is easily available from them [7, 8, 12, 13].

One may obtain many kinds of relevant information from an image. For example, a task may require calculation of intensity from an image, whereas, in another

Soham Ghosh · Sourya Roy · Utkarsh Kumar · Arijit Mallick
Department of Instrumentation and Electronics Engineering, Jadavpur University,
Salt Lake Campus, LB-8, Sector 3, Kolkata, West Bengal, India
e-mail: {sohamghosh1993,souroy099,utkarsh.iee.ju,
     aribryan}@gmail.com

S.M. Thampi, A. Gelbukh, and J. Mukhopadhyay (eds.), *Advances in Signal Processing and Intelligent Recognition Systems*, Advances in Intelligent Systems and Computing 264,
DOI: 10.1007/978-3-319-04960-1_25, © Springer International Publishing Switzerland 2014

problem, we may be interested in highlighting the edges of an image. In such cases, images need to be transformed from one form to another such that the required features are emphasized, whereas those which are not required are either removed or reduced [7, 8, 12, 13]. Image processing enables us to achieve such results. The enhancement of an image is one such category under image processing [7, 8, 12, 13].

One fundamental property of every image, irrespective of what type it is, is that it is always in a discrete form. The smallest undividable part of an image is called a pixel. The pixels are function of spatial coordinates of an image represented as [7, 8, 12, 13]:

$$P=(x,y) \tag{1}$$

Where x and y are spatial coordinates of the image. The value returned by this function is the value of the pixel. This value may be anything from the intensity of the image at that point to grayscale value (in case of monochromatic image). Every image processing technique is essentially an application of the mathematical function at each pixel and/or its surrounding pixels [7, 8, 12, 13]. Every technique aims at drawing information from the pixel and its surrounding pixels, and then generating a new value for the subject pixel.

Image enhancement, as the name suggests, is a class of image processing operations which aims at producing a digital image which is visually more suitable as appearance for its visual examination by a human examiner or machine [7, 8, 12, 13].

A major problem that once use to lie with image processing, in this case with image enhancement, is that a human was needed to judge whether an image was suitable for the required task or not [7, 8, 12, 13]. There was no specific benchmark against which the measurement of enhancement could be drawn and only a human could evaluate a processed image as being suitable for the purpose [7, 8, 12, 13].

The problem of eliminating a human interpreter was finally solved with the introduction of metaheuristic algorithms and genetic programs [1, 2, 3, 4]. The algorithms/programs would improve the enhancement functions and allow the functions to work optimally or as restricted by the user to produce the desired results. In one of the works, C. Munteanu & A. Rosa [2] have demonstrated the application of genetic algorithms to overcome the problem of a human judge to a large extent and also shown the improved capability of enhancement functions in transforming images. Such works have been instrumental in introducing metaheuristic algorithm aided enhancement operations.

As further developments were made in the area of metaheuristic algorithms, it was found that as the algorithms became more capable, the efficiency of the functions increased too. The genetic algorithm (GA) was followed by particle swarm optimization (PSO) and then cuckoo search optimization (CSO).

As we are trying to display the power of CSO based optimization for enhancement function, in this work we have also generated results for some classical enhancement techniques Histogram equalization (HE) and Linear Contrast Stretch (CS) techniques. These algorithms do not include any other metaheuristic algorithms.

In our work we have taken the aid of a novel algorithm known as the Cuckoo Search Optimization (CSO) [5, 6, 9]. The algorithm's efficiency is compared to the other algorithms and conclusions are drawn from the results illustrated. The CSO has shown to be more efficient at tuning the enhancement functions compared to the other metaheuristic algorithms. In this work the resultant images have been shown after subjecting to the algorithms mentioned and the details for comparison of these generated images have been presented. The results clearly display the CSO algorithm's capability.

2 Enhancement Functions and Their Related Equations

The transformation function for the enhancement generates a new intensity value based on the intensity value of each pixel of the original image. These intensity values generated correspond to each pixel of the enhanced image. We also require a function which will determine the quality of the enhanced image, known as the evaluation function or criterion [4, 7, 8, 12, 13].

2.1 Transformation Function

In this case, we use a function that inputs the intensity value of each pixel of a P x Q image to produce an enhanced image. P is the number of columns and Q the number of rows of the image pixels. Local enhancement methods are used to apply transformation functions based on gray-level intensity, in the neighborhood of every pixel in an image.

The transformation equation is given by:

$$g(x,y) = \left(\frac{kM}{\sigma(x,y)+b}\right)\{f(x,y) - c.m(x,y)\} + m(x,y)^a \tag{2}$$

Here $0.5 < k < 1.5$, $a \in \psi_1$, $b \in \psi_2$, $c \in \psi_3$, with $\psi_1, \psi_2, \psi_3 \subset R$

Where $\sigma(x, y)$ and $m(x, y)$ are the local standard deviation and mean computed in the neighborhood centered at (x, y). M is the global mean of the image. $F(x, y)$ And $g(x, y)$ are the input and output grayscale intensities of the image used for enhancement. The ranges of a, b, c, k are given above. For a given pixel of an image, all the variables in g(x,y) are known. The values of a, b, c and k are chosen as needed. So the value of the transformation function depends entirely on them.

The formula for local mean about an n X n window is given by:

$$m(x,y) = \frac{1}{n*n}\sum_{x=0}^{n-1}\sum_{y=0}^{n-1} f(x,y) \tag{3}$$

And the formula for standard deviation is given by:

$$\sigma(x,y) = \frac{1}{n*n} \sqrt{\sum_{x=0}^{n-1} \sum_{y=0}^{n-1} (f(x,y) - m(x,y))^2} \tag{4}$$

The global mean of the image is given by:

$$M = \frac{1}{P*Q} \sum_{x=0}^{P-1} \sum_{y=0}^{Q-1} f(x,y) \tag{5}$$

2.2 Fitness Function

In order to properly obtain the results of image enhancement, a function is required which objectively measures the fitness, i.e. the quality of the enhanced image without human interaction. An enhanced image must have a high intensity of the edges for the optimization technique to be successful. Hence the fitness function must be proportional to the intensity obtained [4, 7, 8, 12, 13]. Also, a good enhanced image must have a high number of edgels, i.e. pixels belonging to that edge. So the number of edgels is also proportional to the function. However, these criteria are not sufficient to determine the degree of enhancement of the image. This is due to the fact that a function that is proportional to number and intensities of edgels might be skewed towards an image that doesn't have a natural contrast. In order to avoid this, the entropy of the image is also taken into consideration. The fitness function described below is seen to be a good fit for evaluation of image enhancement:

$$F(J) = \log\left(\log\left(E(I(J))\right)\right) * \frac{n\_edgels(I(J))}{M*N} * H(I(J)) \tag{6}$$

Here F(J) is the fitness function. I(J) is the image obtained after the transformation function is applied. The parameters a, b, c and k are given by the particle as J= (a b c k). E(I(J)) is the sum of intensity of edges. In this equation, it is obtained using a Sobel edge detector. Other operators such as Canny are also available [4, 7, 8, 12, 13]. The edge detector used in this case is an automatic threshold detector. The Sum of the intensity of the edges is calculated using the following formula:

$$E(I(J)) = \sqrt{\partial u(x,y)^2 + \partial v(x,y)^2} \tag{7}$$

Where,

$$\partial u(x,y) = g(x+1,y-1) + 2g(x+1,y) + g(x+1,y+1)$$
$$- g(x-1,y-1) - 2g(x-1,y) - g(x-1,y+1)$$

$$\partial v(x,y) = g(x-1,y+1) + 2g(x,y+1) + g(x+1,y+1)$$
$$- g(x-1,y-1) - 2g(x,y-1) - g(x+1,y-1)$$

The equation for entropy of the image is given as:

$$H(J) = -\sum_{i=0}^{255} h_i \log_2 h_i \tag{8}$$

Where h_i is the probability of occurrence of ith intensity value of the enhanced image.

3   An Introduction to Cuckoo Search Optimization Algorithm

Metaheuristic algorithms have been developed by drawing inspiration from naturally evolved systems. The very fact that these natural systems have developed after millions of years of evolution and that they are as efficient as possible has tempted researchers to observe the patterns in these systems, and design and simulate algorithms based upon the systems. These algorithms are generally used in solving non-linear problems having huge number of constraints where a linear approach would either be extremely difficult or impossible to implement.

The Cuckoo Search Algorithm is a newly developed metaheuristic algorithm. It was developed and presented by Xin-She Yang and Suash Deb [5, 6, 9]. They were inspired by the unique breeding behavior of cuckoo birds, and used the concept in their development of the optimization algorithm. They also integrated the concept of the Levy flight of certain birds into this algorithm [9].

Most metaheuristic algorithms are imitations of a system already existing in nature, such as Particle Search Optimization. The most important aspects of these systems which are used in developing new algorithms, such as cuckoo search, are intensification and diversification. Intensification can be represented as the ability of the algorithm to generate the best possible outcomes after every major phase and Diversification as the ability of the algorithm to cover the entire breadth of the objective function as quickly as possible[5, 6].

In Cuckoo Search algorithm, the best results, which survive after every phase, are generated by imitating the breeding pattern of the cuckoos. The extent of the range of optimization in the enhanced image for the next iteration is generated using the Levy flight distribution [5, 6]. The salient features of the cuckoo reproduction behavior and Levy flight mechanism and its analogy with the optimization algorithm is described below-

- Each cuckoo lays one egg at a time, and dumps its egg in a randomly chosen nest.
- The top nests containing the highest quality of eggs will carry over to the next iteration.
- The number of vacant host nests is fixed, and the egg laid by each cuckoo may be discovered by the host bird with a probability $p \in (0,1)$. [5, 6]

Here, one of the eggs laid by the cuckoo represents a new solution to the search algorithm. Before advancing to the next stage, a distribution function determines the number of surviving eggs. The new number of eggs serves as the populace for the next iteration. Higher the number of iterations, the better the optimized result obtained. The iterations keep running until some required optimization value is reached [5, 6].

The Cuckoo Search Algorithm uses Levy flight distribution for random-walk search. The distribution is based on the random flight pattern displayed by birds. The flight patterns are random and isotropic in nature and are punctuated with sudden right-angled turns. In mathematical terms, the Levy flight is a succession of random steps having a heavy tailed probability distribution. The next step of the bird utilizing Levy flight pattern depends on its present state and the length of its next step. This distribution gives the input for the next iteration of the search algorithm [5, 6]. Therefore, each iteration gives a different value of the distribution. Mathematically, if f(x) is the number of solutions in the xth stage, then $f(x + 1) = f(x) + \alpha \oplus Levy()$.

The pseudo-code for Cuckoo Search is given by:

— Objective function f(x), $x = (x_1, ..., x_d)^T$;
— Initial a population of n host nests x_i (i = 1, 2... n);
— While (t <MaxGeneration) or (stop criterion);
— Get a cuckoo (say i) randomly by Levy flights;
— Evaluate its quality/fitness F_i;
— Choose a nest among n (say j) randomly;
— If $(F_i > F_j)$,
— Replace j by the new solution;
— End
— Abandon a fraction (p_a) of worse nests
   [And build new ones at new locations via Levy flights];
— Keep the best solutions (or nests with quality solutions);
— Rank the solutions and find the current best;
— End while [5, 6]

4   Procedure

We require the enhanced grayscale image as the output, for a given input grayscale image. For obtaining the enhanced image, the transformation function defined earlier is used on the basis of both local and global parameters of the input image, namely, local and global mean, and standard deviation of the input image. This transformation function contains the parameters a, b, c, k, which are varied in order to obtain the best possible result, that is, a better enhanced image.

The tuning of the parameters a, b, c, k is carried out using the Cuckoo Search Algorithm, in order to maximize fitness of the function. Directly using the transformation without tuning would be possible, but the results obtained might not give more detail. Hence the need for Optimization, which is carried out using CSO. Similarly, we have used other algorithms GA and PSO to optimize the parameters a, b, c, k and used classical HE and CS algorithms to compare the results with CSO. The use of CSO facilitates the optimization of the transformation function, and hence gives various values of a, b, c, k. The algorithm searches the entire breadth of the function and returns the set of values required for the maximization of the fitness evaluation function. The fitness evaluation function is calculated using the sum of intensities of the edges of the image, as well as the entropy and number of edge pixels (edgels). This gives a measure of how well the image enhancement has been done. The results obtained using CSO is then compared to those obtained using different algorithms such as PSO and GA based enhancement and HE and CS enhancements. CSO algorithm has been found to be efficient and less time-consuming than PSO, GA, HE and CS. The comparison is shown in tabular form, along with the input images, and the final enhanced images.

The initial values of the parameters used in the Cuckoo Search Algorithm are as follows:

1.   The number of nests= 25
2.   Discovery rate of Alien eggs per solution= 0.25
3.   Number of eggs per nest= 1

5   Results

The method described previously is applied to various grayscale images. The results obtained for the images are displayed in this paper in Figure 1. In order to judge the capability of this cuckoo-search based enhancement method, we have compared the fitness values and number of edgels obtained, as well as the algorithm computation time, with those generated by using GA and PSO based enhancement, as displayed in Table 2. We have also displayed the fitness scores and number of edgels of the original images in Table 1. The images shown are of different sizes.

Table 1 Characteristics of the Original Images

Image	Size(PxQ)	Number of edgels	Fitness
Circuit	272x280	22254	4.7161
Tire	232x205	9109	2.9952
Forest	1216x1911	39978	3.3595

The Edgels were counted using The Sobel Edge Detector for both the original (Table 1) and enhanced images, with threshold 0.01. Higher edgel values indicate that more detail is present in the picture. The use of enhancement algorithms further increase the value of the edgels, as more details become prominent on use of these. Higher the number of edgels obtained, better is the quality of the image in question, and lesser computational time gives better efficiency, as seen in the table below.

The results generated have been summarized in the following table, and is followed by the actual images.

Table 2 Parameters for comparison after Optimization (A Summary of Results)

Image	Measured	HE	LCS	GA	PSO	CSO
Circuit	Fitness	4.2694	4.7261	7.1799	10.9217	11.0064
	Edgels	23358	22358	29324	28680	29368
	Computation time (in seconds)	0.059	0.067	197.675	1534.66	134.194
Tire	Fitness	3.8531	3.0263	9.7286	10.0824	10.6595
	Edgels	14234	9222	16612	16638	17470
	Computation time (in seconds)	0.051	0.053	176.571	1495.36	127.326
Forest	Fitness	3.3841	3.4208	6.7340	10.0177	10.6278
	Edgels	47052	40646	50086	49097	50237
	Computation time (in seconds)	0.033	0.045	486.699	3762.95	165.341

6   Generated Images

Abbreviations - HE: Histogram Equalization method; CS: Contrast Stretching method

GA: Genetic Algorithm aided enhancement

PSO: Particle Swarm Algorithm aided enhancement

CSO: Cuckoo Search algorithm aided enhancement

6.1 Circuit

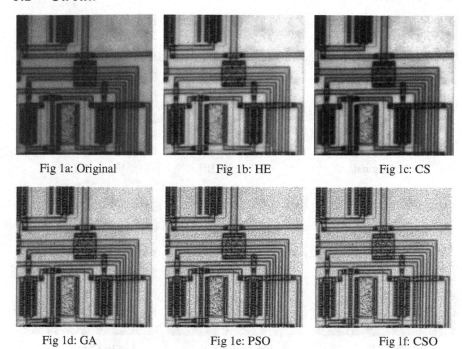

Fig 1a: Original   Fig 1b: HE   Fig 1c: CS

Fig 1d: GA   Fig 1e: PSO   Fig 1f: CSO

6.2 Tire

Fig 2a: Original   Fig 2b: HE   Fig 2c: CS

Fig 2d: GA   Fig 2e: PSO   Fig 2f: CSO

6.3   *Forest*

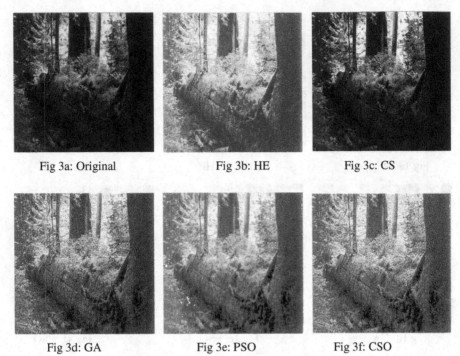

Fig 3a: Original	Fig 3b: HE	Fig 3c: CS

Fig 3d: GA	Fig 3e: PSO	Fig 3f: CSO

7   Discussions and Future Work

The paper may be summarized as follows:

- Three monochromatic images were taken for experimentation
- The enhancement functions were chosen.
- The function was tuned using GA, then PSO and finally with CSO. The images were subjected to the optimized enhancement filter
- The images are also enhanced using HE and CS enhancement methods
- The results were measured in the form of fitness of enhanced image and number of edgels generated for HE, CS, GA aided, PSO aided and CSO aided.

The results clearly depict the capability of Cuckoo Search Algorithm in optimizing the enhancement filters. The images enhanced by CSO enabled enhancement functions look more detailed. The edges have become extremely prominent in the case of CSO. The fitness quotient and the number of edgels are seen to be higher for CSO in all the cases compared to GA, PSO, HE and CS. (Table 2)

The computational time taken by CSO is very low compared to other population based algorithm such as GA and PSO. But in the case of HE and CS we find that the computational time is far below any other metaheuristic algorithm. The results however generated by the HE and CS algorithms show very poor results compared to any other metaheuristic algorithm. (Table 2)

Though the CSO based enhancement as shown in this work may not be able to generate results fast compared to the classical algorithms such as HE and CS, but it very effectively optimizes the enhancement function to generate best possible results.

This work is of immense importance as the improvement over the previous algorithms has been drastic. The algorithm has produced best results in a satisfactory time. The work presented in this paper has very vast range of application areas as mentioned before. Enhanced images find wide applications in various places such as biomedical engineering, fingerprint detection [14, 15] etc.

The work from here may approach two different directions now. As now the power of CSO based optimized enhancement has been displayed, the same method can be applied to other spheres of image processing to generate more efficient tools for analysis. Beyond image processing the CSO algorithm has also been found to be used effectively in various places such as classification [21], in developing data clustering method [16, 17], efficient noise suppression [19], feature selection [18], estimation techniques [20].

The other path that this work may take is optimization of enhancement functions using a more capable metaheuristic algorithm. As more and more developments are made in the field of metaheuristic algorithms, we may find more powerful algorithms at our disposal. These algorithms may optimize enhancement functions better than CSO and yield better results.

Acknowledgements. The authors would like to acknowledge the work of Xin-she Yang. His MATLAB program code (http://www.mathworks.in/matlabcentral/fileexchange/29809-cuckoo-search-cs-algorithm) for Cuckoo search algorithm has been used as original and edited in some cases in the given work.

The authors would also like to bring to the attention of readers that the MATLAB Image Processing toolbox (http://www.mathworks.in/products/image/) has been used in this work.

References

1. Gorai, A., Ghosh, A.: Gray-level Image Enhancement By Particle Swarm Optimization. In: World Congress on Nature & Biologically Inspired Computing, NaBIC 2009, pp. 72–77 (2009) Print ISBN: 978-1-4244-5053-4
2. Munteanu, C., Rosa, A.: Towards automatic image enhancement using Genetic Algorithms. In: Proceedings of the 2000 Congress on Evolutionary Computation, vol. 2, pp. 1535–1542. Inst. Superior Tecnico, Univ. Tecnica de Lisboa, Portugal (2000)
3. Braik, M., Sheta, A.F., Ayesh, A.: Image Enhancement Using Particle Swarm Optimization. In: Proceedings of the World Congress on Engineering, WCE 2007, London, U.K, July 2-4, vol. I (2007) ISBN:978-988-98671-5-7
4. Singh, N., Kaur, M., Singh, K.V.P.: Parameter Optimization In Image Enhancement Using PSO. American Journal of Engineering Research (AJER) 2(5), 84–90, e-ISSN : 2320-0847 p-ISSN : 2320-0936

5. Yang, X.-S., Deb, S.: Cuckoo search via Lévy flights. In: Proc. of World Congress on Nature & Biologically Inspired Computing (NaBIC 2009), India, pp. 210–214. IEEE Publications, USA (2009)
6. Yang, X.-S., Deb, S.: Engineering Optimisation by Cuckoo Search. Int. J. Mathematical Modelling and Numerical Optimisation 1(4), 330–343 (2010)
7. Gonzalez, R.C., Woods, R.E.: Digital Image Processing, 2nd edn. Prentice Hall Publications
8. Gonzalez, R.C., Woods, R.E., Eddins, S.L.: Digital Image Processing using MATLAB, 2nd edn. Prentice Hall
9. Mantegna, R.N.: Fast, accurate algorithm for numerical simulation of Lévy stable stochastic processes. Phys. Rev. E 49(5), 4677–4683 (1994), doi:10.1103/PhysRevE.49.4677 Key: citeulike: 6592204
10. He, Y., Tian, J., Luo, X., Zhang, T.: Image enhancement and minutiae matching in fingerprint verification. Elsevier, Pattern Recognition Letters 24(9-10), 1349–1360 (2003)
11. Sezan, M.I., Tekalp, A.M., Schaetzing, R.: Automatic anatomically selective image enhancement in digital chest radiography. IEEE Trans. Med. Imag. 8, 154–162 (1989)
12. Pratt, W.K.: Digital Image Processing, 2nd edn. John Wiley and Sons (1991)
13. Castleman, K.R.: Digital Image Processing. Prentice Hall (1996)
14. Chaudhary, A., Vatwani, K., Agrawal, T., Raheja, J.L.: A Vision-Based Method to Find Fingertips in a Closed Hand. Journal of Information Processing Systems 8(3), 399–408 (2012)
15. Hong, L., Wan, Y., Jain, A.: Fingerprint image enhancement: algorithm and performance evaluation. IEEE Transactions on Pattern Analysis and Machine Intelligence 20(8) (August 1998)
16. Senthilnath, J.: Clustering Using Levy Flight Cuckoo Search. In: Proceedings of Seventh International Conference on Bio-Inspired Computing, vol. 202, pp. 65–75 (2013)
17. Saida, I.B., Nadjet, K., Omar, B.: A new algorithm for data clustering based on cuckoo search optimization. In: Pan, J.-S., Krömer, P., Snášel, V. (eds.) Genetic and Evolutionary Computing. AISC, vol. 238, pp. 55–64. Springer, Heidelberg (2014)
18. Rodrigues, D., Pereira, L.A.M., Almeida, T.N.S., Papa, J.P., Souza, A.N., Ramos, C.C.O., Yang, X.-S.: BCS: A Binary Cuckoo Search algorithm for feature selection. In: 2013 IEEE International Symposium on Circuits and Systems (ISCAS), May 19-23, pp. 465–468 (May 2013), doi:10.1109/ISCAS.2013.6571881
19. Pani, P.R., Nagpal, R.K., Malik, R., Gupta, N.: Design of planar EBG structures using cuckoo search algorithm for power/ground noise suppression. Progress In Electromagnetics Research M 28, 145–155 (2013), doi:10.2528/PIERM12121108
20. Aly, W.M., Sheta, A.: Parameter Estimation of Nonlinear Systems Using Lèvy Flight Cuckoo Search. Research and Development in Intelligent Systems XXX, 443–449 (2013), doi:10.1007/978-3-319-02621-3_33
21. Goel, S., Sharma, A., Bedi, P.: Journal Title - International Journal of Hybrid Intelligent Systems. Novel approaches for classification based on Cuckoo Search Strategy 10(3), 107–116 (2013), doi:10.3233/HIS-130169 (Issue Cover Date January 1, 2013)

Performance Improvement of Decision Median Filter for Suppression of Salt and Pepper Noise

Vikrant Bhateja, Aviral Verma, Kartikeya Rastogi, Chirag Malhotra, and S.C. Satapathy

Abstract. Integration of decision based schemes with median filtering has been applied previously in numerous works to identify and process only the corrupted pixels during image denoising. However, these approaches are performance limited owing to their dependency upon selection of pre-defined thresholds as a decision measure. This paper presents a novel algorithm for performance improvement of decision median filter for suppression of salt and pepper noise in digital images. The proposed algorithm performs decision (to adaptively increase the window size) by comparing the computed median with the minimum and maximum pixel values within a local window. Thereafter, the pixels are processed with the proposed algorithm; reaching a maximum window size limit of 9x9. Objective analysis of the obtained results is carried out using Peak Signal to Noise Ratio (*PSNR*) and Structural Similarity Index (*SSIM*) as quality parameters. As depicted from the simulations results the proposed algorithm is capable to suppress noise effectively even with the noise contamination levels as high as 90%.

Keywords: Decision median filter, PSNR, salt and pepper noise, SSIM, variable window.

1 Introduction

The process of the image formation, transmission, receiving and processing involves various types of internal as well as external disturbances which are

Vikrant Bhateja · Aviral Verma · Kartikeya Rastogi · Chirag Malhotra
Department of Electronics and Communication Engineering,
Shri Ramswaroop Memorial Group of Professional Colleges (SRMGPC),
Lucknow-227105(U.P.), India
e-mail:{bhateja.vikrant,aviralavi2007,
    kartikeya1991,malhotrachirag1992}@gmail.com

S. C. Satapathy
ANITS, Vizag, (A.P.), India
e-mail: sureshsatapathy@gmail.com

S.M. Thampi, A. Gelbukh, and J. Mukhopadhyay (eds.), *Advances in Signal Processing and Intelligent Recognition Systems*, Advances in Intelligent Systems and Computing 264,
DOI: 10.1007/978-3-319-04960-1_26, © Springer International Publishing Switzerland 2014

unavoidable. There are generally two kinds of noise that corrupt the digital images: one is the additive Gaussian noise and the other is the impulse noise. In course of digital data traffic, impulse noise forms the main cause of error as this noise is present due to bit errors in transmission [1-7]. The two types of impulse noises include: salt and pepper noise and random valued noise. Salt and pepper noise can corrupt images where the corrupted pixel takes either maximum or minimum gray level value; leading to severe degradation of image quality and loss of fine details. The objective of noise suppression in such corrupted images is to filter the impulses (specks of salt and pepper) so that the noise free image is fully restored with minimum signal distortion [8-14]. Several non-linear filters have been proposed for restoration of images contaminated by salt and pepper noise [15-23]. Among them, the conventional filtering approaches include: Standard Median Filter (SMF) [24], Decision based median filtering (DMF) approaches [25-27] and Adaptive median filters (AMF) [28-29]. These approaches have been established to perform reliable filtering when the levels of noise contaminations are fairly low. For impulse noise levels of 50% and above, the major outcomes are edge jitter leading to compromise of fine details as well as prominent blurring [30]-[31]. Chan et al. proposed a new impulse detection and filtering method [32] that uses a 3x3 window for computation. Although the processing time is less but there occurs a heavy blurring at high noise levels. Wang et al. [33] proposed a modified switching median filter which was successful in preservation of edges, only at low noise levels. Vijaykumar et al. proposed an algorithm [34] for detection of high density salt and pepper noise using robust estimation with a variable window of size 17x17. This enhanced the complexity and also the computation time. Jayaraj et al. presented a robust estimation technique [35] which efficiently removed the noise at low densities but reconstructed a poor quality image at higher densities using a maximum window size of 7x7. Recently other variants of median filters in terms of minimization of iterations and window size are also proposed by V. Bhateja et al. [36-40]. Hence, the algorithm proposed in this paper is capable of removing high density of salt and pepper noise by improving the performance of the DMF with minimal complexity (i.e. the maximum window size of 9x9). This algorithm performs robust decision (to adaptively increase the window size) by comparing the computed median with the minimum and maximum pixel values without the requirement of any preset thresholds. The obtained results demonstrated effective suppression of residual noise while satisfactorily preserving the fine details. The remaining paper is structured as follows: Section 2 describes the proposed noise suppression methodology. Section 3 contains the obtained results and its discussion. Section 4 draws the conclusions.

2   Proposed Methodology for Noise Suppression

Standard median filters tend to modify both noisy and noise-free pixels during their non-linear filtering approach. Consequently, effective removal of salt and pepper noise causes blurring and distorted features in the restored image. Thus, to distinguish between noisy and non-noisy pixels (prior to median filtering); DMF are used. However, DMF yielded degradation in performance at high noise densities because of their dependency on pre-defined thresholds; which are generally not standardized but vary with category of images. On the other hand, corrupted pixel values and replaced median pixel values are less correlated in AMF. With an outcome of blurring effect due to failure in estimation of local information as well as smeared edges [41]-[42]. Based on the above idea, the denoising algorithm proposed in this work uses a modified DMF employing a variable window. The key concept of the proposed method is to consider only the corrupted pixels as far as possible and then applying the filtering scheme to those pixels only, leaving the non-noisy pixels unprocessed. The proposed algorithm is initiated by moving a spatially adaptive window (w) of size 3x3 over the noisy image. The statistical parameters within this window are extracted which include minimum (min), maximum (max) and median values for the pixels. Two conditional decisions are applied at this level on the basis of the extracted statistical parameters. Thus, if the median value lies between the minimum and the maximum pixel values; it is further verified if the centre pixel $x(i,j)$ also lies within these limits. Upon fulfillment of the former condition, the centre pixel is left unprocessed; otherwise the centre pixel is replaced by the median value (where: $y(i,j)$ denotes the restored pixel value). However, during the conditional check if the median value do not happens to lie within the minimum and the maximum pixel values; the size of the window is adaptively increased by a factor of 2. Thereafter, again moving to the first step and extracting the parameters within this incremented window. It is worth noting that, the increment in the window size is permissible only to a maximum limit (Wmax) of 9x9; beyond which the centre pixel will be always replaced by the last processed pixel value. To attain effective filtering, the above process is carried out for two iterations for images corrupted with high density noise. The novelty of the proposed algorithm lies in the fact that optimal filtering is obtained without going beyond window sizes of 9x9; unlike the optimization approaches to denoising where the maximum size reaches to 39x39 [43]. In addition, the computational complexity of this algorithm is also minimized as residual filtering can be achieved without iterative application of the filter (beyond 2 iterations). The proposed denoising approach owing to its minimal complexity and feature preservation property can be well employed for prefiltering of medical images. This further catalyzes the enhancement and edge detection process for computer-aided diagnosis of breast cancers [44-54].

3   Results and Discussions

The proposed algorithm is tested using 256×256, 8-bits/pixel gray scale image of Lena as input images. This image is corrupted by salt and pepper noise at various densities and performance is objectively evaluated using the parameters such as Peak Signal to Noise Ratio (*PSNR*) in dB and Structural Similarity Index (*SSIM*) [55]-[59]. Fig. 1. (a)-(g) shows the images which have been corrupted by the salt and pepper noise of different intensities ranging from 10-70% respectively whereas Fig. 1. (h)-(n) shows the denoised images obtained by the proposed denoising

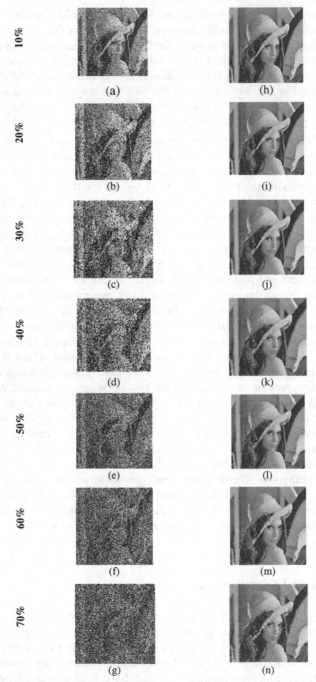

Fig. 1 Shows the results of the proposed noise suppression method for different intensity of salt and pepper noise. (a)-(g) Noisy images. (h)-(n) Images denoised with the proposed (Modified DMF) method.

density (50%-90%), the performance of the proposed algorithm is still able to preserve the details of the image along with suppression of residual noise. It is clear from the table 1(a)-(b) that at low noise densities i.e. from 10% to 20% the performance of the existing [42] as well as proposed DMF methods is almost same but as the noise density increases the proposed (modified) DMF algorithm outperforms the existing one; as validated by the computed values *PSNR* and *SSIM*. Further, for a noise density of 35%, *PSNR* of conventional DMF [26] and the existing DMF are 24.95dB and 25.68dB respectively whereas that of the proposed DMF is 30.26dB; which shows significant leap. The same is method (modified DMF algorithm). Fig. 2 shows comparison among the restored images obtained by standard DMF [26], the recent variant of DMF in work of A. Jourabloo et al. [42] and the modified DMF algorithm proposed in this paper. The quantitative performances in terms of *PSNR* and *SSIM* for all the above mentioned algorithms are shown in Table 1 (a)-(b) and Table 2 respectively. It can be seen that the level of salt and pepper noise is considerably reduced and the visualization of the images is also improved to a great extent in the reconstructed images. It is evident that till 50% noise density, there is very less blurring in the restored image supported with higher *PSNR* and *SSIM* values. Even on increasing the noise also implicit from the visual quality of the restored images. The image recovered from

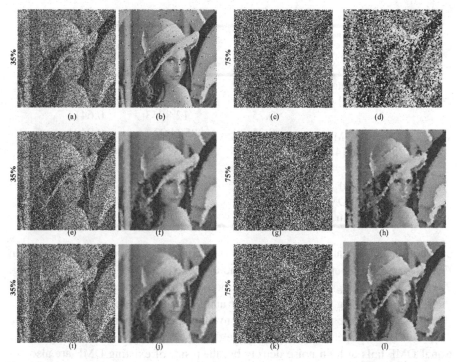

Fig. 2 Images restored by different denoising techniques. (a)-(d) shows the respone of conventional DMF [26] (e)-(h) shows the responseof the existing (recent) DMF algorithm [42]. (i)-(l) shows the response of the proposed denoising method (modified DMF).

Table 1 (a) Computation of *PSNR* (dB) for Proposed and Existing DMF algorithms

Intensity of Noise	Existing DMF Algorithm [42]	Proposed DMF Algorithm
10%	31.06	32.4364
20%	28.45	31.4279
30%	26.60	30.6164
40%	25.12	30.4228
50%	23.89	29.3125
60%	22.71	28.8264
70%	21.65	27.1123
80%	20.49	25.8800
90%	18.94	24.4667

Table 1 (b) Computation of SSIM for Proposed and Existing DMF algorithms

Intensity of Noise	Existing DMF Algorithm [42]	Proposed DMF Algorithm
10%	0.989	0.9927
20%	0.975	0.9909
30%	0.958	0.9892
40%	0.936	0.9875
50%	0.909	0.9845
60%	0.873	0.9813
70%	0.828	0.9746
80%	0.756	0.9647
90%	0.619	0.9491

Table 2 Comparison of Different Denoising Algorithms in terms of *PSNR* (in dB) and *SSIM*

Parameter	Intensity of salt and pepper noise	*PSNR*	*SSIM*
Conventional DMF[26]	0.35	24.9518	0.9555
	0.75	12.8943	0.6411
Existing DMF Algorithm [42]	0.35	25.6802	0.9431
	0.75	21.0813	0.7971
Proposed (Modified) DMF Algorithm	0.35	30.2611	0.9880
	0.75	26.8392	0.9714

conventional DMF (at 35% noise density) still contains traces of noise as shown in Fig. 2(b); on the other-hand the outcome of the existing DMF method as in Fig. 2.(f) shows streaking effect. But, the response obtained with the proposed DMF method shown in Fig. 2. (j) is quite clear and shows detail preservation. Fig. 2(d), (h), (l) shows image recovered from conventional DMF, existing DMF and the proposed DMF respectively at 75% noise density. The results shows that conventional DMF fails at high noise density but the result of existing DMF are also not very promising. However, the response of proposed DMF (at 75% noise density) is promising. Thus, it can be visualized from the obtained results that the proposed

DMF algorithm tends to improves the performance of ordinary DMF approach; yielding above satisfactory results even for high density salt and pepper noise along with preservation of fine image details. The contrast and visualization of the restored images is reasonably maintained in terms of HVS [60-65].

4 Conclusion

This paper focuses upon performance improvement of DMF for suppression of salt and pepper noise; across a wide range of noise densities. The modified DMF algorithm proposed in this work performs decision to adaptively increase the window size depending upon the number of corrupted pixels within the local window; expanding to a maximum window size of 9x9. This limit on window size reduces the computational complexity as well as blurring; which can be observed by improved visual quality coupled with higher *PSNR* and *SSIM* values for restored images. The obtained results are above satisfactory even at high noise densities without usage of any complex optimization approach or iterative application of the filtering algorithm.

References

1. Fabijanska, A., Sankowski, D.: Noise Adaptive Switching Median-Based Filter for Impulse Noise Removal from Extremely Corrupted Images. Image Processing, IET 5(5), 472–480 (2011)
2. Jain, A., Bhateja, V.: An Improved Image Denoising Algorithm using Robust Estimation for High Density Salt and Pepper Noise. In: Proc. of (IEEE) International Conference on Digital Convergence (ICDC 2011), Chennai, India, pp. 25–30 (2011)
3. Jain, A., Bhateja, V.: A Novel Image Denoising Algorithm for Suppressing Mixture of Speckle and Impulse Noise in Spatial Domain. In: Proc. of (IEEE) 3rd International Conference on Electronics & Computer Technology (ICECT 2011), Kanyakumari, India, vol. 3, pp. 207–211 (2011)
4. Garg, A., Shukla, J., Jain, A., Bhateja, V.: An Optimal Spatial Domain Edge Detector for Images Corrupted with Salt and Pepper Noise. In: Proc. of (IEEE) 4th International Conference on Electronics & Computer Technology (ICECT 2012), Kanyakumari, India, vol. 1, pp. 77–81 (2012)
5. Jain, A., Bhateja, V.: A Novel Detection and Removal Scheme for Denoising Images Corrupted with Gaussian Outliers. In: Proc. of IEEE Students Conference on Engineering and Systems (SCES 2012), Alla-habad, U.P., India, pp. 434–438 (2012)
6. Jain, A., Bhateja, V.: A Versatile Denoising Method for Images Contaminated with Gaussian Noise. In: Proc. of (ACM ICPS) CUBE International Information Technology Conference & Exhibition, Pune, India, pp. 65–68 (2012)
7. Gupta, A., Ganguly, A., Bhateja, A.: An Edge Detection Approach for Images Contaminated with Gaussian and Impulse Noises. In: Proc. of (Springer) 4th International Conference on Signal and Image Processing (ICSIP 2012), Coimbatore, India, vol. 2, pp. 523–533 (2012)

8. Han, W., Lin, J.: Minimum-Maximum Exclusive Mean (MMEM) Filter to Remove Impulse Noise from Highly Corrupted Images. Electronics Letters 33(2), 124–125 (1997)
9. Gupta, A., Tripathi, A., Bhateja, V.: De-Speckling of SAR Images via An Improved Anisotropic Diffusion Algorithm. In: Proc. of (Springer) International Conference on Frontiers in Intelligent Computing Theory and Applications (FICTA 2012), Bhubaneswar, India. AISC, vol. 199, pp. 747–754 (2012)
10. Gupta, A., Tripathi, A., Bhateja, V.: Despeckling of SAR Images in Contourlet Domain using a New Adaptive Thresholding. In: Proc. of (IEEE) 3rd International Advance Computing Conference (IACC 2013), Ghaziabad, U.P., India, pp. 1257–1261 (2013)
11. Bhateja, V., Tripathi, A., Gupta, A.: An Improved Local Statistics Filter for Denoising of SAR Images. In: Thampi, S.M., Abraham, A., Pal, S.K., Rodriguez, J.M.C. (eds.) Recent Advances in Intelligent Informatics. AISC, vol. 235, pp. 23–29. Springer, Heidelberg (2014)
12. Bhateja, V., Singh, G.: A. Srivastava, A.: A Novel Weighted Diffusion Filtering Approach for Speckle Suppression in Ultrasound Images. In: Proc. of (Springer) International Conference on Frontiers in Intelligent Computing Theory and Applications (FICTA 2013), Bhubaneswar, India, vol. 247, pp. 459–466 (November 2013)
13. Bhateja, V., Srivastava, A., Singh, G., Singh, J.: A Modified Speckle Suppression Algorithm for Breast Ultrasound Images Using Directional Filters. In: ICT and Critical Infrastructure: Proc (Springer) of the 48th Annual Convention of Computer Society of India (CSI 2013), Vishakhapatnam, India, vol. 2, pp. 219–226 (December 2013)
14. Bhateja, V., Verma, R., Mehrotra, R., Urooj, S.: A Novel Approach for Suppression of Powerline Interference and Impulse Noise in ECG Signals. In: Proc (IEEE) International Conference on Multimedia, Signal Processing and Communication Technologies (IMPACT 2013), AMU, Aligarh, U.P., India, p. xx (November 2013)
15. Gupta, A., Ganguly, A., Bhateja, V.: A Noise Robust Edge Detector for Color Images using Hilbert Transform. In: Proc. of (IEEE) 3rd International Advance Computing Conference (IACC 2013), Ghaziabad, U.P., India, pp. 1207–1212 (2013)
16. Srivastava, A., Alankrita, Raj, A., Bhateja, V.: Combination of Wavelet Transform and Morphological Filtering for Enhancement of Magnetic Resonance Images. In: Snasel, V., Platos, J., El-Qawasmeh, E. (eds.) ICDIPC 2011, Part I. CCIS, vol. 188, pp. 460–474. Springer, Heidelberg (2011)
17. Alankrita, R.A., Shrivastava, A., Bhateja, V.: Contrast Improvement of Cerebral MRI Features using Combination of Non-Linear Enhancement Operator and Morphological Filter. In: Proc. of (IEEE) International Conference on Network and Computational Intelligence (ICNCI 2011), Zhengzhou, China, vol. 4, pp. 182–187 (May 2011)
18. Verma, R., Mehrotra, R., Bhateja, V.: An Improved Algorithm for Noise Suppression and Baseline Correction of ECG Signals. In: Proc (SPRINGER) of the International Conference on Frontiers of Intelligent Computing (FICTA 2013), Bhuvneshwar, India, pp. 733–739 (December 2012)
19. Gupta, A., Ganguly, A., Bhateja, V.: A Novel Color Edge Detection Technique Using Hilbert Transform. In: Satapathy, S.C., Udgata, S.K., Biswal, B.N. (eds.) Proceedings of Int. Conf. on Front. of Intell. Comput. AISC, vol. 199, pp. 725–732. Springer, Heidelberg (2013)

20. Verma, R., Mehrotra, R., Bhateja, V.: A New Morphological Filtering Algorithm for Pre-Processing of Electrocardiographic Signals. In: Proc (SPRINGER) of the Fourth International Conference on Signal and Image Processing (ICSIP 2012), Coimbatore, India, pp. 193–201 (De-cember 2012)
21. Bhateja, V., Verma, R., Mehrotra, R., Urooj, S.: A Non-linear Approach to ECG Signal Processing using Morphological Filters. International Journal of Measurement Technologies and Instrumentation Engineering (IJMTIE) 3(3), 46–59 (2013)
22. Bhateja, V., Urooj, S., Mehrotra, R., Verma, R., Lay-Ekuakille, A., Verma, V.D.: A Composite Wavelets and Morphology Approach for ECG Noise Filtering. In: Maji, P., Ghosh, A., Murty, M.N., Ghosh, K., Pal, S.K. (eds.) PReMI 2013. LNCS, vol. 8251, pp. 361–366. Springer, Heidelberg (2013)
23. Jain, A., Bhateja, V.: An Iterative Non-Linear Filtering Approach for Suppression of High Density Impulse Noise in Mammographic Images. In: Proc. of (IEEE) 3rd International Conference on Machine Learning and Computing (ICMLC 2011), Singapore, vol. 3, pp. 527–531 (February 2011)
24. Pitas, I., Venetsanopoulos, A.N.: Order Statistics in Digital Image Processing. Proc. of the IEEE 80(12), 1893–1921 (1992)
25. Srinivasan, K.S., Ebenezer, D.: A New Fast and Efficient Decision-Based Algorithm for Removal of High-Density Impulse Noises. IEEE Signal Processing Letters 14(3), 189–192 (2007)
26. Sun, T., Neuvo, Y.: Detail-Preserving Median Based Filter in Image Processing. Pattern Recognition Letters 15(4), 341–347 (1994)
27. Ng, P., Ma, K.: A Switching Median Filter with Boundary Discriminative Noise Detection for Extremely Corrupted Images. IEEE Transactions on Image Processing 15(6), 1506–1516 (2006)
28. Chen, P., Lien, C.: An Efficient Edge-Preserving Algorithm for Removal of Salt-And-Pepper Noise. IEEE Signal Processing Letters 15, 833–836 (2008)
29. Hwang, H., Hadded, R.A.: Adaptive Median Filters: New Algorithms and Results. IEEE Transaction on Image Processing 4(4), 499–502 (1995)
30. Wang, S.S., Wu, C.H.: A new impulse detection and filtering method for removal of wide range impulse noises. Pattern Recognition 42, 2194–2202 (2009)
31. Wang, C., Chen, T., Qu, Z.: A Novel Improved Median Filter for Salt-And-Pepper Noise from Highly Corrupted Images. In: IEEE Interna-tional Symposium on Systems and Control in Aeronautics, ISSCAA, Harbin, China, pp. 718–722 (2010)
32. Chan, R.H., Ho, C., Nikolova, M.: Salt-And-Pepper Noise Removal by Median-Type Noise Detectors and Detail-Preserving Regularization. 3rd IEEE Transactions on Image Processing 14(10), 1479–1485 (2005)
33. Wang, G., Li, D., Pan, W., Zang, Z.: Modified switching median filter for impulse noise removal. Signal Processing 90, 3213–3218 (2010)
34. Vijaykumar, V.R., Vanathi, P.T., Kanagasabapthy, P., Ebenezer, D.: High density impulse noise removal using robust estimation based filter. IAENG International Journal of Computer Science 35(3), 259–266 (2008)
35. Jayaraj, V., Ebenezer, D., Aiswarya, K.: High density salt and pepper noise removal in images using improved adaptive statistics estimation filter. International Journal of Computer Science and Network Security 9(11) (November 2009)
36. Mehrotra, R., Verma, R., Bhateja, V.: An Integration of Improved Median and Morphological Filtering Techniques for ECG Processing. In: Proc. of the 3rd International Advance Computing Conference (IACC 2013), Ghaziabad, U.P., India, pp. 1212–1217 (2013)

37. Shukla, A.K., Verma, R.L., Bhateja, V.: Directional Ordered Statistics Filtering for Suppression of Salt and Pepper Noise. In: Proc (IEEE) Int. Conf. on Signal Processing and Integrated Network (SPIN 2014), p. xx(February 2014)

38. Shukla, A.K., Verma, R.L., Bhateja, V.: An Improved Directional Weighted Median Filter for Restoration of Images Corrupted with High Density Impulse Noise. In: Shukla, A.K., Verma, R.L., Bhateja, V. (eds.) Proc (IEEE) Int. Conf. Reliability, Optimization and Information Technology (ICROIT 2014), p. xx (February 2014)

39. Bhateja, V., Rastogi, K., Verma, A., Malhotra, C.: A Novel Approach for Restoration of Images Corrupted with High Density Impulse Noise. In: Proc (IEEE) Int. Conf. on Signal Processing and Integrated Network (SPIN 2014), p. xx (February 2014)

40. Bhateja, V., Rastogi, K., Verma, A., Malhotra, C.: Improved Decision Median Filter for Video Sequences Corrupted by Impulse Noise. In: Proc (IEEE) Int. Conf. on Signal Processinug and Integrated Network (SPIN 2014), p. xx (February 2014)

41. Duan, D., Mo, Q., Wan, Y., Han, Z.: A Detail Preserving Filter for Impulse Noise Removal. In: Proc. of IEEE International Conference on Computer Application and System Modeling, ICCASM, Taiyuan, China, vol. 2, pp. V2-265–V2-268 (2010)

42. Jourabloo, A., Feghahati, A.H., Jamzad, M.: New Algorithms for Recovering Highly Corrupted Images with Impulse Noise. Scientia Iranica Transactions on Computer Science and Engineering and Electrical Engineering 19(6), 1738–1745 (2012)

43. Bakwad, K.M., Pattnaik, S.S., Sohi, B.S., Devi, S., Panigrahi, B.K., Gollapudi, S.V.: Bacterial Foraging Optimization Technique Cascaded with Adaptive Filter to Enhance Peak Signal to Noise Ratio from Single Image. IETE Journal of Research 55(4), 173–179 (2009)

44. Bhateja, V., Urooj, S., Pandey, A., Misra, M., Lay-Ekuakille, A.: A Polynomial Filtering Model for Enhancement of Mammogram Lesions. In: Proc. of IEEE International Symposium on Medical Measurements and Applications (MeMeA 2013), Gatineau, Quebec, Canada, pp. 97–100 (2013)

45. Bhateja, V., Misra, M., Urooj, S., Lay-Ekuakille, A.: A Robust Polynomial Filtering Framework for Mammographic Image Enhancement from Biomedical Sensors. IEEE Sensors Journal, 1–10 (2013)

46. Bhateja, V., Devi, S.: An Improved Non-Linear Transformation Function for Enhancement of Mammographic Breast Masses. In: Proc. of (IEEE) 3rd International Conference on Electronics & Computer Technology (ICECT 2011), Kanyakumari, India, vol. 5, pp. 341–346 (2011)

47. Bhateja, V., Urooj, S., Pandey, A., Misra, M., Lay-Ekuakille, A.: Improvement of Masses Detection in Digital Mammograms employing Non-Linear Filtering. In: Proc. of IEEE International Multi-Conference on Automation, Computing, Control, Communication and Compressed Sensing, vol. 119, pp. 406–408. PalaiKottayam, Kerala (2013)

48. Bhateja, V., Devi, S.: A reconstruction based measure for assessment of mammogram edge-maps. In: Proceedings of the International Conference on Frontiers of Intelligent Computing Theory and Applications (FICTA), pp. 741–746 (2012)

49. Bhateja, V., Devi, S., Urooj, S.: An evaluation of edge detection algorithms for mammographic calcifications. In: Proceedings of the Fourth International Conference on Signal and Image Processing 2012 (ICSIP 2012), pp. 487–498 (2012)

50. Gupta, R., Bhateja, V.: A new unsharp masking algorithm for mammography us-ing non-linear enhancement function. In: Proceedings of the International Conference on Information Systems Design and Intelligent Applications (INDIA 2012) held in Visakhapatnam, India, Visakhapatnam, India, pp. 779–786 (January 2012)

51. Gupta, R., Bhateja, V.: A logratio based unsharp masking (UM) approach for enhancement of digital mammograms. In: Proceedings of the CUBE International Information Technology Conference, pp. 26–31 (2012)
52. Pandey, A., Yadav, A., Bhateja, V.: Volterra filter design for edge enhancement of mammogram lesions. In: Proceedings of the IEEE 3rd International Advance Computing Conference (IACC), pp. 1219–1222 (2013)
53. Pandey, A., Yadav, A., Bhateja, V.: Contrast improvement of mammographic masses using adaptive Volterra filter. In: Proceedings of the Fourth International Conference on Signal and Image Processing (ICSIP 2012), pp. 583–593 (2012)
54. Pandey, A., Yadav, A., Bhateja, V.: Design of new Volterra filter for mammogram enhancement. In: Proceedings of the International Conference on Frontiers of Intelligent Computing: Theory and Applications (FICTA), pp. 143–151 (2012)
55. Gupta, P., Srivastava, P., Bharadwaj, S., Bhateja, V.: A New Model for Performance Evaluation of Denoising Algorithms based on Image Quality Assessment. In: Proc. of (ACM ICPS) CUBE International Information Technology Conference & Exhibition, Pune, India, pp. 5–10 (2012)
56. Gupta, P., Srivastava, P., Bharadwaj, S., Bhateja, V.: A Novel Full-Reference Image Quality Index for Color Images. In: Proc. of the (Springer) International Conference on Information Systems Design and Intelligent Applications (INDIA), Vishakhapatnam, India, pp. 245–253 (2012)
57. Gupta, P., Tripathi, N., Bhateja, V.: Multiple Distortion Pooling Image Quality Assessment. Inderscience Publishers International Journal on Convergence Computing 1(1), 60–72 (2013)
58. Singh, S., Jain, A., Bhateja, V.: A Comparative Evaluation of Various Despeckling Algorithms for Medical Images. In: Proc. of (ACM ICPS) CUBE International Information Technology Conference & Exhibition, Pune, India, pp. 32–37 (2012)
59. Jain, A., Bhateja, V.: A Full-Reference Image Quality Metric for Objective Evaluation in Spatial Domain. In: Proc. of IEEE International Conference on Communication and Industrial Application (ICCIA), Kolkata, W. B., India, pp. 91–95 (2011)
60. Bhateja, V., Srivastava, A., Kalsi, A.: Fast SSIM Index for Color Images Employing Reduced-Reference Evaluation. In: Proc. of (Springer) International Conference on Frontiers in Intelligent Computing Theory and Applications (FICTA 2013), Bhubaneswar, India. AISC, vol. xx, p. xx (2013)
61. Gupta, P., Srivastava, P., Bharadwaj, S., Bhateja, V.: A HVS based Perceptual Quality Estimation Measure for Color Images. ACEEE International Journal on Signal & Image Processing (IJSIP) 3(1), 63–68 (2012)
62. Gupta, P., Srivastava, P., Bharadwaj, S., Bhateja, V.: A Modified PSNR Metric based on HVS for Quality Assessment of Color Images. In: Proc. of IEEE International Conference on Communication and Industrial Application (ICCIA), Kolkata, W.B., India, pp. 96–99 (2011)
63. Jaiswal, A., Trivedi, M., Bhateja, V.: A No-Reference Contrast Measurement Index based on Foreground and Background. In: Proc. of IEEE Second Students Conference on Engineering and Systems (SCES), Allahabad, India, pp. 460–464 (2013)
64. Trivedi, M., Jaiswal, A., Bhateja, V.: A No-Reference Image Quality Index for Contrast and Sharpness Measurement. In: Proc. of IEEE Third International Advance Computing Conference (IACC), Ghaziabad, U.P., India, pp. 1234–1239 (2013)
65. Trivedi, M., Jaiswal, A., Bhateja, V.: A Novel HVS Based Image Contrast Measurement Index. In: Proc. of Springer Fourth International Conference on Signal and Image Processing (ICSIP), Coimbatore, India, vol. 2, pp. 545–555 (2012)

Emotion Recognition from Facial Expressions Using Frequency Domain Techniques

P. Suja, Shikha Tripathi, and J. Deepthy

Abstract. An emotion recognition system from facial expression is used for recognizing expressions from the facial images and classifying them into one of the six basic emotions. Feature extraction and classification are the two main steps in an emotion recognition system. In this paper, two approaches viz., cropped face and whole face methods for feature extraction are implemented separately on the images taken from Cohn-Kanade (CK) and JAFFE database. Transform techniques such as Dual – Tree Complex Wavelet Transform (DT-CWT) and Gabor Wavelet Transform are considered for the formation of feature vectors along with Neural Network (NN) and K-Nearest Neighbor (KNN) as the Classifiers. These methods are combined in different possible combinations with the two aforesaid approaches and the databases to explore their efficiency. The overall average accuracy is 93% and 80% for NN and KNN respectively. The results are compared with those existing in literature and prove to be more efficient. The results suggest that cropped face approach gives better results compared to whole face approach. DT-CWT outperforms Gabor wavelet technique for both classifiers.

Keywords: Frequency Domain, Feature Extraction, Classification, DT-CWT, Gabor Wavelet, Neural network, and KNN.

1 Introduction

Facial expressions are non-verbal signs that play an important role in interpersonal communications. Emotion recognition from facial expression system finds its

P. Suja · Shikha Tripathi
AmritaVishwaVidyapeetham, Amrita School of Engineering, Bengaluru – 560 035, India
e-mail: p_suja@blr.amrita.edu, suja_rajesh@yahoo.co.in,
    shikha.eee@gmail.com

J. Deepthy
CTS, Chennai, India
e-mail: deepthy.swaroop@gmail.com

S.M. Thampi, A. Gelbukh, and J. Mukhopadhyay (eds.), *Advances in Signal Processing*   299
and Intelligent Recognition Systems, Advances in Intelligent Systems and Computing 264,
DOI: 10.1007/ 978-3-319-04960-1_27, © Springer International Publishing Switzerland 2014

application in robotics, automobile, online gaming and text chat applications. The "Humanoid Robots" can carry out intellectual conversation with human beings. In order to enhance the communication between man and humanoid robots the emotion recognition from facial expressions can be used. In automobile industry the driving assistance system can incorporate this system, which in turn helps to reduce the accidents due to driver mistakes. In text chat application the emotion recognition system would automatically insert emoticons based on the user's facial expressions.

Recognition of emotions from facial expressions consists of two steps, viz, feature extraction and classification. Extraction of relevant features from the input image is known as feature extraction and classification refers to the process of classifying the emotions using the extracted features into any one of the six basic expression classes.

Feature extraction can be done either in time-domain or frequency domain. In this proposed work the DT-CWT and Gabor wavelet transform domain techniques are used for feature extraction. Neural network and K-Nearest Neighborhood with Euclidean Distance are used for classification. The results obtained are tabulated and analyzed.

2 Background Work

Automatic facial expression recognition system with high accuracy will help to create humanoid robots and machines that are truly intelligent and facilitate communication with humans. Recent research suggests that frequency domain techniques are giving better results compared to spatial domain techniques.

Extraction of relevant feature is very important in emotion recognition. Several transform based recognition algorithms have been proposed in literature which use Discrete Cosine Transform (DCT), Discrete Fourier Transform (DFT), Discrete Wavelet Transform (DWT) Gabor Wavelet Transform, and Dual Tree Complex Wavelet transform (DT-CWT) to transform the original image into frequency domain. Most of the signal information tends to be concentrated in a few low-frequency components of the DCT. DCT is real valued and provides a better approximation of a signal with fewer coefficients.2-D Discrete Cosine Transform can be used for facial feature extraction [1, 2, 3, 4].

Emotion recognition from facial expression system needs more information from the image. Wavelet transform decomposes the original image without loss of large amount of information. It has good time-frequency localization property and better energy compaction and also has the feature of natural pyramidal decomposition of the data. The 2-D DWT coefficients form the feature vector, which can be used for classification [6, 7].

A good feature extraction technique should have the following features: non redundancy, shift invariance, low-computational complexity and directional sensitivity in many scales [9]. Shift invariance and poor direction selectivity are the main disadvantages of DWT. Literature shows that DT-CWT and Gabor Wavelet methods can overcome the disadvantages of DWT method [5, 7, 10].

In this paper experiments have been carried out to explore the efficiency of DT-CWT [9, 10] and Gabor Wavelet [5, 8]. They are applied on the features extracted from the two approaches viz., whole face and cropped face to form feature vector which are then fed to classifiers to classify it into one of the six basic emotions. The results obtained are tabulated, analyzed and compared with the existing literature.

The rest of the paper is organized as follows: Section 3 gives a description of the proposed methods. Experimental details are discussed in section 4. Results and analysis is presented in section 5. Conclusion and future work is reported in section 6.

3   Proposed Methods for Emotion Recognition

Two techniques for emotion recognition have been proposed in this paper with an assumption that face has been already identified in the given image. In the first approach, the face region is segmented from the input image, resized and the feature vector is formed using the two chosen transform techniques. This approach is called whole face method. In the second approach, the eye along with eyebrow and mouth regions are cropped manually separately from the input image, resized and the feature vector is formed using the two chosen transform techniques. This approach is called cropped face method. These approaches are represented in Fig 1 and Fig 2. The transforms applied for feature vector formation are DT-CWT and Gabor Wavelet Transform. The feature vector formed from both these approaches is classified into one of the six basic emotions using Neural Network and KNN.

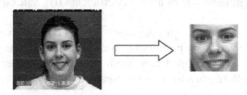

Fig. 1 Whole face obtained from input image

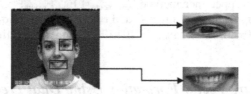

Fig. 2 Cropped Eye & eye brow and Mouth Region obtained from input image

3.1   The Feature Vector Formation Using Gabor Wavelet Transform

A family of Gabor kernel Ψ_k (z) is the product of a Gaussian envelope and a plane wave which is represented by (1) where z = (x, y), is the variable in spatial domain. Different selection of subscript μ and ν gives different Gabor kernel. We have chosen $\mu = 0,1,2,...,7$ and $\nu = 0,1,2, ...,4$ thus totally there are 40 Gabor functions to be used.

For a given image $I_{(x,y)}$, the Gabor wavelet transform is obtained by convolution of image with Gabor kernels $\Psi_{k(x,y)}$ as given in (2). Then the mean (m) and standard deviation (σ) of the magnitude of the transform coefficients are used to represent the image. Feature vector is constructed using m_{ij} and σ_{ij} as feature components, where 'i' (0 to 4) and 'j' (0 to 7) indicate the orientation and scale respectively.

$$\psi_k(z) = k^2_{(\mu,\nu)} \exp(\frac{-k^2_{(\mu,\nu)}}{2\sigma^2} z^2)(\exp(ik^2_{(\mu,\nu)}z) - \exp(-\frac{\sigma^2}{2})) \tag{1}$$

$$O_{k(x,y)} = I_{(x,y)} * \psi_{k(x,y)} \tag{2}$$

The algorithm for feature extraction is as follows:

Step 1: Generate the Gabor kernels using equation (1) for five scales and eight orientations [8].

Step 2: Get the input image from database and crop the mouth and eye region/ whole face region from the image, then resize cropped regions.

Step 3: Convolve the image with Gabor kernels to obtain Gabor wavelets.

Step 4: Find the magnitude part of all the coefficients five sales and eight orientations.

Step 5: Find the mean and standard deviation for 40 Gabor Wavelets.

Step 6: Form feature vector by combining mean and standard deviation of coefficients. In case of cropped face method concatenate the feature vector of eye and mouth regions to form the final feature vector.

Feature vector for whole face method $F_{face}= [m_{00}\sigma_{00} m_{01}\sigma_{01}........ m_{47}\sigma_{47}]$. Size of mean = 40, size of Standard Deviation = 40 and Size of final feature vector = 80.

Feature vector for cropped face method $F_{eye-mouth}=[F_{eye}:F_{mouth}]$

Where $F_{eye}= [m_{00}\sigma_{00} m_{01}\sigma_{01}........ m_{47}\sigma_{47}]$ and $F_{mouth}=[m_{00}\sigma_{00} m_{01}\sigma_{01}........ m_{47}\sigma_{47}]$

Size of $F_{eye}= 40+40=80$, size of $F_{mouth}= 40+40= 80$ and Size of final feature vector = 160.

3.2   The Feature Vector Formation Using Dual Tree-Complex Wavelet Transform (DT-CWT)

DT-CWT was introduced to overcome some drawbacks in DWT. The DT-CWT employs two real DWTs, namely 'Tree-a' and 'Tree-b'. The first DWT gives the real part of the transform while the second DWT gives the imaginary part. The two real wavelet transforms use two different sets of filters, with each satisfying the perfect reconstruction conditions. The two sets of filters are jointly designed so that the overall transform is approximately analytic.

2-D DT-CWT will yield 4 different sub-bands of the source image constituted of complex coefficients. There will be six oriented complex high pass sub-images at each level and a low frequency part of the image which contain most significant information. In 2-level DT-CWT implementation, magnitude part of six sub-images of all 2 levels and the low frequency part are used for representing the cropped eye and mouth regions.

The algorithm for feature extraction is as follows:

Step 1: Input image from database.

Step 2: Extract the mouth and eye region / whole face region from the image, and then resize cropped regions.

Step 3: Perform the 2-level DT-CWT on the normalized images.

Step 4: Extract the low frequency part and magnitude part of complex high pass sub-band image regions

Step 5: Select the highest 500 high pass sub-band image coefficients and all the low frequency coefficients to form the feature vector for whole face approach and in case of cropped face approach concatenate the feature vector of eye and mouth regions to form the final feature vector.

In whole face approach, the whole face region is resized into 112*112 so as to include eye and mouth. The 2-level DT-CWT decomposed of the resized image gives a low frequency image of size 56*56 and 6-high pass sub image of size 56*56 and 28*28 at level-1 and level-2 respectively. To form feature vector, all the low frequency coefficients and the highest 500 high pass sub-band image coefficients are used. So the feature vector size becomes 3636 (56*56 + highest 500 of 56*56*6+28*28*6).

In cropped face approach, the cropped eye and mouth regions are resized into 100*50. The 2-level DT-CWT is performed on the eye and mouth regions separately. The 2-level DT-CWT decomposed of the resized images gives a low frequency image of size 50*26 and 6-high pass sub image of size 50*25 and 25*13 at level-1 and level-2 respectively. To form the feature vector of the eye, all the low frequency coefficients and the highest 500 high pass sub-band image coefficients are used which result in a feature vector of size 1800 (50*26 + highest 500 of 50*25*6+25*13*6). Feature vector of the mouth can be obtained using similar procedure. The final feature vector has been formed by concatenating the eye and mouth region feature vectors. So the final feature vector size becomes 3600.

3.3   Classification Techniques

3.3.1   Neural Network Classifier

The feature vectors are randomly divided into Training, Validation and Test data. They are given to the neural network which consists of input layer, one hidden layer with 20 neurons and one output layer. The number of neurons in the input and output layer depends on the feature vector size and number of emotions. The network is trained with scaled conjugate gradient back propagation (trainscg) algorithm. The network calculates the Mean Squared Error (MSE) and Regression(R) values for the given input data. After training the network with the feature vector inputs and targets, the network is saved and is given to Neural Network Tool along with the testing data. Each test input is applied to the system with the trained network and the output is observed.

3.3.2   KNN Classifier

In KNN classifier suppose we have C classes of N samples each, then the distance between the test sample and each of the sample is calculated. The test data belongs to the class with which the distance is minimum. The distance measure used was 'Euclidean Distance'. Consider two points in the XY-plane; the shortest distance between the two points is along the hypotenuse, which is the Euclidean distance.

Euclidean Distance = $\sqrt[2]{\sum_{i=1}^{n} (x_i - y_i)^2}$ where X = [x1, x2, x3,...... xn] and Y = [y1, y2, y3,...... yn]

4   Experimental Details

The feature extraction and classification algorithms have been implemented using MATLAB R2010a. JAFFE and Cohn-Kanade (CK) database have been considered for the experiments. The main focus was on the applications where the user's expressions were stored in a database with slight change in their expressions and the emotions expressed by the users need to be detected with at most 100% accuracy. Sequence image based approach was adopted for implementing these cases.

4.1   CK Database

CK database consists of video sequence of subjects showing six basic prototypic emotions. But some video sequences do not show all six emotions. The images are of dimension 640*490 and in .png format.

From the CK database 30 female subjects were selected. For each emotion last four most expressive frames in the sequence were used for implementation. Each emotion contains 120 images, so totally 720 images were used. One image from the sequence was used to form the test set and the remaining three were used to form the train set. The experiment was repeated four times by reselecting the test set and the train set and the average recognition rate was found. Then the train set and test set are interchanged and the experiment was repeated four times and the average recognition rate was found. This process was repeated for each emotion.

4.2   JAFFE Database

JAFFE database consists of 219 images taken from 10 Japanese women, where each person has six basic emotions and neutral. The resolution of the image is 256 * 256 and in .tiff format.

10 subjects were taken from JAFFE database for each emotion including neutral and two sequence images were used for implementation. To classify seven emotions a total of 140 images were selected from which 70 were considered for testing and another 70 for training. The experiment was repeated twice by interchanging the test set and the train set and the average recognition rate was determined. The experiment was repeated for six emotions i.e. total 120 images were selected from which 60 was given for training and another 60 was given for testing. This process was repeated for each emotion. DT-CWT and Gabor Wavelet Transform techniques are applied on features extracted from both the datasets to form the feature vectors which are given to the classifiers separately. The results obtained for each transform technique with two different classifiers are tabulated.

5   Results and Analysis

DT-CWT feature extraction technique outperforms Gabor Wavelet feature extraction technique with both the classifiers. The upper part of the face region which includes eye, eyebrow regions and lower part which includes mouth region contains the relevant information while in whole face region unwanted information also exists. So by using the cropped face approach, the feature vector size has been reduced with good recognition rate.

For CK database with 540 train and 180 test images, DT-CWT with Neural Network classifier shows highest accuracy of 99.86% for both the approaches, whereas with 180 train and 540 test images DT-CWT with Neural Network classifier shows highest accuracy of about 98.06% for cropped face approach. With more number of train images than test images, DT-CWT technique with NN gives 100% recognition rate for all the expressions except 'fear'. 'Happiness' and 'surprise' are the two emotions which show the high recognition rates in most cases. 'Fear' gives the lowest recognition rate for the two cases of CK database.

The Gabor wavelet transform method gives comparatively better results with KNN than NN for 540 train and 180 test data, whereas it is reverse for 180 train and 540 test data. This method gives 100% recognition rate for 'surprise' and 'disgust' for cropped face method with NN. It also gives 100% recognition rate for 'anger' and 'happy' for cropped face method when combined with KNN.

In case of JAFFE database for whole face approach DT-CWT with NN shows highest accuracy of about 98.33% and 95.71% for both six class and seven class respectively. This is significantly higher than KNN. The Gabor wavelet transform method gives better results with NN than KNN for both 6 class and 7 class data. 'Neutral' emotion is considered for JAFFE database and only DT-CWT in combination with NN gives 100% recognition rate. JAFFE database results are less efficient compared to CK database due to less number of subjects and some of them being wrongly labeled. Tables 2 to 5 provide the recognition rates for all combinations of extraction, classification techniques and approaches.

Table 1 Comparison of the results of proposed technique and related literature

Reference	Methodologies Used	Database	Feature Extraction Methods	Recognition Rate With Different Classifiers (%)	
Sidra Batool Kazmi, Qurat-ul-Ain and M. Arfan Jaffar[7]	Whole Face	JAFFE 5-CLASS	DWT+PCA	BANK OF NN	
				96.4	
Shishir Bashyal, Ganesh K. Venayagamoorthy [5]	Whole Face	JAFFE 6-CLASS	GABOR WAVELET	LVQ	
				87.51	
Yadong Li, Qiuqi Ruan, Xiaoli Li[10]	Whole Face	JAFFE 7-CLASS	4-LEVEL DT-CWT+PCA	ED	
				88.6	
			GABOR+PCA	78.6	
		CK TRAIN-180 TEST-540	4-LEVEL DT-CWT+PCA	96.83	
Proposed Methods	Whole Face	JAFFE 7-CLASS	2-LEVEL DT-CWT	NN	KNN
				95.71	72.14
			GABOR	80	52.86
		JAFFE 6-CLASS	2-LEVEL DT-CWT	98.33	88.33
			GABOR	82.5	58.33
		CK TRAIN-540 TEST-180	2-LEVEL DT-CWT	99.86	99.45
			GABOR	95.29	95.69
		CK TRAIN-180 TEST-540	2-LEVEL DT-CWT	97.96	87.54
			GABOR	88.97	63.8
	Cropped Eye And Mouth	JAFFE 7-CLASS	2-LEVEL DT-CWT	88.57	65
			GABOR	87.8	54.29
		JAFFE 6-CLASS	2-LEVEL DT-CWT	89.17	63.33
			GABOR	88.33	62.5
		CK TRAIN-540 TEST-180	2-LEVEL DT-CWT	99.86	99.45
			GABOR	98.34	95.69
		CK TRAIN-180 TEST-540	2-LEVEL DT-CWT	98.06	88.52
			GABOR	92.72	78.52

The results are compared with those of existing literature and it is given in table 1. Whole face approach using JAFFE database has been found in majority of existing literature. The performance has improved in the proposed methods with different combination of transform techniques, classifiers and databases. The implemented 2-level DWT gives better results than 4-level DWT for CK database with 180 train and 540 test data [10]. Gabor wavelet technique with NN for JAFFE 6 class using cropped face method yields an average accuracy of 88.83% which is slightly higher than LVQ as reported in [5]. The average accuracy using NN with DT-CWT is again higher than DWT and PCA as reported in [7].

Table 2 Recognition rate using DT-CWT & Gabor Wavelet transform and NN & KNN for CK Database Train-540 Test-180

Expressions	Feature extraction techniques	Neural Network Classifier (%)		KNN Classifier (%)	
		Cropped	Whole face	Cropped	Whole face
Anger	DT-CWT	100	100	99.17	100
	GABOR	99.18	95.85	100	97.50
Disgust	DT-CWT	100	100	100	100
	GABOR	100	93.33	99.17	98.33
Fear	DT-CWT	99.18	99.18	95.84	96.67
	GABOR	95.85	90.85	96.67	88.34
Sad	DT-CWT	100	100	100	100
	GABOR	95.83	96.68	96.67	91.67
Happy	DT-CWT	100	100	99.17	100
	GABOR	99.18	97.5	100	98.34
Surprise	DT-CWT	100	100	100	100
	GABOR	100	97.53	99.17	100
% Average Recognition Rate	DT-CWT	99.86	99.86	99.03	99.45
	GABOR	98.34	95.29	98.61	95.69

Table 3 Recognition rate using DT-CWT & Gabor Wavelet transform and NN & KNN for CK database Train-180 Test-540

Expressions	Feature extraction techniques	Neural Network classifier (%)		KNN Classifier (%)	
		Cropped	Whole face	Cropped	Whole face
Anger	DT-CWT	98.63	98.35	88.33	89.45
	GABOR	98.35	86.13	83.33	74.72
Disgust	DT-CWT	98.08	98.9	86.39	86.67
	GABOR	98.05	90	69.45	60.56
Fear	DT-CWT	96.68	93.6	75.28	74.72
	GABOR	86.38	83.9	66.11	46.11
Sad	DT-CWT	96.38	96.93	91.39	84.17
	GABOR	89.45	88.33	68.89	58.06
Happy	DT-CWT	99.18	100	95.56	94.45
	GABOR	91.4	96.98	85.95	78.06
Surprise	DT-CWT	99.45	100	99.17	95.28
	GABOR	89.73	94.48	96.39	65.28
% Average Recognition Rate	DT-CWT	98.06	97.96	88.52	87.54
	GABOR	92.23	89.97	78.52	63.80

Table 4 Recognition rate using DT-CWT & Gabor Wavelet transform and NN & KNN for JAFFE Database Train-70 Test-70. (7 class)

Expressions	Feature extraction techniques	Neural Network Classifier (%)		KNN Classifier (%)	
		Cropped	Whole face	Cropped	Whole face
Anger	DT-CWT	85	100	80	85
	GABOR	90	85	55	70
Disgust	DT-CWT	80	100	60	100
	GABOR	85	95	30	85
Fear	DT-CWT	80	90	45	55
	GABOR	85	50	50	35
Sad	DT-CWT	90	100	60	55
	GABOR	95	80	45	25
Happy	DT-CWT	90	85	65	50
	GABOR	85	75	55	35
Surprise	DT-CWT	95	100	55	80
	GABOR	95	95	65	30
Neutral	DT-CWT	100	95	90	80
	GABOR	80	80	80	90
% Average Recognition Rate	DT-CWT	88.57	95.71	65	72.14
	GABOR	87.86	80	54.29	52.86

Table 5 Recognition rate using DT-CWT & Gabor Wavelet transform and NN & KNN for JAFFE Database Train-60 Test-60. (6 class)

Expressions	Feature extraction techniques	Neural Network Classifier (%)		KNN Classifier (%)	
		Cropped	Whole face	Cropped	Whole face
Anger	DT-CWT	95	100	65	90
	GABOR	95	85	70	75
Disgust	DT-CWT	70	100	65	100
	GABOR	85	95	45	85
Fear	DT-CWT	90	95	45	70
	GABOR	80	65	50	40
Sad	DT-CWT	90	95	55	70
	GABOR	80	90	70	50
Happy	DT-CWT	90	100	70	80
	GABOR	100	85	75	55
Surprise	DT-CWT	100	100	80	90
	GABOR	90	95	65	45
% Average Recognition Rate	DT-CWT	89.17	98.33	63.33	88.33
	GABOR	88.33	85.83	62.50	58.33

6 Conclusion and Future Work

Extensive tests have been carried out for emotion recognition using different combinations of chosen transform techniques and classification methods on CK and JAFFE databases using the two approaches. This work has been carried out to explore the efficiency of transform techniques. DT-CWT technique outperforms Gabor wavelet technique for both classifiers used. 2-level DT-CWT with normalization of images is computationally more efficient compared to 4-level DT-CWT technique proposed by Yadong et al [10].

Cropped face method reduces the size of feature vector and hence the storage requirement with improvement in time compared to whole face approach. In case of CK database, cropped face method is found to be more effective than the whole face method, while in JAFFE database it is vice versa. 'Fear' and 'sad' gives low recognition rate and the other entire expressions gives 100% recognition rate in CK database with more number of train images, than less number of train images. In JAFFE database surprise, anger and neutral shows high recognition rate whereas sad, happy and fear shows low recognition rate. Selection of the classifier is also important for achieving high recognition rate. Neural Network is found to be better than KNN classifier since the classification efficiency in KNN is greatly affected by the number of training samples.

Research can be continued using CMU PITTSBURG database which contain images with different poses and illumination. In future dimensionality reduction techniques could be applied to further reduce the size of feature vector.

References

1. Kharat, G.U., Dudul, S.V.: Human Emotion Recognition System Using Optimally Designed SVM with Different Facial Feature Extraction Techniques. WSEAS Transactions on Computers 7, 650–659 (2008)
2. Kharat, G.U., Dudul, S.V.: Neural Network Classifier for Human Emotion Recognition from Facial Expressions Using Discrete Cosine Transform. In: International Conference on Emerging Trends in Engineering and Technology, vol. 22, pp. 653–658. IEEE (2008)
3. Thomas, N., Mathew, M.: Facial Expression Recognition System Using Neural Network and MATLAB. In: International Conference on Computing, Communication and Applications (ICCCA). IEEE (2012)
4. Gupta, S.K., Agrwal, S., Meena, Y.K., Nain, N.: A Hybrid Method of Feature Extraction for Facial Expression Recognition. In: Seventh International Conference on Signal Image Technology & Internet-Based Systems, pp. 422–425. IEEE (2011)
5. Bashyal, S., Venayagamoorthy, G.K.: Recognition of Facial Expressions Using Gabor Wavelets and Learning Vector Quantization. Engineering Applications of Artificial Intelligence 21(7), 1056–1064 (2008)
6. Shi, D., Jiang, J.: The Method of Facial Expression Recognition Based on DWT-PCA/LDA. In: International Congress on Image and Signal Processing (CISP), pp. 1970–1974. IEEE (2010)

7. Kazmi, S.B., Ul-Ain, Q., Arfan Jaffar, M.: Wavelets Based Facial Expression Recognition Using a Bank of Neural Networks. In: 5th International Conference on Future Information Technology (FutureTech). IEEE (June 2010)
8. Zhou, S., Liang, X.-M., Zhu, C.: Support Vector Clustering of Facial Expression Features. In: International Conference on Intelligent Computation Technology and Automation, pp. 811–815. IEEE (2008)
9. Selsnick, W., Baraniuk, R.G., Kingsburg, N.G.: The Dual – Tree Complex Wavelet Transform – a coherent framework for multiscale signal and image processing. IEEE Signal Processing, Magazine 22(6), 123–151 (2005)
10. Li, Y., Ruan, Q., Li, X.: Facial Expression Recognition Based on Complex Wavelet Transform. In: IET 3rd International Conference on Wireless, Mobile and Multimedia Networks. IEEE (January 2010)

Systolic Array Implementation of DFT with Reduced Multipliers Using Triple Matrix Product

I. Mamatha, Shikha Tripathi, T.S.B. Sudarshan, and Nikhil Bhattar

Abstract. A generic 2D systolic array for N point Discrete Fourier Transform using triple matrix product algorithm is proposed. The array can be used for a non power of two sized N point DFT where $N=N_1N_2$ is a composite number. It uses an array of size $N_2 \times (N_1+1)$ which requires $(2N+4N_2)$ multipliers. For a DFT of size $4N_2$ (i.e multiple of four), an optimized design which requires $4N_2$ number of multipliers is proposed. It is observed that the proposed optimized structure reduces the number of multipliers by 66.6% as compared to the generic array structure while maintaining the same time complexity. Two examples are illustrated, one with non power of two size DFT and another with a DFT of size $4N_2$. Both the generic and optimized structures use the triple matrix product representation of DFT. The two structures are synthesized using Xilinx ISE 11.1 using the target device as xc5vtx240t-2ff1759 Virtex-5 FPGA. The proposed structure produces unscrambled stream of DFT sequence at output avoiding a necessity of reordering buffer. The array can be used for matrix -matrix multiplication and to compute the diagonal elements of a triple-matrix multiplication.

Keywords: Discrete Fourier Transform, Systolic Array, FPGA.

1 Introduction

Discrete Fourier Transform (DFT) is an efficient tool used in various digital signal processing applications. As it is computationally intensive, it is always desirous to

I. Mamatha · Shikha Tripathi · T.S.B. Sudarshan
AmritaVishwaVidyapeetham, Amrita School of Engineering,Bengaluru, 560 035,India
e-mail : mamraj78@gmail.com, shikha.eee@gmail.com,
        sudarshan.tsb@gmail.com

Nikhil Bhattar
Engineer Staff-II, IC Design,Broadcom India Research Pvt.Ltd, Bangalore, India
e-mail : nikhil_bhattar@yahoo.com

S.M. Thampi, A. Gelbukh, and J. Mukhopadhyay (eds.), *Advances in Signal Processing and Intelligent Recognition Systems*, Advances in Intelligent Systems and Computing 264,
DOI: 10.1007/978-3-319-04960-1_28, © Springer International Publishing Switzerland 2014

compute DFT faster to meet the requirement of real time applications. In addition, low power consumption is an added requirement by portable devices. Several algorithms and architectures are reported in literature to meet this requirement. Several architectures are proposed for FFT algorithm [1,10,11]. The algorithm proposed in [11] for fast computation of DFT though reduces the number of multipliers results in increase in other elements like adders and multiplexers. However, for Very Large Scale Integrated circuit implementation it is desirous to have more local communications rather than global communications which improves the speed of computation. Systolic array architectures [3] are best suited for this purpose due to simplicity, local interconnections, high level of pipelining and its similarity to Field Programmable Gate Array (FPGA) fabric. FFT structure proposed by Cooley and Tukey [4] is not well suited for systolic array implementation due to their complex global communications. Several systolic architectures were proposed in literature for implementing DFT.

Further, FFT based architectures are applicable only for radix 2/4/8 which restricts on the DFT size N which needs to be a power of two, four or eight. There are applications where non power of two sized DFT are used [6], thereby, it is required to develop efficient computational structures for N point DFT where N can be a composite number. Prime factor algorithm, row- column decomposition are few such approaches used to develop an efficient architecture. An architecture developed in [6] based on the decomposition of N as small sized DFTs N_1 and N_2 is proven to be efficient for base-4 (N_2=4) and can compute DFT of size 256n. The cyclic convolution based structures for DFT is another popular and efficient technique to compute DFT [2,8]. These architectures require lot of preprocessing and post processing stages before and after representing the cyclic convolution form. A design for general transform length is also proposed in [9].

Reference paper [5] describes a triple- matrix product algorithm from the theoretical point of view. In this paper, a new approach for realizing the N-point Discrete Fourier Transform where $N=N_1N_2$ is a composite number, is presented. Here, an algorithm is developed to represent N point DFT as triple- matrix product and is mapped onto a 2- D systolic array of processors. The structure hence developed is termed as generic structure for convenience. Further, the architecture is optimized to reduce the number of multipliers when N is a multiple of four. The total time required to compute an N-point DFT is ($3N_2$ - 2)+N time units. To describe the generic architecture implementation, the implementation of a 16-point and 15-point DFT is shown. For computing lager size DFTs, the smaller DFT block can be used repeatedly. The implementation of 16- point and 64- point DFT using optimized structure is illustrated as an example for multiple of four . The architecture can be used for matrix-matrix multiplication and also for the computation of the diagonal terms of triple matrix multiplication.

Rest of the paper is organized as follows. Session 2 describes the triple-matrix product based algorithm for N point DFT which is applicable for power of two and non-power of two sized DFTs. The 2-D systolic array architecture using processing elements for the algorithm proposed is developed in session 3. The performance evaluation of the architecture is also explained in session 3. Session 4 details about the architecture optimization for multiple of four sized DFT. Hardware implementation of both generic and optimized design for different examples and a comparison of the device utilization of optimized design with that of generic design are reported in session 5.Ths session also compares the proposed architecture with that of the other architectures reported in literature. Session 6 concludes the work carried out and briefs about the future scope.

2   Triple Matrix Product Algorithm

In this session, we develop the triple matrix product algorithm for DFT. N point DFT is decomposed into smaller length and a re-indexing technique is used in order to represent as matrix-matrix-matrix product.

N point DFT of a sequence x(n) is

$$X(k) = X_k = \sum_{n=0}^{N-1} x(n) W_N^{kn} \quad for \ k = 0,1,......N-1 \tag{1}$$

where, $W_N^{kn} = e^{-\frac{j2\pi kn}{N}}$

Decomposing $N=N_1N_2$ and re-labeling the indices, $n = N_1 n_1 + n_2$, DFT can be

written as $X_k = \sum_{n_1=0}^{N_2-1} \sum_{n_2=0}^{N_1-1} x(N_1 n_1 + n_2) W_N^{k(N_1 n_1 + n_2)}$

$$\tag{2}$$

Here, the signal sequence x_n can be rearranged to a rectangular array. For example, let N=16= 4x4 (i.e multiple of four) where $N_1=4$, $N_2=4$;

$$X_k = \sum_{n_1=0}^{3} \sum_{n_2=0}^{3} x(4n_1 + n_2) W_{16}^{k(4n_1+n_2)}$$

$$\tag{3}$$

The signal sequence x(n) can be represented by a matrix X.

$$X = \begin{bmatrix} X_{n_1 n_2} \end{bmatrix} = x(4n_1 + n_2) = \begin{bmatrix} x_1 & x_2 & x_3 & x_4 \\ x_5 & x_6 & x_7 & x_8 \\ x_9 & x_{10} & x_{11} & x_{12} \\ x_{13} & x_{14} & x_{15} & x_{16} \end{bmatrix} \text{ is a 4x4 matrix}$$

$$\tag{4}$$

DFT can now be expressed as,

$$X_k = \sum_{n_1=0}^{3} \sum_{n_2=0}^{3} a_{kn_1} x_{n_1 n_2} b_{n_2 k}$$

$$\tag{5}$$

where,

$$A = a_{kn_1} = W_{16}^{4\,kn_1} \quad \text{is a 16x4 matrix ,} \tag{6}$$

$$B = b_{n_2 k} = W_{16}^{kn_2} \quad \text{is a 4x16 matrix.} \tag{7}$$

$C = A.X.B$ is a triple matrix product and the size of C matrix is 16x16

$$X_k = C_{kk}$$

DFT given by X_k is the diagonal elements of matrix C.

Further, the matrix A can be partitioned into four identical matrices. Each of the partitioned matrices is a 4x4 square matrix for N=16.

$$A = [A_1 \quad A_2 \quad A_3 \quad A_4]^T \tag{8}$$

Due to periodicity property of exponential function i.e, $e^k = e^{k+2\pi}$

$$A_1 = A_2 = A_3 = A_4 \tag{9}$$

This property of the matrix based DFT algorithm is significant as it reduces the memory required to store the coefficient matrix A. Only one 4x4 matrix needs to be stored instead of the entire 16x4 matrix. This property is made use of in the DFT implementation by looping each column of the A column matrix so that the same 4x4 matrix can be reused. Fig.3 shows the systolic architecture.

2.1 Triple Matrix Product Form When N is Composite Number

The proposed triple-matrix based algorithm is applicable for non power of two sized DFT also. Consider N=15 = 3 x 5 , where $N_1=3$, $N_2=5$. Using the algorithm, x(n) is written as a 3x5 matrix

$$X_k = \sum_{n_1=0}^{4} \sum_{n_2=0}^{2} x(3n_1 + n_2) W_{15}^{k(3n_1+n_2)} \tag{10}$$

$$X = [X_{n_1 n_2}] = \begin{bmatrix} x_1 & x_2 & x_3 & x_4 & x_5 \\ x_6 & x_7 & x_8 & x_9 & x_{10} \\ x_{11} & x_{12} & x_{13} & x_{14} & x_{15} \end{bmatrix}$$

$$a_{kn_1} = W_{15}^{3n_1 k} \quad \text{is 15x3 matrix}$$

$$b_{n_2 k} = W_{15}^{k n_2} \quad \text{is 3x15 matrix}$$

The systolic array of size $N_2 \times (N_1+1)$ i.e 5x4 is used to compute the DFT.

3   Two-D Systolic Array Architecture for DFT

The triple-matrix based algorithm developed in session 2 is mapped onto a 2-dimensional systolic array composed of an array of processing elements, each of which compute inner-product step of the algorithm.

3.1   Two-D Systolic Generic Architecture

The row-column multiplication (inner product) which involves multiply and accumulates operation (MAC) is given by,

$$D_{out} = D_{in} + A_{in} * X_{in} \tag{11}$$

First step is to carry out the AX matrix multiplication. A systolic array for matrix multiplication is designed which uses an array of processing elements (PE). Each PE does a multiply and an add operation on the individual inputs as shown in Fig.1. This processing element is termed as processing element A, since it uses the A coefficient matrix.

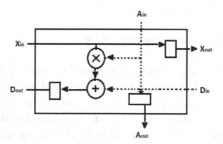

Fig. 1 Processing Element (PE)

The processing element has inputs A_{in} (coefficient A), X_{in} (signal sample from matrix X) and D_{in} (inner product from the previous stage). The new inner product is calculated and is registered, so that it can be used by the next stage in the systolic array. X_{in} and A_{in} are also registered and passed on to the next stage.

As has been discussed earlier, only the diagonal elements of the resulting matrix C have to be computed. For this purpose, another column of PEs namely PE 'B' is used which does the same computation as PE 'A' but on inputs B and D which are complex numbers. Therefore this PE will require a complex-complex multiplier and two adders for adding two complex quantities. Thus PE 'B' requires more hardware than PE 'A'. The complete systolic array for DFT computation is shown in Fig.2. It must be noted that the inputs to the systolic array and the elements of the B coefficient matrix have to be ordered appropriately. This is done by zero padding, as is the case for the elements of the A coefficient matrix. At each clock cycle, the left most column of systolic array A will emit one row of matrix D in a staggered manner. At t=7, A_{11} emits D_{11}, at t=8, A_{21} emits D_{12}, and so on. At t=10, A_{41} emits $D_{14.}$ The first diagonal element of the final result matrix C will be emitted by cell B_{41} at t=11. According to (11) this forms the first element of the DFT, i.e. X_1. At t=12 the second diagonal element of the resultant matrix C will be emitted by B_{41}, which forms the second element of the DFT, i.e. X_2 and so on. It follows that the last element X_{16} will be emitted by B_{41} at time t=26.

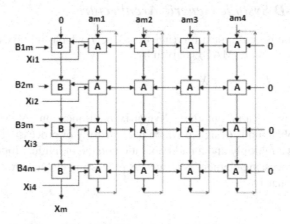

Fig. 2 2D Systolic Array for computing 16 point DFT

When x_{ij} (j=1, 2, 3, 4) reaches the right most PE, the systolic motion of data x_{ij} is stopped. At this point, each cell of array A will hold one value of the input data. The elements of coefficient matrix A keep moving in the vertical direction, while the elements of matrix X do not move any more. As described by (9), the matrix A has a recurrence property. Taking advantage of the same, the coefficients of matrix A are fed back to the top of each column after they are emitted by the bottom most cell of the systolic array. In this manner, it is required to store only the 4x4 matrix A_1 of (8). This leads to reduction in memory required to store the coefficients.

Fig.3 shows a similar structure for a non power of two size DFT considering N=15. Here the array size is 5x4 as stated earlier. The first DFT output is obtained at t=10.

3.2 Performance Evaluation

PE 'A' does complex-real multiplication and PE 'B' does complex-complex multiplication. There are N numbers of A processing elements and $\sqrt{N}$ (if N is a square number) or N_2 number of B processing elements. Hence, the generic architecture requires $(2N+4\sqrt{N})$ number of real multipliers if N is a square number or $(2N+4N_2)$ number of multipliers if N is a multiple of four or any composite number.

The total computation time required to compute the N-point DFT is

Total time units = Initialization time + N time units.

From the foregoing example, 10 time units are required to completely initialize the array before the first result appears at the end of the leftmost column PE (B_{41}). In general, it takes (3n-2) time units to initialize the array. Therefore,

Total time = (3n -2) + N time units.

Where N= Total number of points in the signal sequence

n = total number of rows in the DFT processor

The overhead of initialization is small compared to N, when N in large. Hence, the total time required to compute a N-point DFT is O(N).

The time required to compute an N-point (if N is perfect square) DFT by using a $\sqrt{N}$ ($\sqrt{N}$ +1) DFT processor is

Total time = ($3\sqrt{N}$ -2) + N time units.

If $N=N_1N_2$ is composite number, then total time is

Total time = ($3N_2$ -2) + N time units

Here, N is assumed to be the square of an integer. However, this does not limit the selection of the size of DFT. The 2-D systolic array processor can be designed for any N which can be represented as $N=N_1 \times N_2$.

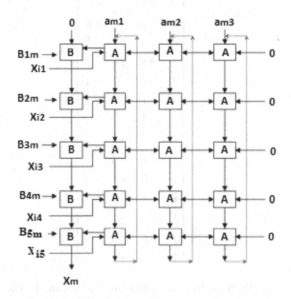

Fig. 3 2D Systolic Array for computing 15 point DFT

4   Architecture Optimization

The above generic architecture is observed to be hardware intensive, primarily because of the use of multipliers. For a $\sqrt{N}$ ($\sqrt{N}$ +1) DFT processor, the number of multipliers required is $2N + 4\sqrt{N}$, which can be prohibitively high for large N.

In this session, we optimize the architecture which dramatically reduces the number of multipliers required for the array processor when N is a multiple of four. Recall that the elements of the coefficient matrix A are given as:

$$a_{kn_1} = W_{16}^{4\,kn_1} = e^{-j\frac{2\pi(4k\,m_1)}{16}} = e^{-j\frac{\pi(k\,m_1)}{2}}$$

(12)

It is observed that the coefficients are of the form $\exp(-j\pi(kn_1)/2)$ and (kn_1) is always an integer. Hence, the coefficient will be either 0, ±1 or ±j. This property allows us to eliminate the multipliers in the AX array. All that is required is one adder/subtractor each for real and complex portions of the running sum D_{in}. Depending on the value of the coefficient, the running sum D_{in} is either left unchanged (when A_{ij} is 0) or the data x_{ij} is added/subtracted to/from it. The coefficient matrix A for 4 x 5 DFT processor (generated with the help of MATLAB) is given below.

$$A = \begin{bmatrix} 1 & 1 & 1 & 1 \\ 1 & -i & -1 & i \\ 1 & -1 & 1 & -1 \\ 1 & i & -1 & -i \\ 1 & 1 & 1 & 1 \\ 1 & -i & -1 & i \\ 1 & -1 & 1 & -1 \\ 1 & i & -1 & -i \\ 1 & 1 & 1 & 1 \\ 1 & -i & -1 & i \\ 1 & -1 & 1 & -1 \\ 1 & i & -1 & -i \\ 1 & 1 & 1 & 1 \\ 1 & -i & -1 & i \\ 1 & -1 & 1 & -1 \\ 1 & i & -1 & -i \end{bmatrix}$$

Hence, the multipliers in the AX array are not required. The PE 'A' has to add/subtract the data value x_{ij} to the running sum D_{in}. Apart from the reduction of hardware and simplification of PE 'A', the real and imaginary parts can be coded in two bits each (signed magnitude format). This means a reduction of about 8x in the amount of memory required to store the coefficient matrix A can be achieved. This also means that fewer flip-flops are needed in each PE of matrix AX to store the coefficient A. Smaller bit-width for A also means lesser routing in between the cells. At the same time, the number of inputs to the systolic array also reduces. Another observation is that, in this scheme, the first and the third column of the coefficient matrix A does not have an imaginary part. Hence, the adder required for the imaginary part in the first and the third columns of the AX array can be eliminated.

Though the above optimization is carried out for the 4x5 array, i.e. 16-point DFT, the same approach can be applied to any other DFT array. The only constraint here is that the re-arranged array X for signal values x_n should have exactly four rows. This will always result in coefficient matrix A having entries only 0, ±1 or ±j. This way, a DFT processor can be designed for any N which is a multiple of 4, i.e. $N = 4N_1$, where N_1 is an integer. As an example, a 64- point DFT requires an array processor of size 16 x 5. The 16 x 4 cells comprising of PE 'A' with no multipliers. Only the left most column of PEs, i.e. PEs 'B' will have multipliers. Hence, the number of multipliers required is N/4 * 4 = N. An N-point DFT now requires only N integer multipliers or N/4 complex multipliers.

The latency for the optimized design will be 7+15 = 22 time units which is the same if a $\sqrt{N}$ ($\sqrt{N}$ +1) array processor would have been used. The generic formula for latency can be derived as:

Latency = ((2x4)-1) + (N/4)-1= N/4 + 6 time units

Total time required is again N time units of processing time + latency

Total time = N + N/4 +6 time units = 5N/4 +6 time units.

Hence, the optimized design has the same computation time as that of the generic architecture.

5   Hardware Implementation

5.1   Proposed Generic 2D Systolic Array Architecture

The 2-D systolic architecture for 16 point DFT and 15 point DFT was implemented on target FPGA, Xilinx VIRTEX V. This FPGA device has DSP48 elements for implementing multiply and accumulate (MAC) operation. The design was coded in Verilog Hardware Description Language (HDL). A 15 bit, two's complement representation was chosen for data as well as coefficients. This facilitates maximum use of the multipliers while maintaining a good accuracy. The range of numbers represented is -1 to $+1-2^{-14}$. The design was tested by feeding different inputs and is observed that the output obtained from ModelSim is matching with the output obtained from MATLAB. The design was then synthesized to xc5vtx240t-2ff1759 Virtex-V FPGA using the Xilinx Synthesis Technology (XST) tool Xilinx ISE 11.1. Out of the 96 DSP48 elements available 48 were used by the generic design for 16 point DFT. Table 1 shows the resource utilization for the generic architecture for 16 point and 15 point DFT and optimized design for 16 point and 64 point DFT.

Table 1 Device Utilization

Device xc5vtx240t-2ff1759	Generic Design	Optimized Design		Generic Design	Available resources
	N=16	N=16	N=64	N=15	
Slice registers	1408	996	3281	1326	149760
Slice LUTs	989	929	3589	1007	149760
Fully used LUT FF Pairs	860	624	2442	840	5128
DSP48s	48	16	64	50	96
Maximum frequency	152.88MHz				

5.2 Proposed Optimized 2D Systolic DFT Architecture

To examine the hardware utilization of the optimized systolic array processor, the designs for 16-point and 64- point DFT processors were coded in Verilog HDL and were synthesized on xc5vtx240t-2ff1759 Virtex-V FPGA. However, the optimized design for 16 point DFT uses only 16 of them by reducing the number of multipliers by 66.6% compared to generic design. In addition, optimized design uses about 27% less number of Look Up Table - Flip Flop (LUT FF) pairs compared to generic design. The data representation was similar to the one used in the case of generic architecture. However, the coefficients a_{kn1} are now coded using only two bits. It is interesting to note that the optimized design is compact and uses much lesser hardware than its generic counterpart. Going by the above results, it has been estimated that a 64 x 5 array for 128-point DFT can fit onto a SPARTAN 3 xc3s2000 device, occupying about 80% of the LUTs available.

5.3 Comparison with the Other Architectures

Table 2 shows the comparison of the number of multipliers used for various other designs and proposed design. The structure in [1, 10, 11] are for Culey & Tukey Butterfly structures for various radix schemes.

Though these structures require less number of multipliers but will restrict on N which should be power of 2, 4 or 8. Design in [4] requires more number of multipliers than our design. The design in [5] reduces multipliers significantly but uses N number of data memories and 3N/8 Read Only Memories (ROM) to store twiddle factors as reported. Convolution based DFT architecture proposed in [8] and 2D array in [7] gives more choice on the DFT size but requires more number of multipliers than the proposed optimized design. Hence, the optimized design not only allows the designer to choose a wide range of DFT size but also reduces the number of multipliers significantly.

Table 2 Comparison

Architecture	No.of complex multipliers			Limits on N
	In general	N=28	N=64	
1D array [1]	(N-1)	27	63	$N=2^n$
Pipelined FFT [11]	$(\log_8 N - 1)$	NA	1	$N=8^n$
Parhi [10]	$3(\log_4 N - 1)$	NA	6	$N=4^n$
Meher [7]	$(N-1)^2$	729	3969	none
Meher[8]	2N-8	48	NA	N=2M, M is prime
Generic Design (proposed)	$((N/2)+N_2)$	18	36	$N=N_1 N_2$
Optimized Design (Proposed)	N/4	7	16	$N=4N_2$

6 Conclusion and Future Work

2D systolic array architecture for implementing DFT is proposed. The architecture uses triple matrix product based algorithm for DFT. The structure is suitable for computing DFT of any size N where N is a composite number. The design is implemented for power of two (16 point) and non power of two (15 point) sized DFTs. Further, an optimized design which considerably reduces the number of multipliers is proposed for N being a multiple of four. The designs are simulated for various inputs and synthesized using Xilinx ISE with xc5vtx240t-2ff1759 Virtex-V as the target FPGA. It is observed that the optimized design uses 66.6% less multipliers and about 26% less memory elements as compared to generic design. Both the generic and optimized designs operate at a maximum frequency of 152.88 MHz. As the number of multiplications is reduced drastically, it reduces the amount of time taken for DFT computation and reduction in power and area. As the multiple of four gives more choice for the selection of DFT size than power of two, the proposed architecture is more suitable for any application as it requires less hardware. The algorithm and the architecture is scalable, hence, 2D DFT implementation is planned for future work. Triple matrix product form of other transforms like discrete sine transform (DST) and discrete cosine transform (DCT) are also needs to be explored.

References

1. Nandi, A., Patil, S.: Performance Evaluation of One Dimensional Systolic Array for FFT Processor. In: IEEE-ICSCN, pp. 168–171 (February 2007)
2. Cheng, C., Parhi, K.K.: Low-Cost Fast VLSI Algorithm for Discrete Fourier Transform. IEEE Transactions on Circuits and Systems-I: Regular Papers 54(4), 791–806 (2007)

3. Kung, H.T.: Why Systolic Architectures? IEEE Computer 15(1), 37–46 (1982)
4. Cooley, J.W., Tukey, J.W.: An Algorithm for the Machine Computation of Comlpex Fourier Series. Math.Comp. 19, 297–301 (1965)
5. Aravena, J.L.: Triple Matrix Product Architecture For Fast Signal Processing. IEEE Trans. Circuits Syst. 35(1), 119–122 (1988)
6. Nash, J.G.: Computationally Efficient Systolic Architecture For Computing The Discrete Fourier Transform. IEEE Tran. Signal Processing 53(12), 4640–4651 (2005)
7. Meher, P.K.: Highly Concurrent Reduced Complexity 2-D Systolic Array For Discrete Fourier Transform. IEEE Signal Proc. Letters 13(8), 481–484 (2006)
8. Meher, P.K.: Efficient Systolic Implementation of DFT Using a Low- Complexity Convolution –Like Formulation. IEEE Transactions on Circuits and Systems-II: Express Briefs 53(8), 702–706 (2006)
9. Meher, P.K., Patra, J.C., Vinod, A.P.: Efficient Systolic Designs for 1 and 2-Dimensional DFT of General Transform –Lengths for High Speed Wireless Communication Applications. Journal of Signal Processing Systems 60(1), 1–14 (2010)
10. Chang, Y.N., Parhi, K.K.: An efficient pipeline FFT architecture. IEEE Transactions on Circuits and Systems-II:Analog and Digital Signal Processing 50(6), 322–325 (2003)
11. Fan, X.: A VLSI-Oriented FFT Algorithm and Its Pipelined Design. In: ICSP 2008 Proceedings, pp. 414–417 (2008)

An Effective Approach Towards Video Text Recognition

Prakash Sudir and M. Ravishankar

Abstract. With the rapid increase in digital video capture and editing technologies in recent times there is a great demand for Semantic-based Video Analysis and indexing algorithms. Video Text is an important semantic clue in video content analysis such as video information retrieval and summarization.In this paper, a Video text detection and localization algorithm is proposed that extracts the Eigen feature embodied in wavelet edge map. Also described is an iterative variance based threshold calculation method for Video text binarization. Experimental results on diverse videos demonstrate the robustness and efficiency of proposed method for detecting and recognizing superimposed text and multi-oriented scene text.

Keywords: Wavelet, Eigen value, Stroke width Transform, Binarization.

1 Introduction

With the advancements in digital media from text, images to videos and rapidly developing internet traffic there is a growing demand for automatic content-based video indexing. Text extracted from video sequence provides natural, meaningful keywords that reflect the videos content.Detection of videotext helps in number of Computer Vision applications, such as computerized aid for visually impaired, automatic geocoding of businesses, and robotic navigation in urban environments. Superimposed text or Caption text usually provides high-level semantic information about video content. News Broadcast and documentary captions annotate information about an event. Scene text is a textual content that was captured as a part of scene, e.g. Road signs, Billboards or Building Names. Although embedded videotexts provide useful information about the scene, it poses numerous difficulties in extracting and recognizing it. A single frame can have multiple sizes, multiple

Prakash Sudir · M. Ravishankar
Department of IS, DSCE, Visvesvaraya Technological University, India
e-mail: {sudirhappy,ravishankarmcn}@gmail.com

S.M. Thampi, A. Gelbukh, and J. Mukhopadhyay (eds.), *Advances in Signal Processing and Intelligent Recognition Systems*, Advances in Intelligent Systems and Computing 264,
DOI: 10.1007/978-3-319-04960-1_29, © Springer International Publishing Switzerland 2014

fonts and Multi-Orientation Characters. Cluttered background, bleeding Colors due to compression and movement of Special effect texts further add to the challenge of video text detection and extraction. During the past decades numerous methods have been applied to text detection in videos and achieved good experimental results. Connected Components (CC) methods [LR00, MK00, CWLK03] which are based on color or intensity homogeneity of characters perform satisfactorily only on high quality videos and are computationally expensive. On the other hand, Edge based methods [CLD05, JX01] give more false alarms and are not robust for complex background images. Texture-based approaches, such as the salient point detection and the wavelet transform, have also been used to detect the text regions. Zhong et al.[YZ00] detect text in JPEG/MPEG compressed domain using texture features from DCT coefficients. Ye et al. [QYZ05] compute the wavelet energy features at different scales and perform adaptive thresholding to find candidate text pixels, which are then merged into candidate text lines.

After the text detection step, the text extraction step should be employed before OCR is applied. The text extraction methods can be classified into color-based [MK00] and stroke based methods [TSS98], since color of text is generally different from that of background, text strings can be extracted by thresholding. Thresholding can be categorized into local [JM00, W.N86] and global [Ots79], both of which have poor accuracy for the superimposed and scene texts as they involve hand tuning of parameters and sensitivity to parameter choice. In [AM11],Anand et al. proposed an automatic graph cut seeding method for binarization of Natural Scene text. Foreground and background colors were modeled in GMMRF (Gaussian Mixture Mode Random Field) to make binarization robust to variations in colors. Saidane and Gracia [SG07] proposed a Video Text Binarization scheme based on Convolutional neural network whose performance depends on training samples. In [MM09], Michele Merler et al. employed local Adaptive Otsu based binarization approach for low quality of Video Scene text which is implemented with integral histograms.

Based on above survey, one arrives at conclusion that although many methods have been proposed for text detection, very few of them address superimposed and scene texts with multi-oriented alignments and also majority of binarization methods are still chained to type of videotexts and limited by text type, contrast and background complexity. Hence in this paper a new approach for Videotext Detection using Wavelet Edge Map and Block Eigen values followed by new iterative variance based threshold calculation method for Video text binarization for both Superimposed and Multi-Oriented Scene text is proposed. The main contribution of this paper lies in the following two aspects: (1) A Novel Text Detection approach for both Superimposed and Scene Video text (2) A Novel Video Text Binarization scheme independent of type of videotext.

2   System Overview

The Block Diagram of the proposed method is shown in Figure.1. Limiting ourselves to only English Characters and resizing video frames to a size of 512×512 for ease in computation, Edge map is first obtained using wavelets because wavelet decomposition generally enhances the high contrast pixels by suppressing low contrast pixels [WW02].Next, a Sliding window of size 8×8 is moved over edge map to obtain Eigen Feature vector. The Fuzzy C-means algorithm is applied to classify the feature vector into two clusters: background and text candidates. SWT (Stroke Width Transform) based verification is adapted to remove the non-text blocks followed by Connected Components (CC) based false positive elimination. Next Text orientation is determined and Alignment corrected Videotext blocks are extracted and then subjected to binarization by iterative variance based threshold calculation method. Recognition of Videotext is accomplished using commercial OCR ABBYY fine reader 9.0.

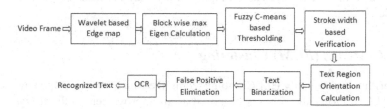

Fig. 1 Block Diagram of the proposed approach

3 Proposed Methodology

3.1 Wavelet based Edge Map generation

Wavelet Analysis is a popular tool for identifying the spatial frequency content of an image and the local regions within the image where those spatial frequencies exist. In [HO00], Li captured wavelet texture property by mean, second-and third order central moments features in Haar-wavelet domain. In [QY03], Ye et al, Captured wavelet texture property by statistical and co-occurrence features in the wavelet domain. Proposed method employs the Eigen feature in the wavelet domain to capture the Texture property. In our case Haar wavelet based decomposition at level 1 is adopted.

In the first level of decomposition of 2D Discrete Wavelet Transform, the image is separated into four parts. Each of them has a quarter size of the original image. They are called approximation coefficients (LowLow or LL), horizontal (LowHigh or LH), vertical (HighLow or HL) and detail coefficients (HighHigh or HH).Approximation coefficients obtained in the first level can be used for the next decomposition level. The principle of the simplest method of edge detection is based on replacing of all approximation coefficients by zeros. This modification removes low frequencies from the image. The image is reconstructed using only the remaining wavelet coefficients. By means of this method the most expressive edges are found which include the text pixels as shown in Figure 2b.

3.2 Eigen Feature Extraction

Edge map obtained by above procedure is divided into k blocks of size m×m where m is power of two (m=8 in our case). For each k block compute Eigen values of the covariance matrix of the blocks as per the equation 1.

$$\Psi = \frac{1}{m} \sum_{j=1}^{m} \left(\left(k_j - \bar{k} \right)^2 \left(k_j - \bar{k} \right) \right) \tag{1}$$

where k_j is the intensity value in every j^{th} column and $\bar{k}$ is the mean of all elements in k^{th} block. The Eigen values λ_{ij} of the covariance matrix Ψ of the block are computed. Maximum Eigen value of each block is stored in a $k \times k$ matrix. The feature vector thus obtained is employed as the input of the Fuzzy C-Means Clustering (FCM) classifier.

3.3 Fuzzy C-Means (FCM) Clustering

The FCM algorithm is essentially Hill-Climbing technique which was developed by Dunn in 1973[J.C73]. This algorithm is proposed as an improvement to the fuzzy k-means clustering. The FCM partition the collection of n vector into C groups and finds a cluster center in each group such that a cost function of dissimilarity measure is minimized. The FCM algorithm assigns pixels to each category using fuzzy membership functions. Let f_{aj} be the text block which is to be clustered into C clusters. The algorithm is an iterative optimization that minimizes the cost function defined as follows

$$\sum_{k=1}^{m \times n} \sum_{i=1}^{C} \mu_{ik}^m \left\| f_{ajik} - V_i \right\| \quad 1 \leq m \leq \infty \tag{2}$$

where 'm' is the any real number greater than $1, \mu_0$ is the degree of membership of f_{ajik} in cluster I, V_i is the i_{th} cluster center, $\|.\|$ is a norm metric. The parameter 'm' controls the fuzziness of the resulting partition. The cost function is minimized when pixels close to the centroid of their clusters are assigned high membership values and low membership values are assigned to pixels far from the centroid. The membership function represents the probability that a pixel belong to a specific cluster. In the FCM algorithm, the probability is dependent on the distance between the pixel and each individual cluster center in the feature domain. Let $E = \{e_1, e_2, ..., e_N\}$ be the dataset of N Eigen Values. Let C=2 be the number of clusters in which the dataset E is to be partitioned. The objective of FCM clustering is to minimize the function

$$X_m = \sum_{i=1}^{N} \sum_{j=1}^{C} \mu_{ij}^m \left\| e_i - v_j \right\|^2 \tag{3}$$

where m=2 is the fuzzification constant. The membership μ_{ij} and cluster centers v_j are updated using

$$\mu_{ij} = \frac{1}{\sum_{k=1}^{C} \left(\frac{\|e_i - v_j\|}{\|e_i - v_k\|} \right)^{\frac{2}{m-1}}} \tag{4}$$

$$where \quad v_j = \frac{\sum_{i=1}^{N} \mu_{ij}^{m} * e_i}{\sum_{i=1}^{N} \mu_{ij}^{m}} \tag{5}$$

Initial membership matrix μ is determined using the formula

$$\mu_{ij} = \begin{cases} 1 & if \ \Delta(e_i, v_j) \\ 0 & \exists k \neq j \ fulfilling \Delta(e_i, v_k) = 0) \\ \left(\sum_{k=1}^{C} \left(\frac{\Delta(e_i, v_j)}{\Delta(e_i, v_k)} \right)^{m} \right)^{-1} & otherwise \end{cases} \tag{6}$$

$\Delta(e_i, v_j)$ is the Euclidian distance between the eigen value e_i and centre v_j. After the clustering process has converged the Eigen value is classified to belong to the cluster with a maximum membership value. The two clusters with the lowest mean value is merged into one and a new cluster is calculated. After the initialization with the new center is done, the pixels are clustered again. The Clustering process is repeated until C=2.Area of the two clusters is computed to separate text and background. A cluster is classified as text if its area is less than to area of other cluster. The sample output of the FCM algorithm is shown by separating the background and the text in Figure 2c.

3.4   Stroke Width Based Text Verification

SWT based Text Verification is next adopted on the text regions obtained from previous steps. Ephstein et al.[BY10] had proved that stroke width determination is a robust approach for detection of text in complex images.SWT is a local image operator which computes per pixel width of the most likely stroke containing the pixel. The output of SWT is a feature map, where each pixel contains the potential stroke width value of input image pixels. Restricting Stroke width to less than 10 probable text pixels are determined as shown in Figure 2d.

3.5   False Positive Elimination

Heuristic filter methods are adopted in order to reject falsely detected regions like too small, too thin CCs, large CCs based on aspect ratio and size of CCs. Figure 2e shows the result after the refinement process, which is the final output of the text detection process. Some samples of text detection are shown in Figure.2,g,h,i.

3.6   Determination of Orientation of Text Regions

Traditional Bounding box (BBs) calculation and plot leads to lot of false Positive as shown in Figure.3a especially for Oriented Scene text because BBs tend to overlap on each other when characters are closely spaced. A method is proposed to prevent overlap wherein oriented BBs are generated as follows

• Each Connected Component (CC) is extracted using the label number, a sample CC is as shown in Figure 3b.

• With every rotation of CC ranging from $i=0°$ to $i=180°$.Mean value of the horizontal projection is calculated.

• Mean of Projection vs. Angle plot is obtained as shown in Figure 3c for a sample CC shown in Figure 3b. From the plot one can observe that least mean value occurs when the alignment of text is perfectly horizontal. Thus angle at which least mean value occurs gives the orientation angle of the CC.

4   Video Text Binarization

Text Binarization Process usually follows text detection and text localization Process in a standard framework of Video OCR. A binarization technique is proposed

Fig. 2 (a) Original Video frame (b) Wavelet based Edge Map (c) Fuzzy C-means threshold output (d) SWT verification output (e) Morphological Operation output (f) Detected Text Region (g),(h),(i) Examples of text detection

that is robust to segment text in Video frames with complex background, arbitrary font, color, orientation and size. The Binarization Process is divided into two steps.

4.1 Back Ground Estimation by Using Median Filter

Background Estimation is a vital step in estimating the foreground and text pixels in the CCs obtained from Localization Process. Initially obtain the Median Filter output of the CC using Window size of 5×5 as shown in Figure.4b.After subtracting the Median Filter Output from the original Grayscale image as shown in Figure.4c. One can observe that the complex background has been almost been removed.

4.2 Iterative Variance Based Threshold Calculation Method

Proposed approach shows that it is convenient to threshold the gray level that is slowly varying function of the position in an image based on global variance value of the CC. Following steps are used to obtain a binarized CC.

i) Initial Threshold $T_0 = \frac{G_{min}+G_{max}}{2}$ where,$T_0 = \{T_k/k = 0\}$ is calculated

ii) Obtain Thresholded Image X by using Threshold value T_0

iii) Obtain Maximum Variance of V_k the thresholded image X of size M×N and mean value μ given by

$$V_x = Max\left\{\frac{1}{M \times N} \sum_{i=1}^{M} \sum_{i=1}^{N} (X_{ij} - \mu)^2\right\} \tag{7}$$

iv) Update the Gray scale values of the thresholded image by adding V_k to them.

v) Repeat Procedure (ii) ,(iii) ,(iv) until variance $V_k = V_{k-1}$

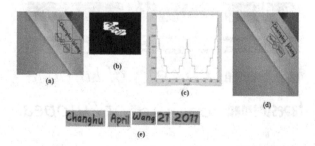

Fig. 3 (a) Erroneous text detection result (b) Sample CC (c) Plot of Mean of projection vs Angle for sample CC (d) Multi oriented text detection result (e) Extracted Text Region

5   Experimental Results

5.1   Datasets Used

Proposed Text Detection and Text Localization methodology has been evaluated on three datasets
1) Since there is no benchmark database, our own dataset is created for the purpose of experimentation which include 50 Indoor and outdoor video clips containing scene texts.
2) **Microsoft Common Dataset**: A Collected 104 Video frame set from German TV news Program and the test set from the Mediaglobe project, which are subsequently referred to as MS testset, TV news testset and MG testset respectively [MZK10].
3) **MSRA Text Detection 500 database (MSRA-TD500)**: It contains 500 natural images. These include indoor (office and mall) and outdoor (street) scenes. The resolutions of the images vary from 1296×864 to 1920×1280. Manual annotations are available at [CYT12].

5.2   Evaluation of Video Text Detection and Localization Procedure

(i) Super Imposed Text: The evaluation procedure proposed in [MZK10] from MS Test set is employed for performance evaluation. The comparisons results on MS test set with existing methods is shown in Table 1 and some experimental results compared to Yang et al.[[HY13] are shown in Figure 5. Table 1 shows that all though our proposed method is not able to outperform the results of [MZK10], it surpasses all the other methods for the MS test set range of datasets provided in the MS test set.
(ii) Multi-Oriented Scene text: The Evaluation procedure proposed in [CYT12] is used for performance evaluation. The Evaluation and Comparison results as illustrated in Table1 shows that proposed method has achieved significantly

Fig. 4 (a),(b),(c) Sample CCs (d),(e),(f) Median Filter output (g),(h),(i) Background eliminated CCs (j),(k),(l) Binarized output (m),(n),(o) OCR output

Fig. 5 (a),(b),(c),(d) Text Detection result for Yang method (e),(f),(g),(h) Text Detection result for proposed method (i),(j)Text Detection result for MSRA Database (k),(l) Text Detection result for own Database

Table 1 Performances of different text detection methods evaluated on the various dataset

Text Type	Method	Precision	Recall	F-Measure
Caption	**Proposed**	**0.95**	**0.95**	**0.95**
	Zhao[MZK10]	0.98	0.94	0.96
	Shivakumar[PS08]	0.9	0.92	0.91
	Gllavata [JGBFST09]	0.87	0.9	0.88
	Yang [HY13]	0.94	0.93	0.93
Scene	**Proposed**	**0.69**	**0.65**	**0.67**
	Ephshtein[BY10]	0.25	0.25	0.25
	Cong yao [CYT12]	0.63	0.63	0.6

enhanced performance over existing methods on texts of arbitrary orientations. Sample
examples are shown in Figure 5.

5.3 Evaluation of Videotext Recognition Procedure

Evaluation of proposed text binarization method is done on the following two test sets using commercial OCR ABBYY fine reader 9.0.

(i)Caption text set: The proposed binarization method is compared with the well-known thresholding based binarization techniques like Otsu [Ots79], Sauvola [JM00], Niblack [W.N86] and Yang [JGBFST09].The Achieved text recognition accuracy is as shown in table 2 which proves that the proposed method outperforms the results of all other reference procedures.

(ii) Scene text set: Since our aim is also to evaluate the Text Recognition procedure with respect to multi-oriented scene text hence 100 sample images (530 words, 3478 characters) are carefully chosen from MSRA text detection 500 database in such a

Input	Proposed	Otsu	Niblack	Sauvola
Simplex	Simplex "Simplex"	Simplex "S\m\>\@1"	Simplex "W LX"	Simplex " __H -ll l' J "
COME TO EC	COME TO EC " gmE; TO EC"	COME TO EC "{EH/1E TO EC"	COME TO EC " QQAEJO EC,"	" "
crocs	crocs "CTOCS"	crocs "@\F\©\@\S"	crocs "L@1T1@1°"	crocs "@w@@%"

Fig. 6 Some Example Binarization and OCR results of the proposed and existing methods

Table 2 Performances of different text detection methods evaluated on the various dataset's database

Method	Correct Characters	Correct Words	Char Accuracy	Word Accuracy
Otsu	2087	301	0.60	0.57
Niblack(k=0.5)	1986	287	0.57	0.54
Sauvola(k=-.01)	1400	213	0.40	0.40
Proposed	**2191**	**323**	**0.63**	**0.61**

way that the set covers most of the challenges of a scene text such as low contrast and Noisy background. Proposed Binarization of Multi-oriented scene text has been compared with Standard thresholding based binarization techniques like Otsu, Sauvola, Niblack and the achieved text recognition accuracy has been tabulated in table. 2. Sample recognition results are shown in Figure.5. Since there is no other binarization evaluation result published from this test set yet hence our recognition results is reported to establish a new baseline for comparison.

6 Conclusion

This paper presents a Text detection and recognition system that detects and recognizes both Caption and Multi-Oriented Video text. The Experimental results reveal that our method is quite robust over texts of unconstrained size and orientation. Future work would focus on detection of text of Curve orientation, the case where our method gives poor result. Our binarization algorithm handles text of arbitrary color superimposed on complex color backgrounds. Our binarization algorithm works well with most font sizes. However in some broadcast videos the stroke width is less than 1, causing our binarization to fail. Future work would focus on binarizing small size text. Proposed work also suggests focusing more on designing OCR modules to handle unique challenges posed by Video texts.

References

AM11.   Jawahar Anand Mishra, C.V., Alahari, K.: An mrf model for binarization of natural scene text. In: International Conference on Document Analysis and Recognition (ICDAR) (2011)

BY10.   Ofek, E., Epshtein, B., Wexler, Y.: Detecting text in natural scenes with stroke width transform. In: Proc. of Computer Vision and Pattern Recognition (2010)

CLD05.   Wang, C., Liu, C., Dai, R.: Text detection in images based on unsupervised classification of edge-based features. In: IEEE ICDAR (2005)

CWLK03.   Jung, K., Lee, C.W., Kim, H.J.: Automatic text detection and removal in video sequences. Pattern Recognition Letters (2003)

CYT12.   Liu, W., Ma, Y., Yao, C., Bai, X., Tu, Z.: Detecting texts of arbitrary orientations in natural images. In: CVPR (2012)

HO00.   Doermann, D., Li, H., Kia, O.: Automatic text detection and tracking in digital video. IEEE Transactions on Image Processing (2000)

HY13.   Yang, H.S.H., Quehl, B.: Text detection in video images using adaptive edge detection and stroke width verification. In: IWSSIP (2013)

J.C73.   Dunn, J.C.: A fuzzy relative of the isodata process and its use in detecting compact, well separated cluster. Cybernatics (1973)

JGBFST09.   Ewerth, R., Gllavata, J., Phan, T.-Q., FreislebenP. Shivakumara, B., Tan, C.-L.: Video text detection based on filters and edge features. In: Proc. of the 2009 Int. Conf. on Multimedia and Expo (2009)

JM00.   Sauvola, J.J., Pietikainen, M.: Adaptive document image binarization. Pattern Recognition (2000)

JX01.   Hua, X.-S., Xi, J.: A video text detection and recognition system. In: IEEE International Conference on Multimedia and Expo (2001)

LR00.   Effelsberg, W., Lienhart, R.: Automatic text segmentation and text recognition for video indexing. In: ACM/Springer Multimedia Sys. (2000)

MK00.   Mariano, V.Y., Kasturi, R.: Locating uniform-colored text in video frames. In: Proc. 15th Int. Conf. Pattern Recognition (2000)

MM09.   Kender., J.R., Merler, M.: Semantic keyword extraction via adaptive text binarization of unstructured unsourced video. In: 16th IEEE International Conference on Image Processing (ICIP) (2009)

MZK10.   Li, S., Zhao, M., Kwok, J.: Text detection in images using sparse representation with discriminative dictionaries. Journal of Image and Vision Computing (2010)

Ots79.   Otsu, N.: A threshold selection method from gray-level histograms. IEEE Trans. Syst., Man, Cybern. (1979)

PS08.   Tan, C.L., Shivakumara, P., Huang, W.H.: Efficient video text detection using edge features. In: ICPR (2008)

QY03.   Wang, W., Zeng, W., Ye, Q., Gao, W.: A robust text detection algorithm in images and video frames. Information, Communications and Signal Processing (2003)

QYZ05.   Gao, W., Ye, Q., Huang, Q., Zhao, D.: Fast and robust text detection in images and video frames. Image and Vision Computing (2005)

SG07.   Saidane, Z., Garcia, C.: Robust binarization for video text recognition. In: International Conference on Document Analysis and Recognition (ICDAR) (2007)

TSS98.   Hughes, E.K., Sato, T., Kanade, T., Smith, M.A.: Video ocr for digital news archive. In: Proc. IEEE International Workshop on Content Based Access of Image and Video Libraries (1998)

W.N86.   Niblack, W.: An introduction to digital image processing. Prentice Hill (1986)

WW02.   Lam, K.K.M., Mao, W., Chung, F., Siu, W.: Hybrid chinese/english text detection in images and video frames. In: ICPR (2002)

YZ00.   Jain, A.K., Zhong, Y., Zhang, H.J.: Automatic caption localization in compressed video. IEEE Transactions on PAMI (2000)

Local Search Method for Image Reconstruction with Same Concentration in Tomography

Ahlem Ouaddah and Dalila Boughaci

Abstract. Image reconstruction in tomography is an attractive research area that has received considerable attention in recent years. The image reconstruction can be viewed as an optimization problem where the main objective is to obtain high quality reconstructed images. In this paper, we proposed a local search (LS) method to improve the quality of reconstructed images in tomography in supposed case of similar concentration of physical phenomena. The proposed method starts with an initial image solution and tries to enhance its quality. A solution is a set of points where each point represents a distribution of a physical quantity resulting by radius emission. Each point is evaluated by a function that estimates the difference between the estimated and the measured projections. The LS makes use of a move operator that permits to generate neighbour solutions and helps in finding the optimal correctness of distribution in each point. The LS is an iterative process that tries to optimize position of the physical parameter on the image in order to obtain a solution corresponding to the reconstructed image. To measure the performance of the proposed approach, we have implemented it and compared it with the Filtered back-projection (FBP). Further, we compared the reconstructed images of LS with the source ones. The numerical results are promising and demonstrate the benefits of the proposed approach.

1 Introduction

The meta-heuristics are optimization techniques used for solving difficult problems and finding optimal solutions when the exact techniques fail to do it. Meta-heuristics have been used successfully in various optimization domains such as scheduling, job

Ahlem Ouaddah · Dalila Boughaci
University of Science and Technology Houari Boumediene USTHB,
Electrical Engineering and Computer Science Department,
El-Alia BP 32 Bab-Ezzouar, 16111 Algiers, Algeria
e-mail: {aouaddah,dboughaci}@usthb.dz

S.M. Thampi, A. Gelbukh, and J. Mukhopadhyay (eds.), *Advances in Signal Processing and Intelligent Recognition Systems*, Advances in Intelligent Systems and Computing 264,
DOI: 10.1007/978-3-319-04960-1_30, © Springer International Publishing Switzerland 2014

shop, industry and son on [2]. However in tomography reconstruction, there is only a few research dealing with meta-heuristic for tomographic images reconstruction [4, 7, 11, 12, 13, 17]. The image reconstruction in tomography is an inverse problem that can be stated as follows: let us consider a set of measures; the problem is to finding which object produced these measures. The ill-posed problem is that the solution isn't unique and the solution does not exist necessary. We can conclude that this problem could be solved as an optimization problem and as we say previously, the meta-heuristics could be used on this case [2].

The medical image reconstruction is a technique used to have representation in image that gives information about a part or all of human body or animals. Such image is used to comprehend function or anatomy of different organs or to detect diseases and for this, the quality of reconstruction has a direct impact on the precision of the diagnostic.

The image reconstruction is an important search area. In this paper, we propose a Local Search (LS) method to reconstruct an image. In this work, we demonstrate that a local search met-heuristic could be used to reconstruct and improve the quality of images from projections and only with restraint number of projections.

The image reconstruction in tomography can be divided into two parts:

1. The acquisition of the projections,
2. from this projections we reconstruct the image that corresponding to distribution of the physical phenomenon [8].

The data acquisition could be modelled analytically by the Radon Transform [10], all measured data are recorded into matrix called sinogram.

In continuous case, using Radon Transform, Projections $P(s,\theta)$ with s is the distance between each point of the ray of projection and the center for an angle θ, such as (1), consist to measure the integral of an infinite domain of all points (x,y) of a function $f(x,y)$ representing an object, these points are the points that contribute in the projection P such as (2).

$$s = x\cos(\theta) + y\sin(\theta) \tag{1}$$

$$P(s,\theta) = \int_{-\infty}^{+\infty} f(x,y)dv \tag{2}$$

By analogy, in discrete case, Radon Transform consist to sum the values of all pixels (x,y) that contribute in each projection P_i also called bin, such as (3); r_{ij} is the value of contribution of pixel j at the projection P_i ; f_j is the value of this pixel [3].

$$P_i = \sum_{j=1}^{m} r_{ij}f_j \tag{3}$$

Added to this transform, John Radon has proved in 1917, that the reconstruction of the object from its projections is possible and could be exact if we have an infinite number of projections, and in reality it's impossible [14].

After publication of this theory and many years later, several methods are proposed to reconstruct image from its projections.

We could divide standard methods of image reconstruction in tomography in two major categories: analytical methods and iterative methods.

i The analytical methods are based on continuous modelling, it consist to inverse measurements equations.

ii The basic principle for iterative methods could be summary as:

  a. Estimate image,
  b. compute projections of estimated image,
  c. compare between estimated projections (b) and measured projections,
  d. use factor of correction to correct error resulted by step (c),
  e. update the projections to create a new estimated image, and go to (a),
  f. The stopping condition could be a fixed number of iterations or when difference is minimized between estimated projections and measured projections.

The goal of iterative method is to minimize the difference between estimated projections and measured projections using a factor of correction.

Filtered back-projection (FBP) [6] is the most frequently used analytical method, and in general, analytical methods are fast but the quality of reconstruction is disputed, for this reason iterative methods have been introduced to improve the quality of reconstructed image, such as maximum likelihood expectation (MLEM) [15] or ordered subset expectation maximization (OSEM) [9]. In general, iterative methods provide a better quality than analytical methods, however these methods need heavy computing power and the noises increase proportionally with the number of iterations. Using them is still limited actually.

In the best of our knowledge, there is few research works about reconstruction using metahueristics approaches. We can cite for example the genetic and fly algorithm in medical imaging reconstruction [4, 7, 11, 12, 13, 17]

The most proposed approach is based on genetic algorithm and these methods are till now, in experimental stage. But in general, some of them improve highly the quality of reconstruction and others less than. We believe that, these approaches can still improve more the reconstruction and it's our goal.

The problem is still open, because there is no exact solution for this problem for all cases and mainly most of them use complete number of projections and gives good results under some conditions.

In this paper, we propose a local search method (LS) in image reconstruction, to improve the quality of reconstruction with a limited number of projections.

The rest of the paper is organized as follow: the second section presents the principle of the proposed method. Then in the third section, we present preliminary results and discussion about the results. Finally, we conclude with the conclusion.

2   Proposed Approach

2.1   Main Principle

The input is a set of measured projections data produced analytically [10] and recorded into sinogram $H(s,\theta)$ where s is the position of the ray of projection according to angle θ, see Fig. 1, each line of the sinogram represent sets of projections for a given angle θ. Different main data types could be used in medical image reconstruction, but the most used data format is sinogram [5], this is why we choose for a first approach a sinogram mode.

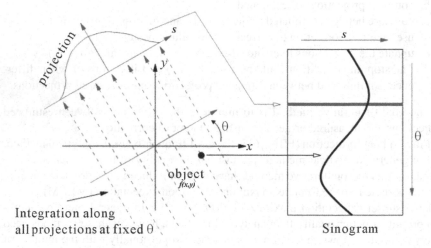

Fig. 1 Data acquisition illustrated. Each projection $P(s,\theta)$ in s position at θ angle is recorded into sinogram $H(s,\theta)[1]$

In summary, the step one consist in producing the projections that we store into sinogram see Fig. 2 and the second step consist to reconstruct the image from its projections (sinogram) see Fig. 3.

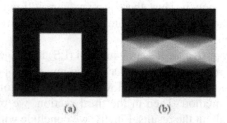

Fig. 2 Example of sinogram (b) estimated for the image source (a)

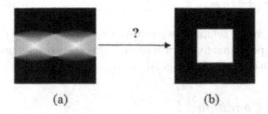

(a)   (b)

Fig. 3 Example that explains the problematic: How reconstruct the image (b) only with its projections (sinogram) (a).

The initial space search represent image generated by the method SBP or by uniform distribution, for each projection in a specific angle, for each point that contribute in this projection, we apply a set of moves and estimate for each move its projections in the same specific angle and using function of evaluation, we estimate the convergence of the estimated projection to the measured projection, than select the best solution corresponding to the solution that containing the changed point which has maximum of convergence, and in other term, minimum of distance between its estimated projection and measured projection, in final we get better quality of reconstructed image, which means, a better distribution of the physical quantity.

2.2 Function of Evaluation

The function of evaluation corresponds at the distance between the measured projections and estimated projections.

Distance of Manhattan [16], is used because it's given a high quality of precision and it's a fast method.

We supposed that the projections were recorded into sinogram $H(s,\theta)$.

We calculate the distance (Dc) between the estimated projection of changed pixel that recorded into sinogram $H_c(s,\theta)$ and measured projection for the same angle θ of the same point recorded into sinogram $H_m(s,\theta)$, such (4):

$$D_c = |H_c(s,\theta) - H_m(s,\theta)| \qquad (4)$$

2.3 Selection

In this algorithm, selection is used to optimize founding of the best distribution and concentration of the physical phenomenon. Selection operator, correspond at selection of the solution that have the lowest value of distance D_c see Section 2.3.

2.4   Operator of Move

For each point of initial solution, two moves are accomplished for finding the best estimation of this point that has improved the convergence of its projection to measured projection in each specific angle θ.

2.5   Stopping Condition

The LS is an iterative process. The boucle stopped when the current solution is the same then the previous one, that means no improvement is noted in the new solution. In other term, when all points are mutated with values having minimum distance between estimated projection and measured projection thus provide a convergence criteria.

In the case where each iteration is always slightly different from the previous, the LS is applied for all points of the images for all measured projections.

The iteration process is stopped if no improvement is noted in the new solution compared to the previous or if LS is applied for all points of the images for all measured projections or until time of computation becomes superior of some fixed value t.

2.6   The LS Algorithm

- **Step0**: **Initialization**: Initial solution corresponding to the initial estimation of the image generated by simple back projection (SBP) or by uniform distribution.
- **Step1**: Number of projections is a restraint and small number comparing to the initial number of projections that always used to reconstruct image.
- **Step2**: Each point of the solution is changed with two values 0 or 1, projection data are estimated for the two cases.
  The operator of move is used to optimize the position of the physical parameter then we can say it allows performing the global optimization.
- **Step3**: **Function of evaluation**: This function that used to select the best solution containing new changed point. It estimates the distance between current projection and measured projection for a specific angle θ for the same point that contributes in this projection.
  For each point that contributes in each projection, we compute this function and proceed to selection of the best new solutions.
- **Step4**: **Selection**: This step concern selection process, we select the solution which its changed point has the function of evaluation correspond to the minimum of distance between the current estimated projection and the measured projection.
- **Step5**: **Stopping condition**: The algorithm is stopped when the algorithm converge to the same solution for a maximum number of iterations or if all points are treated for all projections or if the time of computation became huge.

3 Numerical Results

To validate the proposed method, we developed a set of numerical phantoms. Numerical phantoms are with different small resolutions. The size of different objects is not the same but the concentration of physical phenomena is the same. We can say with a small number of projections, optimization of reconstruction is better than FBP with a large number of projections. The next section presents a series of tests to show the performance of the proposed approach.

3.1 Test1

We test objects with same, different small sizes, different resolutions of the numerical phantom and same concentration of the physical phenomena, with a few numbers of angles of projections. Our goal is to asses, if our approach presents a higher quality of reconstructed image, see Fig. 4.

3.2 Test2

This test case has been introduced to compare our method with FBP method, tested on small resolution of numerical phantom. Applied FBP method is used with all angles of projections ([0..180]). The results show that our method using few number of projections, gives better results for reconstruction than FBP, see Fig. 5.

3.3 Test3

This test case has been introduced to prove the power and efficiency of our method, comparing our results and FBP results on small resolution of numerical phantom case.

Applied FBP method is used with the same numbers and angles degrees of projections used in our method. The results show that LS method is better for reconstruction than FBP despite we have a small number of projections, see Fig. 6.

To show clearly the performance of LS in image reconstruction, we draw the curves in Fig. 7 that compare LS and FBP in term of reconstruction quality point of view.

We estimate the grayscale values of reconstructed image using our method, FBP method and the original one for image test 01. The curves of reconstructed image by LS method and the original one are supcrimposed unlike the reconstructed image by FBP method that is greatly different from the original one.

Table 1 gives a quantitative comparison between LS and FBP on all tested images. The column gap is the difference between the reconstructed image and the original one. The column time is the computation time needed for each reconstruction for each method.

As shown in Table 1, LS succeeds to give high quality of images compared to FBP. However the search process in LS is time consuming compared to the reconstruction process in FBP.

Table 1 Table of gap and computation time needed for each reconstruction using our method and FBP method

Image test	LS		FBP	
	gap	time(s)	gap	time(s)
Image test 01	0	2.2004	3.4981	0.1746
Image test 02	0	1.3640	0.1058	0.1370
Image test 03	0	0.6571	3.5494	0.0860
Image test 04	0	2.1569	6.7300	0.0840
Image test 05	0	3.8657	12.0027	0.0845
Image test 06	0	2.6712	6.7553	0.0835
Image test 07	0	7.6690	9.0000	0.0952
Image test 08	0	6.5062	2	0.0925
Image test 09	0	0.7278	1.6060	0.0830
Image test 10	0	0.4492	2.9482	0.0905

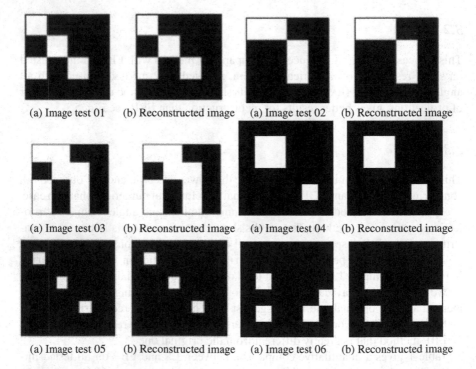

(a) Image test 01   (b) Reconstructed image   (a) Image test 02   (b) Reconstructed image

(a) Image test 03   (b) Reconstructed image   (a) Image test 04   (b) Reconstructed image

(a) Image test 05   (b) Reconstructed image   (a) Image test 06   (b) Reconstructed image

Fig. 4 Reconstruction results for Test 3.1

FBP fails in finding good quality of reconstructed image for all the checked tests. The gap between the images given by FBP is > 0.

Figure 7 confirms the efficiency of LS in reconstruction of images compared to FBP.

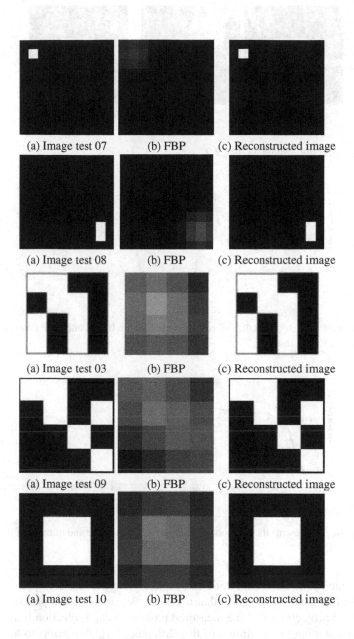

(a) Image test 07   (b) FBP   (c) Reconstructed image

(a) Image test 08   (b) FBP   (c) Reconstructed image

(a) Image test 03   (b) FBP   (c) Reconstructed image

(a) Image test 09   (b) FBP   (c) Reconstructed image

(a) Image test 10   (b) FBP   (c) Reconstructed image

Fig. 5 Reconstruction results for Test 3.2

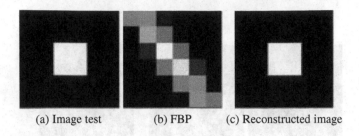

(a) Image test   (b) FBP   (c) Reconstructed image

Fig. 6 Reconstruction results for Test 3.3

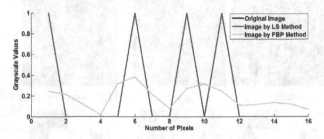

Fig. 7 The graph represents quality of reconstructed image by LS and FBP compared to the original one for image test 01

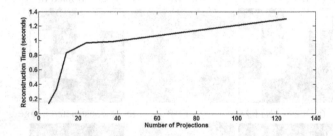

Fig. 8 The graph represents the relation between computation time and number of projections in image test 01

This is due to the good exploration of the search space and using our operators such as the function of evaluation that takes into account the comparison between the estimated projections and the measured projections and selection based on taking the point having the minimum of this difference, which permits to locate each good point that correspond to the position of the physical parameter.

To show clearly how the time of computation depends on the number of projections, we draw the curves in Fig. 8 that represent the evolution of time computation according to the number of projections.

According to the drawn line, the time of computation increase rapidly for the first small number of projections, after a specified value, the time of computation increase slower compared to the beginning, but keeps increasing; we can conclude that the number of projections has a direct impact on time of computation needed for reconstruction, therefore restrict this parameter back to reduce time of computation.

4 Discussion

The data reconstruction is an exact reconstruction, although the number of angles of projections is limited and low compared to those for example to FBP.

Despite the number of angles of projections is wide in FBP, the quality of reconstruction is lower than our method.

According to our research, any proposed approach till now present an exact reconstruction with small numbers of angles of projections.

Using a few numbers of projections is a big advantage because it reduced time of computation principally of reconstruction and in other side reduces time of acquisition.

We think that these preliminary results prove that's a promising approach in tomographic reconstruction, we intend validate this approach in big size data and different concentration of objects.

Because the time of computation is important, despite we use a restraint number of angles of projections for reducing time of computation, we are going to detect the area of the object, and apply our operators only in detected area.

5 Conclusion

In this paper we proposed a local search method (LS) for image reconstruction in tomography. To measure the performance of the proposed method, we have implemented it and compared it with some other well-known method. The numerical results are promising and demonstrate the benefits of LS in image reconstruction. Further refinements will include detection of the area of the objects and apply our algorithm only in the area of the objects. We plan to improve the current work by handling large data size and different concentration of objects with different sizes. We also intend to include evaluation using Shepp-Logan Model, Hoffman simulated phantom data and real phantom data. Further we plan to introduce the parallelism in LS to reduce the computation time. The exploration of the search space could be done by using threads or process working in parallel. It would be nice to compare LS with other iteratives method such as MLEM and/or OSEM.

References

1. Alessio, A., Kinahan, P.: PET Image Reconstruction. In: Henkin, et al. (eds.) Nuclear Medicine, 2nd edn. Philadelphia, Elsevier (2006)

2. Bosman, P.A.N., Alderliesten, T.: Evolutionary algorithms for medical simulations: a case study in minimally-invasive vascular interventions. In: Proceedings of the 2005 Workshops on Genetic and Evolutionary Computation (GECCO 2005), pp. 125–132 (2005)
3. Bruyant, P.: Analytic and iterative reconstruction algorithms in SPECT. J. Nucl. Med. 43, 1343–1358 (2002)
4. Donald, P.L., Kihm, K.D.: Tomographic-image reconstruction using a hybrid genetic algorithm. Opt. lett. 22, 847–849 (1997)
5. Fahey, F.H.: Data acquisition in PET imaging. Journal of Nuclear Medicine Technology 30(2), 39–49 (2002)
6. Fessler, J.: Analytical Tomographic Image Reconstruction Methods, ch. 3 (November 19, 2009)
7. Gouicem, A.M.T., Benmahammed, K., Drai, R., Yahi, M., Taleb-Ahmed, A.: Multi-objective G-A Optimization of Fuzzy Penalty for Image Reconstruction from Projections in X-Ray Tomography. Digital Signal Processing Elsevier 22(3), 486–596 (2012)
8. Herman, G.T.: Image Reconstruction from Projections. Academic Press, New York (1980)
9. Hudson, H.M., Larkin, R.S.: Accelerated image reconstruction using ordered subsets of projection data. IEEE Trans. Med. Imaging 13(4), 601–609 (1994)
10. Jain, A.: Image reconstruction from projections. In: Kailath, T. (ed.) Fundamentals of Digital Image Processing, pp. 431–475. Prentice-Hall, Englewood Cliffs (1989)
11. Li, X., Jiang, T., Evans, D.J.: Medical Image Reconstruction Using a Multi-objective Genetic Local Search Algorithm. International Journal on Computer Mathematics 74, 301–314 (2000)
12. Mou, C., Pen, L., Yao, D., Xiao, D.: Image Reconstruction Using a Genetic Algorithm for Electrical Capacitance Tomography. Tsinghua Science and Technology 10(5), 587–592 (2005) ISSN 1007-0214 12/20
13. Qureshi, S.A., Mirza, S.M., Arif, M.: A Hybrid Continuous Genetic Algorithm for Parallelray Transmission Tomography Image Reconstruction. Journal of Biomedical Informatics (2006) (in process)
14. Radon, J.: On the determination of functions from their integrals along certain manifolds. Math. Phys. Klass. 69, 262–277 (1917) (in German)
15. Shepp, L.A., Vardi, Y.: Maximum likelihood reconstruction for emission tomography. IEEE Trans. Med. Imaging 1(2), 113–122 (1982)
16. Vadivel, A., Majumdar, A.K., Sural, S.: Performance Comparison of Distance Metrics in Content-based Image Retrieval Applications. In: International Conference on Information Technology (CIT), Bhubaneswar, India, pp. 159–164 (2003)
17. Vidal, F.P., Lazaro-Ponthus, D., Legoupil, S., Louchet, J., Lutton, É., Rocchisani, J.-M.: Artificial evolution for 3D PET reconstruction. In: Collet, P., Monmarché, N., Legrand, P., Schoenauer, M., Lutton, E. (eds.) EA 2009. LNCS, vol. 5975, pp. 37–48. Springer, Heidelberg (2010)

A Novel Color Image Coding Technique Using Improved BTC with k-means Quad Clustering

Jayamol Mathews, Madhu S. Nair, and Liza Jo

Abstract. A new approach to color image compression on HSV model with good quality of reconstructed images and better compression ratio using Improved Block Truncation Coding algorithm with k-means Quad Clustering (IBTC-KQ) is proposed in this paper. The RGB plane of the color image is transformed into HSV plane in order to reduce the high degree of correlation between the RGB planes. Each HSV plane is then encoded using IBTC-KQ method. The block sizes are chosen based on the information content of the respective plane. The result of the proposed method is compared with that of other BTC based methods on RGB model and it shows a better performance both in the visual quality and compression ratio. Also the proposed method involves only less number of simple computations when compared with other BTC methods.

Keywords: Color image compression, HSV color model, Block Truncation Coding, k-means quad clustering.

1 Introduction

Digital image processing is a rapidly emerging field with growing applications in Science and Engineering. It has a wide range of applications such as remote sensing data via satellite, medical image processing, radar and robotics. Nowadays, most of the information in computer processing is done online which is either graphical or pictorial in nature and since it is represented by images, the

Jayamol Mathews · Madhu S. Nair
Department of Computer Science, University of Kerala, Kariavattom
Thiruvananthapuram – 695581, Kerala, India
e-mail: jayamolm@gmail.com, madhu_s_nair2001@yahoo.com

Liza Jo
Philips Electronics India Ltd, Bangalore, India
e-mail: liza.jose@gmail.com

S.M. Thampi, A. Gelbukh, and J. Mukhopadhyay (eds.), *Advances in Signal Processing and Intelligent Recognition Systems*, Advances in Intelligent Systems and Computing 264, DOI: 10.1007/978-3-319-04960-1_31, © Springer International Publishing Switzerland 2014

storage and communication requirements are immense. A color image normally contains a great deal of data redundancy and requires a large quantity of storage space. These storage and transmission challenges can be managed by image compression techniques [1]. The objective of an image compression technique is to remove as much redundant information as possible without destroying the image quality. Color image compression using color quantization is used in order to reduce the number of possible colors and thus lower the transmission and storage cost [2]. In this work, moment preserving principle is studied in the image compression process to keep the visual quality of the resulting image good and acceptable for the human visual system. Block Truncation coding (BTC) is a comparatively simple image compression method developed in 1979 [3]. Though it is a simple technique, it has an important role in the history of digital image coding, as many advanced coding techniques have been developed based on BTC. Even though the compression ratio achievable by BTC have long been exceeded by many other image coding techniques such as DCT (JPEG) [4] and wavelet [5], the computational complexity of BTC like image coding techniques has made it attractive in applications where the real time fast implementation is needed. Another feature of this coding is the high processing speed of the compression and decompression algorithms. The compressed file is normally an object of a data transmission operation. Visual comparison of the original and transmitted (decoded) images gives a direct evaluation of the robustness. By using BTC coding, we can eliminate transmission errors and achieve robustness. Some other interesting features of this coding technique are its ease of implementation and minimum use of working storage space.

The standard BTC algorithm is a simple block based lossy image compression algorithm for grayscale images that preserves the block mean and the block standard deviation. Although this method keeps the visual quality of the reconstructed image it creates some artifacts like staircase effect, etc. near the edges. Several modifications on this compression technique have been developed during the last many years and most of the techniques focus on the extraction and retention of visual information of the image. The absolute moments BTC [6], upper and lower mean BTC [7] and adaptive BTC [8] were implemented to improve the quality of coded images. Some other modified or hybrid BTC algorithms have been developed to further reduce the bit rate either regarding the mean vector [9] or the bit-map [10] [11] [12] or both [13] [14]. All of them use the moment preserving principle first and then apply other techniques to truncate the image data. These algorithms were originally implemented on grayscale images but we can apply these methods separately on each color plane of color images, and then merge the resulting triple set mean vectors and bit-maps by using some other method. Therefore, reducing the three planes of bit maps becomes an important factor in the process of compressing color images. Wu and Coll [15] used a single bit map to quantize all the three color planes so that only one out of three bit maps need to be preserved. Kurita and Otsu [16] used the mean vector and the covariance matrix of color vectors to compute the principal score for the pixels in an image block, and classified the pixels in the block into two classes. Two mean vectors and a bit map are preserved.

In this study, a lossy color image compression algorithm giving better compression ratio with good reconstructed image quality is proposed. Most color images are represented in RGB model. In this paper, the compression of color images is done by converting color images from RGB to HSV color space. The RGB color space defines colors as a linear combination of three additive primary colors; red, green and blue. The color space which is commonly used and matches more naturally to human perception is the HSV color space, with the three components hue, saturation and value. Color vision can be controlled using RGB color space or HSV color space. In cases where color description plays an inherent role, the HSV color model is frequently used alternatively of RGB model. The HSV model depicts colors in a similar way to that of how the human eye tends to perceive color. For many applications, processing HSV component images provides superior results [17]. RGB color space is not perceptually uniform. The three components in this color space have equal importance and hence with the same precision these values have to be quantized [18]. The human visual system is more sensitive to information in the luminance component than to information in the chrominance component. Hence, the RGB components are linearly transformed into the HSV components, and the luminance and the chrominance components are treated differently at the time of coding and decoding. When we apply conventional BTC method on each of the three RGB planes of a color image it gives a better compression ratio without much degradation on the reconstructed image. But it shows some artifacts like staircase effect or raggedness near the edges.

In this color image coding technique using Improved Block Truncation Coding Algorithm with k-means Quad Clustering (IBTC-KQ) [19] the image is converted from RGB model to HSV model and then applied IBTC-KQ on each plane. In IBTC-KQ, instead of the bi-clustering technique used in the conventional BTC, quad clusters are formed for each block in each plane using k-means clustering algorithm [20] so that similar pixel values come under the same cluster. And by getting the means of the pixel values of each cluster, the reconstructed image is generated. In the HSV color model, value represents the brightness or luminance of the color, which gives the dominant description. Hence the block size for luminance component (i.e., value) is taken as 4 and variable block sizes are used in other two components. This will yield generally better visual quality for the reconstructed image with better compression ratio.

2   Proposed Method

The proposed method converts the image from RGB model to HSV model and applies IBTC-KQ on each plane of the HSV color space. The algorithm for this method is as follows.

2.1   Encoding

Step 1: Input a color image of size M×N×3 pixels.

Step 2: Convert the input color image in RGB model (*rgb*) to HSV model (*hsv*).

Step 3: Extract the individual planes h, s and v from hsv.

Step 4: Divide the planes h, s and v into blocks, each of size k×k. Value of k can be 4, 8, 16, and so on for h and s planes and for v plane, the value of k is fixed as 4. Then each block in h, s and v planes can be represented as,

$$HW = \begin{bmatrix} hw_1 & hw_2 \cdots & hw_k \\ \vdots & \ddots & \vdots \\ & \cdots & hw_{k^2} \end{bmatrix}$$

$$SW = \begin{bmatrix} sw_1 & sw_2 \cdots & sw_k \\ \vdots & \ddots & \vdots \\ & \cdots & sw_{k^2} \end{bmatrix} \quad \text{and}$$

$$VW = \begin{bmatrix} vw_1 & vw_2 \cdots & vw_k \\ \vdots & \ddots & \vdots \\ & \cdots & vw_{k^2} \end{bmatrix} \quad \text{respectively.}$$

Step 5: Segment the blocks HW, SW and VW into 4 clusters (hc_0, hc_1, hc_2, hc_3), (sc_0, sc_1, sc_2, sc_3) and (vc_0, vc_1, vc_2, vc_3) respectively using k-means algorithm so that similar pixels are grouped into the same cluster.

$$\left. \begin{aligned} hc_{i\,(i=0\,to\,3)} &= \left\{ hw_j \,\middle|\, hw_j \in HW \text{ and } hw_j \text{ closure to } centroid_i \right\} \\ sc_{i\,(i=0\,to\,3)} &= \left\{ sw_j \,\middle|\, sw_j \in SW \text{ and } sw_j \text{ closure to } centroid_i \right\} \\ vc_{i\,(i=0\,to\,3)} &= \left\{ vw_j \,\middle|\, vw_j \in VW \text{ and } vw_j \text{ closure to } centroid_i \right\} \end{aligned} \right\} \;\dots (1)$$

Step 6: Compute the mean ($h\mu_0$, $h\mu_1$, $h\mu_2$, $h\mu_3$), ($s\mu_0$, $s\mu_1$, $s\mu_2$, $s\mu_3$) and ($v\mu_0$, $v\mu_1$, $v\mu_2$, $v\mu_3$) of the pixel values corresponding to each cluster in each plane.

Step 7: Based on these 4 clusters of each plane the bit maps HB, SB and VB are generated.

$$HB = \begin{bmatrix} hb_1 & hb_2 \cdots & hb_k \\ \vdots & \ddots & \vdots \\ & \cdots & hb_{k^2} \end{bmatrix} \quad \text{here} \quad hb_j = \begin{cases} 00, if\ hw_j \in hc_0 \\ 01, if\ hw_j \in hc_1 \\ 10, if\ hw_j \in hc_2 \\ 11, if\ hw_j \in hc_3 \end{cases}$$

$$SB = \begin{bmatrix} sb_1 & sb_2 \cdots & sb_k \\ \vdots & \ddots & \vdots \\ & \cdots & sb_{k^2} \end{bmatrix} \quad \text{where} \quad sb_j = \begin{cases} 00, if\ sw_j \in sc_0 \\ 01, if\ sw_j \in sc_1 \\ 10, if\ sw_j \in sc_2 \\ 11, if\ sw_j \in sc_3 \end{cases} \quad (2)$$

$$VB = \begin{bmatrix} vb_1 & vb_2 \cdots & vb_k \\ \vdots & \ddots & \vdots \\ & \cdots & vb_{k^2} \end{bmatrix} \quad \text{where} \quad vb_j = \begin{cases} 00, if\ vw_j \in vc_0 \\ 01, if\ vw_j \in vc_1 \\ 10, if\ vw_j \in vc_2 \\ 11, if\ vw_j \in vc_3 \end{cases}$$

Only 2 bits per pixel are required to represent each hb_j, sb_j and vb_j.

Step 8: Repeat the steps 5 to 7 for each block in each h, s & v planes. The resultant bit maps represent the encoded planes.

2.2 Decoding

For the reconstruction of each plane of the image, the bit map and the four means of each cluster in every block of each plane are transmitted to the decoder. The decoding procedure is as follows:

Step 1: Divide the bit map of each plane into k×k blocks.

Step 2: Decode bitmap block with the four means of each cluster in such a way that the elements assigned 00 are replaced with mean of cluster 0, elements assigned 01 are replaced with mean of cluster 1, elements assigned 10 are replaced with mean of cluster 2 and elements assigned 11 are replaced with mean of cluster 3. Then the decoded image block Z for each plane h, s & v can be represented as,

$$HZ = \begin{bmatrix} hz_1 & hz_2 & \cdots & hz_k \\ \vdots & & \ddots & \vdots \\ & & \cdots & hz_{k^2} \end{bmatrix} \quad \text{where} \quad hz_i = \begin{cases} h\mu_0, hb_j = 00 \\ h\mu_1, hb_j = 01 \\ h\mu_2, hb_j = 10 \\ h\mu_3, hb_j = 11 \end{cases}$$

$$SZ = \begin{bmatrix} sz_1 & sz_2 & \cdots & sz_k \\ \vdots & & \ddots & \vdots \\ & & \cdots & sz_{k^2} \end{bmatrix} \quad \text{where} \quad sz_i = \begin{cases} s\mu_0, sb_j = 00 \\ s\mu_1, sb_j = 01 \\ s\mu_2, sb_j = 10 \\ s\mu_3, sb_j = 11 \end{cases} \quad (3)$$

$$VZ = \begin{bmatrix} vz_1 & vz_2 & \cdots & vz_k \\ \vdots & & \ddots & \vdots \\ & & \cdots & vz_{k^2} \end{bmatrix} \quad \text{where} \quad vz_i = \begin{cases} v\mu_0, vb_j = 00 \\ v\mu_1, vb_j = 01 \\ v\mu_2, vb_j = 10 \\ v\mu_3, vb_j = 11 \end{cases}$$

Step 3: Repeat Step 2 for each block of each plane and the resultant matrix represents the reconstructed plane of the image.

Step 4: Combine all the three decoded planes into a decoded HSV image (*chsv*).

Step 5: Convert the decoded HSV image (*chsv*) into RGB color image (*crgb*) which results the reconstructed color image.

The flowchart for illustrating the encoding and decoding process is given in Fig. 1.

3   Performance Measures

Once an image compression system has been designed and implemented, performance evaluation based on some image quality measures should be done in order to compare results against other image compression techniques. In this work we focus on measures such as Mean Square Error (MSE), Peak Signal to Noise Ratio (PSNR) and Compression ratio (CR) [14, 15]. MSE and PSNR values give the measure of quality of reconstruction of lossy compression. PSNR is a qualitative measure based on MSE of the reconstructed image. The compression ratio is used to measure the ability of data compression by comparing the size of the compressed image and the size of the original image.

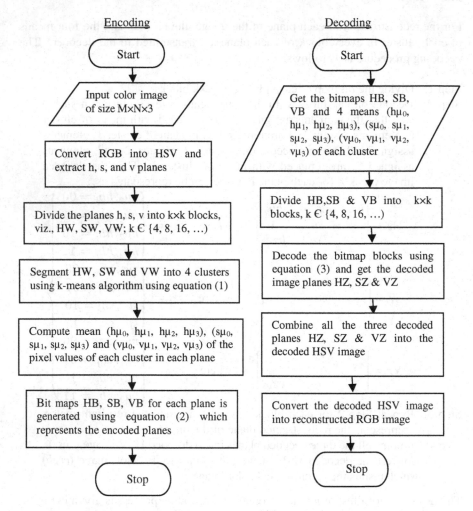

Fig. 1 Flowchart illustrating Encoding and Decoding process

4   Experimental Analysis

Performance of the proposed method has been evaluated for a set of test images of different sizes, viz., 'baboon', 'lena', 'airplane', 'pepper' and 'satellite'. The test images in RGB color space are converted into HSV color space and then extracted each individual components viz., hue (h), saturation (s) and value (v) of the HSV

Table 1 Performance analysis of proposed method on a set of test images for different block sizes for h & s planes and constant block size 4 for v plane

Block size	PSNR in dB				CR			
s ↓ / h →	4	8	16	32	4	8	16	32
lena (256×256)								
4	38.8147	37.5182	36.2483	35.0551	2	2.2857	2.3704	2.3925
8	38.0867	37.0234	35.8597	34.6957	2.2857	2.6667	2.7826	2.8132
16	37.0963	36.2659	35.3122	34.2350	2.3704	2.7826	2.9091	2.9425
32	35.1461	34.8912	33.8618	33.5340	2.3925	2.8132	2.9425	2.9767
lena (512×512)								
4	40.8243	40.0340	39.1380	38.3322	2	2.2857	2.3704	2.3925
8	40.6212	39.8660	39.0592	38.2163	2.2857	2.6667	2.7826	2.8132
16	40.2530	39.5556	38.7624	38.0455	2.3704	2.7826	2.9091	2.9425
32	39.8231	39.2105	38.4457	37.7852	2.3925	2.8132	2.9425	2.9767
airplane (512×512)								
4	40.0848	38.9146	37.8221	36.7505	2	2.2857	2.3704	2.3925
8	39.5928	38.6350	37.6177	36.5840	2.2857	2.6667	2.7826	2.8132
16	38.9921	37.8003	36.9431	36.1053	2.3704	2.7826	2.9091	2.9425
32	37.8115	37.0685	36.4810	35.5037	2.3925	2.8132	2.9425	2.9767
pepper (512×512)								
4	40.8718	39.2747	37.5448	35.7779	2	2.2857	2.3704	2.3925
8	39.1006	37.9665	36.5610	35.0657	2.2857	2.6667	2.7826	2.8132
16	36.6312	35.9277	35.2589	33.9682	2.3704	2.7826	2.9091	2.9425
32	34.6617	33.2834	33.2841	32.7727	2.3925	2.8132	2.9425	2.9767
satellite (1024×1024)								
4	32.3240	32.0592	31.7768	31.5273	2	2.2857	2.3704	2.3925
8	32.2671	31.9815	31.7069	31.4877	2.2857	2.6667	2.7826	2.8132
16	32.1580	31.9038	31.6032	31.3756	2.3704	2.7826	2.9091	2.9425
32	32.0225	31.7531	31.4769	31.2542	2.3925	2.8132	2.9425	2.9767

color image. In RGB model, since (r, g, b) components are correlated, with the same precision these values have to be quantized. Hence, each plane is coded with same block sizes. In the HSV color model, value represents the brightness of the color, which is more sensitive to the human eye. Therefore, for achieving good visual quality in the reconstructed image, the block size for the component v is taken as constant, say, 4 and variable block sizes ranging from 4 to 32 are taken for the other two components h and s. The performance is measured based on the parameters PSNR and CR. Table 1 shows the performance analysis on test images for different block sizes for h & s with the constant block size 4 for v plane. Table 2 shows the performance analysis of IBTC-KQ, AMBTC and BTC on r, g & b planes separately with different block sizes.

When we compare the values in Table 1 & 2 for different test images, we can see the increasing variation in PSNR values with better compression ratio. For example, in the case of 512×512 'lena' image, the PSNR value achieved for block size 4 in (r, g, b) plane is 40.1465dB with a compression ratio of 2 (in Table 2) whereas the PSNR value for block sizes (4, 4, 4) in (h, s, v) plane 40.8243dB with same compression ratio of 2 (in Table 1). And with BTC and AMBTC the PSNR value obtained is only 29.4556dB and 33.1843dB.

Table 2 Performance analysis of IBTC-KQ, AMBTC and BTC on each r, g, b planes with different block sizes

Method	Block size – 4		Block size – 8		Block size – 16		Block size – 32	
	PSNR	CR	PSNR	CR	PSNR	CR	PSNR	CR
lena (256×256)								
BTC	26.7184	4	23.8559	6.4	21.7470	7.5294	19.8585	7.8769
AMBTC	30.9400	4	27.8555	6.4	25.5719	7.5294	23.6503	7.8769
IBTC-KQ	38.2529	2	34.5424	3.2	32.2239	3.7647	30.1786	3.9385
lena512 (512×512)								
BTC	29.4556	4	26.4548	6.4	24.0347	7.5294	22.1325	7.8769
AMBTC	33.1843	4	30.3392	6.4	27.9862	7.5294	25.8793	7.8769
IBTC-KQ	40.1465	2	36.4591	3.2	34.1803	3.7647	32.3672	3.9385
airplane (512×512)								
BTC	26.4967	4	23.8153	6.4	21.8808	7.5294	20.1680	7.8769
AMBTC	31.8545	4	28.5812	6.4	26.4854	7.5294	24.9702	7.8769
IBTC-KQ	39.2827	2	35.5515	3.2	33.5078	3.7647	31.9910	3.9385
babbon (512×512)								
BTC	23.0003	4	21.1752	6.4	19.8415	7.5294	18.5215	7.8769
AMBTC	26.9687	4	25.0069	6.4	23.6425	7.5294	22.6878	7.8769
IBTC-KQ	33.8076	2	30.9736	3.2	29.6361	3.7647	28.5184	3.9385
pepper (512×512)								
BTC	27.7787	4	24.6092	6.4	21.7688	7.5294	19.7993	7.8769
AMBTC	33.3755	4	29.3254	6.4	26.2391	7.5294	23.7962	7.8769
IBTC-KQ	40.5899	2	36.0965	3.2	33.1570	3.7647	30.6388	3.9385
satellite (1024×1024)								
BTC	21.0127	4	18.7240	6.4	17.3198	7.5294	16.5737	7.8769
AMBTC	24.7687	4	22.2407	6.4	20.6722	7.5294	19.7676	7.8769
IBTC-KQ	31.8253	2	28.5514	3.2	26.9108	3.7647	25.9544	3.9385

When we are increasing the block size value for 's' (saturation) by keeping the other two values of block sizes same, the PSNR value is still high when compared with PSNR value in (r, g, b) plane (40.1465dB) and it also achieves a compression ratio greater than 2. For example,

(h, s, v)	PSNR in dB	CR
(4, 8, 4)	40.6212	
2.2852		
(4, 16, 4)	40.2530	
2.3704		

Similarly for other images including the 'satellite image' having large size, we can see the same increase in PSNR value from the two tables. In the case of satellite image, in (r, g, b) plane the PSNR value achieved is 28.5514dB with a compression ratio of 3.2 (in Table 1). With an approximately same compression ratio of 2.9767 the PSNR value achieved is 31.2542dB in (h, s, v) plane with block sizes (32, 32, 4) (in Table 2). This increasing variation in PSNR value shows betterment in the visual quality of the reconstructed image. The variation in PSNR value for different block sizes in HSV and RGB model with IBTC-KQ, AMBTC and BTC on a test image 'lena' (512×512) is shown in Fig. 2. Some of the test images of different sizes and the corresponding reconstructed images are shown in Fig. 3. and Fig. 4.

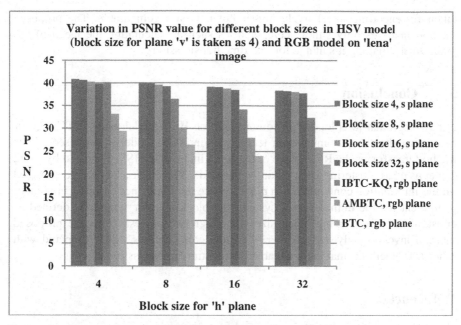

Fig. 2 Variation of PSNR value in HSV model and RGB model

Fig. 3 Original images

Fig. 4 Reconstructed images using IBTC-KQ in HSV plane

The visual quality of the reconstructed images is much more enhanced in HSV model than in RGB model using IBTC-KQ. This improvement is clearer by the variation of PSNR values shown in the graph in Figure 2. Thus the proposed method produces perceptually superior quality reconstructed image with minimum distortions near the edges. Since this method involves quad clustering the time taken for encoding is relatively higher but it is still comparable. The proposed method involves only simple computations such as calculating the mean of the pixel values making it suitable for real time transmission .

5   Conclusion

A new approach to color image compression on HSV model using IBTC-KQ for enhancing the other compression techniques in RGB model is proposed. This method transforms the RGB plane of the input image into HSV plane and by using k-means quad clustering each plane is encoded. The block sizes are given based on the information content of each plane. Since the luminance component is more prominent for the human visual system, the block size for v plane is defined as constant and use varying block sizes for other two planes. Also the proposed method involves only less number of simple computations when compared with other BTC methods making it suitable for real time transmission.

References

1. Sangwine, S.J., Horne, R.E.: The Colour Image Processing Handbook, 1st edn. Chapman & Hall (1998)
2. Gonzalez, R.C., Woods, R.E.: Digital Image Processing, 3rd edn. Prentice Hall (2008)
3. Delp, E.J., Mitchell, O.R.: Image compression using block truncation coding. IEEE Trans. Commun. COM-27, 1335–1342 (1979)
4. Pennebaker, W.B., Mitchell, J.L.: JPEG Still Image Compression Standard Van Nostrand Reinhold, New York (1993)
5. Shappiro, J.M.: Embedded image coding using zero trees of wavelet coefficients. IEEE Trans. Signal Processing 41, 3445–3462 (1993)
6. Lema, M.D., Mitchell, O.R.: Absolute moment block truncation coding and its application to color images. IEEE Trans. Commun. COM-32, 1148–1157 (1984)
7. Udpikar, V.R., Raina, J.P.: A modified algorithm for block truncation coding of monochrome images. Electron. Lett. 21-20, 900–902 (1985)
8. Hui, L.: An adaptive block truncation coding algorithm for image compression. In: Proc. ICASSP 1990, vol. 4, pp. 2233–2236 (1990)
9. Delp, E.J., Mitchell, O.R.: The use of block truncation coding in DPCM image coding. IEEE Trans. Signal Process. 39(4), 967–971 (1991)
10. Arce, G.R., Gallagher, N.C.: BTC image coding using median filter roots. IEEE Trans. Commun. 31(6), 784–793 (1983)
11. Zeng, B., Neuvo, Y., Venetsanopoulos, A.N., Interpolative, B.T.C.: image coding. In: Proc. IEEE ICASSP 1992, vol. 3, pp. 493–496 (1992)
12. Udpikar, V.R., Raina, J.P.: BTC image coding using vector quantization. IEEE Trans. Commun. 35(3), 352–356 (1987)
13. Zeng, B., Neuvo, Y., Interpolative, B.T.C.: image coding with vector quantization. IEEE Trans. Commun. 41(10), 1436–1437 (1993)
14. Wu, Y., Coll, D.C.: BTC-VQ-DCT hybrid coding of digital images. IEEEE Trans. Commun. 39(9), 1283–1287 (1991)

15. Wu, Y., Coll, D.C.: Single bit map block truncation coding for color image. IEEE Trans. Commun. COM-35, 352–356 (1987)
16. Kurita, T., Otsu, N.: A method of block truncation coding for color image compression. IEEE Trans. Commun. COM-35, 352–356 (1987)
17. Baxes, G.A.: Digital Image Processing – Principles and Applications. John Wiley & Sons, Inc. (1994)
18. Vellaikal, A., Jay Kuo, C.C., Dao, S.: Content-Based Retrieval of Color and Multispectral Images Using Joint Spatial-Spectral Indexing. In: SPIE, vol. 2606, pp. 232–243
19. Mathews, J., Nair, M.S., Jo, L.: Improved BTC Algorithm for Gray Scale Images using k-means Quad Clustering. In: Huang, T., Zeng, Z., Li, C., Leung, C.S. (eds.) ICONIP 2012, Part IV. LNCS, vol. 7666, pp. 9–17. Springer, Heidelberg (2012)
20. Kanungo, T., Mount, D.M., Netanyahu, N., Piatko, C., Silverman, R., Wu, A.Y.: An efficient k-means clustering algorithm: Analysis and implementation. In: Proc. IEEE Conf. Computer Vision and Pattern Recognition, pp. 881–892 (2002)
21. Eskicioglu, A.M., Fisher, P.S.: Image quality measures and their performance. IEEE Trans. Communications 34, 2959–2965 (1995)
22. Yamsang, N., Udomhunsakul, S.: Image Quality Scale (IQS) for compressed images quality measurement. In: Proceedings of the International Multiconference of Engineers and Computer Scientists, vol. 1, pp. 789–794 (2009)

Block Matching Algorithms for Motion Estimation – A Comparison Study

Abir Jaafar Hussain, Liam Knight, Dhiya Al-Jumeily,
Paul Fergus, and Hani Hamdan

Abstract. Motion estimation procedures are employed in order to achieve reductions in the amount of resources required for data retention and therefore alleviate transmission demands of steaming digital video sequences across Information System infrastructures. The process comes at high computational expense, however there are a number of techniques in which attempt to expedite processing times for the compression of video streams by increasing the efficiency of resource utilisation through a minimisation of superfluous calculations incurred by imprudent video compression applications. A variety of fast, block-based matching algorithms have been developed to address such issues within the motion estimation process and to also exploit the assumptions made in regards to the behaviours exhibited by distortion distributions. This investigation will analyse a number of such methodologies through a literature review of their accompanying publications and will enable the development comparative analysis based on various performance indicators.

Keywords: Motion estimation, block matching algorithms, video compression.

1 Introduction

The impact of media content provision on the Information Technology and entertainment industries has been hugely significant, the estimated revenue from

---

Abir Jaafar Hussain · Liam Knight · Dhiya Al-Jumeily · Paul Fergus
Liverpool John Moores University, Byroom Street, Liverpool, L3 3AF, UK
e-mail: L.knight@2009.ljmu.ac.uk,
{A.Hussain,P.Fergus,D.Aljumeily}@ljmu.ac.uk ,

Hani Hamdan
SUPÉLEC, Department of Signal Processing & Electronic Systems,
rue Joliot-Curie, 91190 Gif-sur-Yvette, France
e-mail: Hani.Hamdan@supelec.fr

S.M. Thampi, A. Gelbukh, and J. Mukhopadhyay (eds.), *Advances in Signal Processing and Intelligent Recognition Systems*, Advances in Intelligent Systems and Computing 264,
DOI: 10.1007/978-3-319-04960-1_32, © Springer International Publishing Switzerland 2014

advertisement in 2011 reached over three billion dollars and is expected to be a seven billion dollar enterprise by 2015 [1]. The distribution of digital video via the internet has effectively given rise to new source of entertainment media and has shifted the focus of major traditional media outlets such as Channel 4 and ITV, who are currently providing online content in order to capitalise on emerging revenue streams. A collection of these factors have also helped to drive innovation of video compression technologies as competition for user viewing time has increased.

The increasing popularity of the internet as a source for entertainment media consumption and the remarkable development data transfer capabilities have helped video streaming become a viable source for media content provision. Traditionally, the main methods of delivering entertainment media to consumers have been limited to television and radio vendors. Technological advancements have given rise to the online availability of digital video content allowing consumers to stream content directly to their computational devices. The emergence of high-definition video displays with provisional storage media formats such as Blu-Ray and HD discs place an ever greater exertion on the facilitating computational resources. Ubiquitous availability of such applications has become viable thanks to the progressive increase in internet bandwidth speeds and more significantly, the development of video compression techniques.

The video coding environment incorporates a number of different methodologies such as the lossy and lossless techniques, of which demonstrate contrasting operational characteristics and therefore use within practical applications is largely dependent on the requirement of the user. There are several formats associated with the encoding and decoding of digital video streams such as the ISO/IEC MPEG and ITU-U H.26x formats [2], developed with the intention of providing a framework for the standardisation of data compression within this environment. Full Search (FS) motion estimation procedures are able to exploit the redundancies that exist in temporal domains of moving video sequences through the process of motion compensation [3]. This is the activity in which motion differences occurring between digitally sampled frames of video sequences are encoded as opposed to retaining the structural integrity of each. Frame segmentation is applied in order divide a spatial image into a group of sections referred to as Macroblocks, enabling the displacement in which is observed between temporal positions to be predicted. This process is known as block-based motion estimation and whilst a reasonable ratio of compression can be achieved using the FS technique, it is both highly resource intensive and time consuming to accomplish. A variety of fast block matching algorithms have been proposed in order to alleviate the computational demands of FS motion estimation procedures and also obtain comparable levels of fidelity.

Information yielded from our research investigations intend to provide pertinent information in relation to a number of block-based motion estimation techniques so that the optimum method for application within a video coding environment can be determined. Knowledge on this subject is ascertained from a literature review on the publications in which such techniques are proposed and will include a detailed analysis of the experimental data contained within. Conclusions made in respect to which solution provides the least computational complexity, time for processing and levels of distortion will be firmly reinforced by the empirical evidence supplied in the proposal paper simulations. Correlations demonstrated by such techniques in relation to their beneficial characteristics and their operational functionalities will be identified, thus enabling suggestions on their successfulness for implementation within motion estimation procedures.

2   Motion Compensation and Estimation

One of the most effective methods for eliminating temporal redundancy from within digital video streams is through the use of block-based motion estimation techniques in which provide compensation of the movement that occurs between temporal samples (Interframe). The process begins[2] by arranging the pixels within the current frame into a selection of blocks (Macroblocks, commonly 16 x 16 pixels) as shown in Fig.1 (a) , also known as frame segmentation and is represented by Fig.1 (b).

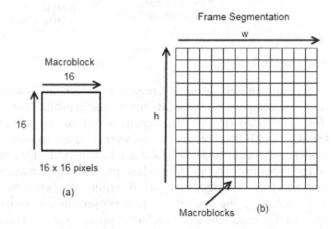

Fig. 1 (a) Macroblock. (b) The process of frame segmentations.

A search is then conducted to identify an area of a reference frame (Motion Vector) that best matches the current macroblock and the displacement between the two is then taken into account, referred to as motion estimation (demonstrated by Fig.2). The source content of certain frames has to be encoded in order to provide reference frames for the motion estimation process. Motion vectors identified within the reference frame are known as residual macroblocks. When each macroblock within the candidate frame is assigned a residual macroblock they are then used to create a residual frame, of which is deducted from the matching areas within the current frame in order to provide motion compensation. The compensated movement between the two matching blocks is used to predict the arrangement of pixels on the current frame. This can provide a significant reduction in data size as the bit sequences for every frame do not have to be retained in the encoded video file.

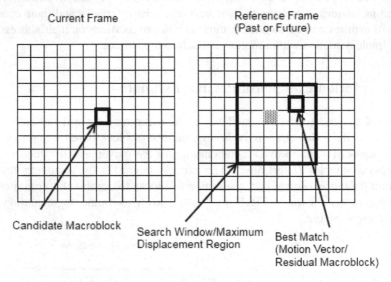

Fig. 2 The process of motion estimation

The block based motion estimation and compensation techniques are utilised by the majority of video compression standards due to their capability for eliminating redundancy from raw video streams. The standards that employ this technology include the MPEG-4 and H.264 formats [4], however there are various methods in which the motion estimation and compensation is achieved. The demand on computational resources from the block matching process can be reduced through the implementation of various algorithms that determine the behaviour of search operations. The block matching algorithms are employed in order to increase the efficiency of the search operations and therefore improve the overall effectiveness of video compression process. These techniques exploit consistencies that are exhibited by the behaviours in the distortion levels of potential matching areas. The suitability of macroblocks within a reference area for their inclusion in the residual frame is determined by the implementation of various distortion metrics such as Mean Squared Error (MSE), Mean Absolute Error (MAE), and Sum of Absolute Errors (SAE).

3   Evaluation Metrics

Distortion metrics are implemented in order to measure the homogeneity between the current macroblock and predicted reference area; this however does not provide an indication for the perceived quality of the resulting decoded video output. Human perception of visual quality is difficult to accurately quantify as this process is largely subjective and evaluation of fidelity can deviate significantly due to individual interpretations of content matter. Endeavours have been undertaken to provide methods in which the performance of the motion estimation process can be measured quantitatively and effectively dehumanise the evaluation of decoded visual quality.

There are a variety of evaluation metrics which are utilised to determine the perceived fidelity of decompressed video output including the Peak Signal-to-Noise Ratio (PSNR) and the Structural Similarity Index (SSIM). The PSNR is the most commonly employed evaluation metric, being the preferred method due to its simplicity and potential for fast calculations. Evaluation metrics are able to supply an objective determination of visual quality and in certain instances this does not correlate with the subjective perceptions of the HVS. A calculation for the MSE is required in order to obtain the PSNR figures for video sequences and the equation used to do so is as follows:

$$PSNR_{dB} = 10log_{10} \frac{(Peak\ signal\ value\ of\ content)^2}{MSE}$$

4   Comparison of Block-Based Matching Algorithms

Primal video coding applications in which the estimation of temporally inhabited motions were conducted, incurred a heavy demand on resource availability due to the appreciable number of computations required in order to achieve such. This is demonstrated by the large number of search point calculations required by the Full Search (FS) motion estimation technique [3] to identify appropriate motion vectors for candidate macroblocks under examination. The vertical scaling of hardware components is a relatively common undertaking within modern day system architectures in order to address these issues. In the formative years of motion estimation application however, commodity physical devices were of limited availability and also of a diminutive potency compared to contemporary device specifications. Thus, the heavy computational cost of FS operations placed further constraints on individuals concerned with compression of digital video sequences in which motion estimation is rendered. The development of fast block matching algorithms sought to address this consequence through a reduction in the computations involved for motion vector assignment and also by increasing the rapidity in which it is achieved.

The inception of fast block matching algorithms commenced in the early Eighties with solutions such as the Three Step Search (TSS) [5] and 2-D Logarithmic Search (TDL) [6] proposed during this era. Intending to address the resource intensity of exhaustive displacement region excavations, the TSS and TDL algorithms are able to significantly reduce the computational expenditures of motion estimation procedures. This however, is not without recourse as the decrease in search point locations is mitigated by a potential for impairments in the perceived fidelity of decompressed video sequences. The major disadvantage of fast block matching algorithms is that the path of search is tempted towards a local minimum point of distortion as opposed to the identification of the true motion vector location. Incorrect specification of motion vector for the candidate macroblock does not provide a true representation of the source information and therefore results in a lossy form of data compression.

Conceptualisation of the both TSS and TDL would not have successfully reduced computational complexity, whilst maintaining reasonable fidelity, unless the dispersion of distortion exhibited a certain level of correlation. Development of such solutions is dependent on valid logical reasoning to facilitate programmable algorithms that can be executed sequentially in order to render a desired outcome. Pattern-based, fast block matching algorithms owe their operational efficiency to the fact that measured distortion levels decrease uniformly on approach to the motion vector position considered as the optimum match.

Supported by the homogeneity of distortion distribution, the TSS and TDL are able to both minimise motion vector calculations and expedite FS search procedures. In a commonly employed displacement region scenario of ±7, the TSS is able reduce permutations of search by a factor of nine. Motion estimation procedures can therefore be achieved in approximately ten percent of the amount of time that is required by the FS method.

Proposal of the New Three Steps Search (NTSS) [7] in the mid-nineties prompted the speculation of an additional assumption in relation to the behaviours exhibited by distortion distributions. This technique relied heavily on the suggestion that the motion vectors assigned from reference frames are in a centralised locality accordant to that of the candidate macroblock coordinates. This reasoning is exploited by the NTSS [7] through the inclusion of an additional 3 x 3 coordinate search pattern during distortion level assessments in order to identify static macroblocks, at which point execution of the algorithm may be terminated. Implementation of two search patterns at the first step will incur seventeen Block Distortion Measures (BDM) calculations and is therefore highly dependent of the validity of said assumption to ensure computational demands are not overly exuberant for motion estimations procedures.

Simulations were conducted by the developers of the NTSS with the TSS placed under identical testing conditions. The findings were particularly significant as the NTSS was shown to be consistently closer to point of global minimum distortion, but more significantly it is the most likely to identify the optimum motion vector match. The study does not however, indicate the required number of search locations for application of the NTSS within the tested video sequences. The NTSS is liable to incur an increased number of BDM calculations per step compared to that of the TSS; however this is mitigated by the innovative half-way stop technique which can also reduce processing times for static macroblocks.

The success demonstrated by the NTSS in adopting centre-biased search functionality stimulated the emergence of a number of similar, block-based matching techniques in the years to follow. This is the framework in which algorithms such as the Four Step Search (4SS) [9] and Block-Based Gradient Decent Search (BBGDS) [8], are configured to exploit in order to increase the efficiency of motion estimation procedures whilst maintaining acceptable levels of fidelity. The two solutions were proposed in the simultaneous year and both incorporate a square conformation of search locations that navigate along the path of least distortion from the displacement region centre. The findings of simulations conducted within each accompanying publication[8][9] were significant in that their suggestions of centralised motion vector residency were affirmed through the performance levels in which were exhibited by both solutions in comparison to previously developed block matching techniques.

Transformations made by the 4SS and BBGDS algorithms during the implementation of block matching, motion estimation were variable in regards to both computational complexity and also levels of observable distortion. Distortion levels incurred by the BBGDS in the tested video sequences were comparable with that of the NTSS, whilst computations were reduced by a factor of six. The ramifications of such, are that the computational resources required to achieve motion vector assignment are considerably less and the time required in order process such is also attenuated.

The 4SS also made reasonable progressions in regards to computational complexity, however the minimisation of motion vector assessments is less remarkable and on average, it eliminated such calculations by ten percent in comparison to that as is observed by the results of the NTSS. Yet, fidelity levels exhibited during the 4SS simulated exercises [9] were a more noteworthy disclosure, as the average MSE results indicate a considerable improvement in the quality of encoded video streams over that of the NTSS. This generation of centre-biased, block matching algorithms helped to reinforce the convictions made by the developers of NTSS, whilst also augmenting their functionality in order to achieve distinguishable levels of improvement. Implementation of either the 4SS or BBGDS within motion estimation procedures is largely conditional based on users' requirements in relation to the rate-distortion trade off as they are beneficial to this process for contrasting reasons.

The beginning of the following decade witnessed the proposal of another method for pattern-based, motion estimation operations called the DS algorithm. The Diamond search (DS) [10] is also heavily reliant on a centre-biased dispersion of true motion vectors as its search pattern conformation is of condensed proportions, compared to that of earlier techniques such as the TSS and TDL. Whilst the DS operates within the same logical foundations as the 4SS and BBGDS, its distribution of search point locations were a novel feature at this time. The majority of preceding solutions apply their points of search using either a square or cross-based distribution. The diamond-like positioning of possible motion vector coordinates results in a reduced number of locations for subsequent operations due to the intersection of points in which occurs. The DS simulations indicated that the technique is successful for achieving such in comparison to the

NTSS and 4SS; however the BBGDS is shown to be the optimum performer in regards to computational complexity.

The distinguishing feature of the DS is that it can provide comparable fidelity levels with the 4SS (Even with the FS in certain scenarios) and also achieve reductions in the calculations necessary to do so. Although the BBGDS can be considered less resource intensive than the DS, this is sacrificed through increased levels of distortion and in certain scenarios is able to achieve a twenty-five percent increase in visual quality. Due to the desirable levels of fidelity observed by this technique compared to the 4SS, would have made the DS the optimum fast block matching solution if high-fidelity, compressed video streams are the preferred requirement.

The performance gains achieved by Zhu et al. [10] in the instance of the DS, provided the motivation for the development of a similar pattern-based technique called the Hexagonal Search (HEXBS). Also developed by Zhu et al. [11], further innovation was evident in this methodology as the distribution of search locations are arranged in a hexagonal formation, of which had not been observed in previous block matching solutions. This search pattern configuration results in three additional assessed motion vector locations for each subsequent stage of execution, therefore increasing the simplicity of motion estimation procedures. The experimental results confirmed such and a twenty-five percent reduction of computational complexity in comparison to that of the DS is demonstrated. This however, is not without consequence as the HEXBS is unable to provide a comparable representation of the source content and fidelity levels are not as desirable as the DS. This is understandable due to the main operational benefit of the DS is its ability accurately identify suitable motion vectors and therefore if efficient processing times are the preferred functionality for motion estimation procedures, the HEXBS would be considered the optimum solution in comparison to the currently available techniques.

The same year in which the HEXBS was proposed saw the introduction of another pioneering block matching solution, named the Adaptive Rood Pattern Search (ARPS) algorithm [12]. The ARPS incorporates, what were at this period in time, two novel features; zero motion prejudgement (ZMP) for stationary or quasi-stationary macroblocks and a predicted motion vector location within the primary search event. ZMP was shown to be effective for eliminating the superfluous computations that are incurred by previous algorithms for macroblocks with zero or minimal amounts of motion in the temporal domain of digital video sequences. The reasoning behind implementation of an additional, predicted motion vector location is that this ensures the path of search originates in an auspicious location and removes the likelihood of becoming trapped into a local minimum as is apparent in earlier techniques. The DS was shown the optimum performer from the analysis of previous experimental results and within the ARPS publication is tested alongside the proposed solution. The two algorithms performed competitively in this regard and the ARPS was even shown to demonstrate near perfect fidelity levels based on FS BDM calculations as the optimum distortion values. More significantly, the results show that the ARPS incurs substantially fewer computations in order to achieve such comparable quality in the assignment of suitable motion vectors. Scenarios in which the ARPS is applied without the implementation of ZMP, the number of motion vector assessments is reduced by a factor of two in comparison to the DS. Instances in which ZMP is utilised are shown to make further decreases in necessary computations, however this is less remarkable for high resolution video sequences due to the increases in paths of movement which obviously result in greater dispersion of motion vector locations between current and reference frames.

DS simulations indicated that it was the most desirable solution to this point in regards to the distortion levels in which it provides, demonstrated by Fig.3. The BBGDS was also shown to be the most suitable solution at this time for video coding environments with significant hardware and processing time constraints as it requires considerably less points of search than that of the DS. A collective review of the simulations conducted to date, show that the reductions made by ARPS in terms of the computational complexity of the DS are more remarkable than the BBGDS and it can therefore be considered the optimum solution for rending such qualities.

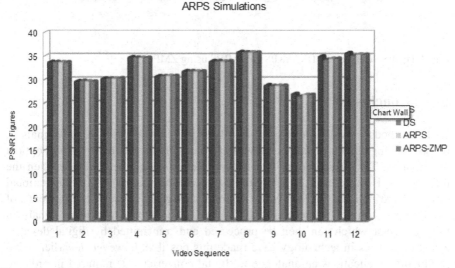

Fig. 3 Simulation results for various fast block matching algorithms.

ARPS experimental data provides empirical evidence showing that its observed distortion levels are comparable with that of the DS (demonstrated by Fig. 4), but also demonstrates that approximately half the calculations are required in order to achieve such. Thus, from our investigation into a variety of fast block matching algorithms it can be determined that the optimum solution for application within motion estimation of video sequences is the ARPS due to its computational simplicity and the desirable levels of fidelity in which are achieved. Simulations conducted within [13] reaffirm our suggestion that the ARPS can be considered the best contemporary solution for such computational activities. Our belief is that the innovative components of the ARPS, such as the ZMP and predicted motion vector locations are extremely effective instruments for increasing the efficiency of block-based matching procedures.

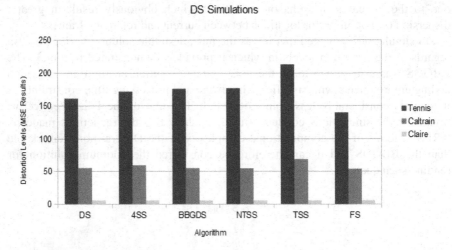

Fig. 4 The PSNR using the DS, ARPS and ARPS using ZMP.

5 Conclusion

Early methods of video production involved the capture of analogue signals to record moving objects or sequences within a particular scene, referred to as optical information. The video coding process can therefore only be achieved when the information is represented in a digital format due to the fact that computational devices are unable to process analogous signals. The development of contemporary digital camcorders has enabled the capture of video signals in binary format which can then be processed and transmitted by digital devices. Recent advances in technology have made this possible, however in earlier video production applications an analogue-to-digital converter was required in order to achieve such undertakings. In this paper a comparison study for various fast block matching algorithms were conducted.

References

1. Social Times (2011) Online Video Revenues Triple in (2011),
   http://socialtimes.com/online-video-revenues-
   triple-in-2011_b85513 (accessed November 19, 2012)

2. Richardson, I.E.G.: H.264 and MPEG-4 Video Compression: Video Coding for Next-Generation Multimedia. Wiley, Chichester (2003)
3. Liyin, X., Xiuqin, S., Zhang, S.: A Review of Motion Estimation Algorithms for Video Compression. In: International Conference on Computer Application and System Modelling (ICCASM 2010) (2010)
4. Weise, M., Weynand, D.: How Video Works: From Analogue to High Definition, 2nd edn. Elsevier Science, U.S.A (2007)
5. Koga, T., et al.: Motion-Compensated Interframe Coding for Video Conferencing. In: Proc. IEEE Telecommunications Conference, New Orleans, New York (1981)
6. Jain, J.R., Jain, A.K.: Displacement Measurement and its Application in Interframe Image Coding. IEEE Trans. Circuits and Systems on Communications COM-29 (1981)
7. Renxiang, L., Bing, Z., Ming, L.L.: A New Three-Step Search Algorithm for Block Motion Estimation. IEEE Trans. Circuits and Systems for Video Technology 4(4), 438–442 (1994)
8. Liu, L.K., Feig, E.: A Block-Based Gradient Decent Search Algorithm for Block Motion Estimation in Video Coding. IEEE Trans. Circuits and Systems for Video Technology 6(4) (1996)
9. Po, L.M., Ma, W.C.: A Novel Four-step Search Algorithm for Fast Block Motion Estimation. IEEE Trans. Circuits and Systems for Video Technology 6 (1996)
10. Zhu, S., Ma, K.K.: A New Diamond Search Algorithm for Fast Block Motion Estimation. IEEE Trans. on Image Processing 6(2) (2000)
11. Zhu, C., Lin, X., Chau, L.P.: Hexagon-Based Search Pattern for Fast Block Motion Estimation. IEEE Trans. Circuits and Systems for Video Technology 12(5) (2002)
12. Nie, Y., Ma, K.K.: Adaptive Rood Pattern Search for Fast Block-Matching Motion Estimation. IEEE Trans. Image Processing 11(12) (2002)
13. Babu, D., et al.: Performance Analysis of Block Matching Algorithms for Highly Scalable Video Compression - Ad Hoc and Ubiquitous Computing. ISA (2006)

An Impact of PCA-Mixture Models and Different Similarity Distance Measure Techniques to Identify Latent Image Features for Object Categorization

K. Mahantesh, V.N. Manjunath Aradhya, and C. Naveena

Abstract. In the current image retrieval systems, there exists a problem of defining and identifying efficient features in order to successfully bridge the gap between low level and high level semantics. In this regard, we propose an approach of efficiently extracting semantic features by combination of EM algorithm and PCA techniques, and thereby exploring PCA-Mixture Model with various similarity techniques for image retrieval system. Firstly, Expectation Maximization (EM) algorithm is applied to learn mixture of eigen values to obtain optimized maximum likelihood clusters. secondly, Principal Component Analysis (PCA) is applied for different mixtures in order to extract efficient features. Further classification is performed using five different distance metrics. Our proposed method reported state-of-the-art classification rate with lesser features and achieved promising results in classifying Caltech-101 object categories compared with other baseline methods performed on the same dataset.

Keywords: Image Retrieval, latent variable, EM algorithm, PCA, Similar distance metrics, Semantic gap.

1 Introduction

The main objective of this paper is to classify the image based on the object category in the large image datasets. Image categorization and retrieval has become increas-

K. Mahantesh
Department of ECE, Sri Jagadguru Balagangadhara Institute of Technology, Bangalore, India
e-mail: mahantesh.sjbit@gmail.com

V.N. Manjunath Aradhya
Department of MCA, Sri Jayachamarajendra College of Engineering, Mysore, India
e-mail: aradhya.mysore@gmail.com

C. Naveena
Department of CSE, HKBK College of Engineering, Bangalore, India
e-mail: naveena.cse@gmail.com

S.M. Thampi, A. Gelbukh, and J. Mukhopadhyay (eds.), *Advances in Signal Processing*
and Intelligent Recognition Systems, Advances in Intelligent Systems and Computing 264,
DOI: 10.1007/978-3-319-04960-1_33, © Springer International Publishing Switzerland 2014

ingly an exigent task for developing methods to archive, query and retrieve color images from the most challenging datasets (eg. Caltech-101 dataset which possess the images belonging to multiple classes with high intensity variations)[1]. One of the early approach of text based image retrieval system searches images through keywords in a manually annotated image database. Since large image databases are context sensitive, prejudiced and incomplete, manual annotation of images are expensive, time consuming and requires much human intervention as well. Some of the interesting works on text based image retrieval can be seen in [2, 3]. Content Based Image Retrieval (CBIR) is an effective and alternative approach of image retrieval, which uses visual features such as color, shape and texture for representation and automatic annotation of images.

Color is the most comprehensively used visual content, color distributions of images are represented using color moments and further color histogram can be used to characterize it into global and local [4], partitioning histogram into coherent and incoherent regions to incorporate spatial information using Color Coherence Vectors (CCV) and color auto-correlograms extracts spatial correlation between identical colors was noticed in [5, 6]. Gabor wavelets are very popular in representing texture features but variant to scale and rotation. Modified Gabor and Circular Gabor Filter (CGF) were proposed in [7] and found to be rotation and scale invariant. Energy distributions of the texture image at different scales obtained from radon wavelet domain [8] and low level texture features extracted from multi resolution subband coefficients which are derived from orthogonal polynomial models [9]. Wang et. al. [10], proposed a statistical approach known as color co-occurrence matrix to identify the relative position of the neighboring pixels which represents color as well as texture correlation to measure the relevancy between two color images. Many color, shape and texture descriptors proposed by MPEG-7 have faced some limitations in object-based image retrieval for large image datasets, to overcome such limitation MPEG-7 DCD (Dominant Color Descriptor) was proposed in [11] which provides condensed representation of colors with in a region of interest of an image.

Although many efforts have been made to extract low level features, evaluating different distance metrics and looking for efficient image retrieval schemes [3], but failed to narrow down the semantic gap between low level features and high level semantics. Machine learning approach is adopted in bridging the semantic gap by using training set to tune the parameters of an adaptive model. Hence the problem of image recognition and retrieval has now become a statistical learning problem [12]. Fei-Fei et al. presented an incremental Bayesian model by incorporating prior information into the category learning process by training very few images, features related to salient regions are extracted and PCA is applied to reduce the dimension of the features. Single mixture component along with first 10 principal components was chosen to obtain prior distribution and obtained better performance compared to both Bayesian and maximum likelihood algorithms [13]. Kristen et al. introduced a new kernel function which extracts desired set of significant features in Eucledian space and mapped to multi-resolution histograms to obtain discriminant features, later classified using weighted histogram intersection between the feature set [14]. In [15], Zhang et al. proposed hybrid classifier model by integrating nearest neighbor classifier and SVM (Support Vector Machine), to find close neighbor of a query sample and to train them using SVM. Spatial pyramids are constructed by partitioning the image into sub-regions and then computing histograms of local features at increasingly fine resolutions yielding high accuracy on Caltech-101 database compared to "bag of features" approach that ignored the locations of individual features [16]. Mutch and Lowe introduced biologically inspired sparsity constraint for learning localized features by matching only the dominant orientation (obtained after applying Gabor filters at different scales and positions) and also discarding features with lesser weights giving significant improvement in generalization performance [17].

Several approaches found in the literature highlighted about constructing efficient feature descriptors and similarity metrics for better retrieval and recognition accuracy. But the data is indeed a latent variables due to the problem with/or limitation of the observation process [21]. Due to inefficient handling of these variables there may be a decline in classification rate. EM algorithm is a powerful tool to effectively handle these latent variables. Also, PCA being very significant in extracting effective features in reduced feature space. Hence motivated by these issues, through this paper we present combination of EM and PCA algorithms developing PCA-Mixture model and also exploring significance of various distance metrics for classification in image retrieval system.

The paper is organized as following: Proposed methodology is explained in section 2, section 3 describes experimental results and performance analysis, finally conclusions are drawn.

2 Proposed Method

Mixture models characterizes object class into number of clusters and its density function of the observed data can be represented as a linear combination of partitioned clusters [23]. The EM algorithm becoming a popular tool in statistical estimation problems involving incomplete data or in problems which can be posed in a similar form, such as mixture estimation [19]. It is an observable fact that low level features are not a local phenomenon and has to be analyzed globally considering different mixtures of good discriminant features in compact domain. Principal Component Analysis (PCA) is a widely used dimensionality reduction technique in data analysis and used as a best feature extractor in recognition. Hence we took the advantages of EM and PCA, to present EM algorithm for PCA and developed PCA mixture model. The algorithm allows a few eigen vectors and eigen values to be extracted from large collections of high dimensional data. It is computationally very efficient in space and time. It also naturally accommodates missing information with a help of estimation-maximization strategy. The details of PCA mixture model is explained in following paragraphs.

A dataset of observation $X = x_1, \ldots \ldots x_n$. The dimensionality of x_n is D. These observed data can be expressed in terms of a marginalization over a continuous latent space z, for each data point x_n there is a corresponding latent variable z_n [19].

Hence, the complete log likelihood function can be expressed in the form:

$$lnp(X,Z|\mu,W,\sigma^2) = \sum_{n=1}^{N} ln(x_n|z_n) + lnp(z_n) \tag{1}$$

Where, the nth row of the matrix Z is given by z_n. W is the general linear function of Z. μ & σ are the mean and covariance of the observed data. By the help of latent and conditional distribution we can obtain expectation with respect to the posterior distribution over latent variables as follows:

$$
\begin{aligned}
E[lnp(X,Z|\mu,W,\sigma^2)] = \ & -\sum_{n=1}^{N}\{\tfrac{D}{2}ln(2\pi\sigma^2) + \tfrac{1}{2}Tr(E[z_n z_n^T]) \\
& + \tfrac{1}{\sigma^2}||x_n - \mu||^2 - \tfrac{1}{\sigma^2}E[z_n]^T W^T(x_n - \mu) \\
& + \tfrac{1}{2\sigma^2}Tr(E[z_n z_n^T]W^T W)\}
\end{aligned}
$$

This posterior distribution depends only through the sufficient statistics of the Gaussian. Therefore in E-step, we use old parameters to find:

$$E[z_n] = M^{-1}W^T(x_n - \bar{x}) \tag{2}$$

$$E[z_n z_n^T] = \sigma^2 M^{-1} + E[z_n]E[z_n]^T \tag{3}$$

In M step, the likelihood function is maximized with respect to W and σ^2, keeping the posterior statistics fixed which can be computed using following equations.

$$W_{new} = [\sum_{n=1}^{N}(x_n - \bar{x})E[z_n]^T][\sum_{n=1}^{N}E[z_n z_n^T]]^{-1} \tag{4}$$

$$\sigma_{new}^2 = \frac{1}{ND}\sum_{n=1}^{N}\{||x_n - \bar{x}||^2 - 2E[z_n]^T W_{new}^T(x_n - \bar{n}) + Tr(E[z_n z_n^T])W_{new}^T W_{new}\} \tag{5}$$

The parameter estimation is conducted by an iterative process using EM algorithm. The process is repeated, and in each iteration E and M step are computed alternatively until the value of W and σ finds better convergence. One of the benefits of the EM algorithm for PCA is computational efficiency for large-scale applications.

Further, for classification purpose we are using five different distance measure techniques such as Minkowski distance, Eucledian distance, Modified Squared Eucledian distance, Correlation coefficient based distance and Angle Based distance.

3 Experimental Results and Performance Analysis

This section reports on PCA Mixture model exhibiting best classification of an image based on convincingly homogeneous visual appearance corresponding to objects. The proposed methodology was experimented on entire Caltech-101 dataset (collected by Fei-Fei et al. [13]) containing 9,197 images of 101 different object categories with varying pose, color and lighting intensity. Sample images of the dataset are shown in Fig 1.

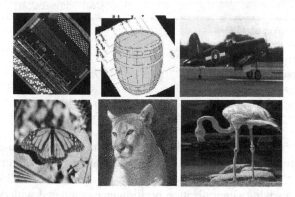

Fig. 1 Sample images of Caltech-101 dataset

The process of the system was initiated by collecting training samples, then EM algorithm is applied to obtain mixture of eigen values and the procedure leads to several iterations until better convergence. The model was trained with 15 and 30 images/category respectively, classified using Minkowski, Eucledian, MSE, Correlation coefficient, Angle Based distance metrics at different mixtures such as K= 2, 3 & 4 respectively. The obtained classification rate for K=2 & 3 mixtures are shown in Table 1. In this, we computed results for our model using 15 and 30 training images per category. The Obtained result is the best of 5 runs, where each runs signifies 5 different distance metrics used for classification.

Table 1 Breakdown of classification rates for K= 2 & 3 mixtures

K	2	3
Classification rate(%)	23.2	41.16

For the purpose of experimental evaluation, we have followed the training procedure i.e. 30 images/category as considered in Grauman et al. [14] and zhang et al. [15]. Remaining images are tested to evaluate the performance with the well known existing techniques [18, 22, 17, 15, 14]. The obtained results summarized in Table 2 is the best of 5 runs, where each run signifies 5 different distance metrics used for classification. So the obtained classification rate with K=4 mixtures for 15 and 30 train images are 36%, and 57.2% respectively. The classification rate with K=4 mixtures is better compared to K=2 and 3 mixtures. Thus, K=4 mixtures proved to yield better classification compare to well known existing methods and the performance evaluation of the system is provided.

Table 2 Performance analysis for Caltech-101 dataset. Experimental results of our model shows best of 5 independent runs as the average recognition accuracy per category in %.

Model	15 Training images/cat	30 Training images/cat
Serrre et al. [18]	35	42
Holub et al. [22]	37	43
Berg et al. [15]	45	-
Grauman & Darrell [14]	49.5	58.2
Mutch & Lowe [17]	33 & 51	41 & 56
Proposed	36	57.2

From the experiment it is noticed that, identifying more latent features are considerably high when K=4 and hence the classification rate increases compared to K= 2 & 3 clusters. In the experiment, The overall classification rate of 57.2% found to be promising in improving generalization performance compared with 56% obtained by recent biologically inspired model mentioned in [17], 42% and 43% of visual cortex inspired features and hybrid object Classifier respectively found in [18, 22]. Fig 2 shows few successful classes such as minaret, butterfly and ant. Least successful classes like panda, car and crocodile.

Fig. 2 Caltech-101 results for some categories, Top: High recognition rate. Bottom: Low recognition rate.

4 Discussion and Conclusion

Various feature descriptors along with different similarity measures are mostly addressed in improving the performance of classification in pattern recognition problems. Kernel based clusters discriminates unordered sets of local features, Bayesian and maximum likelihood methods incorporates prior information in learning process to extract shape features belonging to the object on small training sets, increasing sparsity may also be considered to improve classification rate. Experiments indeed showed that EM is an effective algorithm for fitting a mixture model in unsupervised context to identify hidden structure in unlabeled data and also provides the best reasonable parameter estimates. We also discovered mixing different normal distributions (K=4 mixtures) influences better convergence in assigning object membership for improving classification rate. PCA recognizes average features that reflects global statistical properties in compressed domain. In terms of Gaussian mixture models, the mixture of principal components data is modeled with relatively fewer parameters impacting covariance structure and the number of parameters is controlled by the choice of latent space dimension.

In the proposed work, PCA-mixture models efficiently extracts latent features by capturing covariance structure of observed data in low dimensional feature space for fewer training images and also investigated the impact of PCA-Mixture Model with different similarity metrics and choice of Gaussian mixture component for image retrieval system. First, and rather surprisingly, we have found increasing mixtures as an effective approach in improving generalization performance with multi class dataset and showed significant boost in increasing the classification performance for various distance metrics in comparison with the base line papers considering Caltech-101 benchmark dataset, making it good for attuning newer datasets in order to evaluate further contemporary approaches.

References

1. Enser, P.: Visual image retrieval: seeking the alliance of concept-based and Content-based paradigms. Journal of Information Science 26(4), 199–210 (2000)
2. Enser, P.G.B.: Pictorial information retrieval. Journal of Document 51(2), 126–170 (1995)
3. Datta, R., Joshi, D., Li, J., Wang, J.Z.: Image retrieval: ideas, influences, and trends of the new age. ACM Computing Surveys 40(2), 1–60 (2008)
4. Stricker, M., Orengo, M.: Similarity of color images. SPIE Storage and Retrieval for Image and Video Databases III 2185, 381–392 (1995)
5. Pass, G., Zabith, R.: Histogram refinement for content-based image retrieval. In: IEEE Workshop on Applications of Computer Vision, pp. 96–102 (1996)
6. Huang, J., et al.: Image indexing using color correlogram. In: IEEE Int. Conf. on Computer Vision and Pattern Recognition, Puerto Rico, pp. 762–768 (June 1997)
7. Zhang, J., Tan, T., Ma, L.: Invariant texture segmentation via circular Gabor filters. In: Proceedings of 16th International Conference on Pattern Recognition, vol. 2, pp. 901–904 (2002)
8. Cui, P., Li, J., Pan, Q., Zhang, H.: Rotation and scaling invariant texture classification based on radon transform and multiscale analysis. Pattern Recognition Letters 27(5), 408–413 (2006)
9. Krishnamoorthi, R., Sathiya devi, S.: devi, A multiresolution approach for rotation invariant texture image retrieval with orthogonal polynomials model. J. Vis. Commun. Image R. 23, 18–30 (2012)
10. Xing-Yuan, W., Zhi-Feng, C., Jiao-Jiao, Y.: An effective method for color image retrieval based on texture. Computer Standards & Interfaces 34, 31–35 (2012)
11. Yamada, A., Pickering, M., Jeannin, S., Jens, L.C.: MPEG-7 visual part of experimentation model version 9.0-part 3 dominant color. ISO/IEC JTC1/SC29/WG11/N3914, Pisa (2001)
12. Vapnik, V.: The Nature of Statistical Learning Theory. Springer (1995)
13. Fei-Fei, L., Fergus, R., Perona, P.: Learning generative visual models from few training examples: An incremental bayesian approach tested on 101 object categories. In: IEEE CVPR Workshop of Generative Model Based Vision (WGMBV) (2004)

14. Grauman, K., Darrell, T.: The Pyramid Match Kernel: Discriminative Classification with Sets of Image Features. In: The Proceedings of ICCV, vol. 2, pp. 1458–1465. IEEE (2005)
15. Zhang, H., Berg, A.C., Maire, M., Malik, J.: SVM-KNN: Discriminative Nearest Neighbor Classification for Visual Category Recognition. In: CVPR, vol. 2, pp. 2126–2136. IEEE (2006)
16. Lazebnik, S., Schmid, C., Ponce, J.: Beyond Bags of Features: Spatial Pyramid Matching for Recognizing Natural Scene Categories. In: CVPR, vol. 2, pp. 2169–2178. IEEE (2006)
17. Mutch, J., Lowe, D.G.: Multiclass Object Recognition with Sparse, Localized Features. In: CVPR, vol. 1, pp. 11–18. IEEE (2006)
18. Serre, T., Wolf, L., Poggio, T.: Object recognition with features inspired by visual cortex. In: CVPR, San Diego (June 2005)
19. Bishop, C.: Pattern Recognition and Machine Learning. Springer (2006)
20. Roweis, S.: Em algorithms for pca and spca. Advances in Neural Information Processing Systems 10, 626–632 (1997)
21. Bilmes, J.A.: A Gentle Tutorial of the EM algorithm and its application to parameter estimation for gaussian mixture and HMM. In: Intl. CSI, Berkeley, CA, pp. 1–13 (1998)
22. Holub, A., Welling, M., Perona, P.: Exploiting unlabelled data for hybrid object classification. In: NIPS Workshop on Inter-Class Transfer, Whistler, B.C (December 2005)
23. Kim, H.C., Kim, D., Bang, S.Y.: Face recognition using the mixture-of-eigen faces method. Pattern Recognition Letters 23, 1549–1558 (2002)

Bouncy Detector to Differentiate between GSM and WiMAX Signals

Ammar Abdul-Hamed Khader, Mainuddin, and Mirza Tariq Beg

Abstract. This work considered the differentiation and discrimination between GSM and WiMAX signals in noisy environment. Our new discriminator (Bouncy Detector (BD)) is an extension of the work of Hop Rate Detector (HRD) and it is exploited for differentiation and discrimination process between GSM and Wi-MAX signals. HRD was basically designed to detect and estimate the hop rate of frequency hopping signals only. Other kind of detectors like Matched Filter detector, Energy Detector and Feature Detector cannot give us an indication about the type of signals and also suffer from many drawbacks like complexity in design, determining the value of threshold level and increasing probability of false alarm (Pf) in high noise level etc. The simulation results show that BD can work properly and accurately without confusion between the signals, even in highly noise environment.

Keywords: Bouncy Detector (BD); Hop Rate Detector (HRD); Frequency Hopping (FH); GSM, WiMAX.

1 Introduction

Rapid growth in the development of new wireless devices demands more radio spectrum. The poor utilization of licensed spectrum is noticed by the spectrum regulatory bodies to review their policy so that unutilized spectrum can be utilized by other unlicensed user. Unlicensed user should have ability to sense and get the information from the surrounding environment. One of the important information is to know about the type of the signal (GSM or WiMAX etc.) present in the environment. This will help unlicensed user to sense which portion of the spectrum are available so that best available spectrum can be identified.

Ammar Abdul-Hamed Khader · Mainuddin · Mirza Tariq Beg
Department of Electronics & Communication Engineering,
Jamia Millia Islamia, New Delhi, India
e-mail: ammar_hameed_eng@yahoo.com, moin_s1@rediffmail.com,
  mtbeg@jmi.ac.in

S.M. Thampi, A. Gelbukh, and J. Mukhopadhyay (eds.), *Advances in Signal Processing*   379
and Intelligent Recognition Systems, Advances in Intelligent Systems and Computing 264,
DOI: 10.1007/978-3-319-04960-1_34, © Springer International Publishing Switzerland 2014

In the European Conference of Postal and Telecommunications Administrations (CEPT) the frequency bands 880-915 MHz (Uplink) and 925-960 MHz (Downlink) are allocated to mobile services and are currently used for Global System for Mobile Communication (GSM) and Universal Mobile Telecommunications System (UMTS) networks but also planned for the usage by Long Term Evolution (LTE) and Worldwide Interoperability for Microwave Access (Wi-MAX) and in the future other public mobile networks [1]. Multiple wireless interfaces like (GSM), Wireless Fidelity (Wi-Fi), Bluetooth, and Global Positioning System (GPS) receiver, etc. are being integrated into mobile devices. Wi-MAX, an IEEE802.16-based wireless access technology recently included in the IMT-2000 set of standards by ITU-R. But main problem is that how to operate these radios networks concurrently without interference and hardware conflicts due to congested spectrum allocation and component sharing with radio integration [2].

Today, different types of cellular networks are actively working on the radio links. For instance, the GSM is being used in nearly all of the countries of the world; it has around three billion users all over the world. It is the fully digital system it is considered as a 2G standard and was driven by ETSI (European telecommunication standard institute), which is evolution of (1G) analog system using 900, 1800 MHZ frequency bands. It has become popular very quickly because it provides improved speech quality and, a uniform international standard, makes it possible to use a single telephone number and mobile unit around the world.

The use of Frequency Hopping (FH) in GSM is the most important FH application. FH can introduce frequency diversity and interference diversity. It can be an effective technique for combating Rayleigh fading, reducing interleaving depth and associated delay, and enabling efficient frequency reuse in a multiple access communication   system. Frequency Hopping Spread Spectrum (FHSS) technique has been used widely not only in public safety communications but also in commercial communications such as home RF and Bluetooth, with interference avoidance and multiple access capability [3]. WiMAX Forum is a worldwide organization created to promote and certify compatibility and interoperability of broadband wireless products based on the IEEE 802.16 standard. Evolution and deployment of WiMAX technology rely on cooperative and complementary efforts in the WiMAX forum and IEEE 802.16 standard. The important features of WiMAX are scalable OFDMA, multiple input multiple output (MIMO) antenna, beam forming and adaptive modulation and coding (AMC), support time division duplexing (TDD) and frequency division duplexing (FDD), space time coding, strong security and multiple QoS classes [4-6].

In these cellular technologies and others, we have very limited resources and we have to make best use of them by proper management. The types of signals that presented in the environment are important to be known for the interceptors, cognitive users (CU) and other spectrum's users. This enables us to exploit the unutilized available spectral resources in different manner. Nowadays many commission and researchers are trying to use (800MHz) of TV for WiMAX, LTE and GSM. After switching to this band, the frequencies will become very convergence or very close and it is unavoidable to interfere the neighbouring commercial mobile networks when mobile systems are deployed. However efforts are made by different workers to maximizing the spectrum utilization with minimum interference [7, 8]. The main objective of this work is to scan the spectrum, detect the type of signals and differentiate among them. So our new bouncy detector (BD) can be used to differentiate between GSM and WiMAX signals accurately

2   Literature Survey

The problem of co-existence with other mobile systems is studied by Yi Zhang and et al [9]. They started with the realistic problem, and provided the detailed research method and simulation results. Guihua Piao and David, K. [10] studied Multi-standard radio resource management (MxRRM), based on a user data rate optimised algorithm, for non-real-time (NRT) services in a heterogeneous Wi-MAX/UMTS/GSM scenario. They used an algorithm using a parameter, namely NRT load information, which is mapped from the average user throughput in a respective network. Shaukat R. and Cheema A.R. [11] dealt with the issue of interoperability between heterogeneous networks using Mobile IP. They mentioned some handover solutions based on prior networks between GSM and Wi-MAX. Weiss J. et al. [12] elaborated that by adding elements for signalling of the current spectrum usage of the underlying GSM system to the WiMAX frame, the WiMAX overlay system is able to react very fast to changing utilizations of the spectrum. In 2009, K. Sridhara et al [13] found out the distributed server-based dynamic spectrum allocation (DSA) within liberalized spectrum sharing regulation concept as an alternative to existing regulation based on fixed frequency spectrum allocation schemes towards development of cognitive radio. They investigated a scenario where a block of spectrum is shared among four different kinds of exemplary air interface standards i.e., GSM, CDMA, UMTS and WiMAX. Two mechanisms [14] have been proposed independently by IEEE and 3GPP; namely, Media Independent Handover (MIH) and Access Network Discovery & Selection Function (ANDSF), respectively. These mechanisms enable a seamless Vertical Handover (VHO) between the different types of technologies (3GPP and non-3GPP), such as GSM, Wireless Fidelity (Wi-Fi), WiMAX, UMTS and LTE.

So, from the previous related work, researchers went through co-existence, RRM, interoperability, DSA and VHO problems among different types of technologies. But the novelty in our paper is how to differentiate between GSM and WiMAX signals in noisy environment with little amount of errors.

3   Conventional Signal's Detection Methods

There are a number of methods in which cognitive radios are able to perform spectrum sensing like: energy detection, matched filter detection and cyclostationary detection.

3.1   Energy Detector

Energy detection is a non coherent detection technique in which no prior knowledge of pilot data is required. The detection is based on some function of the received samples which is compared to a predetermined threshold level. The signal detection problem led us to a binary hypothesis-testing problem. In this problem, we need to decide between two hypotheses, the signal is not present (the observation consist of noise only (N[n])) or the signal (X[n]) is present.

$$H0:\ Y[n] = N[n] \qquad \text{signal is not present}$$

$$H1:\ Y[n] = X[n] + N[n] \qquad \text{signal is present}$$

$$n\ (\text{vector element}) = 1, 2,..., N; \qquad \text{where N is observation interval} \qquad (1)$$

The noise is generally assumed to be additive white Gaussian noise (AWGN) with zero mean and variance ($\sigma_w{}^2$) [15]. The detection (P_d) and false alarm (P_f) probabilities are given as in (2) and (3) where (P_r) is the received power, (λ) denotes the threshold level, (v) is the energy detector output, Γ (x) and Γ (x, y) are the complete and incomplete gamma functions respectively, (W) is the signal bandwidth, (T) signal duration and (γ) is defined as $\mu/2$.

$$P_f = P_r\ (v > \lambda:\ H1) = [\ \Gamma(TW/2, \lambda/2) / (\ \Gamma(TW)\ )\ ] \qquad (2)$$

$$P_d = P_r\ (v > \lambda:\ H0) = Q_{TW}\ (\sqrt{2\gamma}, \sqrt{2\lambda}\ ) \qquad (3)$$

$$P_m = 1 - P_d \qquad (4)$$

Where, The expression $Q_x(\alpha, \gamma)$ is known as Marcum's Q function and P_m denotes the probability of missed detection.
The fixed threshold (λ) is determined by the false alarm probability (P_f) and the number of sample points (N). It can be calculated as follows [16]

$$\lambda = Q^{-1}\ (P_f) + N \qquad (5)$$

where

$$Q(x) = 1/ (\sqrt{2\pi}) \int_X^\infty \exp\ (-u^2/2)\ du \qquad (6)$$

An increased sensing time is not the only disadvantage of the energy detector. More importantly, there is a minimum SNR below which signal cannot be detected. This minimum SNR level is referred to SNR_{wall}. There are two very strong assumptions. First, assuming a white noise additive and Gaussian with zero mean and known variance. However, noise is combination of various sources including not only thermal noise at the receiver and underlined circuits, but also interference due to nearby unintended emissions, weak signals from transmitters, etc. Second, assuming that noise variance is precisely known to the receiver, so that the threshold can be set accordingly. However, this is practically impossible as noise could vary over time due to temperature change, environment interference, filtering, etc. Even if the receiver estimates it, there is a resulting estimation error due to limited amount of time [17-18].

3.2   Matched Filter Detector or Coherent Detector

Matched filter detection is the optimal way for any signal detection, since it maximizes received signal-to-noise ratio. It is generally used to detect a signal by comparing a known signal with the input signal. However, a matched filter effectively requires demodulation of a received signal. This means that receiver has a priori knowledge of received signal at both PHY and MAC layers. Such information might be pre-stored in its memory. This is still possible since most transmitted signals have pilots, preambles, synchronization words or spreading codes that can be used for coherent detection. The main advantage of matched filtering is the short time to achieve a certain probability of false alarm or probability of miss detection as compared to other methods that are discussed in this section [17-19].

3.3   Feature Detection or Cyclostationarity-Based Sensing

Cyclostationarity feature detection is a method for detecting transmitted signal by exploiting the cyclostationarity features of the received signals. The transmitted signal generally has a periodic pattern. This periodic pattern (cyclostationarity characteristic), can be used to detect the presence of a signal in the spectrum. A signal is cyclostationary (in the wide sense) if the autocorrelation is a periodic function. With this periodic pattern, the transmitted signal from a licensed user can be distinguished from noise, which is a wide-sense stationary signal. In general, cyclostationary detection can provide a more accurate sensing result and it is robust to variations in noise power. However, the detection is complex and requires long observation periods to obtain sensing result [18, 19].

Conventional signal detection methods have several drawbacks as discussed above and shown in Table 1 [29]. So we propose the extension of the work of Hop Rate Detector (HRD) and exploit it for differentiation and discrimination process between GSM and WiMAX signals.

4   Bouncy Detector (BD)

A block diagram of BD is shown in Fig. 1 [20-22]. The whole input band (W_{ss}) is subdivided into two sequential half bands (an upper band (B.sub.u) and lower band (B.sub.d)), and the signal is collapsed by magnitude squaring. The outputs of the squaring devices are then subtracted to form a bipolar signal. The input signal hops randomly between the two half bands, and thus the first stage output signal is a random direct sequence (DS) waveform with transitions occurring at the hop rate. The BD generates a spectral line at the hop rate with a delay-and-mix circuit, with the delay set to approximately half the hop period ($T_h/2$). However, the BD delay-and-mix circuit generates a square wave with one-half the input signal amplitude, and thus one-fourth the signal power. The delay and multiply circuit involves a delay circuit which feeds a first signal input to the multiplier. The second input to the multiplier comes direct from the LPF, to provide the final output of signal-to-noise ratio SNR_o.

Table 1 Advantages and disadvantages of conventional signal detection methods

Signal's Detection Methods	Advantages	Disadvantages
Energy Detection	Does not need any prior information Low Computational Cost	Cannot work in low SNR Cannot distinguish users sharing the same channel
Matched Filter detection	Optimal Detection Performance Low Computational Cost	Requires a prior knowledge of the primary user
Cyclostationary Detection	Robust in low SNR Robust to Interference	Requires partial information about primary user High Computational Cost

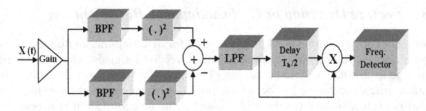

Fig. 1 Bouncy Detector (BD) Simulation Block

The probability of crossing the threshold while the FH signal is "present" is much higher than for the "noise only" case and thus, with L_{th} properly set (L_{th} is the threshold level that depends on the input signal-to-noise ratio), the detector sensitivity can be improved. Though amplitude information is lost by single-bit quantization, it may be understood that, while the signal is present in a particular channel, the signal at the output of the squaring device unit is a DC level. It can thus be seen that information relating to the presence of "signal" is preserved by single-bit quantization of the channel.

The term "false alarm" is used to denote the crossing of a threshold value "L_{th}" when no-signal is present in the channel. The optimal probability of a "false alarm" is designated as p_{opt} [22].

$$p_{opt} = e^{-\frac{L_{th}}{2\sigma L}} \qquad (7)$$

where σL is variance of band pass Gaussian noise present in channel. The "detection probability "q" is the probability that the threshold value L_{th}, will be crossed when a signal is present in the channel. The "first stage" signal-noise ratio, SNR_f can be written:

$$SNR_f = (p - q)^2 / ((L - 1)(p - p^2) + (q - q^2)) \qquad (8)$$

where (p) is the probability of false alarm. Thus given p_{opt}, as a function of the input signal-noise ratio, SNR_i , then the detection probability (q) , can be calculated:

$$q = Q\left(\frac{A}{\sigma L}, \sqrt{-2\log p}\ \right) \qquad (9)$$

where A = signal at output of squaring device, and Marcum Q function is:

$$Q = Q(X, Y) = \int_y^{+\infty} r e^{-(x^2 + r^2)/2} * Io(xr)dr \qquad (10)$$

Where; Io= Modified Bessel function of first kind order 0; x = First variable of Marcum-Q function; y = Second variable of Marcum-Q function; r = Dummy variable for integration; dr = Differential of r [22].

5   Simulation Model

Matlab Simulink has been used to design FH-GSM, WiMAX transmitters and noisy channel as shown in Fig. 2.

5.1   GSM Transmitter

The GSM Air-interface uses two different multiplexing schemes: TDMA (Time Division Multiplexing) and FDMA (Frequency Division Multiplexing). The spectrum is divided into 200 kHz channels (FDMA) and each channel is divided into 8 time slots (TDMA). Each 8 time slot TDMA frame has a duration of 4.6 ms (577 micro sec./ time slot). The GSM system uses slow frequency hopping which means that the frequency changes after each burst (once every 4.6 ms). Where the use of FH is optional in GSM but all phones must support it. All physical channels except the 0 time slot of Broadcast Control Channel (BCCH) can hop.

5.1.1   Hopping Sequence Generation

The hopping sequence use up to 64 different frequencies, with speed of just over 200 hop/s [22]. For a given set of parameters, the index to an Absolute Radio Frequency Channel Number (ARFCN) within the mobile allocation indexing (MAI), (MAI from 0 to N-1, where MAI=0 represents the lowest ARFCN in the mobile allocation (MA), ARFCN is in the range 0 to 7023 and the frequency value can be determined with n = ARFCN and $1 \leq N \leq 64$), is obtained with the following algorithm [23]:

```
If  HSN= 0 (cyclic hopping) then:
MAI, integer (0 ... N-1): MAI = (FN + MAIO)modulo N
    else:
M, integer (0 ... 152) : M = T2 + RNTABLE((HSN XOR
T1R) + T3)
S, integer (0 ... N-1): M' = M modulo (2^NBIN)
       T' = T3 modulo (2^NBIN)
       if M' < N then:
       S = M'
    else:
           S = (M'+T') modulo N
MAI, integer (0 ... N-1): MAI = (S + MAIO) modulo N
```

Where; HSN = Hopping Sequence Number; FN = TDMA Frame Number; T1R= Time parameter T; T3 =Time parameter from 0 to 50 (6 bits); T2= Time parameter from 0 to 25 (5 bits); NBIN= Number of bits required to represent N = INTEGER ($\log_2$ (N) +1); XOR = Bit-wise exclusive OR of 8 bit binary operands.

Table 2 RNTABLE: defined table of 114 integer numbers

Address	Contents
000...009	48, 98, 63, 1, 36, 95, 78, 102, 94, 73
010...019	0, 64, 25, 81, 76, 59, 124, 23, 104, 100
020...029	101, 47, 118, 85, 18, 56, 96, 86, 54, 2
030...039	80, 34, 127, 13, 6, 89, 57, 103, 12, 74
040...049	55, 111, 75, 38, 109, 71, 112, 29, 11, 88
050...059	87, 19, 3, 68, 110, 26, 33, 31, 8, 45
060...069	82, 58, 40, 107, 32, 5, 106, 92, 62, 67
070...079	77, 108, 122, 37, 60, 66, 121, 42, 51, 126
080...089	117, 114, 4, 90, 43, 52, 53, 113, 120, 72
090...099	16, 49, 7, 79, 119, 61, 22, 84, 9, 97
100...109	91, 15, 21, 24, 46, 39, 93, 105, 65, 70
110...113	125, 99, 17, 123

5.2   WiMAX Transmitter and Channel

The WiMAX (802.16-2004) transmitted signal's parameters are: Frequency Band- 2GHz, OFDM carriers -256, Adaptive Modulation -QPSK, 16QAM, 64QAM, Duplexing TDD and Channel Bandwidth 3.5MHz. After these signals crossing noisy channel (Rayleigh and Gaussian Noise), and applying a multipath Rayleigh fading channel model for complex signals, a Bouncy Detector (BD) or Hop Rate Detector is used to differentiate between the shape of these received signals.

5.3   Signals at BD

The signal present at the BD output is a time domain signal which contains only noise components when no FH signal is present at the input of the detector. But when an FH signal is present, the signal at the output will contain an additive sinusoidal component with frequency equal to that of the hop rate of the FH signal. The power spectrum of this signal will, therefore, contain a spectral line at the hop rate. The spectral line is a line which indicates a particular frequency in the frequency spectrum which denotes the presence of an additive sinusoidal component in the signal with the corresponding spectrum.

So the received FH-GSM signal has its own shape. But the case is different in another input signals like WiMAX, where the data and its carrier frequency are fixed in band and couldn't be appear in the two BPFs. So BD gives different shape for GSM and WiMAX signals.

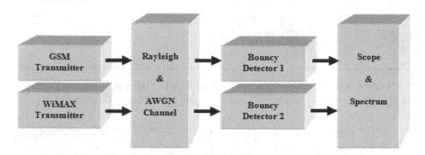

Fig. 2 FH-GSM and WiMAX Transmitters' Block Diagram

6   The Results and Discussion

Fig. 3 shows the two received signals in time domain before entering the detection blocks of BDs, while Fig. 4 shows the spectrum (the signal in frequency domain) where the (X) axis is the frequency and (Y) axis is the power. The detection and differentiation between these two signals can be achieved easily by noticing the output of the BDs in time domain. Where each detected signal has its own shape and it differs from the other signals (they don't look alike).

The performance of the bouncy detector has been tested under different value of signal's parameters with and without noise consideration. Fig. (5, 6 and 7) shows the output of BD in absence of noise. In these results, one of the most important parameter in FH-GSM signal is the hopping rate. In these three figures the hopping rate are considered as 100, 200 and 500 hop/sec. respectively. Another important parameter in WiMAX signal is the cyclic prefix (CP). CP is used in OFDM in order to combat multipath problem by making channel estimation easy. To prevent the ISI (inter-symbol interference) as well as the ICI (inter-channel interference), OFDM symbol is cyclically extended into the guard interval. CP acts as a buffer region where delayed information from the previous symbols can get stored. The OFDM have 256 carriers so the CP ratio used in those three figures is 1/8, 1/16 and 1/32 respectively. We can notice that the envelope shape for each recovered signal is remaining same with slightly small difference when the parameter values are changed.

Fig. (8, 9 and 10) illustrate the shape of the two signals in time domain along with noise consideration. Different cases of signal to noise ratio (SNR) are taken as -10dB, 0dB and 10 dB respectively. Furthermore, respective hop rate of 100, 200 and 500 hop/s has been used for these three cases in FH-GSM whereas respective CP ratio of 1/8, 1/16 and 1/32 are considered for WiMAX signals.

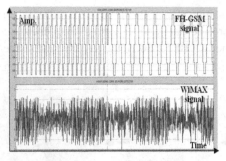

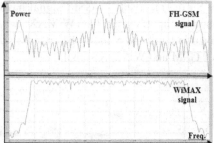

Fig. 3 FH-GSM and WiMAX Transmitted Signals in Time Domain

Fig. 4 FH-GSM and WiMAX Transmitted Signals in Frequency Domain

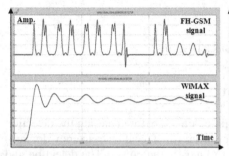

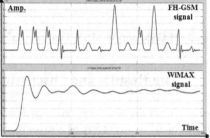

Fig. 5 Signals without Noise, 100hop/s for FH-GSM and 1/ 8cyclic prefix for WiMAX

Fig. 6 Signals without Noise, 200hop/s for FH-GSM and 1/ 16 cyclic prefix for WiMAX

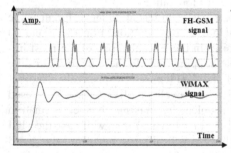

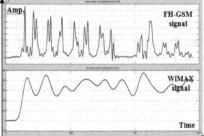

Fig. 7 Signals without Noise, 500hop/s for FH-GSM and 1/ 32cyclic prefix for Wi-MAX

Fig. 8 Signals with SNR= -10 dB, 100hop/s for FH-GSM and 1/8 cyclic prefix WiMAX

It is evident from the above results that the effect of increasing noise power is appeared only in the amplitude of FH-GSM signal and in the tail of the WiMAX signal. Further it is noticeable that the shape of these two signals is clearly distinguishable under different SNR condition. Hence it is concluded from these results that possibility of error in differentiating these signals is negligible.

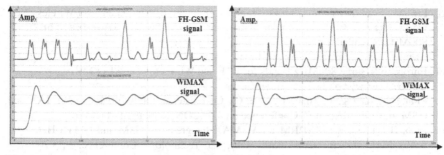

Fig. 9 Signals with SNR = 0 dB, 200hop/s **Fig. 10** Signals with SNR =10 dB, 500hop/s for FH-GSM and 1/16 cyclic prefix WiMAX for FH-GSM and 1/32 cyclic prefix WiMAX

7 Conclusions

With the expected growth of both GSM and WiMAX networks, the number of interference cases will increase. So Coordination carried out between public mobile networks and GSM operators shows that there exist remedies to differentiate them when deployed in adjacent frequencies and in geographical close vicinity. Expanding the function of HRD for differentiation and discrimination between GSM and WiMAX signal is very useful. Matlab software has been used to simulate the two transmitters and the noisy channel (Multipath Rayleigh fading channel plus AWGN channel). The Bouncy Detector's output shows that, it can work to differentiate between FH-GSM and WiMAX signals accurately with negligible error even in low SNR value. This is a big advantage of BD along with its simplicity over conventional methods (matched filter detector, energy detector and feature detector), which suffers from many problems like design complexity, threshold level value and false alarm probability in high noise level.

References

1. ECC REPORT 162, Practical Mechanism to Improve the Compatibility Between GSM-R and Public Mobile Networks and Guidance on Practical Coordination (2011), http://www.erodocdb.dk/docs/doc98/official/word/ECCRep162.doc
2. Zhu, J., Yin, H.: Enabling collocated coexistence in IEEE 802.16 networks via perceived concurrency. IEEE Communications Magazine 47(6), 215–220 (2009)
3. Kennington, J., Olinick, E., Rajan, D.: Wireless Network Design Optimization Models and Solution Procedures. International Series in Operations Research & Management Science, vol. 158, pp. 1–6. Springer Science + Business Media, LLC (2011)
4. Papapanagiotou, I., Toumpakaris, D., Lee, J., Devetsikiotis, M.: A Survey on Next Generation Mobile WiMAX Networks: Objectives, Features and Technical Challenges. IEEE Communications Surveys & Tutorials 11(4), 3–18 (2009)
5. Andrews, J.G., Ghosh, A., Muhamed, R.: Fundamentals of WiMAX Understanding Broadband Wireless Networking. Prentice Hall (2007)

6. Lin, M., Choi, H.: Thomas La Porta: Network Integration in 3G and 4G Wireless Networks, http://www.cse.psu.edu/~molin/network_integration.pdf
7. 800MHz "Cellular" Band,
   http://www.inactivex.net/cellular/800MHz.html
8. Europe to Reserve 800MHz Bands for LTE and WiMAX Networks (2010),
   http://www.cellular-news.com
9. Zhang, Y., Zhang, J., Zhao, T., Xue, C.: Co-Existence Study of Mobile WiMAX and GSM. In: IEEE Region 10 Conference on TENCON 2006, pp. 1–4 (2006)
10. Piao, G., David, K.: MXRRM for WIMAX Integrated to GSM and UMTS Heterogeneous Networks. In: IEEE 65th Vehicular Technology Conference, pp. 768–773 (2007)
11. Shaukat, R., Cheema, A.R.: Mobile IP Based Interoperability between GSM and WiMAX. In: The Fourth International Conference on Wireless and Mobile Communications, pp. 191–195 (2008)
12. Weiss, J., Blaschke, V., Jondral, F.K.: Adapting the WiMAX PHY layer for use in a GSM overlay system. In: The 14th IEEE Mediterranean Electro-technical Conference, MELECON 2008, pp. 114–119 (2008)
13. Sridhara, K., Nayak, A., Singh, V., Dalela, P.K.: Enhanced Spectrum Utilization for Existing Cellular Technologies Based on Genetic Algorithm in Preview of Cognitive Radio. International Journal of Communications, Network and System Sciences 2, 917–926 (2009)
14. Khattab, O., Alani, O.: I AM 4 VHO: New Approach to Improve Seamless Vertical Handover in Heterogeneous Wireless Networks. International Journal of Computer Networks & Communications 5(3), 53–63 (2013)
15. Ÿucek, T., Arslan, H.: A Survey of Spectrum Sensing Algorithms for Cognitive Radio Applications, IEEE Communications Survey & Tutorials. IEEE Communications Survey & Tutorials 11(1), 116–130 (2009)
16. Suresh Babu, R., Suganthi, M.: Review of Energy Detection for Spectrum Sensing in Various Channels and its Performance for Cognitive Radio Applications. American Journal of Engineering and Applied Sciences 5(2), 151–156 (2012)
17. Khader, A.A.-H.: Enhanced Performance of FH Detection System Using Adaptive Threshold Level. Al-Rafidain Engineering Journal 18(5), 107–122 (2010)
18. Tabassam, A.A., Suleman, M.U., Kalsait, S., Khan, S.: Building Cognitive Radios in MATLAB Simulink– A Step Towards Future Wireless Technology. In: UkSim 13th International Conference on Computer Modelling and Simulation, pp. 15–20 (2011)
19. Hossain, E., Dusit, N., Zhu, H.: Dynamic Spectrum Access and Management in Cognitive Radio Networks. Cambridge University Press (2009)
20. Simon, M.K., Omura, J.K.: Robert Scholtz and Barry Levitt: Spread Spectrum Communications Handbook. McGraw-Hill, Inc. (2002)
21. Khader, A.A.H., Shabani, A.M.H., Beg, M.T.: Joint Detection and Discrimination of CDMA2000, WiMAX and Frequency Hopping Signals. American Journal of Scientific Research (89), 116–124 (May 2013)
22. Smith, P.J., Leahy, R.S.: Channelized Binary-Level Hop Rate Detector, United States Patent, http://www.BinaryLevelHopRateDetectorPatrickJ.SmithRonaldLeahy.pdf
23. Technical Specification, GSM 05.02 version 5.1.0. ETSI (August 1996),
   http://www.etsi.org

Modified Gossip Protocol in Wireless Sensor Networks Using Chebyshev Distance and Fuzzy Logic

Raju Dutta, Shishir Gupta, and Mukul K. Das

Abstract. The Flooding is a traditional flat based routing protocol where unlimited broadcasting of the packets in the flooding scheme will cause the huge energy consumption to send the packets from source to sink due to implosion, overlap, resource blindness and consequently creates broadcast storm. However Gossiping routing protocol in WSNs is very much effective due to its simplicity, robustness, distributed and capability to work in noisy and uncertain environments. But due to recirculation of information and repeated data communication of randomized gossip protocol which can lead to a significant energy consumption of the network. This paper proposes an energy efficient routing protocol based on Fuzzy Logic and Chebyshev Distance, which is a modification of gossip protocol. The new protocol determines the optimal routing path from source to destination by selected the best node from candidate nodes in the forwarding paths by favoring highest remaining energy and the lowest distance to the sink. Simulation results shows that the proposed method is efficient to control messages forwarding and improves the performance which minimizes the overall energy consumption and maximize WSNs lifespan.

Keywords: Routing Protocol, Gossiping, Modified Gossiping, Sensor Networks, Fuzzy Logic, Chebyshev Distance.

Raju Dutta
Dept. of Mathematics, Narula Institute of Technology, Kolkata 700109,
West Bengal, India
e-mail: rdutta80@gmail.com

Shishir Gupta
Dept. of Applied Mathematics, Indian School of Mines, Dhanbad 826004, Jharkhand, India
e-mail: shishir_ism@yahoo.com

Mukul K. Das
Dept. of Electronics Engineering, Indian School of Mines,
Dhanbad 826004, Jharkhand, India
e-mail: mkdas12@gmail.com

S.M. Thampi, A. Gelbukh, and J. Mukhopadhyay (eds.), *Advances in Signal Processing*
and Intelligent Recognition Systems, Advances in Intelligent Systems and Computing 264,
DOI: 10.1007/978-3-319-04960-1_35, © Springer International Publishing Switzerland 2014

1   Introduction

A Wireless Sensor Network (WSN) is composed of sensor nodes that are deployed in specified area for monitoring different environmental conditions such as temperature, air pressure, humidity, vibration etc. The sensor nodes are monitoring the environment and usually collect data from environment and send that data information to the sink for remote user access through various communication technologies [1]. In WSNs, the routing protocols are effectively taking part of these data communication and the main objective of the routing protocols is delivery of information between sensors and the sink efficiently as because of energy consumption is the main issue in the development of any routing protocol for WSNs. Due to limited energy resources of sensor nodes, the most energy-efficient manner for data delivery to be adopted by network without compromising the accuracy of the information content [2]. Since, in routing algorithm, the best path is chosen for transmission of data from source to destination and in many routing metrics the shortest path algorithm may not be suitable over a period of time. If all communications is done in the same path for quick transmission, then the battery power of those nodes on this path will get drained fast [3, 4, 5]. Instead, the reasons for energy consumption should be carefully investigated and new energy-efficient routing metrics developed for WSNs. The exchange of information between its neighbors can vary according to the routing techniques. In each case, the overhead of the protocol is increases and nodes consume more energy in exchanging this information through the wireless medium. A simple strategy to disseminate information into a network or to reach a node at an unknown location is to flood the entire network. A sender node broadcasts packets to its immediate neighbors and neighbor nodes are rebroadcasting the packets to its immediate neighbors and so on. This process will repeat until all nodes have received the packets or the packets have traveled for a maximum number of hops. In Flooding routing protocol, if there exists a path to the destination, the destination is sure to receive the data. The main advantage of flooding is its simplicity, while its main disadvantage is that it causes heavy traffic. Gossip protocol is therefore proposed to an enhancement Flooding and to overcome the drawbacks of it, i.e. implosion, overlap and resource blindness [6, 7]. An alternative to forwarding the packet to all neighbors is to forward it to an arbitrary one. Such gossip results in the packet randomly traversing the network in the hope of eventually finding on the destination node. Fig.3, Fig.4 and Fig.5 shows an example of the forwarding packet mechanisms for Flooding, Gossiping and Modified Gossiping protocols. However, using a traditional randomized gossip protocol can lead to a significant waste of energy due to repeated and recirculation of redundant information. Due to this conception, some of researchers have received significant attention to enhancement of gossip performance. The authors [8] proposed an alternative gossip scheme that exploits geographic information. The key issue in this method is the number of iterations for gossip algorithm to converge the network. In [9], the authors using Markov chain to show how to optimize the neighbor selection probabilities for each node. However, for sensor network graphs, even an optimized gossip algorithm can result in excess energy consumption. The authors [10] proposed a Flossiping routing protocol for WSNs. It combines a single branch Gossiping and a controllable low-probability random selective relaying. Although the Flossiping reduced the amount of packets, it suffers longer delay. S. Kheiri et al. in [11] presented LGossiping, an improved Gossiping data distribution technique based on the location of nodes. This protocol assumed that, each node knows the position of the other nodes. This allowed each sensor to send its data reliably to one of known neighbors instead of sending data blindly to one of neighbors. Although in this protocol they deals with delay problem and solved to some extent but there is still problem of many events not reaching the base station. In [12] the authors presented a Energy Lactation based Gossiping protocol (ELGossiping) to determined the optimal path of energy consumption viewpoint for information transfer from sensor nodes to base station. In this protocol when a sensor detects an event, the node that was in transmission radius of sensor and it had a high energy and minimum distance to Sink is selected as next hop. In many traditional optimal path routing schemes each node selects specific nodes to transmit data based on some criteria to maximize network lifespan.

This paper proposes an energy efficient routing protocol based on Fuzzy Logic and Chebyshev Distance to select the neighbor node for the next hop in Modified Gossip protocol. Here node will be selected based on node's residual energy using fuzzy logic and distance. This routing protocol reduces routing messages, balancing energy consumption and maximization of network lifetime for WSNs. The new protocol reduces number of retransmission for the optimal routing path from source to destination by selecting the best node from source nodes by favoring highest remaining energy and the lowest distance to the sink.

2 Fuzzy Logic and Chebyshev Distance Based Routing Protocol

2.1 Fuzzy Rules for Antecedents and Consequents

Fuzzy logic was first introduced by Lotfi-Zadeh in [13]. Recently, its applications have rapidly expanded in adaptive control systems, system identification and so many different fields of Science, Engineering and Technology. It has grate advantages because of its easy implementation, robustness, and ability to approximate to any nonlinear mapping [14]. In fuzzy systems, the dynamic behavior of a system is characterized by a set of linguistic fuzzy rules based on the knowledge of a human expert. The heart of a fuzzy system is the fuzzy rules and may be provided by experts or can be extracted from numerical data. In either case, the rules that we are interested in can be expressed as a collection of IF-THEN statements (IF antecedents THEN consequents). Antecedents and consequents of a fuzzy rule form the fuzzy input space and fuzzy output space respectively are defined by combinations of fuzzy sets. Considering a fuzzy system with p inputs and one output with M rules, then the Lth rule has the form [13]: $R^L : IF\ x_I$ is F_I^L and ....and x_P is F_P^L $THEN\ y$ is G^L Where F_I^L ,........ F_P^L and G^L denote the linguistic variables defined by fuzzy sets and $L=1...M$.

Fig. 1 shows the typical structure of a fuzzy system. It consists of four components namely; fuzzification, rule base, inference engine and defuzzification [13].

The processes of making crisp inputs are mapped to their fuzzy representation in the process called fuzzification. This involves application of membership functions such as triangular, trapezoidal, Gaussian etc. The inference engine process maps fuzzified inputs to the rule base to produce a fuzzy output. A consequent of the rule and its membership to the output sets are determined here. The defuzzification process converts the output of a fuzzy rule into crisp outputs by one of defuzzification strategies [15].

2.2   Proposed Routing Method

In this paper, the topology of a WSN is modeled as a graph G (N, E), where N is the set of sensor nodes, and E is the set of direct edge between the sensor nodes. A sink node is responsible for collecting data from all other nodes within its transmission range. In our protocol when a node detects an event and wants to transmit packets, it selects an optimal neighbor node (height residual energy) using fuzzy logic and the minimum distance within transmission radius as its next hop. The objective of distance based fuzzy logic is to determine the Optimum node $O_N(n)$ that depended on the energy $R(n)$ and the Chebyshev distance $D(n)$ to the Sink of the node n. The proposed method uses three membership functions for each input (R and D) and an output variable (B_N), as shown in Fig. 2. In the fuzzy, the fuzzified values are processed by the inference engine, which consists of a rule base and various methods to inference the rules. Fig.2 shows the IF-THEN rules used in the proposed method, with a total number of $3^3=27$ for the fuzzy rule base. As an example, IF $R(n)$ is *High* and $D(n)$ is *Near* THEN $O_N(n)$ is *Selected*.

Based on these it calculates optimal routing schedule and broadcasts it. The process of finding the optimal path and broadcasting it in the network and sending data from all nodes to the base station by following this routing schedule is repeated in every round. The base station can then determine the routing schedule based on this updated information. The proposed method assumes that: i) all sensor nodes are randomly distributed in the area and every sensor node knows its own position and its neighbor's position and the sink; ii) all sensor nodes have the same transmission range and initial energy.

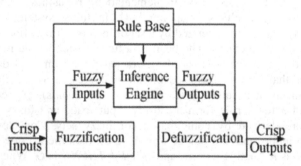

Fig. 1 Typical structure of Fuzzy Inference Engine

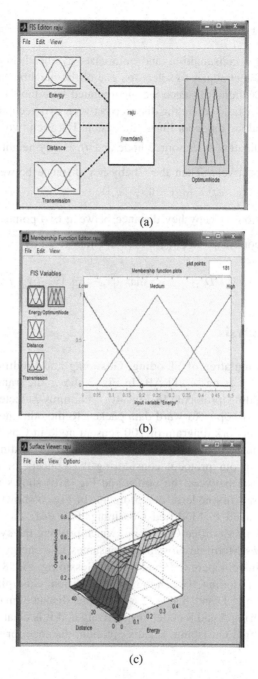

Fig. 2 Member graph for the inputs as (a) Energy, Distance and Transmission range, (b) Energy membership function and (c) Surface view of selection of optimum node based on Remaining Energy and Chebyshev distance.

2.3   Neighbor Node Selection Using Chebyshev Distance

To reduce energy consumption and minimize overhead in WSNs we have implemented new techniques to select the one of the neighbor nodes for the next hop. Instead of Euclidian distance we incorporated here the Chebyshev distance, defined on a vector field where the distance between two vectors is the greatest of their differences along any coordinate dimension. Between two vector or points p and q, with coordinates p_s as source node and q_i as any neighbor node based on our protocol respectively. Then the Chebyshev distance between p_s and q_i is given by $D_{Chebyshev}\ (p_s, q_i) = \max_i\ (|p_s - q_i|)$

In two dimensions, Chebyshev distance between two points p_s and q_i which have Cartesian coordinate (x_s, y_s) and (x_i, y_i), is given by

$$D_{Chebyshev} = \max\ (|x_i - x_s|, |y_i - y_s|) \tag{1}$$

3   Result Analysis

Here extensive simulation of Flooding, Gossiping and Modified Gossiping has been carried out by using MATLAB. In Fig. 3 shows the transmission diagram of nodes in Flooding protocol. In Fig. 4(a) randomly selected node has been considered for next hop to forward data packets to the other nodes and Fig. 4(b) shows the final network diagram of 100 sensors node in Gossiping protocol. In Fig. 5(a) selection of node has been considered for forwarding data packets to the other nodes in our protocol considering the concept of Fuzzy Logic and Chebyshev distance between the nodes and Fig. 5(b) shows the final network diagram of 100 sensors node in our protocol. In Fig. 6 shows more number of retransmissions saved in Modified Gossiping. Fig. 7(a) shows the number of dead node is less in Modified Gossiping, which prolong the system lifetime and in Fig.7 (b) shows Modified Gossiping consumes less energy to broadcast data from source to sink compare to Flooding and Gossiping. Fig. 8 shows the energy consumption of Flooding, Gossiping and Modified Gossiping protocol with hopcount. In Table 1 and Table 2 shows the distance matrix by both the protocol, considering 5 and 8 nodes respectively and it is clear from the distance matrix that Modified Gossiping selects optimum distance for selecting node for next hop.

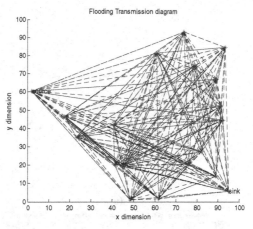

Fig. 3 Flooding Protocol (selected all nodes in forwarding packets)

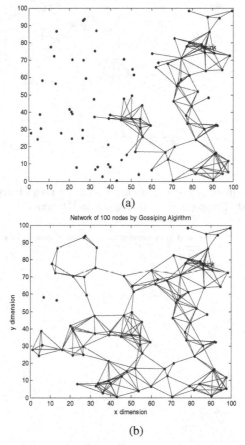

Fig. 4 Gossiping protocol (a) Runtime selection of nodes in forwarding packets, (b) Complete Network diagram of 100 nodes.

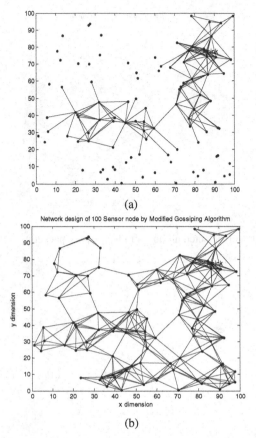

(a)

(b)

Fig. 5 Modified Gossiping protocol (a) runtime selection nodes based on our protocol in forwarding packets, (b) Complete Network diagram of 100 nodes.

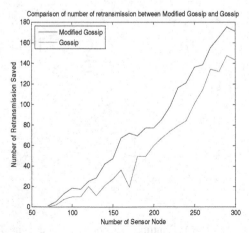

Fig. 6 Number of retransmission saved in forwarding packets

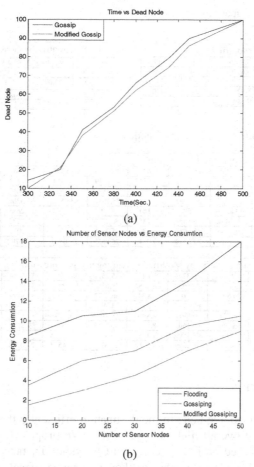

(a)

(b)

Fig. 7 Comparison (a) number of dead node is more in Gossiping (b) Energy consumption with nodes by different protocol.

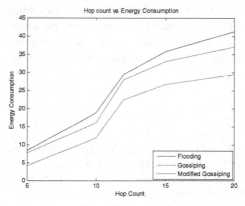

Fig. 8 Comparison of energy consumption with hopcounts.

Table 1. Distance matrix considering 5 nodes.

Distance matrix by Modified Gossiping				
0	39.1125	66.3715	9.2003	6.7908
39.1125	0	70.6039	29.9122	33.2985
66.3715	70.6039	0	65.9247	59.5807
9.2003	29.9122	65.9247	0	6.3440
6.7908	33.2985	59.5807	6.3440	0

Distance matrix by Gossiping				
0	39.3409	69.4521	9.2112	8.9397
39.3409	0	73.0275	30.2760	35.0756
69.4521	73.0275	0	66.8785	61.3533
9.2112	30.2760	66.8785	0	7.1911
8.9397	35.0756	61.3533	7.1911	0

Table 2. Distance matrix considering 8 nodes.

Distance matrix by Gossiping							
0	5.4204	69.8119	29.6477	45.7589	65.1307	34.2662	63.3293
5.4204	0	72.9268	26.6717	49.8719	70.5464	39.1263	68.4212
69.8119	72.9268	0	66.4082	89.5833	61.6653	75.0082	83.5749
29.6477	26.6717	66.4082	0	75.1647	85.7552	62.5266	89.9840
45.7589	49.8719	89.5833	75.1647	0	46.2777	15.2685	26.9855
65.1307	70.5464	61.6653	85.7552	46.2777	0	39.7235	26.0497
34.2662	39.1263	75.0082	62.5266	15.2685	39.7235	0	29.6702
63.3293	68.4212	83.5749	89.9840	26.9855	26.0497	29.6702	0

Distance matrix by Modified Gossiping using expression (1)							
0	4.2325	66.3715	29.6443	45.2522	52.5665	31.5144	54.5844
4.2325	0	70.6039	26.2581	48.6385	56.7990	34.9007	57.9707
66.3715	70.6039	0	65.9247	66.8977	60.1002	53.1599	76.2299
29.6443	26.2581	65.9247	0	74.8965	68.0991	61.1587	84.2288
45.2522	48.6385	66.8977	74.8965	0	45.7758	13.7378	25.3205
52.5665	56.7990	60.1002	68.0991	45.7758	0	39.1125	20.4553
31.5144	34.9007	53.1599	61.1587	13.7378	39.1125	0	23.0700
54.5844	57.9707	76.2299	84.2288	25.3205	20.4553	23.0700	0

4 Conclusion

To send data from source node to the destination efficiently through transmission and to prolong the overall lifetime of a WSNs. We proposed an energy efficient routing protocol based on Fuzzy logic and Chebyshev Distance to consider node for next hop in our protocol. This method is capable of selecting optimal data transmission route from the source node to the sink by favoring highest remaining energy and optimum distance in successive hops within different radio ranges of leading nodes. Based on that it divided different phases where the optimum node has been selected. The performance of the proposed method is evaluated and compared with other protocol under the same criteria. Simulation results demonstrate the effectiveness of our approach with regards to prolong the lifespan of wireless sensor networks.

References

1. Karkvandi, H.R., Pecht, E., Yadid, O.: Effective Lifetime-Aware Routing in Wireless Sensor Networks. IEEE Sensors J. 1112, 3359–3367 (2011)
2. Al-Karaki, J.N., Kamal, A.E.: Routing Techniques in Wireless Sensor Networks: A Survey. IEEE Wireless Commun. 116, 6–28 (2004)

3. Zhang, H., Shen, H.: Balancing Energy Consumption to Maximize Network Lifetime in Data-Gathering Sensor Networks. IEEE Trans. Parallel Distrib. Syst. 2010, 1526–1539 (2009)

4. Akkaya, K., Younis, M.: A survey of routing protocols in wireless sensor networks. Ad Hoc Netw. 33, 325–349 (2005)

5. Ren, F., Zhang, J., He, T., Lin, C., Das, S.K.: EBRP: Energy- Balanced Routing Protocol for Data Gathering in Wireless Sensor Networks. IEEE Trans. Parallel Distrib. Syst. 2212, 2108–2125 (2011)

6. Hedetniemi, S., Liestman, A.: A survey of gossiping and broadcasting in communication networks. Networks 184, 319–349 (1988)

7. Dargie, W., Poellabauer, C.: Network Layer in Fundamental of Wireless Sensor Networks Theory and Practice, pp. 163–204. Wiley, New York (2010)

8. Dimakis, A.G., Sarwate, A.D., Wainwright, M.J.: Geographic gossip: Efficient aggregation for sensor networks. In: Proc. IEEE IPSN, pp. 69–76 (2006)

9. Boyd, S., Ghosh, A., Prabhakar, B., Shah, D.: Gossip algorithms: Design, analysis and applications. In: Proc. IEEE INFOCOM, vol. 3, pp. 1653–1664 (2005)

10. Zhang, Y., Cheng, L.: Flossiping: A New Routing Protocol for Wireless Sensor Networks. IEEE ICNSC 2, 1218–1223 (2004)

11. Kheiri, S., Goushchi, G., Rafiee, M., Seyfe, B.: An Improved Gossiping Data Distribution Technique with Emphasis on reliability and Resource Constraints. In: IEEE CMC 2009, vol. 2, pp. 247–252 (2009)

12. Norouzi, A., Hatamizadeh, A., Dabbaghian, M., Ustundag, B.B., Amiri, F.: An Improved ELGossiping Data Distribution Technique with Emphasis on Reliability and Resource Constraints in Wireless Sensor Network. In: IEEE ICECTECH, pp. 179–183 (2010)

13. Zadeh, L.A.: Soft computing and fuzzy logic. IEEE Software 116, 48–56 (1994)

14. Kulkarni, R.V., Forster, A., Venayagamoorth, G.K.: Computational Intelligence in Wireless Sensor Networks: A Survey. IEEE Commun. Surveys & Tutorials 131, 68–96 (2011)

15. Runkler, T.A.: Selection of Appropriate Defuzzification Methods Using Application Specific Properties. IEEE Trans. Fuzzy Syst. 51, 72–79 (1997)

16. Heinzelman, W.R., Chandrakasan, A., Balakrishnan, H.: Energy- Efficient Communication Protocol for Wireless Microsensor Networks. In: Proc. 33rd Annu. Hawaii Int. Conf. Syst. Sci., vol. 2, pp. 1–10 (2000)

A Novel Cluster Head Selection and Routing Scheme for Wireless Sensor Networks

Dhirendra Pratap Singh, Vikrant Bhateja,
Surender Kumar Soni, and Awanish Kumar Shukla

Abstract. The communication subsystem in Wireless Sensor Networks (WSNs) is primarily responsible for energy consumption; therefore minimization of energy consumption is a challenging issue in these networks. The present work incorporates data reduction using GM (1, 1) as prediction model to propose an efficient routing scheme for WSNs. In this paper, two approaches have been devised involving energy efficient cluster based routing protocol with residual energy and distance based cluster head selection method. This leads to a reduction in overheads during cluster formation. Further, inter-cluster and intra-cluster transmissions are also minimized using proposed prediction based data reduction scheme resulting in an overall reduction in energy consumption of WSNs.

Keywords: Energy Consumption, GM (1, 1) Model, LEACH, Measure of First Node Dead (FND) & Half Node Dead (HND).

1 Introduction

Initially Wireless Sensor Networks were mainly used and motivated by military applications. Later on, their utility has been extended for many other civilian applications, like environmental and species monitoring, smart home automation

Dhirendra Pratap Singh · Vikrant Bhateja
Deptt. of Electronics & Communication Engg., SRMGPC, Lucknow-227105 (U.P.), India

Surender Kumar Soni
Deptt. of Electronics & Communication Engg., NIT Hamirpur (H.P.), India

Awanish Kumar Shukla
Deptt. of Electronics Instrumentation & Control Engg., AIET, Luckonw (U.P.), India
e-mail: {dhirendrapratap.nith,bhateja.vikrant,surender.soni,
    awanishthelibra}@gmail.com

S.M. Thampi, A. Gelbukh, and J. Mukhopadhyay (eds.), *Advances in Signal Processing*   403
and Intelligent Recognition Systems, Advances in Intelligent Systems and Computing 264,
DOI: 10.1007/978-3-319-04960-1_36, © Springer International Publishing Switzerland 2014

etc. [1-5]. In addition, they find applications in biomedical arena in performing computer-aided diagnosis of life-threatening diseases like breast cancer [6-23], brain tumor [24-26] and cardiovascular disorders [27-34]. To optimize the usage of WSNs for various applicative trends, energy consumption forms the major cause of concern due to the energy consumed during transmission of data. This is so, because WSNs are mostly used in such areas where human approach is nearly impossible; and non-rechargeable batteries of sensor nodes cannot be recharged which leading to network failure. Hierarchical routing using multi-hop mode reduces a good amount of energy consumption of WSNs. In this scheme, the entire network is divided into clustered layers and Hierarchical routing can be performed using a basic protocol LEACH. It is self- organizing protocol that incorporates adaptive clustering and uses randomization to distribute the energy consumption evenly among the sensor nodes (in the network). This protocol aims at increasing the lifetime and reducing the latency for transferring the data. To make energy consumption uniform among the nodes in the network, the role of the CH is rotated among all the cluster members. Data sensed by the cluster members is sent to the CH; where CH aggregates the data coming from non-CHs and sends aggregated data to the BS [35]. Lu Tao et al. introduces an energy-efficient protocol [36] that utilizes cluster member threshold to avoid the presence of uneven cluster size at the same time and uses energy rating to prevent uneven distribution of energy. V-LEACH [37] introduces the concept of vice-CH, where a node becomes acting CH if the existing CH dies; ensuring an uninterrupted data delivery to the base station in a particular round. MS-LEACH [38] is another protocol based on the critical value of the cluster area size which analyzes the problem of energy consumption of the single-hop and multi-hop transmissions within a single cluster. In MR-LEACH [39], a multi-hop routing protocol with low energy dynamic cluster hierarchy has been designed to minimize the energy consumption of sensor nodes. In the sequence of improvements made in the existing LEACH protocol; the present work utilizes the role of node residual energy and distance of sensors from sink to develop an efficient CH election and routing scheme. The same has been achieved with the proposed cluster based routing scheme. The remaining part of the paper has been organized as follows: Section 2 presents proposed protocol of cluster-based routing along with two devised approaches for CH selection. Section 3 gives the simulation results along with their analysis and finally, the conclusions are drawn in section 4.

2   Proposed Approach

In the present work, two new clustering approaches have been proposed which are proved to be more energy efficient than LEACH. In these approaches, Approach 1 (A1) and Approach 2 (A2), CH election is done by considering residual energy of sensor nodes and their distance from sink which is different from the election of CHs phase as in LEACH.

2.1   Background

Energy consumption in the WSNs can be calculated according to the set of equation according to which the depleted energy of the sensor nodes to transmit a single bit over a distance 'd' from sensor nodes to BS is given by the expressions as in Eq. (1)-(2) [35].

$$E_{Tx}(l,d) = l \times E_{elec} + l \times \epsilon_{mp} \times d^4, \text{for } d > d_0 \tag{1}$$

$$E_{Tx}(l,d) = l \times E_{elec} + l \times \epsilon_{fs} \times d^2, \text{for } d \leq d_0 \tag{2}$$

where: $d_0 = \left(\dfrac{\epsilon_{fs}}{\epsilon_{mp}} \right)^{\frac{1}{2}}$.

Energy consumed in receiving packet of l bits can therefore be expressed as in Eq. (3).

$$E_{Rx}(l) = E_{Rx-elec}(l) = l \times E_{elec} \tag{3}$$

Energy consumed in computation of Grey model for packet of l bits can be expressed as in Eq. (4),

$$E_{Grey\text{-}computation}(l) = l \times E_{Grey} \tag{4}$$

E_{elec} represents power consumed by launching circuit or receiving circuit, ϵ_{mp} and ϵ_{fs} represent energy consumed by the circuit to launch a single bit of information to 1 square meter in multi-path channel and free space respectively. E_{Grey} is the energy consumption in computation of Grey model for 1 bit while l is the size of data packets in bits.

2.2   Prediction Based Data Reduction Approach

The prediction based data reduction approach relies on the prediction based cooperation between sensor and the sink nodes. Both these nodes will use the same prediction model as well as data for prediction. Sensor node need not to send the data to the sink every time when there is a change in data. Sink node can predict the data in each sensor for that particular round. Thus, a significant amount of energy can be saved by reducing number of transmissions of data if sensor node and sink node efficiently predict the data [40-44].

2.3   GM (1, 1) Model

Grey prediction model is used for predicting the future values of a time series data using past values. GM (1, 1) i.e., a Grey model with first order one variable; this is computationally efficient and less complex to be widely used in many applications. The mathematical equations of GM (1, 1) model which have been used in the proposed approach A2 can be taken from the previous researches of GM (1, 1) model [43-48]. According to the grey system theory, the GM (1, 1) model is defined as:

$$\frac{dx^{(1)}}{dt} + ax^{(1)} = b \tag{5}$$

where: $x^{(1)}$ is defined as the 1st AGO of $x^{(0)}$; the actual sequence of data. 'a' (*developing coefficient*) and 'b' (*driving input*) are the grey model parameters and can be calculated by Least square method in Eq. (5). The predicted value for $(k+1)^{th}$ round corresponding to accumulated data can be expressed as $\hat{x}^{(1)}(k+1)$ using Eq. (6).

$$\hat{x}^{(1)}(k+1) = \left[x^{(0)}(1) - \frac{b}{a} \right] e^{-ak} + \frac{b}{a} \tag{6}$$

Predicted value $\hat{x}^{(0)}(k+1)$ corresponding to original data can be calculated as in Eq. (7).

$$\hat{x}^{(0)}(k+1) = \left[x^{(0)}(1) - \frac{b}{a} \right] e^{-ak} (1 - e^{a}) \tag{7}$$

The percentage prediction error of GM (1, 1) model for k^{th} round is defined as $e(k)$ given in Eq. (8).

$$\% \, e(k) = \left| \frac{x^{(0)}(k) - \hat{x}^{(0)}(k)}{x^{(0)}(k)} \right| \times 100 \tag{8}$$

where: $x^{(0)}(k)$ is the actual value and $\hat{x}^{(0)}(k)$ is the predicted value for k^{th} round.

2.4   Proposed Protocol

For the *proposed hierarchical cluster based routing*, a set of assumptions are made according to which N sensor nodes are distributed randomly in an (D × D) square field. Communication between CHs and BS as well as between sensors and CHs should be reliable; BS knows the location of all the sensor nodes and every node knows the location of other sensor nodes as well as BS. BS is outside the sensor field and has unlimited energy and memory resources and each node has an identity (ID). The proposed hierarchical cluster based routing performs the entire operation in three different phases: Cluster Head Election Phase, Scheduling Phase and Steady State Phase.

1.   In *Cluster Head Election Phase*, first BS divides the sensor field into the desired number of grids. Each grid is called cluster and BS sends a message to all sensor nodes about their corresponding clusters. For the first time, BS will calculate the Residual energy- Distance factor denoted as F expressed as in Eq. (9).

$$F_{ij} = \left[E_{ij} + \frac{1}{D_{ij}} \right] \tag{9}$$

where: F_{ij} is the Residual energy- Distance factor and D_{ij} is the distance of S_{ij} sensor node which is the ith node of jth cluster from BS. BS selects the node having the maximum F_{ij} for all j clusters shown in Eq. (10).

$$CH_j = max[F_{ij}] \tag{10}$$

where: CH_j is the cluster head of jth cluster. After completing first round operation, cluster head election will again take place using the Eq. (9) & (10) the but this time, the last CHs will take the energy information of all the sensor nodes of their corresponding clusters while distance of sensor nodes from BS will be considered as in previous round. Election process will be same as discussed earlier. After electing next CHs, previous CHs will discard the energy information provided by the nodes. Election of CHs will take place in every round.

2. *Scheduling Phase* provides every sensor node a dedicated time slot in which it can send data to the cluster head. After being elected as the CH, node will have the information of all cluster members provided by BS. This information is related to identity (ID) of the particular sensor node and its distance from BS. The CH node sets up a TDMA schedule and assigns each node a time slot when it can transmit. This schedule message is then broadcasted to all the nodes within the cluster. Intra-cluster scheduling is done using TDMA while inter-cluster scheduling is done using CDMA.

3. Now, during *Steady State Phase*, after receiving the TDMA schedule created by their CH nodes, the non-CH nodes start transmitting data depending on the TDMA schedule. To synchronize and start the steady state phase at the same time, the BS can issue the corresponding synchronization pulses to all the nodes. The steady state phase is further divided into frames. During the assigned frame, the sensor node can transmit the data to the CH node. The duration of each frame is constant and depends on the number of nodes in cluster. CHs then start transmitting their data to the sink node within their CDMA schedule.

Using the same protocol mentioned above, two approaches have been designed in this work. In A1, the proposed clustering scheme has been implemented without using prediction based data reduction scheme. This approach reduces the energy consumption by reducing overheads in cluster formation and deciding CHs in each cluster. In A2, the GM (1, 1) prediction model has been implemented to minimize the number of data transmissions thereby reducing the energy consumption by suppressing overheads during cluster formation and deciding CHs (in each cluster). There is a further reduction in energy consumption with A2 by incorporating reduced data transmissions among each CH and BS. As, energy is more consumed during transmission of data; A2 reduces more number of data transmissions in comparison to LEACH and A1depending on the accuracy of the model.

3   Results and Discussions

3.1   Evaluation Parameters

In the setup of simulation, the proposed network model consists of 100 sensor nodes deployed over a field of (D × D) square unit area and simulation has been done for single iteration. In this work, network has been divided into four equal parts; division into optimal number of clusters is an important issue to cover the entire field and all the sensor nodes. Table 1 illustrates the parameters with their values and units which have been used in simulations. Time series prediction based data reduction scheme using GM (1, 1) model is implemented in designing an energy efficient cluster based routing protocol. Performance evaluation at this stage is carried out using objective evaluation measures of signal fidelity [49-52]. The comparisons among LEACH, A1 and A2 (using GM (1, 1) model with prediction error threshold, $\epsilon = 3\%$ and $\epsilon = 5\%$) is made and results have been analyzed in terms of various parameters using the figures 1-3.

Table 1 Parameters used in Simulation of Proposed Protocol

Parameters	Value(s)
Number of Sensor Nodes	100
Network Area (*in Meter²*)	(50×50) ~ (250×250)
BS Location (*in Meters*)	(0,0)
Initial Energy of Nodes (*in Joules*)	0.1
Data Packet Length (*in bits*)	800
Energy Packet Size (*in bits*)	400
Control Packet Size (*in bits*)	200
E_{elec} (*nJ/bit*)	50
ϵ_{mp} (*in nJ/bit/m⁴*)	0.0013
ϵ_{fs} (*in nJ/bit/m²*)	10
E_{DA} and E_{Grey} (*in nJ/bit*)	5
Window Size	3

The various parameters for evaluation and comparison of results are *FND*, *HND* and *Energy consumption* [53]-[55]. FND is measured in terms of number of rounds to which all the sensor nodes of sensor network can serve the desired task whereas HND is measured in terms of number of rounds after which only 50% nodes remain for serving the task. Energy consumption per round shows the depleted energy per round.

3.2   Simulated Results

Figure 1 shows the graphical comparison among the LEACH and proposed approaches (A1, & A2) for different values of network side. From these figures, it can be interpreted that the proposed approaches yields better results for FND in comparison to LEACH.

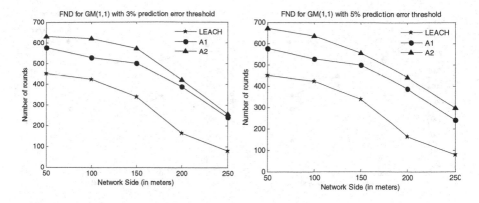

Fig. 1 Comparison of LEACH, A1 and A2 in terms of FND at 3% and 5% threshold prediction error

It can be further seen that A2 with $\epsilon = 5\%$ is giving the superior results. A tabulation of these FND values is made in Table 2.

Table 2 Comparison of LEACH, A1 and A2 in terms of FND parameter

N/W Side(m)	LEACH	A1	A2	
			$\epsilon=3$ %	$\epsilon=5$ %
50	441	565	620	660
100	411	518	610	625
150	326	495	560	545
200	153	376	410	430
250	67	230	245	285

From the results given in table 2, and figure 2, it has been observed that A1 & A2 have shown improvement over LEACH (in terms of FND) as the first node of network dies in LEACH after completing 441 rounds in case of 50 m network side whereas for A1, the first node dies after completing 565 rounds. In approach A2 (with $\epsilon=3\%$ and $\epsilon=5\%$) completes 620 and 660 rounds respectively for FND. Even for the higher network sides, it can be seen that both the approaches (A1 & A2) are showing better performance in comparison to LEACH. Amongst all approaches, A2 with $\epsilon=5\%$, completes more number of rounds for FND. Figure 3 shows the comparison among the LEACH, A1 and A2 based on computed values of HND for different values of network side. From the figures, again it can be seen that A2 with $\epsilon = 5\%$ is giving the superior results (These values are also shown tabulated in Table 3).

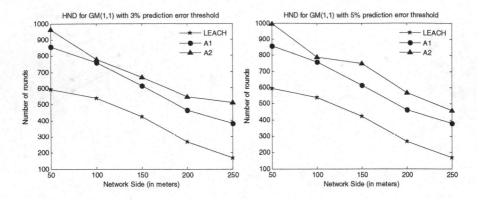

Fig. 2 Comparison of LEACH, A1 and A2 in terms of HND at 3% and 5% threshold prediction error

Proposed approaches (A1 & A2) have also shown improvement over LEACH in terms of network lifetime. Here, A1 is improving network lifetime by a factor of 1.63 while A2 does same by a factor of 1.91 and 2.19 (for ϵ = 3% and 5% respectively). Hence, this comparison shows that the A2 with ϵ = 5% is giving the best results between the two proposed approaches (A1 & A2) in terms of network lifetime making it a robust algorithm for routing in WSNs.

Table 3 Comparison of LEACH, A1 and A2 in terms of HND parameter

N/W Side(m)	LEACH	A1	A2	
			ϵ=3 %	ϵ=5 %
50	584	847	953	985
100	526	749	769	778
150	412	602	655	738
200	258	450	532	555
250	158	368	505	517

In the later part, figure 3 illustrates the comparison among LEACH, A1 and A2 with same values of prediction errors in terms of energy consumption per round. The results with the proposed approaches show energy depletion at a slower rate in comparison to LEACH. In addition, as the network side increases, energy is consumed with very high rate in LEACH whereas this consumption is least with A2 (for ϵ = 5%).

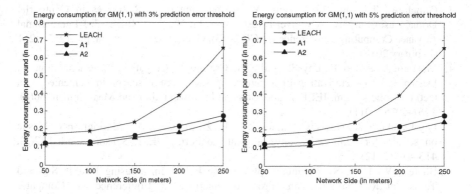

Fig. 3 Comparison of LEACH, A1 and A2 in terms of Energy consumption at 3% and 5% threshold prediction error

4 Conclusion

Most of the WSNs are used in such areas where human approach in nearly impossible. Due to the usage of sensor with non-rechargeable battery having limited power supply, power consumption is the major issue in designing WSNs. Various methods and routing schemes are available to reduce the energy consumption of WSNs so that network can perform the specific task without network failure. It can be concluded from the results obtained with the proposed approaches, a significant amount of energy can be saved by reducing the overheads in CH election process. Further, Time series prediction based data reduction scheme has minimized the energy consumption by reducing the number of data transmissions in intra-cluster and inter-cluster region. GM (1, 1) model serves to provide a reasonably good accuracy by reduction in significant number of data transmissions. The proposed approaches: A1 and A2 have significantly improved the network lifetime as depicted by the computed values of FND and HND; thereby demonstrating improvements over LEACH.

References

1. Gupta, A., Tripathi, A., Bhateja, V.: Despeckling of SAR Images via an Improved Anisotropic Diffusion Algorithm. In: Satapathy, S.C., Udgata, S.K., Biswal, B.N. (eds.) Proceedings of Int. Conf. on Front. of Intell. Comput. AISC, vol. 199, pp. 747–754. Springer, Heidelberg (2013)
2. Bhateja, V., Tripathi, A., Gupta, A.: An Improved Local Statistics Filter for Denoising of SAR Images. In: Thampi, S.M., Abraham, A., Pal, S.K., Rodriguez, J.M.C. (eds.) Recent Advances in Intelligent Informatics. AISC, vol. 235, pp. 23–29. Springer, Heidelberg (2014)

3. Gupta, A., Tripathi, A., Bhateja, V.: Despeckling of SAR Images in Contourlet Domain using a New Adaptive Thresholding. In: Proc. of (IEEE) 3rd International Advance Computing Conference (IACC 2013), Ghaziabad, U.P., India, pp. 1257–1261 (February 2013)

4. Massaro, A., Spano, F., Lay-Ekuakille, A., Cazzato, P., Cingolani, R., Athanassiou, A.: Design and characterization of a nanocomposite pressure sensor implemented in a tactile robotic system. IEEE Transactions on Instrumentation and Measurement 60(8), 2967–2975 (2011)

5. Lay-Ekuakille, A., Davis, C., Kanoun, O., Li, Z., Trabacca, A.: Editorial-Special issue on sensors for noninvasive physiological monitoring. IEEE Sensors Journal 12(3), 413–415 (2012)

6. Bhateja, V., Devi, S.: Mammographic Image Enhancement using Double Sigmoid Transformation Function. In: Proc. of International Conference on Computer Applications (ICCA 2010), Pondicherry, India, pp. 259–264 (December 2010)

7. Bhateja, V., Devi, S.: An Improved Non-Linear Transformation Function for Enhancement of Mammographic Breast Masses. In: Proc. of (IEEE) 3rd International Conference on Electronics & Computer Technology (ICECT 2011), Kanyakumari, India, vol. 5, pp. 341–346 (April 2011)

8. Bhateja, V., Devi, S.: A Novel Framework for Edge Detection of Microcalcifications using a Non-Linear Enhancement Operator and Morphological Filter. In: Proc. of (IEEE) 3rd International Conference on Electronics & Computer Technology (ICECT 2011), Kanyakumari, India, vol. 5, pp. 419–424 (April 2011)

9. Siddharth, Gupta, R., Bhateja, V.: A New UnSharp Masking Algorithm for Mammography using Non-Linear Enhancement Function. In: Satapathy, S.C., Avadhani, P.S., Abraham, A. (eds.) Proceedings of the InConINDIA 2012. AISC, vol. 132, pp. 779–786. Springer, Heidelberg (2012)

10. Bhateja, V., Singh, G., Srivastava, A.: A Novel Weighted Diffusion Filtering Approach for Speckle Suppression in Ultrasound Images. In: Satapathy, S.C., Udgata, S.K., Biswal, B.N. (eds.) FICTA 2013. AISC, vol. 247, pp. 459–466. Springer, Heidelberg (2014)

11. Bhateja, V., Srivastava, A., Singh, G., Singh, J.: A Modified Speckle Suppression Algorithm for Breast Ultrasound Images Using Directional Filters. In: Satapathy, S.C., Avadahani, P.S., Udgata, S.K., Lakshminarayana, S. (eds.) ICT and Critical Infrastructure: Proceedings of the 48th Annual Convention of CSI - Volume II. AISC, vol. 249, pp. 219–226. Springer, Heidelberg (2014)

12. Siddhartha, G.R., Bhateja, V.: An Improved Unsharp Masking Algorithm for Enhancement of Mammographic Masses. In: Proc. of IEEE Students Conference on Engineering and Systems (SCES 2012), Allahabad, India, pp. 234–237 (March 2012)

13. Siddhartha, G.R., Bhateja, V.: A Log-Ratio based Unsharp Masking (UM) Approach for Enhancement of Digital Mammograms. In: Proc. of (ACM ICPS) CUBE International Information Technology Conference & Exhibition, Pune, India, pp. 26–31 (September 2012)

14. Bhateja, V., Devi, S.: A Reconstruction Based Measure for Evaluation of Mammogram Edge-Maps. In: Satapathy, S.C., Udgata, S.K., Biswal, B.N. (eds.) Proceedings of Int. Conf. on Front. of Intell. Comput. AISC, vol. 199, pp. 741–746. Springer, Heidelberg (2013)

15. Pandey, A., Yadav, A., Bhateja, V.: Design of New Volterra Filter for Mammogram Enhancement. In: Satapathy, S.C., Udgata, S.K., Biswal, B.N. (eds.) Proceedings of Int. Conf. on Front. of Intell. Comput. AISC, vol. 199, pp. 143–151. Springer, Heidelberg (2013)

16. Bhateja, V., Devi, S., Urooj, S.: An Evaluation of Edge Detection Algorithms for Mammographic Calcifications. In: Proc. of (Springer) 4th International Conference on Signal and Image Processing (ICSIP 2012), Coimbatore, India, vol. 2, pp. 487–498 (December 2012)

17. Pandey, A., Yadav, A., Bhateja, V.: Contrast Improvement of Mammographic Masses Using Adaptive Volterra Filter. In: Proc. of (Springer) 4th International Conference on Signal and Image Processing (ICSIP 2012), Coimbatore, India, vol. 2, pp. 583–593 (December 2012)

18. Pandey, A., Yadav, A., Bhateja, V.: Volterra Filter Design for Edge Enhancement of Mammogram Lesions. In: Proc. of (IEEE) 3rd International Advance Computing Conference (IACC 2013), Ghaziabad, U.P., India, pp. 1219–1222 (February 2013)

19. Bhateja, V., Urooj, S., Pandey, A., Misra, M., Lay-Ekuakille, A.: Improvement of Masses Detection in Digital Mammograms employing Non-Linear Filtering. In: Proc. of (IEEE) International Multi-Conference on Automation, Computing, Control, Communication and Compressed Sensing (iMac4s-2013), Palai-Kottayam, Kerala, India, vol. (119), pp. 406–408 (March 2013)

20. Srivastava, H., Mishra, A., Bhateja, V.: Non-Linear Quality Evaluation Index for Mammograms. In: Proc. of (IEEE) 2nd Students Conference on Engineering and Systems (SCES 2013), Allahabad, India, pp. 269–273 (April 2013)

21. Bhateja, V., Urooj, S., Pandey, A., Misra, M., Lay-Ekuakille, A.: A Polynomial Filtering Model for Enhancement of Mammogram Lesions. In: Proc. of IEEE International Symposium on Medical Measurements and Applications (MeMeA 2013), Gatineau, Quebec, Canada, pp. 97–100 (May 2013)

22. Jain, A., Singh, S., Bhateja, V.: A Robust Approach for Denoising and Enhancement of Mammographic Breast Masses. International Journal on Convergence Computing, Inderscience Publishers 1(1), 38–49 (2013)

23. Bhateja, V., Misra, M., Urooj, S., Lay-Ekuakille, A.: A Robust Polynomial Filtering Framework for Mammographic Image Enhancement from Biomedical Sensors. IEEE Sensors Journal 13(11), 4147–4156 (2013)

24. Raj, A., Alankrita, S.A., Bhateja, V.: Computer Aided Detection of Brain Tumor in MR Images. International Journal on Engineering and Technology (IACSIT-IJET) 3, 523–532 (2011)

25. Alankrita, R.A., Shrivastava, A., Bhateja, V.: Contrast Improvement of Cerebral MRI Features using Combination of Non-Linear Enhancement Operator and Morphological Filter. In: Proc. of (IEEE) International Conference on Network and Computational Intelligence (ICNCI 2011), Zhengzhou, China, vol. 4, pp. 182–187 (May 2011)

26. Srivastava, A., Alankrita, Raj, A., Bhateja, V.: Combination of Wavelet Transform and Morphological Filtering for Enhancement of Magnetic Resonance Images. In: Snasel, V., Platos, J., El-Qawasmeh, E. (eds.) ICDIPC 2011, Part I. CCIS, vol. 188, pp. 460–474. Springer, Heidelberg (2011)

27. Verma, R., Mehrotra, R., Bhateja, V.: An Integration of Improved Median and Morphological Filtering Techniques for Electrocardiogram Signal Processing. In: Proc. of (IEEE) 3rd International Advance Computing Conference (IACC 2013), Ghaziabad, U.P., India, pp. 1223–1228 (February 2013)

28. Verma, R., Mehrotra, R., Bhateja, V.: An Improved Algorithm for Noise Suppression and Baseline Correction of ECG Signals. In: Satapathy, S.C., Udgata, S.K., Biswal, B.N. (eds.) Proceedings of Int. Conf. on Front. of Intell. Comput. AISC, vol. 199, pp. 733–739. Springer, Heidelberg (2013)

29. Verma, R., Mehrotra, R., Bhateja, V.: A New Morphological Filtering Algorithm for Pre-Processing of Electrocardiographic Signals. In: Proc. of (Springer) 4th International Conference on Signal and Image Processing (ICSIP 2012), Coimbatore, India, vol. 1, pp. 193–201 (December 2012)

30. Bhateja, V., Verma, R., Mehrotra, R., Urooj, S.: A Non-linear Approach to ECG Signal Processing using Morphological Filters. International Journal of Measurement Technologies and Instrumentation Engineering (IJMTIE): An Official Publication of the Information Resources Management Association 3(3), 46–59 (2013)

31. Lay-Ekuakille, A., Vergallo, P., Griffo, G., Conversano, F., Casciaro, S., Urooj, S., Bhateja, V., Trabacca, A.: Entropy Index in Quantitative EEG Measurement for Diagnosis Accuracy. IEEE Transactions on Instrumentation & Measurement x(x), xx (November 2013)

32. Lay-Ekuakille, A., Vergallo, P., Griffo, G., Conversano, F., Casciaro, S., Urooj, S., Bhateja, V., Trabacca, A.: Mutidimensional Analysis of EEG Features using Advanced Spectral Estimates for Diagnosis Accuracy. In: Proc. of IEEE International Symposium on Medical Measurements and Applications (MeMeA 2013), Gatineau, Quebec, Canada, pp. 237–240 (May 2013)

33. Bhateja, V., Urooj, S., Mehrotra, R., Verma, R., Lay-Ekuakille, A., Verma, V.D.: A Composite Wavelets and Morphology Approach for ECG Noise Filtering. In: Maji, P., Ghosh, A., Murty, M.N., Ghosh, K., Pal, S.K. (eds.) PReMI 2013. LNCS, vol. 8251, pp. 361–366. Springer, Heidelberg (2013)

34. Bhateja, V., Verma, R., Mehrotra, R., Urooj, S.: A Novel Approach for Suppression of Powerline Interference and Impulse Noise in ECG Signals. In: Proc (IEEE) International Conference on Multimedia, Signal Processing and Communication Technologies (IMPACT 2013), AMU, Aligarh, U.P., India, p.xx (November 2013)

35. Heinzelman, W.B., Chandrakasan, A.P., Balakrishnan, H.: Application-specific protocol architecture for wireless microsensor networks. IEEE Transactions on Wireless Communications 1(4), 660–670 (2002)

36. Tao, L., Qing-Xin, Z., Luqiao, Z.: An improvement for LEACH algorithm in wireless sensor network. In: 2010 5th IEEE Conference on Industrial Electronics and Applications (ICIEA), pp. 1811–1814 (2010)

37. Bani Yassein, M., Al-zou'bi, A., Khamayseh, Y., Mardini, W.: Improvement on leach protocol of wireless sensor network (VLEACH). International Journal of Digital Content Technology and its Applications 3(2), 132–136 (2009)

38. Qiang, T., Bingwen, W., Zhicheng, W.C.: MS-Leach: A Routing Protocol Combining Multi-hop Transmissions and Single-hop Transmissions. In: Pacific-Asia Conference on Circuits, Communications and Systems, pp. 107–110 (2009)

39. Abdulsalam, H.M., Kamel, L.K.: W-LEACH: Weighted Low Energy Adaptive Clustering Hierarchy Aggregation Algorithm for Data Streams in Wireless Sensor Networks. In: 2010 IEEE International Conference on Data Mining Workshops, pp. 1–8 (2010)

40. Farooq, M.O., Dogar, A.B., Shah, G.A.: MR-LEACH: Multi-hop Routing with Low Energy Adaptive Clustering Hierarchy. In: Fourth International Conference on Sensor Technologies and Applications (SENSOR COMM), pp. 262–268 (2010)

41. Anastasi, G., Conti, M., Di Francesco, M., Passarella, A.: Energy conservation in wireless sensor networks: A survey. Ad Hoc Networks 7(3), 537–568 (2009)
42. Katiyar, V., Chand, N., Soni, S.: Grey System Theory-Based Energy Map Construction for Wireless Sensor Networks. In: Abraham, A., Mauri, J.L., Buford, J.F., Suzuki, J., Thampi, S.M. (eds.) ACC 2011, Part III. CCIS, vol. 192, pp. 122–131. Springer, Heidelberg (2011)
43. Soni, S.K., Singh, D.P.: Energy Map Construction using Adaptive Alpha Grey Prediction Model in WSNs. International Journal of Computer and Information Engineering 6(3), 107–111 (2012)
44. Soni, S.K., Chand, N., Singh, D.P.: Reducing the data transmission in WSNs using time series prediction model. In: Proc. of 2012 IEEE International Conference on Signal Processing, Computing and Control (ISPCC), pp. 1–5 (March 2012)
45. Singh, D.P., Agrawal, Y., Soni, S.K.: Energy optimization in Zigbee using prediction based shortest path routing algorithm. In: Proc. of 2012 3rd National Conference on Emerging Trends and Applications in Computer Science (NCETACS), pp. 257–261 (March 2012)
46. Deng, D.: Introduction of Grey System Theory. Journal of Grey System 1(1), 1–24 (1989)
47. Kayacan, E., Ulutas, B., Kaynak, O.: Grey system theory-based models in time series prediction. Expert Systems with Applications 37(2), 1784–1789 (2010)
48. Kumar, U., Jain, V.K.: Time series models (Grey-Markov, Grey Model with rolling mechanism and singular spectrum analysis) to forecast energy consumption in India. Energy 35(4), 1709–1716 (2010)
49. Gupta, P., Srivastava, P., Bhardwaj, S., Bhateja, V.: A Novel Full Reference Image Quality Index for Color Images. In: Satapathy, S.C., Avadhani, P.S., Abraham, A. (eds.) Proceedings of the InConINDIA 2012. AISC, vol. 132, pp. 245–253. Springer, Heidelberg (2012)
50. Gupta, P., Tripathi, N., Bhateja, V.: Multiple Distortion Pooling Image Quality Assessment. Inderscience Publishers International Journal on Convergence Computing 1(1), 60–72 (2013)
51. Bhateja, V., Srivastava, A., Kalsi, A.: Fast SSIM Index for Color Images Employing Reduced-Reference Evaluation. In: Satapathy, S.C., Udgata, S.K., Biswal, B.N. (eds.) FICTA 2013. AISC, vol. 247, pp. 451–458. Springer, Heidelberg (2014)
52. Jaiswal, A., Trivedi, M., Bhateja, V.: A No-Reference Contrast Measurement Index based on Foreground and Background. In: Proc. of IEEE Second Students Conference on Engineering and Systems (SCES), Allahabad, India, pp. 460–464 (2013)
53. Wen, K.L., Chang, T.C.: The research and development of completed GM (1,1) model toolbox using Matlab. International Journal of Computational Cognition 3(3), 41 (2005)
54. Ray, A., De, D.: Energy efficient cluster head selection in wireless sensor network. In: Proc. of 2012 1st International Conference on Recent Advances in Information Technology (RAIT), pp. 306–311 (March 2012)
55. Singh, D.P., Bhateja, V., Soni, S.: Prolonging the Lifetime of Wireless Sensor Networks using Prediction based Data Reduction Approach. In: Proc. of IEEE Int. Conf. on Signal Processing & Integrated Networks (SPIN 2014), Noida, U.P., India, p. xx (2014)

Identifying Sources of Misinformation in Online Social Networks

K.P. Krishna Kumar and G. Geethakumari

Abstract. The importance of online social networks as a media for dissemination of news has increased in the last decade. The real time nature of the contents and the speed and volume of propagation have posed great challenges to assess the quality of information in an acceptable time frame. Collusion of users to spread false information and simultaneous spread of multiple false messages have made their detection a challenging task. In this paper we propose a methodology based on principles of cognitive psychology for detecting and monitoring sources who collude with each other to spread misinformation. We use social network as a social computing platform to classify sources as credible or non-credible based on the level of acceptance of their messages by other users and patterns of propagation. The proposed methodology could form a framework for an effective social media monitoring system. We have implemented our algorithm in the online social network 'Twitter'.

1 Introduction

Online Social Networks (OSN) have become an important source of information for a large number of people in the recent years. As the usage of social networks is increased, the abuse of the media to spread disinformation and misinformation has also been rapidly increased. The spread of information or misinformation in OSNs is context specific and studies have revealed topics such as health, politics, finances and technology trends are prime sources of misinformation and disinformation in different contexts to include business, government and everyday life [8]. The detection of misinformation in large volumes of data is a challenging task. Methods using machine learning and Natural Language Processing (NLP) techniques exist to automate the process to some extent (ref Section 2). However, because of the semantic nature of the contents, the accuracy of automated methods is limited and quite often

K.P. Krishna Kumar · G. Geethakumari
BITS-Pilani, Hyderabad Campus, Hyderabad, India
e-mail: kpkrishnakumar@gmail.com,
     geetha@hyderabad.bits-pilani.ac.in

S.M. Thampi, A. Gelbukh, and J. Mukhopadhyay (eds.), *Advances in Signal Processing*   417
and Intelligent Recognition Systems, Advances in Intelligent Systems and Computing 264,
DOI: 10.1007/978-3-319-04960-1_37, © Springer International Publishing Switzerland 2014

require manual intervention. The amount of data generated in OSNs is huge. This makes the task computationally expensive to be done in real time.

In this paper, we have proposed a framework for detecting sources of misinformation in OSNs using social networks as social computing systems. In particular we are interested in identifying sources of disinformation, where the sources collude with each other in deliberately spreading false information in the network. Deliberate spread of false information with a view to affect the behaviour and perceptions of others is part of semantic attacks. Once the false information has been inserted, the flow of misinformation and disinformation would not be able to be distinguished. We identify the sources of misinformation by measuring the level of acceptance of messages of each source user by the others. We have proposed a novel method of measuring acceptance based on gini coefficient. The proposed method is also computationally efficient for use by any social media monitoring system. The rest of the paper is organized as follows. In Section 2 we summarise the latest work done in the field. We describe our proposed methodology and the implementation in Twitter data sets in Section 3. In Section 4, we show the results obtained and in Section 5 we conclude and outline our future work.

2 Related Work

Social computing systems like Recommender systems and Reputation systems are being increasingly used in the web. OSNs are also social computing systems. A detailed survey of the various techniques used in Recommender systems can be found in [1] [2]. The collaborative filter techniques used in Recommender systems to include user based systems, item based systems and hybrid systems have been described in [15]. As per the authors, the collaborative filter systems provide much better recommendations than content based systems and are more scalable when the volume of data is very large. Manipulation of recommender systems in the form of *Shilling attacks* and their possible counter measures are described in [12] [13]. A study of the various types of Shilling attacks on collaborative filter systems like bandwagon attack, average attack, sampling attack, random attack etc would reveal that the manipulation of the ratings are done to either improve the ratings of the target items - called push attacks, or pull down the ratings of items - called nuke attacks. In all these cases, for the attack to become successful, the attackers inject false profiles into the systems in sufficient numbers to achieve their aim. These attacks prove that collusion of users could result in altering the correct behavior of the systems, if not detected.

The use of collaborative filter mechanisms to predict information propagation trends in social networks has been recently studied. The role of underlying core-periphery structure and community structure in the diffusion of information in social networks was brought out in [5]. The authors have proposed algorithms for early warning analysis of large-scale protests and other such events in social networks.

The collective intelligence of the web in deciding the quality of the web pages by means of PageRank algorithm has been demonstrated by Google [14]. Google's

PageRank algorithm uses information which is external to the web pages themselves. The back links proposed by them are a sort of peer review. Hence, collective evaluation of quality of an item is bound to produce better results than expert evaluation in most of the social computing systems. In our work, we use a similar idea to estimate the quality of information in social networks based on a peer review process of using re-propagation links.

Twitter has emerged as one of the most popular micro-blogging platforms. The information propagation process in Twitter has been a subject of research in the recent years. The use of Twitter as a social filter has been established. The credibility of tweets propagated through Twitter has been analyzed in [4]. While validating the best features to be used for automatic determination of credibility of tweets, the propagation based features were ranked the best. The measure of influence as given by the retweet mechanism offers an ideal mechanism to study large scale information diffusion in Twitter [9]. The text and author based features alone are not sufficient to determine the credibility of tweets. The use of Twitter as a collaborative-filter mechanism has been studied for critical events in [6] [11]. Reliability of Twitter under extreme circumstances was also investigated in [11]. The authors confirm that the propagation of rumours in Twitter is different from spread of credible news as rumours tend to be questioned more. The studies validate our approach of using the retweet mechanism in evaluating the spread of information and determining its quality. However, our work differs from them in devising a methodology to use the collective intelligence of social network to determine the credibility of source of messages as well as detect patterns of manipulation of the diffusion process.

Our contributions in this paper are:-

1. We have proposed a methodology for automatic detection of collusion between sources to spread misinformation.
2. We have proposed an algorithm based on principles of cognitive psychology and acceptance level of messages in OSNs to detect sources of misinformation.
3. Our algorithm would enable developing an effective social media monitoring system.

3 Proposed Algorithm

3.1 Basic Idea

Our aim is to use social network as a social computing platform to discern the level of acceptance of messages and then segregate possible sources of misinformation. Initially, we analyse the data sets using principles of cognitive psychology to identify cues of deception. We model the network as a graph which would enable us to measure the difference in acceptance levels. We use appropriate metrics like the gini coefficient to segregate sources of misinformation and also estimate the spread of their messages. The steps involved are shown in Fig 1.

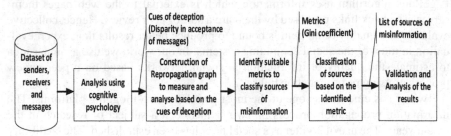

Fig. 1 Process of determination of sources of misinformation

3.2 Identifying Cues of Deception Using Cognitive Psychology

The presence of misinformation in the society and real world social networks have been studied from the psychological point of view extensively. An excellent review of the mechanisms by which misinformation is propagated and how effective corrective measures can be implemented based on cognitive psychology can be found in [10]. As per the authors, the spread of misinformation is a result of a cognitive process by the receivers based on their assessment of the truth value of information. This decision by the receiver is based on a set of parameters which can be characterised by asking four relevant questions. These questions are given below and illustrated in Fig. 2.

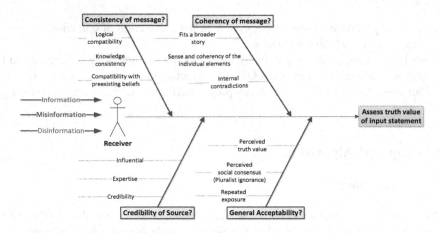

Fig. 2 Cognitive process of assessing cues of misinformation or deception

1. **Consistency of message.** Is the information compatible and consistent with the other things that he believes?
2. **Coherency of message.** Is the information internally coherent without contradictions to form a plausible story?

3. **Credibility of source.** Is the information from a credible source?
4. **General Acceptability.** Do others believe this information?

Information is more likely to be accepted by people when it is **consistent** with other things that they believe is true. If the logical compatibility of a news item has been evaluated to be consistent with their inherent beliefs, the likelihood of acceptance of the misinformation by the receiver increases and the probability of correcting the misinformation goes down. Preexisting beliefs play an important part in the acceptance of messages. Stories are easily accepted when the individual elements which make them up are **coherent** and without internal contradictions. Such stories are easier to process and easily processed stories are more readily believed. The familiarity with the sender of a message, and the sender's perceived **credibility** and expertise ensure greater acceptance of the message. The acceptability of a news item increases if the persons are subjected to repeated exposure of the same item. Information is readily believed to be true if there is a perceived social consensus and hence **general acceptability** of the same. Corrections to the misinformation need not work all the time once misinformation is accepted by a receiver. Efforts to use collaborative filtering to estimate the quality of information in social networks should keep the above aspects in mind. Any proposed algorithm should be able to evaluate the credibility of the source of information as well as measure the extent of spread of information.

An analysis of the four aspects would reveal that the aspects of *coherency* and *consistency* of messages are decided based on the intrinsic characteristics of the messages. To typically understand this, content analysis of the messages using NLP techniques would have to be used. While feasible, we would like to filter the messages based on the cues of *credibility* of messages revealed by the other two factors viz *credibility of source* and *general acceptability*. By using successive filters based on the above principle, we intend to develop a framework for a practical social media monitoring system.

3.3 Analysis of Credibility of Sources of Information

The credibility of sources is decided by the level of acceptance of messages by the users. In order to study this aspect, we obtained data from Twitter. The classification of information as genuine information or misinformation is with respect to the context in which they are studied. We obtained data from Twitter for different contexts as defined in Table 1. We used Twitter API to collect the tweets. The spreadsheet tool TAGS v5 used for collection of tweets using the Search API was provided by Martin Hawskey [7].

- **Bodhgaya.** The spread of information about terrorist attacks on 7 Jul 2013 at 'Bodhgaya' temple in India was tracked for a period of nineteen days from 07 Jul 2013 to 25 Jul 2013. The tweets were collected using the keyword 'bodhgaya'.
- **MyJihad.** We tracked a particular hashtag 'MyJihad' which we observed had contents which were controversial and the frequency of tweets were quite high.

The tweets were collected over a period of eight days between 20 Jul 2013 and 27 Jul 2013.

- **Telangana.** The spread of politically sensitive information in India over the demand for a separate state of Telangana was studied using the keyword 'telangana'. The tweets were collected over a period of eight days between 23 Jul 2013 and 30 Jul 2013 prior to the government decision being announced.
- **Egypt.** We investigated the spread of news related to the political unrest and massive protests in Egypt during the period from 13 Aug 2013 to 23 Sep 2013. The tweets were collected using the keyword 'egypt'.
- **Syria.** We tracked the events of use of chemical agents in Syria and all news related to it using the keyword 'syria'. The tweets were collected over the period between 25 Aug 2013 and 21 Sep 2013.

Table 1 Details of data sets

Data set	Users	Tweets	Sources	Retweets	Period	Type
Bodhgaya	4573	8457	660	4230	07 Jul 2013 to 25 Jul 2013	Terrorism
MyJihad	1166	5925	140	3232	20 Jul 2013 to 27 Jul 2013	Religious
Telangana	2671	6787	464	2177	23 Jul 2013 to 30 Jul 2013	Political
Syria	25415	104867	11452	44671	25 Aug 2013 to 21 Sep 2013	Political
Egypt	27532	141682	10850	51723	13 Aug 2013 to 23 Sep 2013	Civil Movement

We used the 'retweet' mechanism in Twitter to study the patterns of spread of messages of each source. Users propagate any news item they receive by retweeting it to their followers. This could be considered as a positive affirmation by the retweeting node about the acceptance of the message. It also enables us to monitor the spread of a particular news item in the network.

3.4 Establishment of Ground Truth

We used human annotation to classify the tweets and their sources. We concentrated only on tweets which have been retweeted, as we were interested only in misinformation which has the potential to spread in the network. We classified each of the retweeted tweets as possible misinformation or genuine information. Each of the original source was also classified as a potential source of misinformation even if one of his tweets were retweeted and classified as misinformation. We repeated the procedure with all the data sets. Finally, we had a list of sources which are considered spreading misinformation as well as a list of messages.

3.5 Analysis of Sources

Having classified the data sets, we performed a detailed analysis of all the sources. We plot graphs for each source of tweets - both misinformation and genuine information. For each source we plotted the percentage of tweets made by the source and the percentage of users who have retweeted them out of the total users who have retweeted at least a single tweet of the source. Fig.3 is for four sample sources whose messages have been classified as normal users and Fig. 4 are for a sample set of four users classified as misinformers.

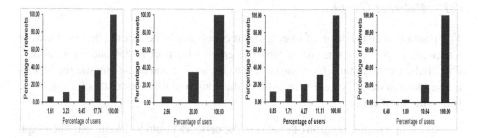

Fig. 3 Distribution of retweets of four different sources of genuine information amongst users retweeting their tweets

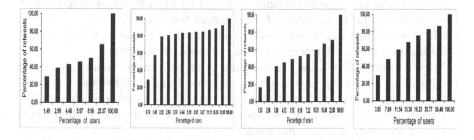

Fig. 4 Distribution of retweets of four misinforming sources amongst users retweeting their tweets

The difference in types of acceptance levels of the two types of sources are clear from Fig 3 and Fig.4. In the case of misinforming sources, there seems to be a smaller set of users who are retweeting most of the tweets of the source. The disparity in retweeting between the retweeting nodes is much sharper in the case of sources who misinform as compared to sources who do not. This also points towards the fact that there could be collusion of users who retweet each others tweets in order to ensure their spread in the network. This in turn would result in greater communication links between the colluding nodes as compared to their communication with others. The presence of such colluding nodes and disparity in their retweet behaviour can be detected using suitable network algorithms like core and community detection algorithms in graphs.

4   Results and Discussion

The difference in levels of acceptance of different sources were measured by drawing a retweet graph and calculating the gini coefficients of the distribution of retweets of the sources.

4.1   Retweet Graph

The pattern of distribution of tweets and retweets could be studied by constructing a retweet graph. We construct a bipartite graph with user nodes and tweet nodes. Directed edges are made from the tweets to the source and also from the retweeters to the tweet nodes. The representation of the complete retweets in this form has the following implications:-

1. The complete propagation of tweets in the network could be identified. The graph would depict as to who 'infected' whom and the tweets involved.
2. The graph would enable the use of standard PageRank algorithm [14] to calculate the general acceptance level of the user nodes and the message nodes. The rating of the source nodes would depend on the rating of their tweets which in turn would be decided by the rating of the retweeter nodes
3. The disparity between the retweeters of a particular source node could be calculated using Gini coefficient as would be explained subsequently.
4. The possibility of collusion between users in propagating misinformation in the network by retweeting each others messages could be identified using community detection and core identification algorithms. Frequent retweet behaviour between the nodes would result in formation of cycles and the nodes falling in the same community or core of the graph.

4.2   Gini Coefficient as a Measure of Disparity

Gini coefficient is often used as a measure of disparity in the distribution of any measurable quantity in a target population. Often used to measure the distribution of incomes in a target population, we measure how the retweets are distributed amongst the retweeters of a particular source node. We calculate the Gini coefficient, G using the equation given below. Let X_k be the cumulative proportion of the population variable, for k=0,...n and X_0 =0 and X_n=1 and let Y_k be the cumulative proportion of the re-propagated news items, for k=0,..,n and Y_0=0 and Y_n=1. If X_k and Y_k are indexed such that $X_{k-1} < X_k$ and $Y_{k-1} < Y_k$, the Gini coefficient, G is given by

$$G = 1 - \sum_{k=1}^{n}(X_k - X_{k-1})(Y_k + Y_{k-1}) \tag{1}$$

The Gini coefficients calculated for the four normal users shown in Fig 3 and misinforming users in Fig 4 are given in Fig 5. The gini coefficients vary between 0 and 1. Perfect equality of distribution is denoted by gini coefficient of 0 and perfect inequality by 1. The actual values vary in between with normal users having a value closer to 0. A high value of gini coefficient for a source node would indicate greater difference in acceptance levels of his tweets and consequent reduction in credibility. The gini coefficients of misinforming sources had higher values as compared to the normal users and indicated their suitability for classification of sources.

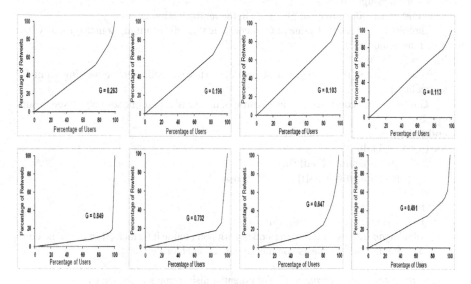

Fig. 5 Gini coefficients for four sample users spreading genuine information (top set of figures) and for four sample users spreading disinformation (bottom set of figures).

4.3 Construction of a Binary Classifier

The classification of sources as credible or non-credible is based on a binary classifier. The binary classifier bases its decision on the threshold value of gini coefficient for the sources in the data set. This value is normally above 0.4. However, the exact value needs to be determined by evaluating the sources which have higher gini values. Following the heavy tailed distribution, these numbers are small. When we decrease the threshold values, false positives would increase. Similarly, an increase in threshold value result in some misinforming sources being left out. We arrive at a threshold figure by starting with the maximum gini value in the graph and slowly reducing them till we continue to get sources of misinformation.

The collusion of sources also enables their detection. Even if one non credible source is identified, the others could be detected by using standard community detection algorithms based on modularity [3] in the retweet graph. We present the whole methodology of the binary classifier for detection of colluding nodes in Algorithm 1.

4.4 Validation of Results

We validated the proposed methodology based on its ability to detect all the colluding nodes, i.e, True positives and the computation requirements for the same. The

Algorithm 1. Methodology for detection of colluding nodes to spread misinformation

---

Input: Details of users and messages (tweets) involved in the spread of information for specific 'context'

Preprocessing Step: repropagationgraph ← re-propagation bi-partite graph with user nodes and message nodes

sourcelist ← List of sources of messages in the repropagation graph

ginithreshold ← Threshold value of Gini coefficient for disinforming sources (usually 0.4)

k ← length(sourcelist)

for (i = 1 → k) **do**

    sourcesubgraph[i] ← subgraph of messages (tweets) and users retweeting them for sourcelist[i]

    Gini[i] ← Gini coefficient of distribution of degree of all user nodes only in sourcesubgraph[i]

end for

for (i = k → 1) **do**

    if (Gini[i] > ginithreshold) **then**

        potentialdisinformers[i] ← sourcelist[i]

    end if

end for

for (i = 1 → length(potentialdisinformers)) **do**

    colludingnodes[i] ← Interacting nodes with potentialdisinformers[i] in the same community using modularity based community detection algorithm

end for

Output: *potentialdisinformers:* List of potential disinformers in the data set.

Output: *colludingnodes:* List of colluding nodes for each of the potentialdisinformers.

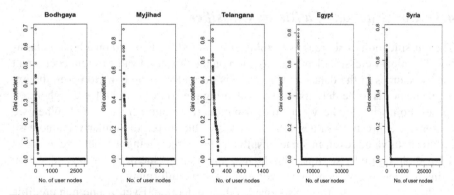

Fig. 6 Distribution of Gini coefficients in all the data sets

distribution of gini coefficients of degree distribution of all retweeting nodes of the source nodes in all the data sets being studied is given at Fig 6. The graph shows a heavy tailed distribution with few source nodes having higher gini coefficients. This also is intuitive in that the social network is mainly used by genuine users to spread information rather than misinformation.

The number of source nodes with gini coefficients above the threshold value (taken as 0.4 in our data sets) varied from 0.5% to 1.6% of the total users nodes. All the sources of disinformation had high gini coefficients, while all nodes with high gini coefficients were not spreading misinformation. The use of semi automated means of segregating the disinforming sources is limited to less than 2 % of the source nodes and hence the algorithm is computationally very efficient. The identification of colluding nodes using community detection algorithm [3] with these sources nodes enabled detection of all misinforming user nodes. Thus the algorithm could identify all sources spreading misinformation through collusion by inspecting less than 2% of the user nodes. The results validate the suitability of the algorithm for any social network monitoring system. The worst case complexity of the algorithm is proportional to the number of nodes in the graph. However, we are interested in monitoring the social media and once the Gini coefficient values are initially calculated, we would need to monitor only the changes.

5 Conclusion

In this paper we have proposed a methodology for automated detection of sources of misinformation in a social network. Based on the principles of cognitive psychology and using gini coefficient as a measure of acceptability of a source, we could achieve good results in identifying sources of misinformation especially deliberate efforts to spread false information. The methodology adopted by us in identifying the sources of misinformation rather than the messages seem more robust and implementable for large social media monitoring systems. Integrating the proposed algorithm with NLP techniques would help us to further improve the results obtained.

References

1. Adomavicius, G., Tuzhilin, A.: Toward the next generation of recommender systems: A survey of the state-of-the-art and possible extensions. IEEE Transactions onKnowledge and Data Engineering 17(6), 734–749 (2005)
2. Almazro, D., Shahatah, G., Albdulkarim, L., Kherees, M., Martinez, R., Nzoukou, W.: A survey paper on recommender systems. arXiv preprint arXiv:1006.5278 (2010)
3. Blondel, V.D., Guillaume, J.L., Lambiotte, R., Lefebvre, E.: Fast unfolding of communities in large networks. Journal of Statistical Mechanics: Theory and Experiment 2008(10), P10,008 (2008)
4. Castillo, C., Mendoza, M., Poblete, B.: Information credibility on twitter. In: Proceedings of the 20th International Conference on World Wide Web, pp. 675–684. ACM (2011)
5. Colbaugh, R., Glass, K.: Early warning analysis for social diffusion events. Security Informatics 1(1), 1–26 (2012)
6. Gupta, A., Kumaraguru, P.: Credibility ranking of tweets during high impact events. In: Proceedings of the 1st Workshop on Privacy and Security in Online Social Media, p. 2. ACM (2012)

7. Hawksey, M.: Twitter Archiving Google Spreadsheet TAGS v5. JISC CETIS MASHe: The Musing of Martin Hawksey, EdTech Explorer (2013), http://mashe.hawksey.info/2013/02/twitter-archive-tagsv5/ (accessed September 2013)

8. Karlova, N.A., Fisher, K.E.: "Plz RT": A social diffusion model of misinformation and disinformation for understanding human information behaviour. Information Research 18(1), 1–17 (2013)

9. Kwak, H., Lee, C., Park, H., Moon, S.: What is twitter, a social network or a news media? In: Proceedings of the 19th International Conference on World Wide Web, pp. 591–600. ACM (2010)

10. Lewandowsky, S., Ecker, U.K., Seifert, C.M., Schwarz, N., Cook, J.: Misinformation and its correction continued influence and successful debiasing. Psychological Science in the Public Interest 13(3), 106–131 (2012)

11. Mendoza, M., Poblete, B., Castillo, C.: Twitter under crisis: Can we trust what we rt? In: Proceedings of the First Workshop on Social Media Analytics, pp. 71–79. ACM (2010)

12. Mobasher, B., Burke, R., Bhaumik, R., Sandvig, J.: Attacks and remedies in collaborative recommendation. Intelligent Systems 22(3), 56–63 (2007)

13. Mobasher, B., Burke, R., Bhaumik, R., Williams, C.: Effective attack models for shilling item-based collaborative filtering systems. In: Proceedings of the 2005 WebKDD Workshop (2005)

14. Page, L., Brin, S., Motwani, R., Winograd, T.: The pagerank citation ranking: bringing order to the web (1999)

15. Su, X., Khoshgoftaar, T.M.: A survey of collaborative filtering techniques. Advances in Artificial Intelligence 2009, 4 (2009)

Modified Teacher Learning Based Optimization Method for Data Clustering

Anoop J. Sahoo and Yugal Kumar

Abstract. Clustering is an important task in engineering domain which can be applied in many applications. Clustering is a process to group the data items in the form of clusters such that data items in one cluster have more similarity to other clusters. On the other side, Teaching–Learning Based Optimization (TLBO) algorithm is a latest population based optimization technique that has been effectively applied to solve mechanical design problems and also utilized to solve clustering problem. This algorithm is based on unique ability of teacher i.e. how the teacher influence the learners through its teaching skills. This algorithm has shown good potential to solve clustering problems but it is still suffering with some problems. In this paper, two modifications have been proposed for TLBO method to enhance its performance in clustering domain instead of random initialization a predefined method previously used to exploit initial cluster centers as well as to deal the data vectors that cross the boundary condition. The performance of proposed modified TLBO algorithm is evaluated with six dataset using quantization error, intra cluster distance and inters cluster distance parameters and compared with K-Means, Particle Swarm Optimization (PSO) and TLBO. From the experimental results, it is clearly obvious that the proposed modifications have shown better results as compared to previously ones.

Keywords: Clustering, K-Means, PSO, TLBO.

1 Introduction

Clustering has important applications in pattern recognition, data mining and machine learning domain. It is NP Complete problem when number of cluster is

Anoop J. Sahoo
Infosys Technologies, Chennai, India
e-mail: anoop.jyoti88@gmail.com

Yugal Kumar
Department of Information Technology, Birla Institute of Technology, Mesra, Ranchi
Jharkhand, India
e-mail: yugalkumar.14@gmail.com

S.M. Thampi, A. Gelbukh, and J. Mukhopadhyay (eds.), *Advances in Signal Processing and Intelligent Recognition Systems*, Advances in Intelligent Systems and Computing 264, DOI: 10.1007/978-3-319-04960-1_38, © Springer International Publishing Switzerland 2014

greater than three. It can be defined to find out unknown patterns, knowledge and information from a given dataset D which was previously undiscovered using some criterion function [1]. In clustering problem, a dataset is divided into K number of sub-groups such that elements in one group are more similar to another group. In last two decades large number of algorithms have been developed based on swarms, insects and natural phenomena's to solve optimization problems. Some of these are Artificial Bee Colony (ABC) [2], Ant Colony Optimization (ACO) [3], Genetic Algorithm (GA) [4], Particle Swarm Optimization (PSO) [5], Cat Swarm Optimization (CSO) [7], Black Hole (BH) [6], Gravitational Search Algorithm (GSA) [8], TLBO [9] and many more. These algorithms have shown tremendous potential over traditional algorithms such as K-Means algorithm [10]. Naik et al. [11] has applied the Teacher Learning based optimization (TLBO) method to solve the data clustering problem and compared it to K-Means and PSO method in which TLBO provides better results than other two. From the study of [9, 11, 12],first it has been found that initial population for TLBO algorithm is randomly generated and experimentally it has also been proved that random initialization of population does not exploit the data vectors uniformly from dataset. But in clustering task, the performance of clustering techniques depends on the initial cluster centers. If initial cluster center has not been generated well then, there is a chance that the TLBO algorithm of trapping in local optima and convergence of algorithm is also affected. Second, there is no mechanism to deal with data vectors which crosses the boundary values of a dataset (especially in clustering task). These are the two main drawbacks of TLBO method. So, in this paper we have proposed a modified TLBO algorithm to overcome the above mentioned limitations.

2   Teacher Learning Based Optimization (TLBO) Method

Rao et al. [9] have developed Teacher Learning based optimization (TLBO) method and applied this algorithm to solve constrained mechanical design optimization problems. It is based on influence capability of a teacher on learners i. e. how a teacher influence the students through its teaching capability. As all we know, a good teacher has great impact on students. Thus, the TLBO method sketches out this relationship of teacher and student. TLBO method contains two phases: Teacher phase and Learner phase. Teacher phase correspondents to learning from the teacher while learning phase correspondent to learning among learners through interaction. A   mathematical model has been formed to show the influence of a teacher on the learners. This mathematical model contains a probability density function which illustrates the performance of learners as well as influence of teachers on learners. Similar to other nature-inspired algorithms; TLBO is also a population based method that uses a population of solutions to proceed to the global solution. For TLBO, the population is considered as a group of learners or a class of learners. In optimization algorithms, the population consists of different design variables. In TLBO, different design variables will be analogous to different subjects offered to learners and the learners' result is analogous to the 'fitness', as in other population-based optimization techniques.

The teacher is considered as the best solution obtained so far. The process of TLBO is divided into two parts. The first part consists of the Teacher Phase and the second part consists of the 'Learner Phase'. The 'Teacher Phase' means learning from the teacher and the learner Phase means learning through the interaction between learners.

2.1 Teacher Phase

As shown in Fig. 2, the mean of a class increases from M_A to M_B depending upon a good teacher. A good teacher is one who brings his or her learners up to his or her level in terms of knowledge. But in practice this is not possible and a teacher can only move the mean of a class up to some extent depending on the capability of the class. This follows a random process depending on many factors.

Let M_i be the mean and T_i be the teacher at any iteration i. T_i will try to move mean M_i towards its own level, so now the new mean will be T_i designated as M_{new}. The solution is updated according to the difference between the existing and the new mean given by

$$Difference\_Mean_i = r_i * (M_{new} - TF * M_i) \tag{1}$$

where TF is a teaching factor that decides the value of mean to be changed and r_i is a random number in the range [0, 1]. The value of TF can be either 1 or 2, which is again a heuristic step and decided randomly with equal probability as

$$TF = round[1 + rand(0,1)\{2 - 1\}]$$

This difference modifies the existing solution according to the following expression

$$X_{new,i} = X_{old,i} + Difference\_Mean_i \tag{2}$$

2.2 Learner Phase

Learners increase their knowledge by two different means: one through input from the teacher and the other through interaction between themselves. A learner interacts randomly with other learners with the help of group discussions, presentations, formal communications, etc. A learner learns something new if the other learner has more knowledge than him or her. Learner modification is expressed as

For i = 1 : P_n
Randomly select two learners Xi and Xj, where i/= j
If f (X_i) < f (X_j)
$X_{new, i} = X_{old,i} + r_i(X_i - X_j)$
Else
$X_{new,i} = X_{old,i} + r_i(X_j - X_i)$
End If
End For
Accept X_{new} if it gives a better function value.

3   Modified TLBO

TLBO is the latest population based meta-heuristic algorithm which has been applied to various mechanical design problems as well as clustering problem. But this algorithm mostly suffered with initialization and boundary values problems. In TLBO, initial cluster centers are defined randomly and there is no mention of mechanism to deal with the data vectors which cross the boundary values of a dataset. Hence, to overcome these problems, two modifications have been proposed in TLBO method.

- First, instead of random initialization of initial cluster centers, the equation 1 is used to find the value of initial cluster centers in a given dataset D.

$$C_k - minimum(D_j) + \frac{(k-1)\left(maximum(D_j) - minimum(D_j)\right)}{K}$$

$$where, k = 1,2,...K \ and \ j = 1,2...M \qquad (3)$$

Where, K is total number of cluster centers, M represents total features, minimum (D_j) is minimum value of jth feature in dataset D, maximum (D_j) represents the maximum value of jth feature in dataset D.

- Second, a data vector which crosses the boundary of dataset has been killed and regenerated using equation 2. Equation 2 can be defined as

$$C_k = D_{min}^j + rand(0,1)\left(D_{max}^j - D_{min}^j\right) \qquad (4)$$

$$where \ j = 1,2 ... . M$$

Where, M represents the total features, D_{min}^j is minimum value of jth feature in dataset D, D_{max}^j represents the maximum value of jth feature in dataset D.

Hence, in original TLBO method, these two modifications have been done to overcome the limitations of TLBO. Now, the steps of TLBO algorithm can be given as

Pseudo Code of Modified TLBO algorithm for clustering
Algorithm (Input (training dataset, K), Output (cluster centers))

Step 1: Load the dataset and specify the number of initial cluster centers i.e. Teachers (K).
Step 2: for each k=1 to K
          for each j=1 to m

               Determine the initial position of Cluster Centers (C_k) using equation 1;
               end

       end
       Iteration =0;
Step 3: Calculate the value of objective function (SSE).
Step 4: Find the best solution with minimum objective function value called teacher.

Step 5: Update the solution based on the best solution using given equation.

$$X_{new} = X_{old} + r(X_{teacher} - (T_F) * Mean)$$

Step 6: if (SSE(X_{new})< SSE(X_{old}))
        then Q → X_{new}
        else go to step 5
Step 7: Randomly pick any two solutions from Q
        If (SSE (X_i) ≤ SSE (X_j))
            then $X_{new} = X_{old} + r(X_i - X_j)$
            else $X_{new} = X_{old} + r(X_j - X_i)$
Step 8: if (SSE(X_{new})< SSE(X_{old}))
        then Q → X_{new}
        else go to step 7
        Iteration = iteration+1
Step 9: if {X_{new} > maximum (D)||X_{new} < minimum (D)}
        then generate new cluster center using
            for each j=1 to m
                $$C_k = D_{min}^j + rand(0,1)(D_{max}^j - D_{min}^j)$$
        end
Step 10: Repeat the steps 5 to 10 until the termination condition satisfied.
Step 11: Obtain final cluster centers.

4   Experimental Results

The performance of proposed modified TLBO method is investigated on six datasets out of which five datasets are real and one is artificial dataset using quantization error, intra cluster distance and inters cluster distance parameters. The five real datasets are iris, glass, wine, cancer, haberman survival which are downloaded from UCI repository. Synthetic dataset 1 is two dimensional artificial dataset, generated in matlab to test the effectiveness of proposed algorithm. This dataset includes 300 instances with the two attributes and three classes. The classes in dataset are disseminated using μ and λ where μ is the mean vector and λ is the variance matrix. The data has generated using given μ_i and λ_i values where i= 1,2,3.

$$\mu1 = [4,6], \mu2 = [1,3], \mu3 = [7,9]$$
$$\lambda1 = [0.3,0.5], \lambda2 = [0.7,0.4], \lambda3 = [0.4,0.6]$$

The Figure 1(a) depicts the distribution of data into synthetic dataset 1 and Figure 1(b) shows the clustering of same data using proposed method. Table 1 show the characteristics of all these datasets which are used to investigate the performance of proposed modified TLBO method. The proposed algorithm is implemented in Matlab 2010a environment on a core i5 processor with 4 GB using window operating system. The quality of solutions are measured using intra cluster distances, inter cluster distances and quantization error. Intra cluster distances can be defined as average distance between the instances placed in a cluster to the corresponding cluster center while minimum distance between any two instances of cluster centers known as inter cluster distance. Quantization error can be described as cost function for a clustering algorithm. It can be defined as

$$Q_E = \frac{\sum_{k=1}^{K} [\sum_{\forall x_i \in c_k} dist(x_i, m_k)]/|C_k|}{K}$$

where K is total number of cluster centers, C_k is k^{th} cluster center, m_k represents center points of kth cluster and x_i represents i^{th} data instance of a dataset.

To check efficiency and adaptability of proposed modification in TLBO algorithm, the investigational results are compared with K-Means, PSO and TLBO algorithms which are given in Table 2.

Table 1 Characteristics of dataset

Name of data set	No. of classes	No. of features	Total instance in dataset	Number of classes in dataset and Instance in each classes
Synthetic dataset 1	3	2	300	3 and (100, 100, 100)
Iris	3	4	150	3 and (50, 50, 50)
Glass	6	9	214	6 and (70,17, 76, 13, 9, 29)
Cancer	2	9	683	2 and (444, 239)
Haber man Survival	2	3	306	2 and (225,81)
Wine	3	13	178	3 and (59, 71, 48)

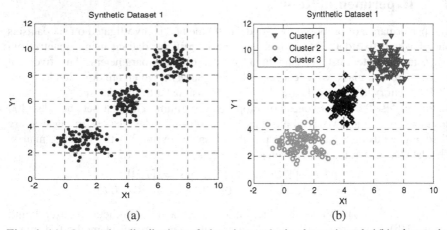

Fig. 1 (a) shows the distribution of data in synthetic dataset1 and 1(b) shows the corresponding cluster centers using proposed method.

Table 2 summarizes experimental results obtained from all four clustering algorithms. From this table, it is concluded that modification in TLBO algorithm is enhanced its performance as compare to original TLBO and all other methods. Modified TLBO provides low quantization error among all these techniques and lowest inter and intra cluster distance expect Haberman's Survival datasets

(50.386 and 16.4236). While the value of quantization error K-Means algorithm is high as compare to others except wine dataset.

Table 2 Shows the comparative results with six different datasets

Dataset Name	Algorithms Name	Parameters		
		Quantization Error	Intra Cluster Distance	Inter Cluster Distance
Iris Data	K-Means	0.6505± 0.0124	2.5147±0.0064	1.7928±0.0045
	PSO	0.3023±0.0263	3.2806±0.2314	1.5403±0.1693
	TLBO	0.2845±0.0047	3.1350±0.2187	1.6939±0.2173
	Modified TLBO	0.2561	2.8343	1.4162
Wine Data	K-Means	101.9587±4.1863	458.6488±3.6020	327.0094±58.2466
	PSO	342.2986±9.5387	456.6758±4.8859	279.7534±56.4985
	TLBO	345.0649±20.7124	450.7811±23.9000	278.1714±29.5149
	Modified TLBO	334.216	443.3276	275.9876
Breast Cancer Data	K-Means	5.2318±0.0016	18.5776±0.00	13.8972±0.0060
	PSO	0.4719±0.2333	20.665±.0.00	13.4433±0.00
	TLBO	0.0514±0.0171	20.6651±0.00	13.4433±0.00
	Modified TLBO	0.0496	19.6783	13.4256
Haberman's Data	K-Means	8.5926±0.0199	50.8811±2.7782	17.5622±0.3637
	PSO	6.8734±0.7443	50.6146±2.2313	17.6723±3.8535
	TLBO	6.1354±0.3461	50.2354±1.5903	16.2889±2.8530
	Modified TLBO	6.0321	50.386	16.4236
Glass Data	K-Means	0.8912±0.0156	3.8371±1.5920	3.6852±1.5902
	PSO	0.0574±0.0037	6.6325±0.8561	0.5290±0.2149
	TLBO	0.0507±0.0106	5.9905±0.6239	0.4924±0.1411
	Modified TLBO	0.0502	5.6187	0.4673
Artificial Data	K-Means	4.2526±0.2154	11.6157±0.6482	5.9458±1.2367
	PSO	3.8675±0.3256	10.4561±1.2650	5.6480±2.1464
	TLBO	3.0238±0.1623	10.2135±0.2352	5.2134±1.2610
	Modified TLBO	2.8783	9.8374	4.3486

Figure 2 demonstrates the convergence behavior of proposed algorithm for iris dataset. From Figure 2, it is observed that K-means algorithm is faster but it exhibits premature convergence such as large quantization error, while the Modified TLBO algorithm provides slow convergence rate among all these methods and low quantization errors. From the experimental setup, it is found that K-means. PSO, TLBO and Modified TLBO algorithms have converged after 15, 90, 130 and 140 function evaluations respectively.

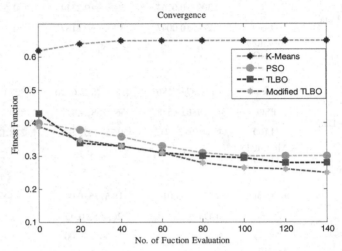

Fig. 2 Convergence of proposed algorithm with iris dataset

5   Conclusion

This paper focuses on the drawbacks of teaching learning based optimization (TLBO) to cluster the data and their solutions. To overcome the drawbacks of TLBO algorithm, two modifications have been proposed- First, a method is proposed to exploit the initial cluster centers without random initialization because in clustering, the performance of an algorithm is highly depends on initial cluster centers and it is also found that random initialization does not explore the data vectors uniformly from a dataset. Second, a mechanism is introduced to deal with the data vectors that cross the boundary values. In our approach, data vectors that cross the boundary values are killed and generated using equation 2. These modifications not only enhance the performance of TLBO algorithm but also improve the convergence rate of TLBO. In addition to it, the proposed modification in TLBO can produce high quality clusters and there is no need to run the algorithms multiple times. In future, the TLBO method may be exploited in other applications, such as image segmentation, scheduling problem, classification and feature selection. The combination of the TLBO with other heuristic approaches and their application to data clustering is also another research direction to the curious academicians and researchers.

References

1. Garey, M.R., Johnson, D., Witsenhausen, H.: The complexity of the generalized Lloyd-max problem. IEEE Transactions on Information Theory 28(2), 255–256 (1982)
2. Karaboga, D.: An idea based on honey bee swarm for numerical optimization, technical report-TR06, Erciyes University, Engineering Faculty, Computer Eng. Dep. (2005)
3. Dorigo, M., Di Caro, G., Gambardella, L.M.: Ant algorithms for discrete optimization. Artificial Life 5(2), 137–172 (1999)
4. Holland, J.H.: Adaption in natural and artificial systems (1975)
5. Kennedy, J., Eberhart, R.: Particle Swarm Optimization. In: Proc. of IEEE international Conference on Neural Networks (ICW), Perth, Australia, vol. IV, pp. 1942–1948 (1995); Santosa, B., Ningrum, M.K.: Cat swarm optimization for clustering. In: International Conference on Soft Computing and Pattern Recognition (SOCPAR 2009), pp. 54–59 (2009)
6. Hatamlou, A.: Black hole: A new heuristic optimization approach for data clustering. Information Sciences 222, 175–184 (2013)
7. Chu, S.-C., Tsai, P.-w., Pan, J.-S.: Cat swarm optimization. In: Yang, Q., Webb, G. (eds.) PRICAI 2006. LNCS (LNAI), vol. 4099, pp. 854–858. Springer, Heidelberg (2006)
8. Rashedi, E., Nezamabadi-Pour, H., Saryazdi, S.: GSA: a gravitational search algorithm. Information Sciences 179(13), 2232–2248 (2009)
9. Rao, R.V., Savsani, V.J., Vakharia, D.P.: Teaching–learning-based optimization: A novel method for constrained mechanical design optimization problems. Computer-Aided Design 43(3), 303–315 (2011)
10. MacQueen, J.: Some methods for classification and analysis of multivariate observations. In: Proceedings of the Fifth Berkeley Symposium on Mathematical Statistics and Probability, vol. 1(281-297), p. 14 (1967)
11. Satapathy, S.C., Naik, A.: Data clustering based on teaching-learning-based optimization. In: Panigrahi, B.K., Suganthan, P.N., Das, S., Satapathy, S.C. (eds.) SEMCCO 2011, Part II. LNCS, vol. 7077, pp. 148–156. Springer, Heidelberg (2011)
12. Rao, R.V., Savsani, V.J., Vakharia, D.P.: Teaching–learning-based optimization: an optimization method for continuous non-linear large scale problems. Information Sciences 183(1), 1–15 (2012)

A Novel Disease Outbreak Prediction Model for Compact Spatial-Temporal Environments

Kam Kin Lao, Suash Deb, Sabu M. Thampi, and Simon Fong

Abstract. One of the popular research areas in clinical decision supporting system (CDSS) is Spatial and temporal (ST) data mining. The basic concept of ST concerns about two combined dimensions of analyzing: time and space. For prediction of disease outbreak, we attempt to locate any potential uninfected by the predicted virus prevalence. A popular ST-clustering software called "SaTScan" works by predicting the next likely infested areas by considering the history records of infested zones and the radius of the zone. However, it is argued that using radius as a spatial measure suits large and perhaps evenly populated area. In urban city, the population density is relatively high and uneven. In this paper, we present a novel algorithm, by following the concept of SaTScan, but in consideration of spatial information in relation to local populations and full demographic information in proximity (e.g. that of a street or a cluster of buildings). This higher resolution of ST data mining has an advantage of precision and applicability in some very compact urban cities. For proving the concept a computer simulation model is presented that is based on empirical but anonymized and processed data.

1 Introduction

Clinical information system is in imperative need for the human society, especially when people experienced some epidemic diseases like severe acute respiratory

Kam Kin Lao · Simon Fong
University of Macau, Taipa, Macau SAR
e-mail: ccfong@umac.mo

Suash Deb
Cambridge Institute of Technology, Ranchi, India
e-mail: suashdeb@gmail.com

Sabu M. Thampi
Indian Institute of Information Technology & Management, Kerala, India
e-mail: sabu.thampi@iiitmk.ac.in

S.M. Thampi, A. Gelbukh, and J. Mukhopadhyay (eds.), *Advances in Signal Processing*   439
and Intelligent Recognition Systems, Advances in Intelligent Systems and Computing 264,
DOI: 10.1007/978-3-319-04960-1_39, © Springer International Publishing Switzerland 2014

syndrome (SARS), swine flu and enterovirus, etc., which has a high prevalence rate. They outbreak at a very rapid speed, and spread wide and far. There are research papers which advocate developing the clinical decision support system which predicts the time series and space area. However, the efficacy of clinical decision support system is based on the underlying analysis model. Some data related challenges are like: what kind of data attributes the system need to use? How about the scope of data? Is the data useful or not? Which analyzing method is efficient and effective? Any other parameter need to be concerned? What is the trend of disease outbreak? Etc.

Many researchers suggested embedding the clinical decision support system into the GIS (Geographic Information System) as it seems to be more accurate to detect the area whether is in a high prevalence rate and their adjacency areas [1], or even using the ST analyzing method to focus on the analyzing risk of the disease outbreak [2]. Actually these research papers assume the field of analysis is of large terrain or vast piece of land. It is useful for large countries. However, it may not be so applicable for compact urban cities like Macao, Hong Kong, Taipei etc. where the human population is very dense, but they are not necessarily evenly distributed. In the words, the radius approach might not work well in estimating the next infested areas. Different from the other researchers, we introduce a new and simple approach by dividing the city into respective regular polygons, each polygon which is square cell as assumed in this case, is of equal size; and they form grid over the coverage of the city regardless what shape the city is. The number of the square cells to be defined can be selected arbitrarily by the user that depends on the land area and the resolution required.

As a demonstrative case in this paper we use the data simulation of the entorvirus as the experiment part of our research, Macao land and the Tapai land will be separately divided by various numbers of cells and combined for analysis. At the start, the risk of the virus would be evaluated in order to find co-relationship among the areas, the analyzing model will predict how risky of each zone of the city, depends on the risk analyze (some factors may need to reference the previous disease record of the zone and the risk analyzing in the surrounding zones). After locating the high risk areas, the analyzer can group these zones and focus on the relations and/or correlations of them as try to know more about virus and its spread. The associated relationships among the areas are those that have the disease outbreak simultaneously. Technically it will involve using various classifiers of the decision tree and association rules analyzing model.

After applying these two models, some high risk areas and the relationship which the areas almost have the disease outbreak in the same time will be found. The analyzers can concentrate on analyzing the specific characteristic of the areas and deciding which attributes will have the significant relationship between the inflected areas. As the demographic information changes and the risk evaluation suggest, the experiment will vary for different time-series. Finally the analyzers can trace back the source of virus and identify the "flow" of various attributes; and investigate whether the virus has been mutated. This analyzing method will be novel as the part of risk evaluation for detecting among various areas, especially when the forecast of the disease outbreak is changing obviously, the risk index and the associated relationship can reflect the status of the virus extension. On the other hand, this method could be quite effective and efficient that the users can refer to the spatial and temporal analyzing model to adjust the whole analyzing model, as the learning process to develop a more accurate schema for detecting the disease outbreak.

2 Related Work

Spatial and temporal (Spatio-temporal analyzing, ST) analyzing model is a hot research topic in the last decade, concerning the time and geographic factors to predict the result. The ST analyzing should be based on the spatial analyzing, with an extension part of analyzing the geographic phenomena, which combined with the time sequences. Its purpose is on tracing back or trending the future result. Many researchers advocated their method and framework for the ST analyzing, most of them have a great contribution through their experiment to prove their ST analyzing model is feasible [3, 4, 5]. The details of the classification of the ST data mining task and techniques can be found in [6], ST analyzing can be engaged in various classifiers as clustering [7] and the association rules for ST analyzing [8]. Just like for the traffic jam detection [9], the users can use the ST data about the traffic conditions to simulate the real time traffic surveillance system, warning the drivers which road occurred traffic jam. ST analyzing can also be engaged as the utilization of the land cover change [10], combined with the association rules method, the analyzers can find out which demographic information will impact the utilization of the land reasonably, the relationship between antecedent and consequence can be determined, and the analyzers are able to utilize the result to make the land allocation more scalable.

One of the most popular topics for ST analyzing is applying in the disease outbreak. Many researchers issue various ideas and conducting experiments about it, for their report they are more concerning the serious disease outbreak occurred in a big country, different formulas and external factors like the demographic information, natural disaster like hazard, exposure and vulnerability [11]. As abovementioned, the direction of the ST analyzing is inclined towards the geographic information with the polygon pattern as the original GIS of the country. In fact, it will be bias as the virus extension should not only be analyzing the geographic framework like the predefined map, may be the serious outbreak area is in the edge of the various regions. Nevertheless, how about when the disease outbreak occurred in small urban city like Macao? So far there is no literature on about ST analysis for small cities.

3 Our Proposed Model

Our proposed analysis framework has two parts: locating the risk areas and studying the association between those areas of high risks.

3.1   Risk Analysis via a Spiral Model

The risk of disease outbreak for each specific small cell is computed in a spiral fashion. The computation in terms of risk grades or indices is taken into account of the distance between the adjunct cells and the active cell, the timing, and the strength of the virus dissemination, and other information. For example, a quantitative risk index for a particular cell (called active cell) in a grid is calculated based on the facts about its risk in the previous years, demographic (density, age, race, visitors' traffic) and geographic (climate, number of buildings) factors that will affect the virus dissemination. The risk index would be normalized between 0 and 1 where 0 means it has not ever been infested before. For consideration of risks over certain years, an overall index can be defined by I where:

$$I = (r_{i-1} + r_{i-2} + r_{i-3+\ldots} + r_1) / \text{number of referenced years} \tag{1}$$

i is the concurrent year need to evaluate, and r is the risk factor of the year. This index I of the particular cell can illustrate the "virus record per history" of the specific zone. It consists of the temporal factors in the analysis when combined with the spatial information. Here, for doing the spatial risk evaluation of each cell in the grid, the coordinates of each grid cell, for the cell i (with coordinate $X = n$, $Y = m$), for estimating the risk index of its adjacency area.

Zone 1 $(n-1,m+1)$	Zone 2 $(n,m+1)$	Zone 3 $(n+1,m+1)$
Zone 4 $(n-1,m)$	Active Zone (Zone 5) (n,m)	Zone 6 $(n+1,m)$
Zone 7 $(n-1,m-1)$	Zone 8 $(n,m-1)$	Zone 9 $(n+1,m-1)$

Fig. 1 Illustration of grids

The risk index of an active cell in consideration of its neighbor adjacent is defined by:

$$R_i = I \times [(R_{z1} + R_{z2} + R_{z3} + R_{z4} + R_{z6} + R_{z7} + R_{z8} + R_{z9}) / 9] \tag{2}$$

The computation starts at the top left corner and goes in a spiral fashion around each zone, updating the corresponding risk indices along the way. Considering the history of risk of each area is very important. As an infested area is likely to be infested again, given the same conditions arise. In this spiral model, the factor of distance among the zones would impact the dissemination rate and the outspread. It is assumed that the closer where it is to the infested zone, the more likely the disease will propagate over.

Since square grid is used in partitioning the region of a city into square cells, the level or the gravity of the contagiousness is abstracted into some estimates of distance effects. An example is shown in Fig. 3 where concentric rings are logically laid over the grid, with each ring position at the outer area decreases in contagiousness proportionally.

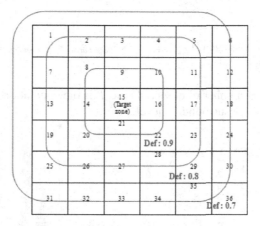

Fig. 2 The effects of the disease are represented by different rings around the zones

In the spiral approach, the "rings" represent various levels of effects measuring from the center (target zone) to how close the designated area by the ring are. In computation, the spiral model generalizes the concept of contagiousness over the distance apart. The spiral index (S_i) to represent this "distance effect" of each area. Moreover, the analyzers can reference from the previous disease outbreaks and calibrated the distance impact index of each level. As in level 1, the distance impacted index is be defined as 0.8 by the user as the previous result illustrated it is not really impacted the specific area significantly. So a moderate factor 0.8 is assumed this time.

In addition to the spatial factors, spatial and temporal analyzing is combining the elements of the time-series with a purpose of predicting the result more reasonable and sensible in consideration of space-and-time. Time series factors are being considered in our model, and it is coined as Seasonality. For instance, Enterovirus is the virus that recognized as it will be disseminated in the middle or later of summer to the beginning of autumn. Actually, for analyzing specific virus, its cycle time should be checked and the seasonal index should be estimated. Given an example as shown in Fig. 3, for the target area 3, we would want to calculate the adjacency areas' risks as well. By considering the time-series factor, the risk (R) value is calculated as

$$D_{(area)} = \{N_{(t,area)}, N_{(t+1,area)}, N_{(t+2,area)}, N_{(t+3,area)} .. N_{(12,area)}\} \tag{3}$$

where $D_{(area)}$ is the data collection of specific area's patients number, and

$$I_{(area)} = \{I_{(t,area)}, I_{(t+1,area)}, I_{(t+2,area)}, I_{(t+3,area)}.... I_{(12,area)}\} \tag{4}$$

where $I = N_{(specific\ month)}/N_{(year)}$. Therefore, the seasonality index is computed as:

$$S_{(area)} = \{S_{(t,area)}, S_{(t+1,area)}, S_{(t+2,area)}, S_{(t+3,area)}.... S_{(12,area)}\} \tag{5}$$

And,

$$R_{(Adjacency\ area)} = Average\ (D_{(Adjacency\ area)} \times I_{(target\ area)} \times S_{(target\ area)} \times P_{(Adjacency\ area)} \times (A_{(Adjacency\ area)})) \tag{6}$$

where N is the total number of observed case, P is the population density, and A is the spatial index of the surrounding level.

Month	Season	M_1	M_2	M_3	M_4	M_5	M_6
1	Winter	0	1	0	1	0	0
2	Winter	0	1	7	1	0	0
3	Spring	0	0	0	2	0	0
4	Spring	0	0	0	5	5	0
5	Spring	0	0	7	5	13	0
6	Summer	0	0	21	15	13	0
7	Summer	0	3	25	12	18	0
8	Summer	0	4	16	22	7	0
9	Autumn	0	2	27	18	8	0
10	Autumn	0	1	18	0	0	0
11	Autumn	0	0	0	0	8	0
12	Winter	0	1	2	3	0	0

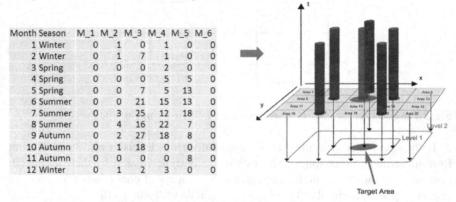

Fig. 3 Illustration of how the seasonality of the Enterovirus outbreaks is incorporated into the spatial-temporal computation model

3.2 Data Mining the Relationships among the Infested Areas

In order to extract the potential rules of co-relationship of inflected areas in a small city, some tasks are needed: 1.) Finding decision rules on the likeliness of having the disease outbreaks that happened simultaneously across multiple areas. 2.) Based on the decision rules, applying our pre-defined formulas to the evaluation of each area, so to determine whether the areas belong to those serious areas of disease outbreak or otherwise. 3.) Through the ranking of the evaluation, analyze the demographic information of different ranked areas across various periods. 4.) Predicting the trend line of disease outbreak and the prevalence rate of each zone, with options of mining deeper for the demographic information of co-effected areas.

3.2.1 Decision Rules Generation of Co-inflected Areas

Extracting decision rules from the disease cases is necessary at the beginning of our method. Above all, the city will be divided by the equal cells as a grid.

The disease outbreak in the small city will be simulated and using the data mining software to find out the co-relationship of this area. Two analyzing methods are used to extract the useful rules. They are decision tree and association rule analyzing models. They estimate the degrees of co-relations of the related areas and provide the measures on how reliable the rules are, in terms of Accuracy rate, lift, Confidence, Leverage, and Conviction. Moreover, for each analyzing model, various classifiers and associates rule miner will be applied, (J48, RadomTree, Apriori and HotSpot, etc.) for ensuring the fairness. Default parameters are assumed in each method.

After applying the various models and classifiers, many sets of rules are extracted. The subsequent step is to rank them, and judge on which rules are useful for further processing. In the decision tree model we can use the accuracy rate to decide which rules are acceptable or not. But for the association rules model, four performance results (Lift, Conf, Lev, Conv, in short) are considered. Assuming Lift = L, Conf = Cf, Lev = Lv, and Conv = Cv, *Max* is the maximum number of the total number of rules we selected, for the number i of Rule, a referenced Score can be calculated as:

$$Si = (L_i/L_{Max})/4 + (Cf_i/Cf_{Max})/4 + (Lv_i/Lv_{Max})/4 + (Cv_i/Cv_{Max})/4 \qquad (7)$$

As we are concerning the co-relationship of related inflected area, we opt to filter out any rule disqualified. The workflow of the whole is shown in Fig. 4.

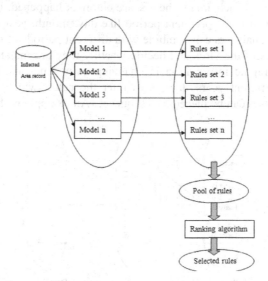

Fig. 4 Concept of rule generation

3.2.2 Analysis with Demographic Information

The likelihood of an area being infested, in addition to its adjacent neighbor areas, is determined by the demographic factors such as transportation, population density, mobility of the residents, and the vulnerability of the age group of those resides and travelers in the vicinity.

In the previous step, the analyzers can find which areas are relatively more directly impacted and their co-inflected areas as the same group of the disease outbreak. The inflected zone will be extended as like as they have some identical characteristics of these areas (it can also applied in the risk evaluation part of this research paper as finding the characteristics of the high risk areas), just like demographic information by evening the units of the habitant buildings/homes. Combing the analysis results of the demographic and geographic data, we merge information such as how many residents they are born in Macao, their age groups, how long they have stayed in the city, how many schools, human traffic flows etc. in each various area.

After that we will apply various classifiers of the decision tree model for predicting whether the area is highly impacted or potentially a highly impacted area. By using this method, some important attributes of the areas which may significantly impact the prevalence rate will be shown.

3.2.3   Trend-Line for the Prediction of the ST Data

Analyzers can apply this model into various time units as it is flexible to recognize the occurred cases of each area if the disease outbreak happened. Users can define various time unit of the experiment period like day, month, year or even decade. Nevertheless the analyzers can combine with different periods of result to identify the updated status of the virus and decide whether they are mutated or not. Likewise the demographic information is changing over time, and the system can combine with different years' factors to infer the rules, which might have appeared and the cases can be pieced as a sequence. An example is shown in Fig. 5.

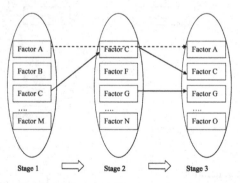

Fig. 5 An example of factors that form a trend

In this example, the factor G is found to have a strong co-relationship of the inflected areas no matter how much the time goes by. It can be illustrated that this factor should be paid more attention as it may be as important as the core factors to impact the disease outbreak. But for factor A it skips one stage and there exists an association relationship between stage 1 and stage 3. As for this kind of relationship, we call it the "gap" as the interval among different stages for evaluating how strong of these factors through frequency counting in the referenced years. We would observe if such trend has a relation to the prevalence of the virus over the years.

4   Experiment

In order to valid our proposed model, empirical data obtained from the Health Bureau of Macao Special Administrative Region of Macao, are used. The URL to the website is: http://www.ssm.gov.mo/statistic/2012/index.html. A total of 9 years of records which are the number of enterovirus infected patients are obtained in the past nine years. They are, chronologically, 218, 1023, 144, 822,1678, 1023, 1188, 2030 and 1017. A general regression trend indicates that the number critically increased over the years. The r-square value of the regression is 0.4114 which is not a bad fitting of forecast trend and the actual data. The rapid increase in recent year is quite uncommon. Allegedly it is concerned that the virus might have mutated and the disseminated rate probably escalates.

For the population, we assume a parameter called human mobility that is the sum of averaged tourists number over that area and the number of original habitants. In 2012, Macao population is 582,000. On the other hand, the total tourists' number is 28,082,292, especially for 13,577,298 tourists will stay overnight. In average there are at least 37,198 tourists stay in Macao every day. Such numbers are approximately divided into different areas in the grid model.

The objective is by conducting an experiment for the enterovirus surveillance, to predict the likely areas to be infested in the near future. Refer to the infected case of various areas in referenced period and its related population, the index is called the 'basic risk index" (Rbi) which calculates by the formula we develop. It illustrates how risky an area will be infected in accordance to their inflected history. The formula of Rbi is defined as:

$$Rbi = (N_i / N) \times (I / P) \tag{8}$$

where N_i is the number of years (can be months, or even years) that there are detected of infected cases in the target area. N is the surveillance period of time for the analysis. I is the number of infected cases that had been observed in the target area. P is the target area population in terms of human mobility; basically it will be assumed as how many people spend most of time here, regardless of tourist or permanent residents. After Rbi is calculated, the spiral model (as in Fig. 2) will be applied to calculate the new risk index which concerning the geographic position and other factors. The idea of spiral model is centered on the target area as the center point, then the effects of the adjacency areas which surround the target area are estimated accordingly with various levels of effect. Each level is inputted with a user-defined value as a parameter, such as level 1=0.9 or 0.7, level 2=0.6 and so on. Later on the level parameter will be multiplied by the risk index of the corresponding area that it associates with. Here assuming there are some residents who are already infested with enterovirus infection in the surveillance period, N_i. Further, each area will be set as a target area in turn, and the spiral computation traverses to re-calculate the risk index with reference to the other adjacent areas. In our case there are 30 areas in the grid of the city. There will be 30 target areas to be considered in turn by the spiral model to analyze the related risk index (S_r). Finally all the 30 risk indices would be computed in preparation for the next step – that is to judge according to some user-defined rules on whether this area is deemed as risky area.

For example, area 1 is set as the target area. By utilizing the spiral model to calculate the risk index in relation to its adjacent areas, we use the data set $D_r = \{ S_{r1}, S_{r2}, S_{r3}, S_{r4}, S_{r5} ... etc \}$. Then we use the ranking method to select the several highest risk area. The types of risks are predefined as the following Table 1.

Table 1 Descriptions of the levels of risks

	Over average Risk	Under average Risk	
		Target risk > 0	Target risk = 0
Over or reach 4/6	Breaking Point, High disseminated rate	Safe but potentially risky	Safe and observed area
Under 4/6	Breaking Point, Low disseminated rate	Safe and observed area	Safe and observed area

The computed results assist the user to determine this area is a center point of disease outbreak or potentially risky areas. Moreover, the risk standard as assumed in Table 1 is calculated by the averaging the risk index (A_R) of the area in the selected level and the mean of this list of number (M_R), the formula goes like this: the risk standard (S_R) = ($A_R + M_R$) / 2. From the computed results, analyzer can study specific target areas by their disease breaking point and the propagation rate, potential intermediate or even safety zone.

The operation steps of the spiral approach are as follow:

Step 1: Set each area as the target area,

Step 2: Compute the influential risk from the adjacent areas for the target area.

Step 3: By referring to the Table 1, identify the area type.

Step 4: Move to the next target area. Repeat step 1 until all the area is coved.

Through the spiral model the risk index for each area of the city is calculated. The corresponding color code of risk by Table 1 applies. The left hand side of Fig. 6 is an extract of Macao map over which is a translucent layer. The colored boxes simply indicate the number of residents contracted with enterovirus infection. However, the right side of Fig. 6 is a predicted risk distribution computed by the spiral model. It shows also the potential areas which will likely be infested, the next potential disease out-breaking point or safe zones, etc. Therefore personnel from CDC can well fine-tune their resources and pay attention to the predicted next outbreak areas.

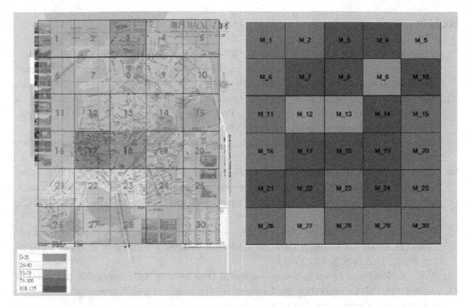

Fig. 6 Left: Observed cases of diseases over a city map. Right: Predicted next outbreak locations

5 Conclusion

In this paper we present two major processes for analyzing the risk evaluation and finding the co-relationship of areas which have occurrences of disease outbreak at the same time. We proposed a framework which combines in use of spatial and temporal factors for predicting disease outbreak in a small city. The method can be extended to detecting the earth quake, risk of the hotel occupancy rate and the other analyzing of finding the evaluation of which will concern the spatial and temporal factors. In order to validate the proposed model, an experiment is conducted for analyzing the spatial and temporal factors on empirical data of enterovirus infection in Macao. Through this method, certain areas can be identified as risky zones that have high spreading rate. Our model is quite different from the traditional spatial-temporal analyzing methods, such as it will not do the surveillance for the area which is not polygon as the city size like to be but for the equivalent "grid", Without the limitation of the polygon of the city size, the analysis is more flexible as the analyzer can do the investigation to evaluate the risk relationship among the target area and its adjacent areas. By using this grid scheme and the spiral calculation method, we can calculate the risk index of each area, and be able to identify areas that are of high or low risks.

References

1. Boulos, M.: Towards evidence-based, GIS-driven national spatial health information infrastructure and surveillance services in the United Kingdom. International Journal of Health Geographics 3(1), 1 (2004)
2. Raheja, V., Rajan, K.S.: Risk Analysis based on Spatio-Temporal Characterization - a case study of Disease Risk Mapping. In: Proceedings of the First ACM SIGSPATIAL International Workshop on Use of GIS in Public Health, pp. 48–56 (2012)
3. Ward, M.P.: Spatio-temporal analyzsis of infectious disease outbreaks in veterinary medicine: clusters. Hotspots and Foci, Vet Ital 43(3), 559–570 (2007)
4. Si, Y.L., Debba, P., Skidmore, A.K., Toxopeus, A.G., Li, L.: Spatial and Temporal Patterns of Global H5N1 Outbreaks. The International Archives of the Photogrammetry. In: Remote Sensing and Spatial Information Sciences, Beijing, vol. XXXVII (B2), pp. 69–74 (2008)
5. Pathirana, S., Kawabata, M., Goonatilake, R.: Study of potential risk of dengue disease outbreak in Sri Lanka using GIS and statistical modeling. Journal of Rural and Tropical Public Health 8, 8–17 (2009)
6. Yao, X.: Research Issues in Spatio-temporal Data Mining. In: Geographic Information Science (UCGIS) Workshop on Geospatial Visualization and Knowledge Discovery, Lansdowne, Virginia, pp. 1–6 (2003)
7. Sato, K., Carpenter, T.E., Case, J.T., Walker, R.L.: Spatial and temporal clustering of Salmonella serotypes isolated from adult diarrheic dairy cattle in California. J. Vet Diagn Invest. 13(3), 206–212 (2001)
8. Shu, H., Dong, L., Zhu, X.Y.: Mining fuzzy association rules in spatio-temporal databases. In: Proc. SPIE 7285, International Conference on Earth Observation Data Processing and Analysis (ICEODPA), 728541 (2008), doi:10.1117/12.815993.
9. Jin, Y., Dai, J., Lu, C.-T.: Spatial-Temporal Data Mining in Traffic Incident Detection. In: SIAM Conference on Data Mining, Workshop on Spatial Data Mining, pp. 1–5. Bethesda, Maryland (2006)
10. Mennis, J., Liu, J.W.: Mining Association Rules in Spatio-Temporal Data: An Analysis of Urban Socioeconomic and Land Cover Change. Transactions in GIS 9(1), 5–17 (2005)
11. Raheja, V., Rajan, K.S.: Risk Analysis based on Spatio-Temporal Characterization a case study of Disease Risk Mapping. In: Proceedings of the First ACM SIGSPATIAL International Workshop on Use of GIS in Public Health, pp. 48–56 (2012)

An Exploration of Mixture Models to Maximize between Class Scatter for Object Classification in Large Image Datasets

K. Mahantesh, V.N. Manjunath Aradhya, and C. Naveena

Abstract. This paper presents a method for determining the significant features of an image within a maximum likelihood framework by remarkably reducing the semantic gap between high level and low level features. With this concern, we propose a FLD-Mixture Models and analyzed the effect of different distance metrics for Image Retrieval System. In this method, first Expectation Maximization (EM) algorithm method is applied to learn mixture of Gaussian distributions to obtain best possible maximum likelihood clusters. Gaussian Mixture Models is used for clustering data in unsupervised context. Further, Fisher's Linear Discriminant Analysis(FLDA) is applied for $K = 4$ mixtures to preserve useful discriminatory information in reduced feature space. Finally, six different distance measures are used for classification purpose to obtain an average classification rate. We examined our proposed model on Caltech-101, Caltech-256 & Corel-1k datasets and achieved state-of-the-art classification rates compared to several well known benchmarking techniques on the same datasets.

Keywords: Image Retrieval System, Maximum likelihood clusters, EM algorithm, FLD, Similarity measures, Gausssian Mixtures.

K. Mahantesh
Department of ECE, Sri Jagadguru Balagangadhara Institute of Technology, Bangalore, India
e-mail: mahantesh.sjbit@gmail.com

V.N. Manjunath Aradhya
Department of MCA, Sri Jayachamarajendra College of Engineering, Mysore, India
e-mail: aradhya.mysore@gmail.com

C. Naveena
Department of CSE, HKBK College of Engineering, Bangalore, India
e-mail: naveena.cse@gmail.com

S.M. Thampi, A. Gelbukh, and J. Mukhopadhyay (eds.), *Advances in Signal Processing and Intelligent Recognition Systems*, Advances in Intelligent Systems and Computing 264, DOI: 10.1007/978-3-319-04960-1_40, © Springer International Publishing Switzerland 2014

1 Introduction

In this paper, we address the problem of statistical inference of classifying and recognizing the relevant images in the large image datasets. Images considered as high dimensional data subjected to random variation of intensity, scales, poses, and layouts has heaved many challenges in pattern recognition and machine learning applications. The crisis of developing good visual descriptors to reduce the semantic gap between low level and high level features for image retrieval system seems to be continuous and rigid [1]. Multiple Kernel Learning (MKL) in [2], combined set of diverse and complementary features based on color, shape and texture information to learn correct weighting of different features from training data thereby improving the classification rate.

Color, shape and texture features are the predominant visual features used for image representation and classification. Color plays a vital role in describing image content, the idea of color histograms along with histogram intersection and color moments can be seen in [3, 4]. Color Coherence Vector (CCV) partitions image into coherent and incoherent regions of histogram bins [5]. Color correlograms holds spatial information between pair of colors along with color distribution of the pixels [6]. Some remarkable color feature based Content Based Image Retrieval (CBIR) techniques can be found in [7, 8, 9]. Wang et al., computed structure elements histogram in quantized HSV color space creating spatial correlation between color & texture and derived structural elements descriptor for CBIR [10]. Texture features such as energy, inertia, entropy and uniformity are extracted using 8 connected neighbors in gray level co-ocurrene matrix [11]. Orthogonal polynomial model coefficients are restructured into multi resolution sub bands, edge features are extracted using gradient magnitudes of the sub bands and Pseudo Zernike moment was computed to generate global shape feature vector and are used for image retrieval [12].

A popular Bag of Features(BoF) model failed to identify the spatial order of local descriptor for image representation and was conquered by Spatial Pyramid Matching (SPM) by partioning image into segments and computing BoF histograms, thereby increasing the descriptive power for image representation and classification [13]. Jianchao et al., extended SPM using sparse codes thereby generalizing Vector quantization showed considerable improvement over linear SPM kernel on histogram in categorizing the sparsity of image data [14]. Boiman et al. [16] proposed Naive Bayes Nearest Neighbor (NBNN) and proved decrease in discriminativity of the data due to the quantization of descriptors in the BoF. The significant contribution made by Local NBNN in representing classes with greater reliability towards their posterior probability estimates can be seen in [17]. Turk and pentland were first to explore the use of PCA (Principal Component Analysis) in obtaining significant features for recognition by computing eigen vectors over high dimensional feature space which describes the data at the best but suffers at the varying intensity directions [15]. Since PCA operates in an unsupervised context with the absence of class label information, it maximizes within-class scatter which is considered to be redundant for the purpose of classification. To overcome this drawback, Fisher's Linear Discriminant Analysis (FLDA) was proposed [18]. To improve the performance of

PCA and LDA based recognition models, Gaussian Mixture Model is used to extract multiple set of features [19].

With the motivation by the aforesaid facts and also due to intrinsic benefit of integrating subspace methods with mixture models, we have explored FLD-Mixture Models to outline the scatter in order to make it more reliable in improving classification rate. The rest of the paper is structured as: Section 2 explains proposed FLD-Mixture Models. Section 3 presents experimental results and performance analysis. Conclusion and future work are drawn at the end.

2 Proposed Method

In this section, we propose a method based on FLD and mixture models. This is driven by the idea that the classification rate is improved by representing each class into a mixture of several features and by executing the classification in the compressed feature space [19]. To obtain this, each object class is partitioned into several clusters and each cluster density is estimated by a Gaussian distribution function in the FLD transformed space. The parameter estimation is performed by an iterative EM algorithm. Detail description is given in the following sub sections.

2.1 Fisher's Linear Discriminant Analysis(FLDA)

Linear Discriminant Analysis (LDA) with an objective to reduce the dimension of the data by preserving class discriminatory feature information by projecting the samples on to the line [20]. Fisher suggested maximizing the difference between the means, normalized by a measure of the within-class scatter leads to better classification [21]. let $x_1, x_2, \ldots, x_k$ be samples of $'k'$ set of images of $'c'$ classes. In order to select projection that gives large separation between the class and small variance within class we compute between-class scatter matrix and within-class scatter matrix as given in equations 1 & 2.

$$S_B = \sum_{i=1}^{c} N_i (\mu_i - \mu)(\mu_i - \mu)^T \tag{1}$$

$$S_W = \sum_{i=1}^{c} \sum_{x_k \in X_i} (x_k - \mu_i)(x_k - \mu_i)^T \tag{2}$$

where μ_i is the mean image of class X_i and N_i is the number of samples in class X_i. Recall that we are looking for the projection that maximizes the ratio of between class to within class, it has $C - 1$ dimensions and the objective function is defined as

$$W = argmax_w \frac{|W^T S_B W|}{|W^T S_W W|} = [w_1 w_2 \ldots w_m] \tag{3}$$

Where $w_1, w_2, \ldots, w_m$ is the set of generalized eigen vectors of S_B and S_W corresponding to the m largest generalized eigen values. Since the obtained within class covariance is singular, PCA is applied to reduce the dimension and later FLDA is used to achieve resulting nonsingular within-class scatter matrix S_w.

2.2   Gaussian Mixtures

The centers and variances of the Gaussian components along with the mixing co-efficients are considered to be as regulating parameters in the learning process of classification. Analysis of complex probability distribution is formulated in terms of discrete latent variables. Gaussian mixtures are used to find clusters in a set of data points [22]. Gaussian mixture distribution defined as weighted sum of $'M'$ Gaussian component densities written as:

$$p\left(\frac{X}{\lambda}\right) = \sum_{i=1}^{M} W_i * g\left(\frac{X}{\mu_i}, \Sigma_i\right) \tag{4}$$

where $W_i, i = 1, 2, 3, \ldots, M$ are the mixture weights and $g\left(\frac{X}{\mu_i}, \Sigma_i\right)$ are component Gaussian densities of D-variate Gaussian function.

2.3   FLD-Mixture Models

As mentioned in section 2.1, FLD projects high-dimensional data onto a line and perform classification in this one-dimensional space among overall classes. The transformed matrix (space) failed to clasp adequate amount of features for classifying complex data with many classes which are subjected to high intensity variations. To overcome this drawback, we use subspace mixture models [23] that uses several transformation matrices (also known as mixtures) among over all classes and developed FLD-Mixture Models for better image representation and classification.

In particular, we apply PCA to the set of mean m_i of each category with k-different mixtures and obtained cluster mean C_k, transformed matrix T_k & diagonal matrix U_k along with eigen values $'\lambda_{k_d}'$ as diagonal elements which is the dth largest eigen value of co-variance matrix. With these results we tend to obtain between-class scatter matrix and within-class scatter matrix for the kth mixture component which can be formulated as:

$$S_{Bk} = T_k U_k T_k^T \tag{5}$$

$$S_W = \sum_{l \in L_k} \frac{1}{n_l} \sum_{x \in c_i} (x - m_l)(x - m_l)^T \tag{6}$$

By considering equations 5 & 6, we compute transformation matrix W_k for the kth mixture component with an objective to maximize the following criterion function to obtain generalized eigen vectors corresponding to largest eigen values.

$$SJ_k(U) = \frac{|U^T S_B U|}{|U^T S_{W_k} U|} \tag{7}$$

In order to measure the similarity between feature vectors of trained and query (test) images for classification purpose, we have used six different distance metrics such as Manhattan, Eucledian, Minkowski, Modified Squarred Eucledian, Correlation coefficient based, Angle based distance measures.

3 Experimental Results and Performance Analysis

This section presents the results based on three widely used benchmarking image datasets: Caltech-101 [24], Caltech-256 [25] and Corel 1K [26] consisting of various object categories. Gray scale images are used as input and scaled to give a sensible number of features per image. Based on our preliminary experiments, we have witnessed that selecting $k = 4$ Gaussian mixtures are optimal in order to achieve competitive classification rates. We initialized the system by gathering training samples and vectorizing each image from 2-D to 1-D clusters.The parameter estimation is performed by an iterative EM algorithm and obtained estimated weights, estimated mean vectors, estimated covariance matrices, log likelihood of estimates of Gaussian mixtures for $k = 4$ mixtures and constructed feature vectors in reduced feature space. Further, for classification six different distance measures are used and considered as an individual run under each distance metric to measure the similarity between train feature dataset and query feature vector.

We followed standard experimental procedure [25] by dividing the entire dataset into train & test, with two different settings i.e. by randomly choosing 15 & 30 image/category separately as train dataset and the remaining considered to be as test. Experiment is repeated for six times with different distance metrics, and recorded an average of per class recognition under each category for all runs. classification rate for benchmarking datasets are given in the following subsections.

3.1 Caltech-101

Caltech-101 dataset consists of 9,144 images of 101 different categories (animals, aeroplanes, cars, flowers, human face, etc.), ranging from 31 to 800 images per category collected by Fei-Fei et al.[24]. In this dataset, most of the images are centered, few of them are occluded and affected by corner artifacts, and mostly one of the challenging dataset available with large intensity variations. We demonstrated our method on an entire Caltech-101 dataset by randomly selecting and training n_{labels} = 15 & 30 images per category individually and remaining unlabeled images as test. Our highest classification rate was obtained at $k = 4$ mixtures, where as Breakdown of classification rates for K= 2 & 3 mixtures for 15 train images are as shown in Table 1. In our Experiment, totally 11 classes and 16 classes reached 100% classification rate for 15 & 30 train image per category respectively and found to be the best compared to that of Locality-constrained Linear Coding technique mentioned in [27]. Figire 1 illustrates three out of those 16 classes which has atleast 10 test images. An experimental result exhibited in Table 2 proves that the proposed model is superior compared with the most popular techniques found in the literature considering similar dataset and experimental procedures.

Table 1 Breakdown of classification rates for K= 2 & 3 mixtures

K	Classification rate(%)
2	28.6
3	45.12

With reference to the Table 2, we noticed that the proposed FLD-Mixture Models in compressed feature space efficiently preserve the neighborhood structure of the dataset. Proposed mixture model conferred the leading result when compared to the conventional classifiers mentioned in [28, 29, 30] and found competitive with Bags of features [13] and nearest neighbor classifier methods [16],

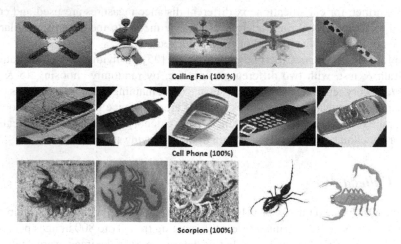

Ceiling Fan (100 %)

Cell Phone (100%)

Scorpion (100%)

Fig. 1 Few sample images of Caltech-101 dataset with 100 % classification rate

3.2 Caltech-256

Griffin et al. introduced a challenging set of 256 object categories containing a total of 30607 images [25]. Caltech-256 is collected by downloading google images and manually screening out which does not fit with the category. It presents high variations in intensity, clutter, object size, location, pose, and also increased the number

Table 2 Performance analysis for Caltech-101 dataset(Average of per class recognition for six runs in %)

Method	15 Training images/cat	30 Training images/cat
Serrre et al. [28]	35	42
Holub et al. [29]	37	43
Berg et al. [30]	45	-
Lazebnik et al. [13]	56.4	64.6
Griffin et al. [25]	59	67.5
Boiman et al. [16]	59	67.5
Proposed	**54.6**	**62.2**

of category with atleast 80 images per category. We have considered the standard experimental procedure mentioned in [25], trained 15 & 30 images per category respectively and rest are considered to be as the test images, and the overall classification rate is computed as the average of six runs. Each run signifies the diverse distance metric used for classification considering $k = 4$ mixtures(optimal value of K based on previous section results) and obtained result is summarized in Table 3. It is evident that, we have considered the entire set of images for testing purpose and seems to be competitive and promising compared to Local NBNN and vector quantization benchmarking techniques which uses not more than 50 images per category in testing phase [17, 14]. Figure 2 shows few samples of Caltech-256 dataset with high classification rate.

Fig. 2 Few classes of Caltech-256 dataset with high classification rate

Table 3 Performance analysis for Caltech-256 dataset (Average of per class recognition for six runs in %)

Method	15 Training images/cat	30 Training images/cat
Griffin et al. [25]	28.30	34.10
Sancho et al. [17]	33.5	40.1
Jianchao et al. [14]	28	34.02
Proposed	**27.2**	**33.56**

Note: we have considered the entire set of images for testing purpose, where as benchmarking techniques mentioned in [17, 14] uses not more than 50 images per category in testing phase.

3.3 Corel-1K

Corel-1K was initially created for CBIR applications comprising 1000 images classified into 10 object classes and 100 images per class. Corel-1K dataset containing natural images such as African tribal people, horse, beach, food items, etc [26]. We experimented the proposed method with $k = 4$ mixtures and considered 50 images per category as training and remaining 50 images for testing. With these the performance of the proposed method is evaluated and compared with the existing methods. Table 4 exhibits the results on Corel-1K dataset and highlighted improved discriminativity in terms of high classification rate of 89.20% in comparison with 77.90%, 86.20% & 88.40% mentioned in [31], [14] and [32] respectively. Figure 3 illustrates few classes of Corel-1K dataset which has achieved 100 % classification rate.

Table 4 Result performed on Corel-1K dataset

Method	Classification rate (%)
Lu et al. [31]	77.90
Jianchao et al. [14]	86.20
Gao et al. [32]	88.40
Proposed	**89.20**

Fig. 3 Few sample images of Corel-1K with 100 % classifiaction rate

4 Discussion and Conclusion

Despite the fact that there are wide variety of feature extraction techniques available in literature of image retrieval systems, the complex distributions over transformed space of observed and latent variables has gained lot of importance in pattern recognition and machine learning. In this paper, we have introduced FLD-Mixture Models to investigate the efficient training method and influence of linear combination of Gaussian mixture distributions to observe latent variables in finding the posterior distribution over objects. An effort to integrate over all possible positions and orientations of an object in reduced feature space and the choice of $K = 4$ mixtures is highly prejudiced and noteworthy to achieve better classification rate. Since Mixture models contains smaller number of components compared to training data, it results in faster evaluation during testing phase with increase in numerical stability. The use of different similarity distance measure techniques to obtain an average classification rate is first of its kind in the literature and affirmed to yield progressive result.

Experiments were conducted on different datasets with diverse distance metrics for improving classification rate and obtained leading results for correlation and angle based distances. Experimental results proves that the proposed model extracts significant features in compressed domain which outperforms generalized vector quantization model [14], spatial pyramid matching technique which conquered local descriptors for image representation and recognition [13], and NN classifier model for visual category recognition [30].In general, our method achieves extremely promising and state-of-the-art classification rate on benchmarking image datasets. In the near future, we would like to examine more about feature based transformation models with the impact of different distance metrics for conventional classifiers and also it is very crucial to analyze image dataset.

References

1. Bo, L., Ren, X., Fox, D.: Multipath Sparse Coding Using Hierarchical Matching Pursuit. In: IEEE International Conference on Computer Vision and Pattern Recognition (2013)
2. Lanckriet, G.R.G., Cristianini, N., Bartlett, P., Ghaoui, L.E., Jordan, M.I.: Learning the kernel matrix with semidefinite programming. JMLR 5, 27–72 (2004)
3. Swain, M.J., Ballard, D.: Indexing via color histograms. In: Proceedings of 3rd Int. Conf. Comput. Vision, pp. 11–32 (1991)
4. Stricker, M., Orengo, M.: Similarity of color images. In: Proc. SPIE, Storage Retriev Image Video Database, pp. 381–392 (1995)
5. Pass, G., Zabih, R., Miller, J.: Comparing images using color coherence vectors. In: Proceedings of 4th ACM Multimedia Con., pp. 65–73 (1997)
6. Huang, J., Kumar, S.R., Mitra, M.: Combining supervised learning with color correlograms for content-based image retrieval. In: Proceedings of 5th ACM Multimedia Conf., pp. 325–334 (1997)
7. Chen, L.T., Chang, C.C.: Color image retrieval technique based on color features and image bitmap. Inf. Process. Manage. 43(2), 461–472 (2007)
8. Ma, W.Y., Zhang, H.J.: Benchmarking of image features for content-based retrieval. In: Proceedings of the 32nd Asilomar Conference on Signals, Systems and Computers, vol. 1, pp. 253–257 (1998)
9. Lin, C.H., Huang, D.C., Chan, Y.K., Chen, K.H., Chang, Y.J.: Fast color-spatial feature based image retrieval methods. Expert Systems with Applications 38(9), 11412–11420 (2011)
10. Xingyuan, W., Zongyu, W.: A novel method for image retrieval based on structure elements' descriptor. J. Vis. Commun. Image R. 24, 63–74 (2013)
11. Xing-yuan, W., Zhi-feng, C., Jiao-jiao, Y.: An effective method for color image retrieval based on texture. Computer Standards & Interfaces 34, 31–35 (2012)
12. Krishnamoorthy, R., Sathiya Devi, S.: Image retrieval using edge based shape similarity with multiresolution enhanced orthogonal polynomials model. Digital Signal Processing 23, 555–568 (2013)
13. Lazebnik, S., Schmid, C., Ponce, J.: Beyond bags of features: Spatial pyramid matching for recognizing natural scene categories. CVPR 2, 2169–2178 (2006)
14. Yang, J., Yuz, K., Gongz, Y., Huang, T.: Linear Spatial Pyramid Matching Using Sparse Coding for Image Classification. In: IEEE CVPR, pp. 1794–1801 (2009)
15. Turk, M., Pentland, A.: Eigenfaces for recognition. Journal of Cognitive Neuroscience 3, 7186 (1991)
16. Boiman, O., Shechtman, E., Irani, M.: In defense of nearest-neighbor based image classification. In: IEEE CVPR, pp. 1–8 (2008)
17. Cann, S.M., Lowe, D.G., Bayes, L.N.: Nearest Neighbor for Image Classification. In: IEEE CVPR, pp. 3650–3656 (2012)
18. Belhumeur, P.N., Hespanha, J.P., Kriegman, D.J.: Eigenfaces vs fisherfaces: Recognition using class specific linear projection. IEEE Transactions on Pattern Analysis and Machine Intelligence 19, 711–720 (1997)
19. Kim, H.C., Kim, D., Bang, S.Y.: Face recognition using LDA mixture model. Pattern Recognition Letters 24, 2815–2821 (2003)
20. Martinez, A.M., Kak, A.C.: PCA versus LDA. IEEE Transactions on Pattern Analysis and Machine Intelligence 23, 228–233 (2001)
21. Fisher, R.A.: The statistical utilization of multiple measurements. Annals of Eugenics 8, 376–386 (1938)
22. Bishop, C.: Pattern Recognition and Machine Learning. Springer (2006)

23. Jordon, M., Jacobs, R.: Hierarchical mixtures of experts and the EM algorithm. Neural Computing 6, 181–214 (1994)
24. Fei-Fei, L., Fergus, R., Perona, P.: Learning generative visual models from few training examples: An incremental bayesian approach tested on 101 object categories. In: IEEE CVPR Workshop of Generative Model Based Vision (2004)
25. Griffin, G., Holub, A., Perona, P.: Caltech 256 object category dataset. Technical Report UCB/CSD-04-1366, California Institute of Technology (2007)
26. Wang, J.Z., Li, J., Wiederhold, G.: SIMPLIcity: Semantics-sensitive Integrated Matching for Picture LIbraries. IEEE Trans. on Pattern Analysis and Machine Intelligence 23(9), 947–963 (2001)
27. Wang, J., Yang, J., Yu, K., Lv, F., Huang, T., Gong, Y.: Locality-constrained Linear Coding for Image Classification. In: IEEE CVPR, pp. 3360–3367 (2010)
28. Serre, T., Wolf, L., Poggio, T.: Object recognition with features inspired by visual cortex. In: CVPR, San Diego (June 2005)
29. Holub, A., Welling, M., Perona, P.: Exploiting unlabelled data for hybrid object classification. In: NIPS Workshop on Inter-Class Transfer, Whistler, B.C. (December 2005)
30. Zhang, H., Berg, A.C., Maire, M., Malik, J.: SVM-KNN: Discriminative Nearest Neighbor Classification for Visual Category Recognition. In: CVPR, IEEE-2006, vol. 2, pp. 2126–2136 (2006)
31. Lu, Z., Ip, H.H.: Image categorization by learning with context and consistency. In: CVPR (2009)
32. Gao, S., Tsang, I.W.H., Chia, L.T., Zhao, P.: Local features are not lonely? laplacian sparse coding for image classification. In: CVPR (2010)

An Efficient Variable Step Size Least Mean Square Adaptive Algorithm Used to Enhance the Quality of Electrocardiogram Signal

Thumbur Gowri, P. Rajesh Kumar, and D.V. Rama Koti Reddy

Abstract. The main aim of this paper is to present an efficient method to cancel the noise in the ECG signal, due to various sources, by applying adaptive filtering techniques. The adaptive filter essentially reduces the mean-squared error between a primary input, which is the noisy ECG, and a reference input, which is either noise that is correlated in some way with the noise in the primary input or a signal that is correlated only with ECG in the primary input. The Least Mean Square (LMS) algorithm is familiar and simple to use for cancellation of noises. However, the low convergence rate and low signal to noise ratio are the limitations for this LMS algorithm. To enhance the performance of LMS algorithm, in this paper, we present an efficient variable step size LMS algorithms which will provide fast convergence at early stages and less misadjustment in later stages. Different kinds of variable step size algorithms are used to eliminate artifacts in ECG by considering the noises such as power line interference and baseline wander. The simulation results shows that the performance of the variable step size LMS algorithm is superior to the conventional LMS algorithm, while for sign based, the sign regressor variable step size LMS algorithm is equally efficient as that of variable step size LMS with additional advantage of less computational complexity.

Keywords: Adaptive filtering, Variable Step Size LMS, Signal to Noise Ratio.

Thumbur Gowri
Dept. of ECE, GIT, GITAM University, Visakhapatnam-530045, A.P, INDIA
e-mail: gowri3478@yahoo.com

P. Rajesh Kumar
Dept. of ECE, AUCE, Andhra University, Visakhapatnam-530003, A.P, INDIA
e-mail: rajeshauce@gmail.com

D.V. Rama Koti Reddy
Dept. of Inst. Tech., AUCE, Andhra University, Visakhapatnam-530003, A.P, INDIA
e-mail: rkreddy_67@yahoo.co.in

S.M. Thampi, A. Gelbukh, and J. Mukhopadhyay (eds.), *Advances in Signal Processing and Intelligent Recognition Systems*, Advances in Intelligent Systems and Computing 264, DOI: 10.1007/978-3-319-04960-1_41, © Springer International Publishing Switzerland 2014

1   Introduction

Biological signals originating from human body are usually very weak and so easily surrounded by background noise. To detect the heart disease problem, one need to know how best approach we are diagnosing the electrocardiogram (ECG) signal. The ECG signal gives important information related to the functionality of the heart condition. The electrical activity of the heart muscles can be measured by placing the electrodes on the surface of a body. The predominate artifacts of an electrocardiogram are: Power Line Interference (PLI), Muscle Artifact (MA), Electrode Motion artifact (EM), and Baseline Wander (BW). Along with these, the channel noise is added during transmission which will mask the microscopic features of ECG signal. These artifacts reduce the signal quality, frequency resolution and strongly effect the ST segment of the ECG. Sometimes these artifacts may produce large amplitude signals in ECG that can favour PQRST waveform and disguise tiny features in ECG, which are important for proper clinical monitoring and diagnosis. Hence it is very important to eliminate these noises for better clinical diagnosis.

Telecardiology is transmission of electrocardiogram data to a remote health monitoring centre that has the capability to interpret the data and affect decision-making. These wireless ECG biotelemetry implantable systems play a vital role when the measuring point is far from the monitoring place. For example, on the emergency medical helicopter, doctor scan hardly hear the heartbeat and respiratory sound of a patient due to blade flapping noise. For better diagnosis, it is necessary to eliminate artifacts in ECG signal to improve the accuracy and reliability. There are several methods have been implemented to reduce the noises in the ECG signal. The basic methods are non-adaptive filtering methods; these are static filters, because their coefficients are fixed which cannot reduce the time varying behaviour of the instrumentation noise. Different adaptive filtering methods have been developed to overcome the limitations in static filter.

Several approaches are reported in the literature regarding the filtering techniques that include both adaptive and non-adaptive; to denoise the power line interface and baseline wander in the ECG signal [1, 2, 3, 4, 5, and 6]. The adaptive filtering techniques are used to detect the periodic interference in the ECG signal. In [2], Salvador Olmos et al. proposed an Mean Square Error (MSE) convergence of the least mean square (LMS) algorithm using the adaptive orthogonal linear combiner when the reference input is a deterministic signal. In [3], Karthik et al. proposed a high resolution cardiac signal using different adaptive Least Mean Forth (LMF) algorithms, which will result a better steady state error compared to LMS.

The gradient based adaptive step size LMS algorithms and zero-attracting variable step size LMS algorithm leads to less mean square error which is due to the changing of the step size parameter [7, 8]. In [9, 10, 11], Rahaman et al. presented variations on LMS algorithm. Namely Block Leaky LMS, Block Normalized LLMS, Block Error Normalized LLMS and Normalized Sign Regressor LMS etc. algorithms. These algorithms are used to improve the capability of filtering and to eliminate artifacts in ECG in time domain as well as frequency domain with fixed step size. For less computational complexity

Normalized Sign Regressor LMS algorithm was used. In [12] Abadi *et al.* proposed Variable step size block least mean square (VSSBLMS) adaptive filter which will give the fast convergence rate than the BLMS adaptive filter. Hyun-Chool Shin *et al.* [13] proposed Variable Step Size Normalized LMS and affine projection algorithms, these algorithms led to improved convergence rate and lower misadjustment error.

In this paper, we propose an efficient Variable Step Size LMS (VSSLMS) algorithm for analysing the ECG signal. It gives not only a better convergence rate but also high signal to noise ratio when compared to LMS, Normalised LMS (NLMS) and Sign regressive normalized LMS (SRNLMS) algorithms; because in these algorithms the step size is fixed to that of the maximum Eigen value of the auto correlation matrix for the given reference input, but in the case of variable step size, step size is varied. To eliminate artifacts in ECG signal, we have presented various VSSLMS based adaptive structures such as Error Normalized Variable Step Size LMS (ENVSSLMS), Sign Variable Step Size LMS (SVSSLMS), Sign Regressor Variable Step Size LMS (SRVSSLMS) and Error Normalized Sign Regressor Variable Step Size LMS (ENSRVSSLMS) algorithms. Sign based algorithms provides the right solution with less complexity. When the order of the filter is increased, it gives better signal to noise ratio. The rest of the paper is organised as follows. In Section 2, we discuss about the basic LMS adaptive filter theory and then we propose efficient variable step size algorithms. In Section 3, we present the performance results for the proposed algorithms. Finally, conclusions are given in Section 4.

2 Adaptive Noise Cancelling

The block diagram of adaptive noise canceller system is depicted in Fig.1. In this system, primary input D of the adaptive filter is the combination of a desired signal S and a noise N. The reference input X_1 is equal to noise X. Noise N is derived from X after passing an unknown (channel) system.

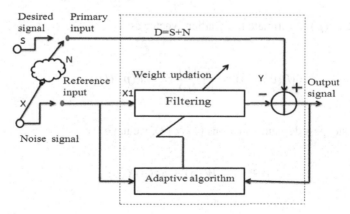

Fig. 1 Adaptive FIR filter

The reference input X_1 is filtered to produce an output Y, this output is then subtracted from the primary input to produce the system output (error signal), equals to S+N-Y. The main aim of adaptive algorithm is to estimate the transmission characteristics of the unknown system when X passes. After convergence, the weight vector of filter to be close to that of the unknown system, then Y becomes a close replica of N, and system output error signal is nearly equals to the source signal S.

2.1   Least Mean Square Algorithm

The most popularly used Least Mean Square algorithm has been applied to a large number of filtering applications such as biomedicine, speech processing, signal modeling and beam forming [14, 15]. The LMS algorithm is easy to compute and was introduced by Widrow and Hoff in 1959; they used a gradient-based method of steepest decent [16]. Consider a input vector $X(p)$, the output $d(p)$, the corresponding set of adjustable weight vector $W(p)$, and the estimated output $y(p)$ as $y(p) = X^T(p) W(p)$ the output estimated error $e(p)$ is $e(p) = d(p) - y(p)$. From the method of steepest decent, weight update equation for the adaptive noise canceller is

$$W(p + 1) = W(p) + \frac{1}{2}\mu[-\nabla(E\{e^2(p)\}].\tag{1}$$

Where μ is the step size parameter which controls the convergence rate, and $E\{.\}$ is the expectation value.

In order to find the minimum value of $e^2(p)$, choose R as $R = E\{X(p)X^T(p)\}$ and P_1 be defined as $P_1 = E\{d(p)X^T(p)\}$. Now, $E\{e^2(p)\}$ can be written as

$$E\{e^2(p)\} = E\{d^2(p)\} - 2P_1^T W(p) + W^T(p)RW(p).\tag{2}$$

So we use $e^2(p)$ to estimate the gradient vector as

$$\nabla[E\{e^2(p)\}] = \frac{\partial e^2(p)}{\partial W(p)} = -2e(p)X(p)\tag{3}$$

From the steepest descent equations (1) & (3), we have

$$W(p + 1) = W(p) + \mu e(p)X(p).\tag{4}$$

The equation (4) is the LMS algorithm, where $W(p)$ the present weight vector and $W(p + 1)$ is the updated weight vector. The parameter μ is a positive constant called step size of the input signal that controls both the rate of convergence and the stability of the algorithm. The value of the step size μ has to be smaller than $2/\lambda_{max}$ (or $2/tr[R]$), where λ_{max} is the highest Eigen value of the autocorrelation matrix R, and $tr[R]$ is denotes the trace of R. The step size rate μ can be selected as small as possible so that it gives a better result, but the training sequence takes longer time (learning rate).

2.2 Variable Step Size LMS Algorithms

It can be observed that the LMS algorithm updates the adaptive filter coefficients, by choosing a suitable μ which is of great importance to the convergence as well as the stability margin of the LMS algorithm. The results shows that, when smaller step size is chosen then the convergence rate is slower, which leads to low steady state misadjustmnet (SSM); on the other hand, if the larger step size is chosen, it leads to fast convergence rate and larger SSM. To solve this problem, the Variable step size LMS (VSSLMS) is used. VSSLMS algorithm uses a large step size in the initial stage to speed up the convergence rate and later the step size is gradually adjusted to a smaller value during the convergence, so that both the convergence rate and the misadjustment are better.

There are many Variable step-size LMS-based algorithms have been proposed in [17, 18], to improve the fundamental trade-off between convergence rate with less mean square error (MSE). This algorithm converges, as the step size decreases in value with the decreasing of MSE. But these algorithms results a high SSM and low signal to noise ratio (SNR). To improve this, an efficient variable step size LMS algorithm has been used. The weight update equation (5) for the VSSLMS algorithm can be written as

$$W(p + 1) = W(p) + \mu(p)e(p)X(p). \tag{5}$$

Where $\mu(p)$ is the variable step-size parameter and it is constant in LMS. In the Kwong's [17] the step size parameter is given as $\mu(p + 1) = \alpha\mu(p) + \gamma e^2(p)$, where α and γ are positive control parameters. The main motivation for this algorithm is strongly dependent on the additive noise, which reduces its performance for a low signal-to-noise ratio (SNR) environment. Aboulnasr's [18] algorithm includes a change in Kwong's Algorithm, by adjusting the step-size parameter considering the autocorrelation between $e(n)$ and $e(n - 1)$, instead of the square error $e^2(p)$, but Aboulnasr's algorithm, lag(1) error autocorrelation function could reduce the step-size value too early in some situations, resulting in a slower convergence. The proposed efficient variable step size is

$$\mu(p) = \frac{(1-\mu)}{2(1-(\mu)^{p+1})}, 0.5 < \mu < 1, \text{for } p=1, 2, 3, \ldots \ldots \tag{6}$$

where p is the occurrence number in the iteration filter. The equation (6) produces good estimation of fast convergence for the beginning signal occurrences, because the variable step size is large (for $p=1$ then $\mu(1) \cong 0.5$), and decreasing step size in the latter cases leads (for $p=1000$, $\mu=0.7$ then $\mu(1000) \cong 0.15$) to a less misadjustment error ratio, and it finally reaches to a steady state value. This decreasing step size is the main variation in the VSSLMS which is not possible in LMS. If μ value is nearer to one, then variable step size LMS gives better result.

Sign based VSSLMS algorithms that make use of the signum (polarity) of either the error or the input signal, or both have been derived from the LMS algorithm for the simplicity of implementation, and less computational complexity. The Sign Regressor Variable Step Size LMS (SRVSSLMS) weight updating vector (7) which will give less computational complexity than VSSLMS is given as

$$W(p + 1) = W(p) + \mu(p)e(p)sgn(X(p)). \tag{7}$$

Where 'sgn' is the well-known signum function, i.e.

$$sgn\{x(p)\} = \left\{ \begin{array}{l} 1: x(p) > 0 \\ 0: x(p) = 0 \\ -1: x(p) < 0 \end{array} \right\} \tag{8}$$

The Sign VSSLMS weight updating vector is

$$W(p + 1) = W(p) + \mu(p)sgn(e(p))(X(p)). \tag{9}$$

When the variable step size is normalised with reference to error, it gives good filtering capability, and this algorithm is called as Error Normalised Variable Step Size LMS (ENVSSLMS). This algorithm enjoys less computational complexity if we include signum function. The Error Normalized VSSLMS (10) and Error Normalised Sign Regressor Variable Step Size (ENSRVSSLMS) (11) weight update equation is as follows

$$W(p + 1) = W(p) + \frac{\mu(p)}{c + e(p)^T e(p)} e(p)X(p). \tag{10}$$

$$W(p + 1) = W(p) + \mu_s e(p) sgn(X(p)),\qquad(11)$$

where $\mu_s = \frac{\mu(p)}{c + e(p)^T e(p)}$.

The main drawback of adaptive noise cancellers in LMS based algorithms is the large value of excess mean square error which will produce signal distortion in the noise cancelled signal. In the ENVSSLMS and ENSRVSSLMS algorithms, the time-varying step-size is inversely proportional to the squared norm of error vector rather than the input data vector as in the Normalised LMS algorithm. This algorithm provides a significant improvement in reducing the mean squared error and consequently minimizing the signal distortion.

3   Performance Results

The performances of the LMS algorithm and different VSSLMS algorithms have been studied in the presence of an external reference signal. These methods have been validated using ECG recordings from MIT-BIH arrhythmia data base [19] with various morphologies. We have taken the non-stationary (BW) and stationary (PLI) noises, from the benchmark MIT-BIH arrhythmia database, as the reference for our work. The BW real noise is obtained from MIT-BIH Normal Sinus Rhythm Database (NSTDB). The arrhythmia database contains 23 records of the '100 series' and 25 records of the '200 series' ECG signals.

The MIT-BIH database consists of ECG recordings, which were obtained from 47 subjects, including 25 men aged 32-89 years, and 22 women aged 23-89 years (Records 201 and 202 came from the same male subject). Each record is slightly over 30 minutes of duration in length. Each signal file contains two signals sampled at 360 Hz (samples per second) with 11-bit resolution over a 10 m V range. The header file contains information about the leads used, the patient's age, sex, and medications.

3.1   Power Line Interference Cancellation

In our simulations we used the data base records of 102, 103, 104,105 and 108 to demonstrate power line interference (PLI) cancellation with a amplitude of 1mv and frequency 60Hz, and sampled at 200Hz. Using different algorithms, the simulation filtered results for record 105 are shown in Fig.2 and Fig.3 (due to space constraint only one record output is shown). For all the figures, number of samples is taken on x-axis and amplitude on y-axis, unless stated. In our experiment we used 4500 samples of ECG signal with random noise variance of 0.001 is added for evaluating different factors using different algorithms with length of the filter is five chosen.

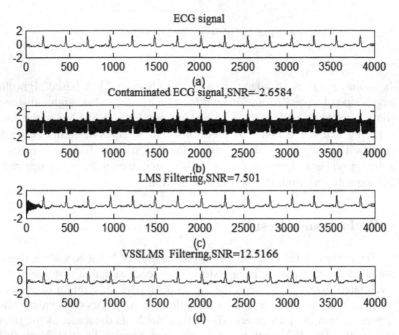

Fig. 2 (Rec. No.105) Typical filtering results of PLI cancellation (a) ECG signal, (b) ECG signal with real noise, (c) filtered signal with LMS algorithm, (d) filtered signal with VSSMS algorithm.

For evaluating the performance of the proposed adaptive filters, we measured the Signal to Noise Ratio (SNR) for different database records as shown in Table 1. In SNR measurements, it is found that VSSLMS algorithm gets better SNR with of 12.5166dB compared to all other algorithms; when less computations with closer SNR of VSSLMS are considered then we choose SRVSSLMS algorithm with SNR of 12.3055dB for the record No. 105 as shown in Table 1. Convergence characteristics of various algorithms are as shown in Fig 4. It is clear that VSSLMS algorithm gives better convergence rate.

The reference input signal and noise present in the primary input signal are perfectly correlated in the adaptive filter as shown in Fig. 1, then the output error signal is called as Minimum mean square error (MMSE) [20]. The steady state Excess mean square error (EMSEss) is the state when the adaptive filter converges and the filter coefficients no longer have significant changes [9]. The misadjustment (M) is a common factor which is a normalized measure of the noise in the filter output in the steady-state due to the fluctuations in the filter coefficients. The MMSE, MSE, EMSEss and misadjustment (M) factor, which all are compared for different algorithms and for different ECG records as shown in Table 2. For a VSSLMS the misadjustment is of 2.9871e-06, which is less value compared to all other algorithms.

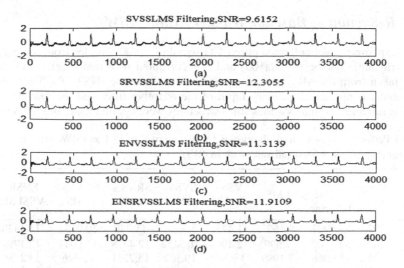

Fig. 3 (Rec. No.105) Typical filtering results of PLI cancellation (a) filtered signal with SVSSLMS algorithm, (b) filtered signal with SRVSSLMS algorithm, (c) filtered signal with ENVSSLMS algorithm, (d) filtered signal with ENSRVSSMS algorithm.

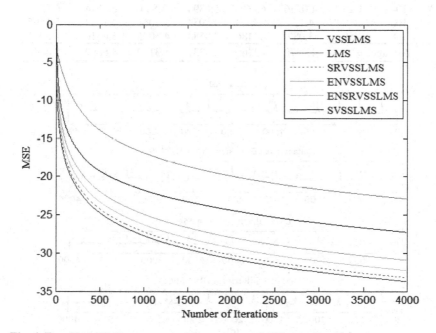

Fig. 4 (Rec. No.105) Convergence characteristics of different algorithms

3.2   Reduction of Baseline Wander Artifact (BW)

In our experiment, we collected pure ECG signal of 4500 samples from the MIT-BIH arrhythmia database (data105) and it is corrupted with real baseline wander (BW) taken from the MIT-BIH Noise Stress Test Database (NSTDB). The noisy ECG signal is applied at primary input of the adaptive filter and the real BW is given as reference signal. Various filter structures were applied using the LMS and

Table 1 Performance of various algorithms for the removal of PLI and BW artifacts from ECG signals interms of SNR (all values are in dB's)

Noise	Rec. No.	SNR Before Filtering	LMS	VSS-LMS	SVSS-LMS	SRVSS-LMS	ENVSS-LMS	ENSR-VSSLMS
	102	-3.9922	6.1605	11.4751	8.5652	11.749	10.9484	11.4328
	103	-2.5302	7.599	12.4904	10.3452	12.4478	11.6043	12.1799
PLI	104	-3.0066	7.1999	13.346	10.363	13.7211	12.626	13.2789
	105	-2.6584	7.501	12.5166	9.6152	12.3055	11.3199	11.9109
	108	-3.1827	7.0179	13.0331	7.9753	13.4234	12.3772	12.9233
	Average		7.0956	12.5733	9.3727	12.7293	11.7751	12.3451
	102	1.2501	4.7583	4.446	5.5043	5.7964	3.9073	5.754
	103	1.2499	4.2512	6.1088	4.5097	5.5705	5.4215	5.4943
BW	104	1.25	4.0326	6.663	5.6894	5.894	6.2574	6.5271
	105	1.2499	4.156	4.8663	4.5034	5.2064	4.1083	5.1595
	108	1.2501	3.087	4.0486	3.9584	4.1025	3.6516	4.2971
	Average		4.0570	5.2265	4.8330	5.3139	4.6692	5.4464

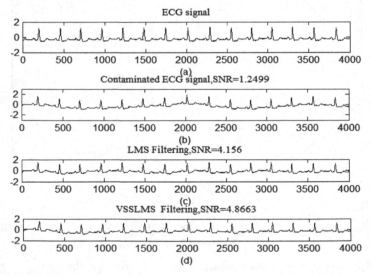

Fig. 5 (Rec. No.105) Typical filtering results of BW cancellation (a) ECG signal, (b) ECG signal with real noise, (c) filtered signal with LMS algorithm, (d) filtered signal with VSSMS algorithm.

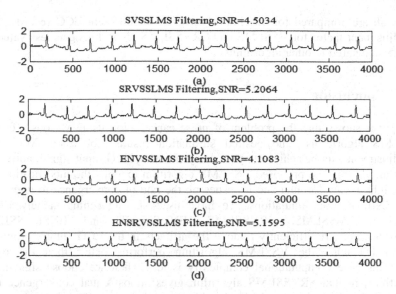

Fig. 6 (Rec. No.105) Typical filtering results of BW cancellation (a) filtered signal with SVSSLMS algorithm, (b) filtered signal with SRVSSLMS algorithm, (c) filtered signal with ENVSSLMS algorithm, (d) filtered signal with ENSRVSSMS algorithm.

Table 2 (Rec. No.105) Comparison of MMSE, MSE, steady state Excess MSE (EMSEss) in dB, and Misadjustment (M) factor for various algorithms with a variance of 0.001

Noise	Algorithm	MMSE	MSE	EMSEss	M
PLI	LMS	0.1456	0.0046	-46.6643	1.4803e-04
	VSSLMS	0.1456	3.7823e-04	-63.6187	2.9871e-06
	SVSSLMS	0.1456	0.0017	-33.8374	0.0028
	SRVSSLMS	0.1456	4.2550e-04	-59.5205	7.6687e-06
	ENVSSLMS	0.1456	7.2023e-04	-53.4830	3.0794e-05
	ENSRVSSLMS	0.1456	5.2705e-04	-58.9693	8.7064e-06
BW	LMS	0.1456	0.0216	-17.8437	0.1128
	VSSLMS	0.1456	0.0156	-22.1879	0.0415
	SVSSLMS	0.1456	0.0184	-19.2846	0.0810
	SRVSSLMS	0.1456	0.0133	-22.1876	0.0415
	ENVSSLMS	0.1456	0.0221	-18.9664	0.0871
	ENSRVSSLMS	0.1456	0.0136	-22.6695	0.0371

different VSSLMS algorithms to study the relative performance and results are plotted in Fig.5 and Fig.6. As shown in Table 1, on an average SRVSSLMS algorithm gets SNR value as 5.3139dB, where as ENSRVSSLMS gets 5.4464dB for less computational complexity equivalent SNR choose SRVSSLMS. As shown in Table 2, Mean square error (MSE), Minimum mean square error (MMSE), Steady state Excess mean square error (EMSEss) and misadjustment (M) factor, which all are compared for different algorithms for different ECG records. The misadjustment factor for VSSLMS and SRVSSLMS algorithms gets less value of 0.0145, compared to other algorithms.

4 Conclusion

This paper shows that the problem of noise cancellation using various efficient VSSLMS algorithms. The desired simulation results, for SNR, MSE and misadjustment, can be achieved by properly choosing ECG input signal, noise and better algorithm. The proposed VSSLMS algorithm exploits the modifications in the weight update formula and it pushes up the rate of convergence to high with high SNR. In our simulations, we have used six algorithms namely LMS, VSSLMS, SVSSLMS, SRVSSLMS, ENVSSLMS and ENSRVSSLMS algorithms, and are efficiently remove the artifacts in the ECG signal. From the simulated results, the VSSLMS algorithm performs better than the other algorithms. The computational complexity is less when we choose sign based algorithms, in that SRVSSLMS algorithm gives almost equal convergence rate with that of variable step size LMS algorithm.

References

1. Li, M., Zhao, J., Zhang, W., Zheng, R.: ECG signal base line filtering and power interference suppression method. In: 2nd International Conference on Information Science and Engineering (ICISE), pp. 207–209 (2010)
2. Olmos, S., Laguna, P.: Steady-state MSE convergence of LMS adaptive filters with deterministic reference inputs with applications to biomedical signals. IEEE Transactions on Signal Processing 48(8), 2229–2241 (2000)
3. Karthik, G.V.S., Sugumar, S.J.: High resolution cardiac signal extraction using novel adaptive noise cancellers. In: 2013 International Multi-Conference on Automation, Computing, Communication, Control and Compressed Sensing, pp. 564–568 (2013)
4. Ku, C.T., Hung, K.C., Wu, T.C., Wang, H.S.: Wavelet-Based ECG Data Compression System With Linear Quality Control Scheme. IEEE Transactions on Biomedical Engineering 57(6), 1399–1409 (2010)
5. Wang, X., Meng, J.: A 2-D ECG compression algorithm based on wavelet transform and vector quantization. Digital Signal Processing 18(2), 179–188 (2008)
6. Koike, S.: Analysis of adaptive filters using normalized signed regressor LMS algorithm. IEEE Transactions on Signal Processing 47(10), 2710–2723 (1999)
7. Ang, W.P., Farhang-Boroujeny, B.: A new class of gradient adaptive step-size LMS algorithms. IEEE Transactions on Signal Processing 49(4), 805–810 (2001)
8. Salman, M.S., Jahromi, M.N.S., Hocanin, A., Kukrer, O.: A zero-attracting variable step-size LMS algorithm for sparse system identification. In: International Symposium, I.X. (ed.) IX International Symposium on Telecommunications (BIHTEL), pp. 1–4 (2012)

9. Rahman, M.Z.U., Karthik, G.V.S., Fathima, S.Y., Lay-Ekuakille, A.: An efficient cardiac signal enhancement using time-frequency realization of leaky adaptive noise cancellers for remote health monitoring systems. Measurement 46(10), 3815–3835 (2013)

10. Rahman, M.Z.U., Ahamed, S.R., Reddy, D.V.R.K.: Efficient and simplified adaptive noise cancellers for ECG sensor based remote health monitoring. IEEE Sensors Journal 12(3), 566–573 (2012)

11. Rahman, M.Z.U., Shaik, R.A., Reddy, D.V.R.K.: An Efficient Noise Cancellation Technique to Remove Noise from the ECG Signal Using Normalized Signed Regressor LMS Algorithm. In: 2009 IEEE International Conference on Bioinformatics and Biomedicine, pp. 257–260 (2009)

12. Abadi, M.S.E., Mousavi, S.Z., Hadei, A.: Variable Step-Size Block Least Mean Square Adaptive Filters. In: First International Conference on Industrial and Information Systems, pp. 593–595 (2006)

13. Shin, H.C., Sayed, A.H., Song, W.J.: Variable step-size NLMS and affine projection algorithms. IEEE Signal Processing Letters 11(2), 132–135 (2004)

14. Lalith kumar, T., Soundara Rajan, K.: Noise Suppression in speech signals using Adaptive algorithms. International Journal of Engineering Research and Applications (IJERA) 2(1), 718–721 (2012)

15. Gorriz, J.M., Ramirez, J., Cruces-Alvarez, S., Puntonet, C.G., Lang, E.W., Erdogmus, D.: Novel LMS Algorithm Applied to Adaptive Noise Cancellation. IEEE Signal Processing Letters 16(1), 34, 37 (2009)

16. Haykin, S.: Adaptive Filter Theory, 4th edn. Prentice Hall, Upper Saddle River (2002)

17. Kwong, R.H., Johnston, E.W.: A Variable size LMS algorithm. IEEE Trans. Signal Proc. 40(7), 1633–1642 (1992)

18. Aboulnasr, T., Mayyas, K.: A robust variable size LMS type algorithm: analysis and simulation. IEEE Trans. Signal Processing 45(3), 631–639 (1997)

19. MIT-BIH noise stress test database,
http://www.physionet.org/physiobank/database/nstdb

20. Diniz, P.S.: Adaptive Filtering: Algorithms and Practical Implementation, 4th edn., pp. 57–60. Springer, Kluwer Academic Publisher, Norwell (2002)

Human vs. Machine: Analyzing the Robustness of Advertisement Based Captchas

Prabaharan Poornachandran, P. Hrudya, and Jothis Reghunadh

Abstract. A program that can ensure whether the response is generated by a human, not from a machine, is called CAPTCHA. They have become a ubiquitous defense to protect web services from bot programs. At present most of the visual CAPTCHAs are cracked by programs; as a result they become more complex which makes both human and machine finds it difficult to crack. NLPCAPTCHAs, an acronym for Natural Language Processing CAPTCHA, are introduced to provide both security and revenue to websites owners through advertisements. They are highly intelligent CAPTCHAs that are difficult for a machine to understand whereas it's a better user experience when compared to traditional CAPTCHA. In this paper we have introduced a method that is able to analyze and recognize the challenges from three different types of NLPCAPTCHAs.

Keywords: NLPCAPTCHA, CAPTCHA, Tesseract, OCR.

1 Introduction

The word CAPTCHA is an abbreviation of Completely Automated Public Turing Test to Tell Computers and Humans Apart .They act as an interrogator in the Turing test [1] to conclude whether the person sitting in the front of the computer is person or machine. Any user entering the correct answer is considered to be human. They are used to provide security functionality in online polls, prevent email spam, automatic account registration etc [2].

Prabaharan Poornachandran · P. Hrudya · Jothis Reghunadh
Amrita Center for Cyber Security, Amrita Vishwa Vidyapeetham,
Kollam-690525,India
e-mail: praba@amrita.edu,
       {hridyathul,jothisreghu}@gmail.com

S.M. Thampi, A. Gelbukh, and J. Mukhopadhyay (eds.), *Advances in Signal Processing*   477
and Intelligent Recognition Systems, Advances in Intelligent Systems and Computing 264,
DOI: 10.1007/ 978-3-319-04960-1_42, © Springer International Publishing Switzerland 2014

But, a question arises why we need to distinguish human from machine. Consider an online poll that selects the best institute in the state. If a hacker is able to come up with a program that can generate thousands of vote within a minute to a specific institution then that institute will win the poll [2]. So to avoid such invalid votes from bot program a challenge is given to the person (or machine) to crack CAPTCHA to ensure that the vote is not generated by a machine since machines are unable to solve image challenges.

Most of the Class 1, 2, 3 and 4 CAPTCHAs were cracked using different segmentation and recognition techniques [3, 4], as a result CAPTCHAs are designed with more distortions and slanted characters that makes harder to both machines and humans. Since CAPTCHAs are designed in an annoying way, people are unable to access the web services which results in increasing the conversion rate of the website [4]. It is also found that non-native speakers are slower and less accurate in solving traditional CAPTCHAs [13]. NLPCAPTCHAs (Natural Language Processing CAPTCHA) are highly intelligent CAPTCHA that are easily understood by human but extremely hard for an AI. Instead of giving distorted characters as in traditional CAPTCHA, NLPCPATCHAs provide user friendly puzzles to users which includes identify the colored word or the underlined word from the image or it can be an game [18] etc. Fig 1 [5] shows an example for NLPCAPTCHA.

Fig. 1 Example of NLPCAPTCHA

The main idea behind this paper is to compromise NLPCAPTCHA using image processing techniques. Here we have introduced a method to analyze and recognize the challenges from three different types of NLPCPATCHAs. The rest of the paper is organized as follows related work about CAPTCHA is explained in Section 2. Section 3 deals with NLPCAPTCHA and why it is preferred over traditional CAPTCHA. The methodology used for cracking NLPCAPTCHA is explained in Section 4. Section 5 describes about the experimental results and finally conclusion is discussed in section 6.

2   Related Work

CAPTCHAs are security features that are added by web services to prevent it from bot programs. To prevent bot programs CAPTCHAs are made more complex, but most of the visual CAPTCHAs are broken using different methods. Some of the works related to CAPTCHAs are explained as follows.

In [3] our previous work we have classified CAPTCHA as Class 1 CAPTCHAs (CAPTCHA with complex background), class 2 (BotBlock and Botdetect CAPTCHAs), Class 3 and 4 contains much more complex CAPTCHAS, like google and gimpy CAPTCHAS. We mainly focused on breaking overlapped CAPTCHAs using DFT with 80% success ratio.

In [4] Han-Wei Liao, Ji-Ciao Lo, Chun-Yuan Chen proposed a method to crack ReCAPTCHA. They implemented two algorithms, bisect ReCAPTCHA and overlap template for preprocessing. They used Template Matching algorithm to recognize characters. They also provide examples of complex CAPTCHA.

In [7] Jason Ma, Bilal Badaoui, Emile Chamoun, proposed a Generalized Method to Solve Text-Based CAPTCHAs. They used learning algorithms to classify each character. And also they developed an algorithm to separate the individual characters using support vector machines and optical character recognition algorithms. Jiquing Song, Zuo Li, Michael R.Lyu and ShijieCai in [8] proposes a merged character segmentation and recognition method based on forepart prediction, necessity-sufficiency matching and character adaptive matching.

In [9] Manisha B. Thombare, Parigha M. Derle, ArpitaK.Wandre explains about designing complex CAPTCHA which are harder for OCR (Optical Character Recognition) to recognize. In [10] Ahmad S El Ahmad, Jeff Yan, MohamadTayara proposed a method to crack those CAPTCHA that are resistant to segmentation mechanism. The attack was able to exploit shape pattern found in individual characters, as well as connection patterns of adjacent characters.

Ryan Fortune, Gary Luu, and Peter McMahon in [11] proposed methods for line removal, noise removal and clustering based segmentation techniques. In [12] Philippe Gole proposed a SVM to crack Asirra CAPTCHA which relies on the problem of distinguishing images of cats and dogs. The classifier was able to solve 12 Asirra images with probability 10.3%.

In [13] ElicBursztein, Steven Bethard, Celine Fabry, John C. Mitchell, Dan Jurafsky, they found that CAPTCHAs are often difficult for humans, with audio captchas being particularly problematic. They also found that non-native speakers of English are slower in general and less accurate on English-centric CAPTCHA schemes. In [14] Jennifer Tam, Sean Hyde, Jiri Simsa, Luis Von Ahn analysed the security of audio CAPTCHA and achieved correct solution for the test samples with accuracy upto 71%.

Jeff Yan, Ahmad Salah El Ahmad in [15] explained several methods to crack visual CAPTCHA. They explained the method of dictionary attack and snake segmentation algorithm in their paper. PWNtcha [16] is an excellent web site which demonstrates the inefficiency of many CAPTCHA implementations. It briefly comments on the weaknesses of some simple CAPTCHAs, which were claimed to be broken with a success ranging from 49% to 100%.

In [18] SuhasAggarwal ,Indian Institute Of Technology, Guwahati, India, discussed about Animated CAPTCHAs which can be used for advertising.

According to the paper animated CAPTCHAs are more fun and an interesting medium for advertisements. These CAPTCHAs were easy for people from different age groups and increased positive advertisement impressions.

In[19] DagaoDuan, MengXie ,Qian Mo, ZhongmingHan,Yueliang Wan proposed a method for implementing Hough transform that employs the "many-to-one" mapping and sliding window neighborhood technique to reduce the computational and storage load. The proposed method had a better performance than the previous variations of Hough transform. In[20] Takimoto, R.Y, Chalella das Neves, A. ; Mafalda, R. ; Sato, A.K. ; Tavares, R.S. ; Stevo, N.A. ; de Sales Guerra Tsuzuki, M. presented a modified Hough transform for detecting function patterns implemented with interval arithmetic. The results obtained were much better when compared with the conventional Hough transform implementation. Bing Wang and Shaosheng Fan in [21] proposed an improved Canny Edge Detection by replacing the Gaussian filter with a self-adaptive filter. The results of experiment showed an improved Canny edge detection algorithm.

Breaking all kinds of NLPCAPTCHAs is still need more research. We have tried to break three major NLPCAPTCHAs and could do it successfully.

3   NLPCAPTCHA

The Internet, as a global communications medium, provides advertisers with unique and often cost-effective ways of reaching advertising audiences. Internet advertising offers a large number of possible formats, including pop-ups, ads including audio and video, classified advertising and e-mail marketing etc. But these advertisements lack attention due to clustering of other advertisements [5]; as a result we don't notice the advertisement of our favorite brand when we visit the website. In addition to that we close the pop up advertisements even before it loads. One solution to grab attention is through NLPCAPTCHAs. NLPCAPTCHAs are CAPTCHAs with advertisement along with security features. They are highly intelligent CAPTCHAs that are easy for a human to understand whereas they are hard for a machine (software) to crack.

The advantages [5] of using NLPCAPTCHAs include 1.Gureented attention 2. New stream of revenue 3.better enhanced security 4. Better user experience. 5. Quickest way to know what people think about a brand or in other words it provides a free survey to know what people think about a brand (fig 2 [5]).

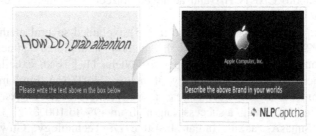

Fig. 2 Using NLPCAPTCHA to know what people think about a brand

4   Methodology

In the following sections we have explained methods to break three different NLPCAPTCHA. NLPCAPTCHA that asks users to enter 1.Specific colored word from the image (figure 3) 2. Underlined word from the image (figure 4) 3. Blinking character from the image (figure 5).

Fig. 3 Find word with specific color

 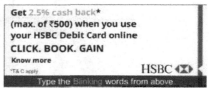

Fig. 4 Find the underlined word   **Fig. 5** Find the blinking word

4.1   *Methodology-To Break NLPCAPTCHA of type1*

The steps include different phases such as pre-processing, splitting, color recognition and text extraction. This method focuses to crack NLPCAPTCHA of certain type that is shown in figure 1. The methodology is implemented in C++ with the aid of Open Computer Vision (OpenCV) version 2.4.5 and Tesseract [6]. Figure 6 is a diagrammatic representation of different phases.

4.1.1   Preprocessing

In preprocessing phase, the input NLPCAPTCHA image is enlarged. The enlarged image may be blurred therefore the image is converted into smoothened image using Gaussian filter [17].

4.1.2   Splitting Image

The enlarged image is then separated into two images type.jpgand nlp.jpg (figure 10) .The former contains how to solve NLPCAPTCHA and latter contains the advertisement. The splitting is done using ROI (Region of Interest) function in Opencv.

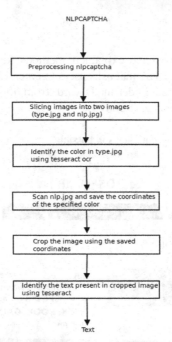

Fig. 6 Breaking NLPCAPTCHA type 1

4.1.3 Color Recognition

In this phase, type.jpg is converted to grey scale image [2].This makes the Tesseract, an open source OCR (Optical Character recognition),   easy to recognize the text that is present in the image [6].

Using the extracted text, we then searched for the specified color. Once the color is identified we call the corresponding procedure to identify the color. Following algorithm identifies yellow.

Algorithm 1 Algorithm to identify yellow pixels

$Pixels = \{p_1, p_2...p_n\}$
for all p_i in $Pixels, i = 1..n$ **do**
 if $p_i.red\_component \geq 212$ **then**
  if $p_i.green\_component \geq 193$ **then**
   if $p_i.blue\_component \leq 110$ **then**
    if $minposition \leq i$ **then**
     $minposition = i$
    end if
   end if
  end if
 end if
end for
$maxposition = loc(p_i)$
Crop the image using minposition and maxposition

4.1.4   Text Extraction

Using the procedure, the rectangular coordinates of the specified color is saved which are then used to crop the portion of the image containing the specified color. The cropped image is then converted to grey scale image before it is given to Tesseract. The OCR extracts the text from the cropped image which is then displayed as plaintext.

4.2   Methodology to Break NLPCAPTCHA of Type 2

The preprocessing steps that are discussed in 4.1.1 and 4.1.2 are same for type 2. After preprocessing, the image is converted to binary image then Canny edge detection algorithm is applied to detect edges in figure 4.The figure 7 shows the result after applying Canny edge detection.

Fig. 7 Canny edge detection

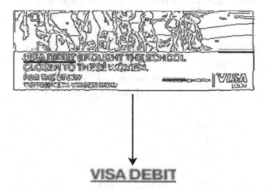

Fig. 8 Identifying target line

The Hough Transform [19, 20] method is useful for detecting any parametric curves (eg lines, conics etc.). It can detect line segments of different angles from the given edge points in the image. But we are concerned only about the underlined words (horizontal lines) which can be detect by setting Θ to 90° in the polar coordinates of the line given by the equation $\rho = x\cos\Theta + y\sin\Theta$,where ρ is the distance of the line from the origin. Instead of this method, we did a simple approach to detect the group of collinear points along the horizontal axis. For that, a horizontal scan is performed on figure 7. From our understanding we found that, we need at least 40 collinear points of black pixels to underline a single character and based on this assumption we have identified the lines in the image(figure 8),which is obtained after applying Canny edge Detection algorithm. The problem that can occur is that, it may classify some fake lines (eg: line at the top of a rectangle) as target line. In the figure 8 the blue lines represent the lines that are recognized by the method. Now we need to identify which line among the identified line is the actual line. For that we performed a horizontal search for black pixel (text pixels) starting from the line to top of the image. For example, if two end points of the line are (10, 30) and (60,30) then we perform a horizontal scan from (10,29) to (60,29),(10,28) to (60,28) up to (10,20) to (60,20). If there are no black pixels occur within 10 pixels above the line, then the line under consideration is a not the target line. Once we identify the target line we have obtained the coordinate position (two end points of the line) of the line. The coordinates of the line form the bottom two coordinates of the region of interest. Next we need to identify the remaining two coordinates to identify the region of interest (target word).For that, we executed a horizontal scan for white pixels (background pixels) from the line to the top like we did for identifying the target line. The red line in the fig 8 shows the portion where we have identified all pixels of same pixel values (white) that corresponds to the background of the image. Now we have obtained the four coordinates of the region (ends points of red and blue line) which then can be used to crop the region of interest from the image. This is explained as follows

---

Algorithm 2 Algorithm to identify target line

$LineCoord = \{line_1, line_2..line_n\}$
$Targetlines\{..\} = NULL$
for all $line$ in $LineCoord$ **do**
　　repeat
　　　　$Temp = LineCoord_y - 1$
　　　　if $LineCoord_x \& Temp \neq white$ **then**
　　　　　　$Targetlines = LineCoord$
　　　　　　break
　　　　end if
　　until $LineCoord_y - 10$
end for
if $Targetlines \neq NULL$ **then**
　　for all $line_i$ in $Targetlines$ **do**
　　　　repeat
　　　　　　$line\_start = line_i\_startcoord$
　　　　　　$line\_finish = line_i\_endcoord$
　　　　　　$line\_endcoord = line\_finish - 1$
　　　　until There is no black pixles from line_start
　　　　and line_finish
　　end for
end if

Methodology to Break NLPCAPTCHA of Type 3

To identify the blinking word from the image, we took screenshots of the image at different time intervals (Figure 9). The images are compared and stored the pixel positions where intensity value varies. Using the coordinates, the region (blinking) is extracted from the image containing more number of black pixels in the region of interest. For example, suppose in t1 at (20, 10) the pixel value corresponds to white whereas it is black in t3. Now we have obtained the first coordinate of region of interest. Similarly we have found the remaining three coordinates. Using the four coordinates we cropped the portion of the image (either t3 or t4) that has more black pixels (text pixels) in the region of interest.

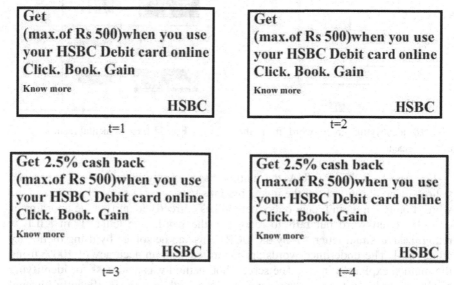

Fig. 9 Screen shots at different time interval

5 Experimental Results

Figure 10 shows the method of identifying the specified colored word. In this example, the challenge is to find the green word from the advertisement[5]. The method extracted the word with 100% accuracy. The table(figure 11) shows the experimental result of some NLPCAPTCHAs. Image 1 asks the user to identify yellow colored word from the image, images 4,5,6 and 7,8 asks the user to identify red and green word from the image ,respectively. Image 2 and 3 corresponds to type 2 and type 3 NLPCAPTCHA.

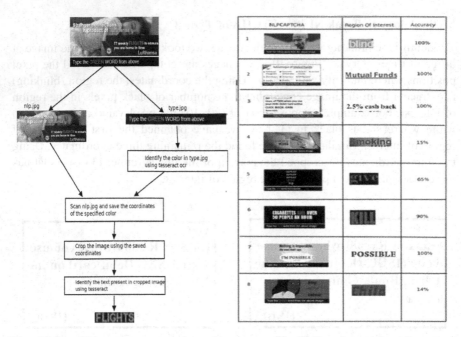

Fig. 10 Identifying green word from the   **Fig. 11** Experiemental results
advertisement

For type 1 NLPCAPTCHA, the method has correctly identified red, green and yellow colored region from the image but fails to recognize some characters in the word. For example, in image 6 the method has correctly identified the coordinates of red colored word but fails to recognize the word. The letter 'I' in KILL is recognized as small letter 'L' by the OCR. This can be solved by using dictionary attack [15]. The underlined words are identified with an accuracy of 100% using the method explained in 4.2. The screenshot method was a success for identifying blinking words. In type 1, the accuracy of the method was significantly lowered when the specified color is also present in the background. For example in figure 13(a), the user is asked to identify blue color word from the image but the blue pixels are also present in the background. The region of interest obtained from the method (figure 12) is difficult for the OCR to recognize the target word.

The accuracy is calculated based on the formula:

$$Accuracy = \frac{number\ of\ corrrectly\ recognized\ characters}{Total\ number\ of\ characters}$$

Fig. 12 Region of interest obtained from fig.12(a)

6   Conclusion and Future Work

CAPTCHA can ensure whether the response is generated by a human, not from a machine. They have become a ubiquitous defense to protect web services from bot program. But most of the Class 1, 2, 3 CAPTCHAs are cracked by programs; as a result CAPTCHAs are designed with more distortions and slanted characters that makes harder to both machines and humans. NLPCAPTCHA are CAPCTHA with advertisements. They are complex CAPTCHA for a machine but it's easy for a human to crack .A method for attacking NLPCAPTCHA like in fig 1 that we have discussed in this work can be summarized into following procedure: 1) Preprocessing NLPCAPTCHA which includes enlarging the image followed by smoothing 2) Splitting the image 3)Color recognition using OCR 3) Extracting the text with specified color and displaying the result.

To identify a line, the edges of the image are identified using Canny edge detection method. A line is identified based on the assumption that if the horizontal scan finds 40 continuous black pixels. NLPCATPCHA to identify blinking character is obtained by comparing different screenshot of the image taken at different time. Though NLPCAPTCHAs are introduced for better security, they are vulnerable to bot attacks using the methods that are discussed in the paper.

The future works include NLPCAPTCHA with complex background (figure 13(a)). There are NLPCAPTCHAs like figure 13(b) that require better methods to recognize the target word. Some NLPCAPTCHA asks users to enter the brand name of a particular product from the advertisement (figure 2), which requires complex machine learning algorithms.

Fig. 13 Complex NLPCAPTCHA

Acknowledgement. We are so grateful to many a people who have helped and supported us for the completion of this project. We express our deepest sense of gratitude to Dr.Krishnasree Achuthan, Director of TBI, for believing in our potential and providing us an opportunity to work for Amrita Center for Cyber Security with adequate facilities. Also we would like to thank our Institution, Amrita School of Engineering, Amrita Vishwa Vidyapeetham, Amritapuri Campus for their help and support.

References

1. Cui, J.S., Wang, L.J., Mei, J., Zhang, D., Wang, X., Peng, Y., Zhang, W.Z.: CAPTCHA design based on moving object recognition problem. In: 2010 3rd International Conference on Information Sciences and Interaction Sciences, ICIS (2010)
2. Chandavale, A.A., Sapkal, A.M., Jalnekar, R.M.: Algorithm To Break Visual CAPTCHA. In: Second International Conference on Emerging Trends in Engineering and Technology, ICETET 2009 (2009)
3. Hrudya, P., Gopika, N.G., Poornachandran, P.: Forepart based captcha segmentation and recognition using dft
4. Liao, H.-W., Lo, J.-C., Chen, C.-Y.: NTU CSIE CMLAB, NTU CSIE CMLAB, T NTU CSIE CMLAB "GPU catch breaker"
5. Official website of nlpcaptcha,
   http://www.nlpcaptcha.in/en/index.html
6. Patel, C., Patel, A., Patel, D., Patel, C.M.: Optical Character Recognition by Open Source OCR Tool Tesseract: A Case Study Institute of Computer Applications (CMPICA) Charotar University of Science and Technology(CHARUSAT). International Journal of Computer Applications 55(10), 975–8887 (2012)
7. Ma, J., Badaoui, B., Chamoun, E.: A Generalized Method to Solve Text-Based CAPTCHAs
8. Song, J., Li, Z., Lyu, M.R., Cai, S.: Recognition of merged characters based on forepart prediction, necessity-sufficency matching and character adaptive masking. IEEE Transactions on Systems, Man, and Cybernetics, Part B: Cybernetics 35(1) (February 2005)
9. Thombare, M.B., Derie, P.M., Wandre, A.K.: Integrating-Random-Character-Generation-to-Enhance-CAPTCHA-using-Artificial-Intelligence. International Journal of Electronics, Communication and Soft. Computing Science and Engineering 2(1) (April 2012)
10. El Ahmad, A.S., Yan, J., Tayara, M.: The Robustness of Google CAPTCHAs. School of Computer Science Newcastle 4University, UK
11. Fortune, R., Luu, G., Mc Mahon, P.: Cracking Captcha
12. Golle, P.: Machine Learning Attacks against the Asirra CAPTCHA. In: Proceeding, C.C.S. (ed.) 2008 Proceedings of the 15th ACM Conference on Computer and Communications Security, pp. 535–542. ACM, New York (2008)
13. Bursztein, E., Bethard, S., Fabry, C., Mitchell, J.C., Jurafsky, D.: How Good are Humans at Solving CAPTCHAs? A Large Scale Evaluation
14. Tam, J., Hyde, S., Simsa, J. Von, L.: AhnComputer Science Department Carnegie Mellon University,"Breaking Audio CAPTCHAs"
15. Yan, J., El Ahmad, A.S.: Breaking Visual CAPTCHAs with Naïve Pattern Recognition Algorithms. In: Twenty-Third Annual Computer Security Applications Conference, ACSAC 2007, School of Computing Science, Newcastle University, UK, December 10-14 (2007)
16. Yan, J., El Ahamad, A.S.: A low-cost attack on a Microsoft Captcha. In: Proceedings of the 15th ACM Conference on Computer and Communications Security, CCS 2008. ACM, New York (2008)

17. Deng, G., Cahill, L.W.: An adaptive Gaussian filter for noise reduction and edge detection. In: Nuclear Science Symposium and Medical Imaging Conference, IEEE Conference Dept. of Electron. Eng., La Trobe Univ., Bundoora, Vic., Australia (1993)
18. Aggarwal, S.: Animated CAPTCHAs and games for advertising. In: Proceeding WWW 2013 Companion Proceedings of the 22nd International Conference on World Wide Web, International World Wide Web Conferences Steering Committee Republic and Canton of Geneva, Switzerland, pp. 1167–1174 (2013)
19. Duan, D., Xie, M., Mo, Q., Han, Z., Wan, Y.: An Improved Hough Transform for Line Detection. In: 2010 International Conference on Computer Application and System Modeling (ICCASM), vol. 2 (October 2010)
20. Takimoto, R.Y., Chalella das Neves, A., Mafalda, R., Sato, A.K., Tavares, R.S., Stevo, N.A., de Sales Guerra Tsuzuki, M.: Detecting function patterns with interval Hough transform. In: 2010 9th IEEE/IAS International Conference Industry Applications, INDUSCON (November 2010)
21. Wang, B., Fan, S.: An Improved CANNY Edge Detection Algorithm. In: Second International Workshop on Computer Science and Engineering, WCSE 2009, Department of Electrical and Information Changsha Univ. of Sci. & Technol., vol. 1 (October 2009)

Content-Based Video Copy Detection Scheme Using Motion Activity and Acoustic Features

R. Roopalakshmi and G. Ram Mohana Reddy

Abstract. This paper proposes a new Content-Based video Copy Detection (CBCD) framework, which employs two distinct features namely, motion activity and audio spectral descriptors for detecting video copies, when compared to the conventional uni-feature oriented methods. This article focuses mainly on the extraction and integration of robust fingerprints due to their critical role in detection performance. To achieve robust detection, the proposed framework integrates four stages: 1) Computing motion activity and spectral descriptive words; 2) Generating compact video fingerprints using clustering technique; 3) Performing pruned similarity search to speed up the matching task; 4) Fusing the resultant similarity scores to obtain the final detection results. Experiments on TRECVID-2009 dataset demonstrate that, the proposed method improves the detection accuracy by 33.79% compared to the reference methods. The results also prove the robustness of the proposed framework against different transformations such as fast forward, noise, cropping, picture-in-picture and mp3 compression.

1 Introduction

In the present Internet era, due to the exponential growth of multimedia sharing activities, enormous amount of video copies are proliferating on the Internet and causing huge piracy issues. For instance, according to Canadian Motion Picture Distributors Association (CMPDA)-2011 report [1], 133 million pirated movies are watched in Canada in 2010. Therefore, video copy detection techniques are essential in order to solve piracy as well as copyright issues. Two standard methodologies are used in the literature for detecting the copies of a video content namely, digital watermarking and Content-Based video Copy Detection (CBCD) [2]. The primary goal of any CBCD system is, to detect video copies by utilizing the content based features of the media [3].

R. Roopalakshmi · G. Ram Mohana Reddy
NITK, Surathkal, Mangalore, India-575025
e-mail: {roopanagendran2002,profgrmreddy}@gmail.com

S.M. Thampi, A. Gelbukh, and J. Mukhopadhyay (eds.), *Advances in Signal Processing*   491
and Intelligent Recognition Systems, Advances in Intelligent Systems and Computing 264,
DOI: 10.1007/978-3-319-04960-1_43, © Springer International Publishing Switzerland 2014

In the CBCD literature, existing works can be broadly classified into global and local descriptor techniques. For example, Hua et al. [4] proposed ordinal measure based video fingerprints for detecting the duplicate video sequences. Hoad et al. [5] introduced color-shift and centroid based descriptors to detect the illegal video contents. Though global descriptors are effective for frame level transformations; yet, they are less robust against region-based attacks such as picture-in-picture.

Due to the limited capability of global descriptors, many researchers introduced local features, which compute a set of interest points to facilitate local matching. For instance, SIFT descriptor [6] is widely popular in the CBCD domain, which is partially invariant to lighting changes. Although local descriptors are more robust against region based transformations; still, their computational cost is high when compared to that of global features.

Though motion activity features contribute an essential information about a video content; yet, in the CBCD literature, motion signatures are considered as poor descriptors [7], due to these reasons: (a) They are very close to zero values, when captured at normal frame rates (25-30fps); (b) Noisy nature of raw motion vectors. However, Tasdemir and Cetin [8] attempted to employ motion features for their CBCD task, by capturing frames at a lower rate (5fps) and using motion vector magnitudes. Though, the proposed signatures are resistant to illumination and color changes; still, they fail to describe the spatio-temporal motion activity of the video.

On the other hand, audio content is a significant information source of a video; yet, only very few attempts are made to detect video copies using acoustic signatures. For instance, Itoh et al. [9] utilized acoustic power features for their copy detection task. Although this method is efficient; still, its high computational cost degrades the performance of the system. In [10], the authors utilized both the visual and audio features for detecting duplicate videos, but the performance of their method is limited to global transformations.

2 Motivation and Contributions

State-of-the art CBCD methods are primarily utilizing visual features to detect the video copies. However, it is stated in CBCD literature that, copy detection using only visual features may not be efficient against various video transformations [11]. From another perspective, exploiting audio fingerprints for the CBCD task is compulsory, because of the these reasons: i) audio content constitutes an indispensable information source of video; ii) From copy detection point of view, in most of the illegal camcorder captures, the audio content is less affected compared to the visual data [10]. Furthermore, exploiting several complementary features of the video for the copy detection task, not only improves the detection performance, but also widens the coverage to more number of video transformations. Based on these aspects, this paper proposes a new copy detection framework, which integrates motion activity features and audio spectral descriptors to detect the video copies.

3 Proposed CBCD Framework

The proposed copy detection framework is shown in Fig. 1 and the relevant notations are described in Table 1. In the proposed framework, first the motion activity features and audio spectral signatures are extracted from the master and query video frames. Then the resultant signatures are further processed and the corresponding Motion Activity (MA) and Spectral Descriptive (SD) words are computed. The MA words comprehensively represent the overall motion activity, whereas the SD words summarize audio profile of the given video sequence. After this step, K-means clustering technique is employed in order to obtain the low-dimensional representation of MA and SD words of video contents. The resulting cluster centroids are stored as fingerprints of video sequences. Finally, clustering based pruned similarity search is performed and consequently the detection results are reported.

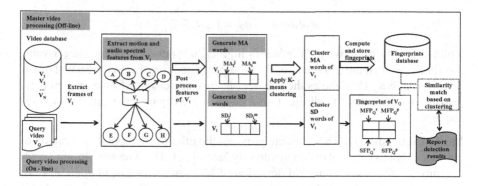

Fig. 1 Block diagram of the proposed copy detection framework

Table 1 Description of notations used in Fig. 1

Notation	Description	Notation	Description
N	Number of master videos	G	Audio spectral roll-off value
V_i	i-th master video in database	H	Spectral flux of audio signal
A	Intensity of motion activity	MA_i^j	j-th MA word of V_i, $j = \{1,2,3,...,m\}$
B	Number of active regions	SD_i^j	j-th SD word of V_i, $j = \{1,2,3,...,m\}$
C	Dominant direction of activity	SFP_Q^r	r-th spectral fingerprint of V_Q
D	Mean motion vector magnitude	MFP_Q^r	r-th motion fingerprint of V_Q
E	Audio spectral centroid	m	Number of MA/ SD words of V_i
F	Energy of spectrum	p	Number of video signatures of V_Q

3.1   Problem Definition

Let $R = \{x_i | i = 1, 2, ..., m\}$ be the reference video, where x_i is the i-th frame of reference video. Let $Q = \{y_j | j = 1, 2, ..., n\}$ be the query clip, where y_j is the j-th frame of the query clip and $m >> n$. Here Q is derived by applying different types of video transformations such as blurring and cropping to one or more subsequences of R. Our goal is to quickly and accurately identify the reference video (Here it is R), by comparing the features of R and Q. In the proposed copy detection framework, the motion activity and audio spectral features are extracted as follows.

3.2   Motion Activity Features Extraction

MPEG-7 Motion activity descriptor captures the intensity of activity or pace of action in a video segment [13], which includes the following attributes:

$$Motion\ activity = \{I, D, S, T\} \tag{1}$$

where intensity of activity (I) indicates high or low intensity by an integer value and direction attribute (D) specifies the dominant direction of activity. Spatial distribution of activity (S) denotes the number of active regions in a frame and temporal distribution attribute (T) indicates the variation of activity over the duration of video sequence.

Motion Intensity (I): This attribute provides the effective temporal description of the video shot in terms of different intensity levels [12]. The Average Motion Vector magnitude (AMV) and Standard deviation of Motion Vector magnitude (SMV) of the frame can be calculated as given by,

$$AMV = \frac{1}{MN} * \sum_{i=1}^{M} \sum_{j=1}^{N} mv(i, j) \tag{2}$$

$$SMV = \sqrt{\frac{1}{MN} * \sum_{i=1}^{M} \sum_{j=1}^{N} |mv(i, j) - AMV|^2} \tag{3}$$

Where $mv(i, j)$ indicate the motion vector magnitude of (i, j)-th block and MN is the frame size. The proposed framework utilizes SMV of blocks to compute the motion intensity. In [14], the motion activity values are quantized into 1-5 range, as per the MPEG-7 standard [12]. However, the quantization thresholds are suggested mainly for MPEG-1 videos. But, our video database includes various file formats (i.e. MPEG-1,2,4); hence, we experimented different threshold values and finally utilized the new threshold values given in Table 2 for implementing the proposed copy detection task.

Spatial Distribution of Activity (S): This attribute indicates, whether the activity is spread across many regions or confined to one region [14]. The segmentation of

Table 2 New quantization thresholds used in the proposed CBCD task

Activity value	Range of SMV
1	$0 \leq SMV < 2.5$
2	$2.5 \leq SMV < 9.7$
3	$9.7 \leq SMV < 16.1$
4	$16.1 \leq SMV < 24.4$
5	$24.4 \leq SMV$

frame into $n \times n$ regions, plays a significant role in predicting the accurate number of active regions in a given frame. In order to handle this discrepancy, we experimented our data set with different n values ranging from 2-5. Maximum accuracy(85.36%) is achieved at $n=3$; thus, we computed spatial distribution of activity of frames by segmenting it into 3×3 regions.

In the proposed CBCD framework, the algorithm described in [15] is enhanced, so that it includes the detailed description of *Spatial Activity Matrix (SAM)* and *Mean Motion Distribution (MMD)* calculations. More specifically, the proposed algorithm, which calculates the spatial distribution of activity in the frame in terms of number of active regions is described as follows.

Algorithm: Calculation of Spatial Distribution of Activity

1: Calculate Spatial Activity Matrix (*SAM*) of frame F using the equation,

$$SAM(F) = \begin{cases} mv(i,j) & if \ mv(i,j) \geq AMV \\ 0 & otherwise \end{cases} \qquad (4)$$

where $mv(i,j)$ is the motion vector magnitude of block (i,j), such that i = { 1, 2, 3, ..., M} and j ={ 1, 2, 3, ..., N}. This SAM computation of F retains only high activity blocks of F.

2: Segment $SAM(F)$ into 3×3 non overlapping blocks of size $W \times H$, where $W = \frac{M}{3}$ and $H = \frac{N}{3}$. The motion activity of k-th region R_k of F is computed as,

$$R_k(F) = \sum_{x=1}^{W} \sum_{y=1}^{H} B_m(x,y) \qquad (5)$$

where $k \in \{1, 2, 3, ..., 9\}$, B_m is the m-th block of R_k and $m \in \{1, 2,.., W \times H\}$.

3: Extract Mean Motion Distribution (*MMD*) of R_k of F as follows,

$$MMD(R_k(F)) = \frac{\sum_{x=1}^{W} \sum_{y=1}^{H} B_m(x,y)}{W \times H} \qquad (6)$$

4: Sort the *MMD* values of all regions of a frame in the ascending order.

5: Regions with higher *MMD* values ($MMD \geq \alpha \times AMV$, where α is set as 2.4) are considered as the active regions of the given frame.

Dominant Direction of Activity (D): Here, the goal is not to calculate the exact direction of motion of all objects, but to compute the approximate dominant direction of activity in the frame, so that the robustness of the CBCD task can be improved. Precisely, in the proposed CBCD framework, the dominant direction of activity is calculated as illustrated in [15]. More precisely, the proposed method calculates the *direction vector (DIR)*, that indicates the total amount of motion in four major directions including up, down, left and right as given by,

$$DIR = \{Up, Down, Left, Right\} \tag{7}$$

The highest value of *DIR* represents the dominant direction of activity in the given frame. Direct processing of resultant motion activity features in terms of MA words is computationally expensive. In order to solve this issue, compact representation of MA words are generated using clustering technique. Precisely, we experimented MA words with two widely used clustering methods, namely K-means and Fuzzy C-means techniques. Table 3 compares the representative MA words generated by the two clustering techniques for 1 to 60 minutes video sequences respectively. Results from Table 3 indicate the better performance of K-means clustering technique, compared to the other method; hence, it is utilized in the proposed CBCD task.

Table 3 Comparison of total amount of extracted MA words

Duration of video files (in minutes)	Dimension of resultant MA words			Total reduction using KC(in %)
	No clustering	K-means(KC)	Fuzzy C-means(FC)	
1 - 15	1515	74	83	95.11
16 - 30	3732	154	142	95.87
31 - 60	4419	197	229	95.54

3.3 Audio Spectral Descriptors Extraction

In the proposed framework, the audio spectral descriptors including Spectral Centroid, Energy, Roll-off and Flux are extracted from the short term power spectrum of the audio signals, as described in [15]. More precisely, the audio signal is down sampled to 22050Hz and segmented into 11.60ms windows with an overlap factor of 86% using Hamming window function [15]. Then the spectral descriptors are extracted as follows.

Spectral Centroid (SC): This centroid is a timbral feature, which describes the brightness of a sound signal [16]. In the proposed framework, mean frequency distribution values are computed as described in [15], and consequently utilized as spectral centroid features for the CBCD task.

Spectral Energy (SE): This descriptor describes the envelope of the input signal in terms of amplitude values. In the proposed copy detection framework, we employed the squared magnitude of audio samples to extract the spectral energy descriptors, as illustrated in [15].

Spectral Roll-off (SR): This feature is referred as the skew present in the shape of power spectrum i.e. frequency boundary, where 85% of the total energy of spectrum resides [16]. Precisely, SR is calculated as follows,

$$SR = \sum_{k=0}^{R} x^d[k] = 0.85 \sum_{k=0}^{N-1} x^d[k] \tag{8}$$

where $x^d[k]$ indicates the magnitude of frequency components.

Spectral Flux (SF): This descriptor is popularly known as the delta magnitude spectrum, which defines the amount of energy differences between the consecutive analysis frames [16]. The SF descriptors is extracted as follows [15],

$$SF = |x_f^d[*] - x_{f-1}^d[*]| \tag{9}$$

where $x^d[*]$ represents magnitude of frequency components and $f, f-1$ indicate current and previous frames respectively. In order to obtain the compact representation of SD words, we experimented our dataset with K-means and Fuzzy C-means clustering techniques. Table 5 proves the better performance of K-means compared to Fuzzy C-means technique; hence, it is used in the proposed CBCD task.

Table 4 Comparison of total amount of extracted SP words

Duration of videos (in minutes)	Dimension of resultant SP words			Total reduction using KC(in %)
	No clustering	K-means(KC)	Fuzzy C-means(FC)	
1 - 15	30308	531	547	98.24
16 - 30	74645	615	624	99.17
31 - 60	88379	686	672	99.22

3.4 Fingerprints Matching

In the proposed system, motion activity and spectral descriptors of video sequences are grouped into clusters using K-means technique. The value of K ranges from 55-317 based upon the individual video contents. Let $R(k)$ and Q be the kth reference and query video clips, then mp_r and mp_q are their corresponding motion activity features. Then, the similarity Sim_{MO} between the motion activity features of $R(k)$ and Q is computed using Comparative Manhattan distance metric as follows,

$$Sim_{MO}(R_k, Q) = \sum_{i=1}^{m} \sum_{j=1}^{n} \frac{|mp_r(i) - mp_q(j)|}{|mp_r(i)| + |mp_q(j)|} \tag{10}$$

where m and n indicate the size of motion fingerprints of $R(k)$ and Q respectively. The Sim_{MO} scores are evaluated against the predefined confidence measure(CM_1), ranging between 0.50-0.75. We observed better detection results for 0.60 threshold, which noticeably reduces false positive rates; thus, it is used in the proposed CBCD task.

Consider sp_r, sp_q be the spectral fingerprints of videos $R(k)$ and Q respectively. Then, the similarity Sim_{SP} between spectral features of $R(k)$ and Q is computed using squared Euclidean distance measure as follows,

$$Sim_{SP}(R(k), Q) = \sum_{i=1}^{a} \sum_{j=1}^{b} (sp_r(i) - sp_q(j))^2 \tag{11}$$

where a and b indicate the size of spectral fingerprints of $R(k)$ and Q respectively. The Sim_{SP} scores are evaluated against the confidence threshold (CM_2), which is set us 0.69 based on experimental results. After this step, the final similarity score $Final_{SS}$ between $R(k)$ and Q is computed as,

$$Final_{SS} = \begin{cases} 1 & if\ Sim_{MO} \geq CM_1\ \&\ Sim_{SP} \geq CM_2 \\ 0 & otherwise \end{cases} \tag{12}$$

Based upon $Final_{SS}$ scores, the copy detection results are reported.

4   Experimental Setup and Results

4.1   Reference Database and Query Dataset

The proposed framework is evaluated on TRECVID-2009 Sound & Vision dataset [17]. More precisely, the video database includes 200 hours of video covering a wide variety of content. Since, motion vectors are efficient when they are captured at lower frame rates [8]; hence, we experimented different frame rates ranging from 4-10. In the proposed scheme, 6fps is used for extracting motion features because of its high detection accuracy. Table 5 represents the list of visual and audio transformations used in the proposed CBCD task.

For experimentation purpose, 45 video clips are selected from the reference database. Different types of transformations listed in Table 5 are applied to the 45 video clips, where the duration of clips vary between 30-45s. The resulting 765 (45×17) video sequences are used as query clips for the proposed copy detection task. Each query video is used to detect the corresponding video sequence in the

Table 5 List of transformations used in the proposed framework

Type	Category	Description
T1	Fast forward	Double the video speed
T2	Slow motion	Halve the video speed
T3	Color change	Changing color spectrum
T4	Blurring	Blurring by 20%
T5	Brightness change	Increase brightness by 25%
T6	Noise addition	Add 15% random noise
T7	Pattern insertion	Insert text pattern into selected frames
T8	Moving caption insertion	Insert moving titles into entire video
T9	Cropping	Crop top & bottom frame regions by 15% each
T10	Picture-inside-picture	Insert smaller resolution picture into selected frames
T11	Mp3 compression	Change audio file format
T12	Single band compression	Compress only specific frequency band
T13	Multi band compression	Compress different frequency bands independently
T14	Combination of 3	Cropping by 18%, 20% of noise & moving caption

reference database. To measure the detection accuracy of the proposed scheme, we used standard performance metrics given by,

$$Precision(P) = TP/(TP + FP) \tag{13}$$

$$Recall(R) = TP/(TP + FN) \tag{14}$$

$$F\text{-}Measure = (2 \times P \times R)/(P + R) \tag{15}$$

where, True Positives (TP) refer positive examples and false Positives (FP) are negative examples. False Negatives (FN) indicate the positive examples incorrectly labeled as negatives. F-measure indicates the robustness and discrimination ability of the system.

4.2 Copy Detection Results

We implemented the following five methods for performance evaluation:

(1) Ordinal measure [4] (abbreviated as 'OM');
(2) Tasdemir et al.'s method [8] ('KT');
(3) Motion activity features ('MA');
(4) Audio spectral descriptors ('SD');
(5) Combination of motion activity and audio spectral features ('MA+SD').

Our methods [methods-(3), (4) and (5)] include different combinations of the proposed techniques. More specifically, in method (3) motion activity features including motion intensity, dominant direction and spatial distribution of activity are used for the copy detection task. In method (4), we considered the four spectral descriptors including spectral centroid, signal energy, roll-off and flux for detecting video copies. In method (5), both the proposed motion activity and spectral features are jointly utilized for identifying the duplicate video sequences.

Ordinal measure (method (1)) is one of the widely used global signatures in the CBCD domain [4], which is extracted as follows. Each video frame is partitioned into 4×4 blocks and normalized average intensity of blocks are calculated and consequently the resultant ranking order of blocks are known as Ordinal measures. Then, Euclidean distance metric is used to calculate the similarity between the master and query video sequences. Tasdemir et al.'s method (method (2)) is based on mean motion vector magnitudes of video frames. First frames are sampled at 5fps and average motion vector magnitudes of macro blocks are stored as video signatures. The similarity between the master and query clips are computed using L2-norm distance measure.

Table 6 shows the copy detection results of the five compared methods for T1-T5 transformations. The results demonstrate the improved performance of method (5) by 36.19%, when compared to the reference methods. Methods (4) and (5) generally perform well for all five transformations. Specifically, for T2 transformation, method (2) yields poor recall rate compared to method (5). The reason is, spectral features are much affected by temporal attacks such as slow motion. Method(2) scores low precision rate for blurring transformation, because lot of false positives are retrieved from the dataset. However, the global descriptive nature of Ordinal measure results in better performance for blurring transformation when compared to that of method (2).

Table 6 Copy detection results (in %) for T1-T5 transformations

Transformations		OM (1)	KT (2)	MA (3)	SD (4)	MA+SD (5)
Fast forward	P	60.10	62.71	65.85	97.89	99.81
	R	61.54	69.64	84.37	96.37	97.03
	FM	60.81	65.99	73.96	97.12	98.40
Slow motion	P	71.35	71.87	75.00	**100.00**	**100.00**
	R	59.18	70.35	87.80	79.48	92.76
	FM	64.69	71.10	80.89	88.56	96.24
Color change	P	59.26	60.01	63.63	99.39	**100.00**
	R	67.79	69.27	84.83	99.40	97.62
	FM	63.23	64.30	72.71	99.39	98.79
Blurring	P	56.86	55.81	57.14	99.69	99.92
	R	79.14	72.56	92.30	**100.00**	**100.00**
	FM	66.15	63.09	70.58	99.84	99.95
Brightness change	P	56.93	70.15	82.85	90.14	**100.00**
	R	70.19	68.09	75.00	89.92	91.93
	FM	62.86	69.10	78.72	90.02	95.79

Table 7 lists the detection results of the five compared methods for T6-T10 transformations. The results demonstrate the enhanced performance of method (5) by 35.02%, compared to the reference methods. Precisely, method (1) performs poor for noise addition type, because adding random noise severely affects the intensity values of blocks. Methods (2) and (3) also score poor PR rates for this category due to the noisy nature of raw motion vectors. However, methods (4) and (5) using spectral descriptors are less affected by this category and thus provide better detection results. For cropping attacks, Ordinal measure scores low results, because the surrounding black borders on frame regions noticeably increase the false positive rates.

Table 7 Copy detection results (in %) for T6-T10 transformations

Transformations		OM (1)	KT (2)	MA (3)	SD (4)	MA+SD (5)
	P	41.69	42.17	50.00	88.56	**100.00**
Noise addition	R	40.48	41.18	46.15	89.72	94.74
	FM	41.07	41.66	47.99	89.13	92.79
	P	79.68	79.82	82.85	99.09	99.96
Pattern insertion	R	80.24	79.47	85.29	98.80	**100.00**
	FM	79.95	79.64	84.05	98.94	99.97
	P	70.64	74.58	82.50	97.42	**100.00**
Moving caption insertion	R	71.15	72.94	84.61	98.86	99.92
	FM	70.89	73.75	83.54	98.13	99.95
	P	68.29	65.83	80.00	99.00	99.98
Cropping	R	53.58	69.98	82.75	92.66	93.50
	FM	60.04	67.84	81.35	95.72	96.63
	P	61.54	60.17	**100.00**	94.44	**100.00**
Picture-inside-picture	R	43.67	66.73	74.54	90.26	99.01
	FM	51.09	63.28	85.41	92.30	99.50

Table 8 shows the detection results of the five compared methods for T11-T14 transformations. The results prove the improved accuracy of method (5) by 30.18% compared to the other methods. Ordinal measure gives very poor detection rates for T14 type, since the global descriptors are less robust against region based attacks such as cropping. Method (4) provides good PR rates for T14, when compared to methods (1)-(3). Utilization of the robust spectral descriptors for copy detection task is the exact reason for the good performance of method (4). On the other hand, method (5) outperforms method (4) by yielding better PR rates for T14 type.

Summary: Experimental results prove that, the proposed techniques improve the detection accuracy by 30-35% compared to reference methods. Specifically, method (5) gives consistently good performance for all fourteen types of video transformations. The reason for this improved performance of method (5) is that, the joint utilization of robust audio spectral signatures and spatio-temporal motion activity features for the CBCD task.

Table 8 Copy detection results (in %) for T11-T15 transformations

Transformations		OM (1)	KT (2)	MA (3)	SD (4)	MA+SD(5)
	P	73.65	79.91	83.72	67.22	99.17
Mp3 compression	R	72.58	68.29	70.51	57.29	97.35
	FM	73.11	73.64	76.54	61.85	98.25
	P	80.06	79.62	83.62	73.44	98.37
Single-band compression	R	73.64	75.36	80.61	50.28	97.82
	FM	76.71	77.43	82.08	59.69	98.09
	P	66.28	69.16	85.05	70.36	91.74
Multi-band compression	R	61.15	61.19	81.64	52.22	93.82
	FM	63.61	64.93	83.31	59.94	92.76
	P	58.33	60.93	80.65	98.96	**100.00**
Combination of 3	R	51.29	64.74	79.83	98.21	99.29
	FM	54.58	62.77	80.23	98.58	99.64

4.3 Computational Cost Comparison

To evaluate the proposed method, we implemented the code in MATLAB using a PC with 2.8GHz CPU and 3GB RAM. The total computational cost of all five methods including signatures extraction and matching are shown in Table 9. They are measured using 35s query clip and 50h of master dataset for detecting copies. The total computational cost of method (5) is slightly high compared to the methods (2)-(4). Although method (4) is the most cost effective method, its detection results are poor for audio transformations, when compared to that of method (5). Thus, the results prove that, method (5) significantly improves the detection accuracy and also widens the coverage to more number of transformations at the cost of slight increase in computational cost. If feature extraction and matching procedures are implemented in parallel, then the computational cost of the proposed method can be substantially reduced.

Table 9 Comparison of computational cost

Computational Cost	OM	KT	MA	SD	MA+SD
	(1)	(2)	(3)	(4)	(5)
Signature extraction	170.95	223.04	187.98	111.56	196.41
Signature matching	57.43	42.02	34.38	1.27	35.68
Total cost	228.38	265.06	222.36	112.83	232.09

5 Conclusion

This paper proposes an efficient CBCD method for detecting video copies, which integrates acoustic and motion activity features. Our future work will be targeted at improving the robustness of the proposed CBCD system against complex transformations such as camcording.

Acknowledgment. This research work is supported by Department of Science & Technology (DST) of Government of India under research grant no. SR/WOS-A/ET-48/2010.

References

1. CMPDA- Feb 2011 report, "Economic consequences of movie piracy" (2014)
2. Sarkar, A., Singh, V., Ghosh, P., Manjunath, B.S., Singh, A.: Efficient and Robust Detection of Duplicate Videos in a Large Database. IEEE Trans. Circuits & Sys. for Video Tech. 20(6), 870–885 (2010)
3. Chiu, C.Y., Wang, H.M.: Time-Series Linear Search for Video Copies Based on Compact Signature Manipulation and Containment Relation Modeling. IEEE Trans. Circuits & Sys. for Video Tech. 20(11), 1603–1613 (2010)
4. Hua, X.S., Chen, X., Zhang, H.J.: Robust video signature based on ordinal measure. In: proc. of IEEE Int. Conf. on Image Proc. (ICIP), pp. 685–688 (2004)
5. Hoad, T.C., Zobel, J.: Detection of video sequence using compact signatures. Proc. of ACM Trans. on Inf. Sys. 24, 1–50 (2006)
6. Lowe, D.G.: Distinctive image features from scale-invariant key points. Int. Journal of Computer Vision, 91–110 (2004)
7. Hampapur, A., Hyun, K.H., Bolle, R.: Comparison of Sequence Match- ing Techniques for Video Copy Detection. In: Proc. of IEEE Int. Conf. on Multimedia & Expo, pp. 737–740 (2001)
8. Tasdemir, K., Cetin, A.E.: Motion Vector Based Features for Content Based Video Copy Detection. In: Proc. of IEEE Int. Conf. on Pattern Recog., 2010, pp. 3134–3137 (2010)
9. Itoh, Y., Erokuumae, M., Kojima, K., Ishigame, M., Tanaka, K.: Time-space Acoustical Feature for Fast Video Copy Detection. In: Proc. of 2010 IEEE Int. Workshop on Multimedia Sig. Proc. (MMSP-2010), pp. 487–492 (2010)
10. Saracoğlu, A., Esen, E., Ateş, T.K., Acar, B.O., Zubari, E.C., Ozan, E., özalp, A., Alatan, A., Çiloglu, T.: Content Based Copy Detection with Coarse Audio-Visual Fingerprints. In: 7th Int. Workshop on Content-Based Multimedia Indexing, pp. 213–218 (2009)
11. Küçüktunç, O., Baştan, M., Güdükbay, U., Ulusoy, O.: Video copy detection using multiple visual cues and MPEG-7 descriptors. Comp. Vision & Image Understanding 21, 125–134 (2010)
12. Jeannin, S., Divakaran, A.: MPEG-7 Visual Motion Descriptors. IEEE Trans. on Circ. & Sys. for Video Tech. 11(6), 720–724 (2001)
13. Sun, X., Ajay, D., Manjunath, B.S.: A Motion Activity Descriptor and Its Extraction in Compressed Domain. In: IEEE Pacific-Rim Conf. Multimedia (PCM), pp. 450–453 (2001)
14. Roopalakshmi, R., Reddy, G.R.M.: A Novel CBCD Approach Using MPEG-7 Motion Activity Descriptors. In: Proc. of IEEE Int. Symp. on Multimedia, USA, pp. 179–184 (2011)

15. Roopalakshmi, R., Reddy, G.R.M.: A Novel Approach to Video Copy Detection Using Audio Fingerprints and PCA. Published in Elsevier Procedia Computer Science, vol. 5, pp. 149–156 (2011)
16. Park, T.H.: Introduction to digital signal processing- Computer musically speaking. World Scientific Press (2010)
17. TRECVID (2010), Guidelines, http://www.nlpir.nist.gov/projects/tv2010/tv2010.html

Redundant Discrete Wavelet Transform Based Medical Image Fusion

Rajiv Singh and Ashish Khare

Abstract. In this work, we propose redundant discrete wavelet transform (RDWT) based fusion for multimodal medical images. The shift invariance nature of RDWT shows its usefulness for fusion. The proposed method uses maximum scheme for fusion of medical images. We have experimented with several sets of medical images and shown results for three sets of medical images. The effectiveness of fusion results has been shown using edge strength, and mutual information fusion metrics. The qualitative and quantitative comparison of the proposed method with spatial domain fusion methods (Linear, Sharp, and principal component analysis (PCA)) and wavelet domain fusion methods (discrete wavelet transform (DWT), lifting wavelet transform (LWT), and multiwavelet transform (MWT)) proves the superiority of the proposed fusion method.

Keywords: Medical Image Fusion, Spatial and Transform Domain Fusion, Shift Invariance, Redundant Discrete Wavelet Transform, Fusion Metrics.

1 Introduction

The development of multimodality medical imaging sensors for extracting clinical information has influenced to explore the possibility of data reduction and having better visual representation. X-ray, ultrasound, MRI (Magnetic resonance imaging) and CT (Computed tomography) are a few examples of biomedical sensors. These sensors are used for extracting clinical information, which are generally complementary in nature. For example, X-ray is widely used in detecting fractures and abnormalities in bone position, CT is used in tumor and anatomical detection and MRI is used to obtain information about tissues. Thus, none of these modali-

Rajiv Singh · Ashish Khare
Department of Electronics & Communication,
University of Allahabad, Allahabad, India
e-mail: jkrajivsingh@gmail.com, ashishkhare@hotmail.com

S.M. Thampi, A. Gelbukh, and J. Mukhopadhyay (eds.), *Advances in Signal Processing and Intelligent Recognition Systems*, Advances in Intelligent Systems and Computing 264,
DOI: 10.1007/978-3-319-04960-1_44, © Springer International Publishing Switzerland 2014

ties is able to carry all complementary and relevant information in a single image. Medical image fusion [1-3] is the only possible way to combine and merge all relevant and complementary information from multiple source images into single composite image which facilitates more precise diagnosis and better treatment.

The basic requirements for image fusion [4] are: first, fused image should possess all possible relevant information contained in the source images; second, fusion process should not introduce any artifact or unexpected feature in the fused image.

The medical images are generally of poor contrast and may be corrupted by noise and blur due to imperfections of image acquisition systems. Thus, fusion of medical images becomes challenging and the quality of fused image depends on the fusion strategy. Spatial domain and transform domain are the two most common fusion strategies used for medical image fusion. The performance and visual representation of spatial domain fusion methods (Linear [5], Sharp [6], and principal component analysis (PCA) [7]) are limited and visual artifacts are generated during fusion process of medical images. This visual artifact has been observed in Sharp fusion [6] method for medical images and reported in [2].

Therefore, transform domain fusion techniques have been used for fusion to overcome the limitations of spatial domain fusion methods. Pyramid and wavelet transforms [8] have been used for multiscale fusion under category of transform domain methods. However, artifacts such as blocking effects, lack of directional information and low signal to noise ratio in pyramid transforms [9] motivated to use wavelet transforms for medical image fusion.

The discrete wavelet transform (DWT) [10] is the most commonly used wavelet transform for medical image fusion. It provides spectral as well as increased directional information with three spatial orientations: vertical, horizontal and diagonal. Although, DWT has been proved beneficial over pyramid transform based fusion, it needs huge memory and computation steps for convolution of input signals. To provide faster computation and efficient reconstruction, lifting wavelet transform (LWT) [11] has been used for fusion. LWT reduces the computation complexity by a factor of two and inversion of the LWT is obtained by inverting the lifting steps (split, predict and update). Multiwavelet transform (MWT) [12] has overcome the disadvantages of scalar wavelets and possesses orthogonality, symmetry and smoothness for better image analysis. It provides flexible filtering through which high frequency energy can be transferred into low frequency energy and is beneficial to improve compression ratio.

Wavelet transforms, discussed so far, have been able to capture information efficiently from images, yet they are shift variant and failed to provide shift invariant fusion algorithms. The shift invariance [13, 14] is highly desirable for robust fusion of medical images. Thus, to provide shift invariant fusion of medical images, we have used redundant discrete wavelet transform (RDWT) [15, 16] for fusion and proposed maximum fusion scheme for multimodal medical images. To show the effectiveness of the proposed fusion scheme, we have compared the proposed method with wavelet domain (DWT [10], LWT [11], and MWT [12]) and spatial domain (Linear [5], Sharp [6], and PCA [7]) fusion methods. The mutual information and edge strength fusion metrics have been used for quantitative evaluation of the simulated fusion results.

The rest of the paper is organized as follows: Section 2 briefly describes the redundant discrete wavelet transform (RDWT). In Section 3, the proposed fusion method is elaborated. Results and discussions are given in Section 4. Finally, conclusions of the work are given in Section 5.

2   The Redundant Discrete Wavelet Transform (RDWT)

The use of real valued wavelet transforms provides an efficient way for medical image fusion. Yet, these are critically sampled, hence shift sensitive. For robust fusion of medical images and to avoid merging artifacts of wavelet coefficients [17], shift invariant wavelet transform is highly desired. Further, it has been shown in [18] that RDWT has better performance for fusion as it avoids aliasing and ringing artifacts.

The RDWT is an undecimated version of the classical DWT and provides shift invariance by removing down samplers from DWT. This fact can be easily verified from the Fig. 1, where a step function is decomposed by DWT and RDWT up to three levels and plotted [19]. This diagram illustrates that the DWT is far from being shift invariant.

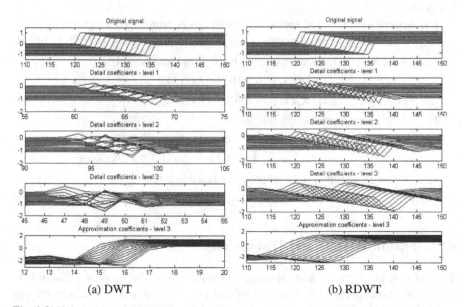

(a) DWT   (b) RDWT

Fig. 1 Shift invariance of RDWT

The implementation of RDWT [15] eliminates the decimation step from Mallat's algorithm of DWT. Therefore, size of wavelet subbands and original signal and/or image is same. The upsampling operator for RDWT is defined as:

$$x[k] \uparrow 2 = \begin{cases} x[k/2], & k \text{ even} \\ 0, & k \text{ odd} \end{cases} \tag{1}$$

The scaling and wavelet filters at scale $j+1$ are given as

$$h_{j+1}[k] = h_j[k] \uparrow 2$$
$$g_{j+1}[k] = g_j[k] \uparrow 2 \tag{2}$$

where

$$h_0[k] = h[k]$$
$$g_0[k] = g[k] \tag{3}$$

3   The Proposed Method

The proposed method uses RDWT for medical image fusion. The effectiveness and relevancy of the maximum fusion rule [2, 20] have compelled to use it for fusion of medical images. The high valued wavelet coefficients carry salient information about images such as edges, corners, and boundaries etc. Therefore, merging of wavelet coefficients by maximum fusion scheme provides high quality fused image. The proposed method is comprised of three steps: decomposition of source images, followed by application of fusion rule, and lastly, reconstruction of fused image.

The steps of the proposed fusion algorithm can be summarized as follows:

1.   Decompose source images $I_1(x, y)$ and $I_2(x, y)$ using RDWT to obtain wavelet coefficients $R_1(x, y)$ and $R_2(x, y)$ respectively.

$$R_1(x, y) = RDWT [I_1(x, y)]$$
$$\text{and } R_2(x, y) = RDWT [I_2(x, y)] \tag{4}$$

2.   Apply maximum fusion rule as:

$$R(x, y) = \begin{cases} R_1(x,y), & \text{if } |R_1(x,y)| \geq |R_2(x,y)| \\ R_2(x,y), & \text{if } |R_2(x,y)| > |R_1(x,y)| \end{cases} \tag{5}$$

3.   Reconstruction of $R(x, y)$ provides fused image $F(x, y)$.

$$F(x, y) = Inverse\ RDWT [R(x, y)] \tag{6}$$

4   Fusion Results and Discussions

This section gives visual and quantitative evaluation of the proposed fusion method. The experiments have been performed over several sets of medical images and fusion results have been shown for three sets of medical images in Figs. 2-4 respectively. The first and third sets of medical images are CT and MRI, and second set of medical images is T1-MR and MRA (magnetic resonance angiogram). To show the effectiveness of the proposed method, we have compared the proposed method with wavelet domain (DWT [10], LWT [11], and MWT [12]), and spatial domain (Linear [5], Sharp [6], and PCA [7]) fusion methods. The maximum fusion scheme has been performed for DWT, LWT, and MWT fusion methods and results have been shown in Figs. 2-4 for three different sets of medical images. Similarly, the comparison results for spatial domain (Linear, Sharp, and PCA) fusion methods have been shown in Figs. 2-4 for better analysis of proposed fusion method.

For objective evaluation [21] of the proposed fusion method, we have used non-reference fusion metrics; mutual information (MI) [22] and edge strength (Q_{AB}^F) [23].

For two source images A, B, and fused image F, edge strength (Q_{AB}^F) is defined as:

$$Q_{AB}^F = \frac{\sum\limits_{n=1}^{N} \sum\limits_{m=1}^{M} Q^{AF}(n,m)w^A(n,m) + Q^{BF}(n,m)w^B(n,m)}{\sum\limits_{i=1}^{N} \sum\limits_{j=1}^{M} (w^A(i,j) + w^B(i,j))} \qquad (7)$$

Similarly, for the source images A, B, and fused image F, mutual information (MI) is given by:

$$MI = MI_{AF} + MI_{BF} \qquad (8)$$

where MI_{AF} and MI_{BF} are the mutual information of source images A and B fused image F respectively.

High values of these metrics show the effectiveness of the fused image. To compare the proposed fusion method with other wavelet domain (DWT [10], LWT [11], and MWT [12]), and spatial domain (Linear [5], Sharp [6], and PCA [7]) fusion methods, we have computed the values of mutual information and edge strength, and tabulated them in Tables 1-3 for the three sets of medical images.

By observing Table 1, one can easily found that fused image obtained by the proposed fusion method has the second highest value of edge strength for first set of medical images. However, qualitatively, the proposed method is found superior over Sharp fusion and PCA fusion methods as Sharp fusion method shows the merging artifact (reported in [2]) and PCA fusion is not able to capture the edge information presented in CT image. This fact can be easily verified by looking the Figs. 2(a), (b) and (i).

Similarly, the observations of Tables 2 and 3 clearly indicate that the proposed fusion method has the highest values of mutual information and edge strength for second and third sets of medical images. This comparison shows the goodness and effectiveness of the proposed fusion method over other wavelet domain (DWT [10], MWT [11], and LWT [12]) and spatial domain (Linear [5], Sharp [6], and PCA [7]) fusion methods.

Therefore, it can be concluded that the proposed fusion approach using RDWT is able to provide better fusion results than any of the existing wavelet domain (DWT [10], LWT [11], and MWT [12]) and spatial domain (Linear [5], Sharp [6], and PCA [7]) fusion methods.

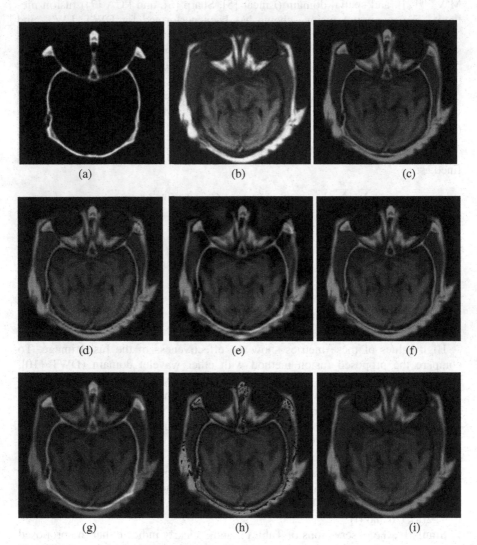

(a)   (b)   (c)

(d)   (e)   (f)

(g)   (h)   (i)

Fig. 2 Fusion results for first set of medical images

(a) CT image, (b) MRI image, (c) Fused image by the proposed fusion method, (d) DWT maximum fused image, (e) LWT maximum fused image, (f) MWT maximum fused image, (g) Linear fused image, (h) Sharp fused image, and, (i) PCA fused image.

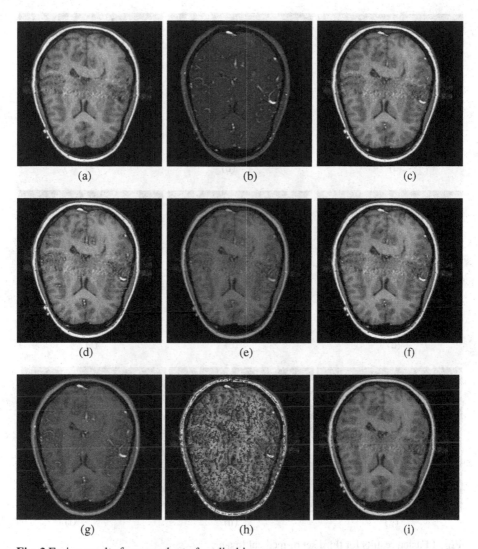

Fig. 3 Fusion results for second set of medical images

(a) T1-MR image, (b) MRA image, (c) Fused image by the proposed fusion method, (d) DWT maximum fused image, (e) LWT maximum fused image, (f) MWT maximum fused image, (g) Linear fused image, (h) Sharp fused image, and, (i) PCA fused image.

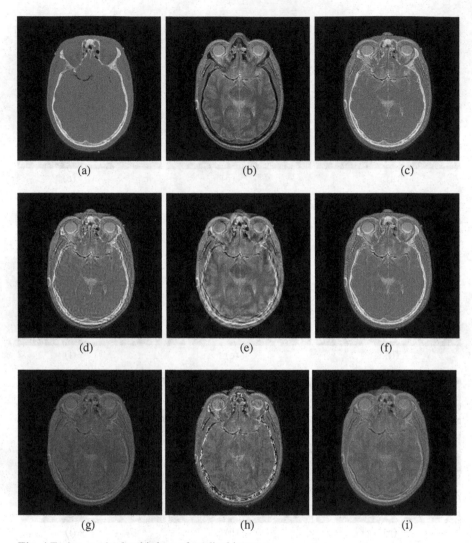

Fig. 4 Fusion results for third set of medical images

(a) CT image, (b) MRI image, (c) Fused image by the proposed fusion method, (d) DWT maximum fused image, (e) LWT maximum fused image, (f) MWT maximum fused image, (g) Linear fused image, (h) Sharp fused image, and, (i) PCA fused image.

Table 1 Fusion performance evaluation for first set of medical images

Fusion Method	Mutual Information	Edge Strength
The proposed fusion method	1.7031	0.6805
DWT maximum fused [10]	2.2454	**0.7403**
LWT maximum fused [11]	1.0883	0.5809
MWT maximum fused [12]	1.7130	0.6651
Linear fused [5]	**2.2821**	0.6409
Sharp fused [6]	1.4048	0.5602
PCA fused [7]	1.8034	0.5278

Table 2 Fusion performance evaluation for second set of medical images

Fusion Method	Mutual Information	Edge Strength
The proposed fusion method	**4.9597**	**0.6534**
DWT maximum fused [10]	4.1804	0.4791
LWT maximum fused [11]	3.8142	0.4564
MWT maximum fused [12]	4.8002	0.6306
Linear fused [5]	3.8227	0.3897
Sharp fused [6]	3.7095	0.3704
PCA fused [7]	4.6493	0.6402

Table 3 Fusion performance evaluation for third set of medical images

Fusion Method	Mutual Information	Edge Strength
The proposed fusion method	**3.5746**	**0.5094**
DWT maximum fused [10]	3.2282	0.5087
LWT maximum fused [11]	2.8899	0.4641
MWT maximum fused [12]	3.4864	0.4713
Linear fused [5]	3.0443	0.2216
Sharp fused [6]	3.3191	0.3953
PCA fused [7]	3.1848	0.3315

5   Conclusions

In the present work, we have proposed a new medical image fusion method using RDWT which is based on maximum fusion scheme. The role of shift invariance in fusion algorithms made RDWT suitable for medical image fusion. The experimental results have been shown for three different sets of medical images and fusion results have been compared with wavelet domain (DWT [10], LWT [11], and MWT [12]) and spatial domain (Linear [5], Sharp [6], and PCA [7]) fusion methods. The quantitative comparison of the proposed method with above mentioned fusion method has been performed with mutual information and edge strength metrics. The qualitative and quantitative comparisons showed the effectiveness of the proposed method with maximum fusion rule over existing wavelet domain (DWT [10], LWT [11], and MWT [12]) and spatial domain (Linear [5], Sharp [6], and PCA [7]) fusion methods.

References

1. Dasarathy, B.V.: Information fusion in the realm of medical applications – A bibliographic glimpse at its growing appeal. Information Fusion 13(1), 1–9 (2012)
2. Singh, R., Khare, A.: Fusion of multimodal medical images using Daubechies complex wavelet transform- A multiresolution approach. Information fusion (Article in Press), http://dx.doi.org/10.1016/j.inffus.2012.09.005
3. Singh, R., Srivastava, R., Prakash, O., Khare, A.: Multimodal medical image fusion in dual tree complex wavelet domain using maximum and average fusion rules. Journal of Medical Imaging and Health Informatics 2(2), 168–173 (2012)
4. Rockinger, O., Fechner, T.: Pixel level fusion: the case of image sequences. In: Signal Processing, Sensor Fusion, and Target Tracking (SPIE), vol. 3374, pp. 378–388 (1998)
5. Clevers, J.G.P.W., Zurita-Milla, R.: Multisensor and multiresolution image fusion using the linear mixing model. In: Stathaki, T. (ed.) Image Fusion: Algorithms and Applications, pp. 67–84. Academic Press, Elsevier (2008)
6. Tian, J., Chen, L., Ma, L., Yu, W.: Multi-focus image fusion using a bilateral gradient-based sharpness criterion. Optics Communications 284(1), 80–87 (2011)
7. Naidu, V.P.S., Raol, J.R.: Pixel-level image fusion using wavelets and principal component analysis. Defence Science Journal 58(3), 338–352 (2008)
8. Hamza, A.B., He, Y., Krim, H., Willsky, A.: A multiscale approach to pixel-level image fusion. Integrated Computer-Aided Engineering 12(2), 135–146 (2005)
9. Li, H., Manjunath, B.S., Mitra, S.K.: Multisensor image fusion using the wavelet transform. Graphical Models and Image Processing 57(3), 235–245 (1995)
10. Cheng, S., He, J., Lv, Z.: Medical images of PET/CT weighted fusion based on wavelet transform. In: The Second International Conference on Bioinformatics and Biomedical Engineering (ICBBE), pp. 2523–2525 (2008)
11. Kor, S., Tiwary, U.S.: Feature level fusion of multimodal medical images in lifting wavelet transform domain. In: 26th Annual International Conference of the IEEE Engineering in Medicine and Biology Society (EMBS 2004), vol. 1, pp. 1479–1482 (2004)
12. Liu, Y., Yang, J., Sun, J.: PET/CT medical image fusion algorithm based on multi-wavelet transform. In: Second International Conference on Advanced Computer Control (ICACC), vol. 2, pp. 264–268 (2010)
13. Zhang, Q., Wang, L., Li, H., Ma, Z.: Similarity-based multimodality image fusion with shiftable complex directional pyramid. Pattern Recognition Letters 32(13), 1544–1553 (2011)
14. Singh, R., Khare, A.: Multimodal medical image fusion using Daubechies complex wavelet transform. In: ICT 2013, pp. 869–873 (2013)
15. Fowler, J.: The redundant discrete wavelet transform and additive noise. IEEE Signal Processing Letters 12(9), 629–632 (2005)

16. Singh, R., Vatsa, M., Noore, A.: Multimodal Medical Medical Image Fusion using Redundant Wavelet Transform. In: Proceedings of Seventh International Conference on Advances in Pattern Recognition, pp. 232–235 (2009)
17. Yockey, D.A.: Artifacts in wavelet merging. Optical Engineering 35(7), 2094–2101 (1996)
18. Stathaki, T. (ed.): Image Fusion Algorithms and Applications. Elsevier (2011)
19. Adam, I.: Complex wavelet transform: application to denoising. PhD Thesis, Politehnica University of Timisoara Universite DE RENNES (2010)
    http://www.tc.etc.upt.ro/docs/cercetare//
    teze_doctorat/tezaFiroiu.pdf
20. Li, S., Yang, B., Hu, J.: Performance comparison of different multi-resolution transforms for image fusion. Information Fusion 12(2), 74–84 (2011)
21. Singh, R., Khare, A.: Objective evaluation of noisy multimodal medical image fusion using Daubechies complex wavelet transform. In: Proceedings of the 8th Indian Conference on Vision, Graphics and Image Processing (ICVGIP-12), IIT Mumbai, India (2012)
    http://dx.doi.org/10.1145/2425333.2425405
22. Guihong, Q., Dali, Z., Pingfan, Y.: Information Measure for Performance of Image Fusion. Electronics Letters 38(7), 313–315 (2002)
23. Xydeas, S., Petrovic, V.: Objective Image Fusion Performance Measure. Electronics Letters 36(4), 308–309 (2000)

Hybrid Shadow Restitution Technique
for Shadow-Free Scene Reconstruction

Muthukumar Subramanyam, Krishnan Nallaperumal, Ravi Subban,
Pasupathi Perumalsamy, Shashikala Durairaj, S. Selva Kumar, and S. Gayathri Devi

Abstract. Shadows are treated as a noise in computer vision scenario, even though it may found useful in many applications. This research focuses the insignificant shadow restitution methodology to improve the scene visibility and to support the dynamic range reduction. The Hybrid technique combines the physical, geometric, textural, spatial and photometric features for shadow detection. Using feature importance statistics the appropriate criteria is chosen and applied. The experiments over wide dataset prove that the proposed hybrid technique outperforms peer research proposals with the expense of computational cost and time. The output results in a shadow-free, visually plausible high quality image.

Keywords: Shadow removal, Shadow detection, Shadow reconstruction, De-Shadowing, Shadow Enhancement, Hybrid Technique, Texture, Gradient, Chromaticity, Shadow Enhancement, Image Reconstruction.

1   Introduction

Human vision is a wonderfully complex system that analyzes the world, and under normal circumstances, is able to make sense of it. Images (scenes) are the input to

Muthukumar Subramanyam
Dept of CSE, NIT, Puducherry, India
e-mail: sm.cite.msu@gmail.com

Krishnan Nallaperumal · Pasupathi Perumalsamy · Shashikala Durairaj ·
S. Selva Kumar · S. Gayathri Devi
CITE, MS University, Tirunelveli, India
e-mail: {Krishnann1751968,pp.cit.msu,shashikalait85,
        gayathri.s7,smartselva2}@gmail.com

Ravi Subban
Dept of CSE, Pondicherry University, Pondicherry, India
e-mail: sravicite@gmail.com

S.M. Thampi, A. Gelbukh, and J. Mukhopadhyay (eds.), *Advances in Signal Processing*   517
and Intelligent Recognition Systems, Advances in Intelligent Systems and Computing 264,
DOI: 10.1007/978-3-319-04960-1_45, © Springer International Publishing Switzerland 2014

the visual system, providing a constant stream of information about the world surrounding us. One of the most fundamental tasks for any visual system is that of separating the changes in an image which are due to a change in the underlying imaged surfaces. Shadows are often one of the largest problems and have become a topic worthy of much research [1]. Such processes largely rely on similarities within natural scenes, whether natural is taken to mean scenes with no manmade structures, or natural in the more general sense of scenes that exist in the real world. These similarities are indeed striking once images are analyzed with suitable tools. The area of natural image statistics provides a toolset for quantitatively analyzing natural scenes and the regularities that appear among them. Shadows in images are typically affected by several phenomena in the scene, such as lighting conditions, type and behavior of shadowed surfaces, occluding objects etc. Additionally, shadow regions may undergo post-acquisition image transformations, e.g., image contrast enhancement, which may introduce noticeable artifacts in the shadow-free images. Shadows are crucial for enhancing realism and provide important visual cues [2]. This can be beneficial for the purposes like revealing information about the object's shape, orientation, size, position, and intensity, physical characteristics of the screen and even about the light source, type and behavior of shadowed surfaces. Despite many attempts, the problem remains largely unsolved, due to several inherent challenges. Dark regions are not necessarily shadow regions, since foreground object or background can be dark too.

A commonly used assumption is that, shadow falls only on ground plane, but it is not valid for all the general cases. This research tries to serve as a guide to better understand limitations and failure cases, advantages and disadvantages and suitability of the shadow rectification algorithms for any kind of application scenarios. Even in algorithms focusing on artistic representations of scenes, some degree of realism is desired. The main focus is on real-time automatic solutions for the shadowing problem but also approaches scene for a better understanding of natural scenes by the analysis of various visually plausible scene reconstruction methodologies [3]. This research provides a concise, yet comprehensive review of the literature on shadow perception from across the cognitive sciences, including the theoretical information available, the perception of shadows in human and machine vision, and the ways in which shadows can be used.

Shadows offer important visual cues for human perception of shape, displacement, depth, and subtleties of contact between stacked objects, but the process of deriving these cues remains elusive [4]. A shadow occurs when an object partially or totally blocks the direct light source. Shadow can take any size and shape. In general, shadow can be divided into two major classes: self and cast shadows. A self shadow occurs in the portion of an object that is not illuminated by direct light. Cast shadows are the areas projected on a surface in the direction of direct light. Cast shadow can be further classified into umbra and penumbra. The region where the direct light source is totally blocked is called the umbra, while the region where it is partially blocked is known as the penumbra [5]. If there is a clear distinction between shadow and lit area it is referred as dark or thick shadows else if there is a smooth transition contains progressive shadow edges are called soft or shallow shadows.

Despite of the fact that many of shadow detection has been published during the last years, only few works in the literature address these three major problems: chromatic shadow identification, moving shadows and shadow detection in camouflaged areas. Shadow in image is generally divided into static and dynamic shadows. Static shadows are shadows due to static objects such as building, parked cars, trees, etc. In contrary, dynamic (moving) shadows appear due to moving object such as animals, vehicles, pedestrians, etc. For example, typical problems caused by shadows in surveillance scenarios, shadows may cause merging of multiple vehicle objects. Clearly, in many image analysis applications, the existence of shadows may lead to inaccurate object segmentation. Consequently, tasks such as object description and tracking are severely affected, thus inducing an erroneous scene analysis [6]. The main goal is to contribute to the design of a framework for automated shadow-free image perception. More specifically the focus is on extraction and compensation of insignificant shadow areas. This work is also being motivated from a practical point of view by the shortcomings of the current approaches [7].

2 Literature Review

Shadows in a scene can be quite complex, and the changes of the occluding objects, inter-reflections, reflectance variation, as well as the intensity and color variation of the illumination, all have effects on the shadowed scene [8]. Thus, shadow removal using a single shadow image is a difficult task. In addition, the computational complexity of shadow removal calculations can be prohibitively high for high resolution images. In spite of these extensive studies, more researches focus on providing a general method for arbitrary scene images and thereby obtaining "visually pleasing" shadow free images. In contrast with the rapidly expanding interests on shadow removal, no comprehensive survey is completely reported on this particular topic.

Liu et al., [9] use a human-guided method to select areas of shadow before relighting the shadow areas using a texture-based method. Traditionally, cast shadow detection algorithms have been mostly developed based on the single-frame approach or reference-frame approach. Scanlan et al., [10] presented a shadow removal algorithm that employs a simple histogram modification function on the image intensity. Jiang et al., [11] presented a shadow identification and classification method for real images in a laboratory environment. In their method, the shadow intensity and shadow geometry are analyzed. Funka-Lea et al., [12] presented an active shadow recognition method by combining color and geometric properties of the image. Salvador et al., [13] presented a method that is based on the use of invariant color models to identify and classify shadows in color images. Mikic et al., [14] presented an algorithm that statistically classifies pixels into the shadow, object, and background classes. Horprasert et al., [15] presented an algorithm for detecting moving objects from a static background scene that contains shading and shadow. They developed a background subtraction algorithm that is able to cope with local illumination changes and global illumination changes.

Nadimi et al., [16] proposed a physical model for moving shadow and object detection in video. Wu et al., [17] use a Bayesian technique where the user selects four different areas of the image: shadow regions, shadow-free regions, items to be excluded or unknown items, for a specific image. Leone et al., [18] use a multilayered approach to remove shadows from objects in a surveillance application. Arbel et al., [2] describe a method for removing shadows on curved surfaces. Their method requires the shadow regions to be detected prior to analysis, after which they detect the penumbra region and re-light the shadowed region accordingly. They are specifically looking to relight soft shadows on curved surfaces only. Withagen et al., [19] provide an interesting method for shadow detection that uses band-ratios (e.g., values of red/green and blue/green) to move pixels into a plane for segmentation of shadowed regions of an image. They make a significant assumption in that the color response of the camera is at across each color band. Xiao et al., [20], who detect outlines of objects (vehicles) and shadows on highways and subsequently attempt to determine whether the components causing the outlines are vehicles or shadows. The methods exist that use only grey-scale images for shadow and scene-change detection, such as the method by Ibrahim et al., [21].

Finlayson et al., [22] compare edges in the original RGB image to edges found in an illuminant-invariant image. This method can work quite well with high-quality images and calibrated sensors, but often performs poorly for typical web-quality consumer photographs. To improve robustness, others have recently taken a more empirical, data-driven approach, learning to detect shadows based on training images. In monochromatic images, Zhu et al., [23] classify regions based on statistics of intensity, gradient, and texture, computed over local neighborhoods, and refine shadow labels using a conditional random field (CRF). Cucchiara et al., [24] use shadow properties in the HSV color space to distinguish shadows from moving objects. These properties show that cast shadows darken the background in the luminance component, while the hue and saturation components change within certain limits. Prati et al., [25], present two layers taxonomy (algorithm-based taxonomy). The first layer classification considers whether the decision process introduces and exploits uncertainty.

3 Proposed Methodology

This section enumerates the various problems and challenges found in t. It is worth, noting that a given shadow image may have different phenomena.

However, in order to develop a robust shadow restitution methodology which can effectively handle shadow images taken under different conditions and of different scene types, any shadow extraction algorithm should account for the various types of possible phenomena [2]. The researcher applies a multi-stage approach inspired by the use of color and gradient information, together with known shadow properties [15]. In this research, the properties of shadow (spatial, geometrical, photometric, statistical and textural) are analyzed in a scene and prominent features based taxonomy is fused with tricolor attenuation model.

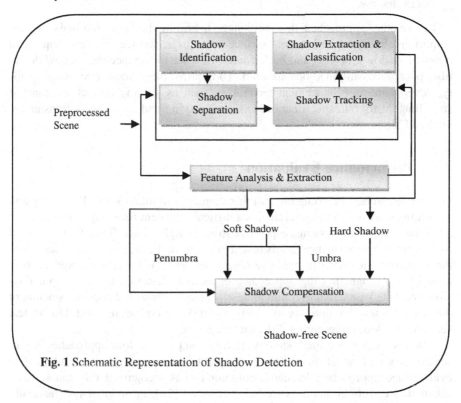

Fig. 1 Schematic Representation of Shadow Detection

The stages of the proposed approach for shadow detection has stated in the following five important steps:

- Apply Tricolor Attenuation Model (TAM) [5, 26] intensity model to identify shadows in the input scene.
- Combining scene adaptive feature selection and successive thresholding to improve the detection of shadows [29].
- Uses gradient, boundary, edges, smoothness and invariant properties for shadow tracking and accurate shadow refinement [28].

- Extending the shadow detection to cope with the categories, fuzzy support vector machine and hybrid classifier based clustering techniques are applied. The shadow map produces a refined shadow [27].
- The shadow classification stage segregates hard/soft shadow and umbra/penumbra regions with the automatic adaptation of scene knowledge from features.
- Finally, the shadow reduction is done with natural shadow matting or Gradient Similarity patch in-painting approach based on the category of soft or dark shadow.

Unlike other approaches, this multi-stage hybrid method does not make any assumptions about camera location, surface geometries, surface textures, shapes and types of shadows, objects and background. In order to enhance the shadow detection, post processing methods are used. To improve the shadow-free image quality, several image enhancement techniques such as gamma correction, contrast stretching, histogram specification, unsharp masking and smoothness transfer are applied [26].

4   Performance Evaluation

Despite researches are going on, still this domain is in infant stage. Extracting and restituting shadows in images remains a difficult problem for computer vision and artificial intelligence systems, especially from a single image. Therefore, it is hard to measure the performance in this task. However, in the area of shadow detection, the methodologies used to evaluate object detection and tracking could be borrowed [8]. Recognizing the importance of shadow detection to the computer vision and video processing communities, a unique shadow detection benchmark dataset collected that consists of nearly 7,00,000 natural scenes and 250 set test sequences(video) representing different categories.

In this section, the methodology used to compare the four approaches is presented. A set of novel quantitative and qualitative metrics has been adopted to evaluate the approaches. Research community has recognized this and serious, substantive efforts in this area are being reported [28]. In order to systematically evaluate various shadow detection, it uses to two important quality measures: good detection (low probability to misclassify a shadow point) and good discrimination (the probability to classify non-shadow points as shadow should be low, i.e. low false alarm rate). In addition to the above quantitative metrics, the authors also consider the following qualitative measures in our evaluation: robustness to noise, flexibility to shadow strength, width and shape, object independence, scene independence, computational load, detection of indirect cast shadows and penumbra [25], shadow camouflage and movement blur effect, limits the object / shadow detection performance. Further feedback of specific task/scene domain knowledge improves the performance.

This section provides the evaluation of various methodologies using different criteria across several categories of scenes. Table 1 specifies the experimental environment. The selected properties of benchmark dataset are given in Table 2. Normally, performance evaluation analyzes the performance in quantitative and qualitative criteria. Table 3 and table 4 explore the qualitative analysis of feature based and model based taxonomies. The performance outcomes of different algorithms are evaluated and are presented in Table 5.

Figures 2, 3, 4, 5 and 6 show the performance comparison of proposed multistage hybrid approach over the-state-of-the techniques. It is observed that, improved accuracy in shadow detection and discrimination rates. Also, the result shows reduced false detection rate and improved ROC rate (Higher True Positive Rate and Lower False Positive Rate). Therefore the proposed approach is effective and efficient. The performance is good comparable with the other proposals.

Table 1 Experimental Environment

Natural Strategy		Categories
Scene Type	Indoor, Outdoor	
Surface	Textures	Grass, Road, Wall, Floor, Steps, Concrete, Sand, Wood, Carpet, Reflective , Clothes, Pathway, Flower, Asphalt, Bricks and Other Natural Surfaces
	Non Textures	Single Color (White, Black, Grey, Green, Red, Blue, Yellow, Brown, Purple, Pink, Orange)
Noise	High, Medium, Low, Very Low, No Noise	
Object Type	Human, Trees, Ball, Boat, Cycle, Bike, Car ,Van, Building, Baby, Can, Cat, Cup, Bridge, Bag, Mouse, Street Lamp, Beam, Mug, Boy, Wood, Vegetable, Fruit, Flower, Others	
Shadow Direction	Horizontal, Vertical, Multiple, Complex	
Shadow Strength	Strong, Medium, Weak, Very Weak	
Shadow Size	High, Medium, Low, Very Low	
Lighting	Sunlight, Overcast, Dawn, Dusk, Artificial (Single Light Source)	
Surface Orientations	Horizontal, Vertical, Sloping	
Timing	Morning, Evening, Night	
Weather	Sunny, Cloudy, Windy	
Distance	Casual, Aerial, Satellite	

Table 2 Characteristics of Benchmark Dataset

Sequence Types	Frames			Scene			Object		Shadow				
	No	Hl	S	Y	B	N	Cl	S	S	Vi	Di	F	E
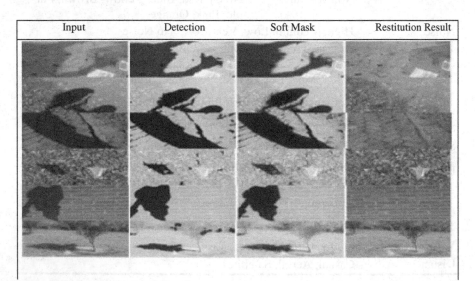CVC	800	12	A	O	T	Lo	P	L	L	H	Sh	Lo	M
PETS	795	16	C	O	V	Lo	P	V	V	Lo	Sh	Lo	L
Hall	1800	13	A	I	T	M	P	V	V	Lo	Sh	Lo	L
Lab	887	14	A	I	W	Lo	Po	M	M	H	Mi	L	H
Caviar	1388	45	D	I	R	M	P	M	M	Lo	Vr	H	L

Note : Hl-Hand –Labeled, B-Background, Vi-Visibility, F- Camouflage, E- Chromatic Effect, O-OutDoor, I-InDoor, A- Asphalt, W-White, C- Carpet, R- Reflective, M- Medium, Po- People/Other, P- People, L-Large, V- Variable, H- High, Sh- Single horizontal, Mi- Multiple, Vr- Vertical, T-Texture, P-People, Di-Direction, Cl-Class, Lo-Low, S-Size, Y-Type, A-320x240,D-384x288,C-720x576.

Input	Detection	Soft Mask	Restitution Result

Fig. 2 Hybrid Shadow Restitution Results

Table 3 Qualitative evaluation comparison of Feature based Techniques

Mehods/Criteria	SI	OI	SID	ID	PD	RN	DT	CC	CS	SC	ST
Geometry	M	L	L	H	M	M	L	L	M	M	M
Chromacity	M	H	H	H	L	L	H	L	M	M	M
Physical	M	H	H	M	M	M	M	L	M	M	H
SRTextures	M	H	H	M	H	H	H	H	L	M	H
LRTextures	H	H	H	M	H	H	L	M	M	H	H
Proposed	H	H	H	H	H	H	M	H	L	L	L

Note: SI-Scene Independence, OI-Object Independence, SID-Shadow Independence, ID-Illumination Independency, PD-Penumbra Detection, RN-Robustness to Noise, DT-Detection/Discrimination Trade Off, CC-Computational Complexity, CS-Chromatic Shadows, SC-Shadow Camouflage, ST-Surface Topology, L- Low , M- Medium, and H- High.

Table 4 Qualitative Evaluation of Model based Taxonomy

Methods/Criteria	SI	OI	SID	ID	PD	RN	DT	CC	CS	SC	ST
SNP	H	H	H	H	H	H	L	L	L	L	L
SP	L	H	M	M	L	H	H	L	L	M	L
DNM1	H	H	H	H	L	H	H	L	H	H	H
DNM2	M	L	H	H	H	L	H	H	H	H	H
Proposed	H	H	H	H	H	H	M	H	L	L	L

Table 5 Qualitative Evaluation of Techniques proposed by significant Authors

Criteria/Research Approach	SI	OI	SID	ID	PD	RN	DT	CC	CS	SC	ST
Horprasert et al.,	L	L	L	M	M	M	L	H	H	H	L
Stauder et al.,	L	L	M	M	M	L	L	M	H	H	L
Mikic et al.,	L	L	M	M	M	M	M	M	H	H	H
MC Kenna et al.,	L	L	L	M	M	M	M	M	H	H	H
Cucchiara et al.,	L	M	M	M	M	M	M	M	H	H	L
Kim et al.,	L	L	L	M	M	L	M	M	H	H	L
Yao et al.,	L	L	M	L	M	M	M	M	L	M	H
Siala et al.,	L	M	M	L	M	L	L	M	L	H	L
Salvador et al.,	L	M	M	L	M	L	M	M	H	H	M
Nadimi et al.,	M	L	M	M	L	M	L	M	L	H	M
Tian et al.,	L	M	M	M	L	M	L	M	L	M	H

Table 5 (*continued*)

Wong et al.,	L	L	M	L	M	M	L	M	L	H	H
M-Brisson et al.,	L	L	M	M	L	M	L	H	H	H	M
Leone et al.,	L	L	M	L	M	M	L	M	H	H	M
Joshi et al.,	L	H	M	M	M	M	M	M	L	H	H
Finlayson et al.,	L	L	L	M	M	M	M	M	L	M	H
Sanin et al.,	H	H	L	L	M	M	M	H	H	H	M
Qin et al.,	L	L	L	L	M	M	M	L	L	H	M
Chen, Agarawal et al.,	M	L	M	L	M	M	L	M	L	L	M
Al-Najdawi et al.,	H	M	M	M	M	H	L	H	H	H	M
Amato et al.,	H	H	H	M	H	M	H	M	H	H	H
Proposed	H	H	H	H	H	H	M	H	L	L	L

Table 6 Quantitative Comparison of Decision based Taxonomy

Data Set →	CVC		PETS		Hallway		Lab		Caviar	
Para meter ↓	Dr %	Dc %	Dr %	Dc %	Dr %	Dc %	Dr %	Dc %	Dr %	Dc %
SNP	81.14	82.34	83.14	76.77	69.28	79.1 20	84.03	92.35	76.93	87.46
SP	64.34	79.44	69.34	82.17	64.74	74.9 30	64.85	95.39	81.17	83.74
DNM1	68.47	82.14	72.14	73.44	80.01	69.6 6	76.26	89.87	69.25	75.89
DNM2	74.76	87.44	87.43	80.11	85.41	82.1 40	60.34	82.57	78.69	82.15
Proposed	94.14	96.23	97.66	97.43	88.17	95.3 20	90.24	96.17	96.24	93.79

Table 7 Quantitative Evaluation of Method based Taxonomy (Shadow Detection Rate (Dr) and DisCrimination (Dc) rates)

Data Set →	CVC		PETS		Hallway		Lab		Caviar	
Parameter ↓	Dr %	Dc %	Dr %	Dc %	Dr %	Dc %	Dr %	Dc %	Dr %	Dc %
Geometry	82.31	36.42	79.44	47.62	71.29	24.67	38.36	28.86	78.66	45.82
Chromacity	87.62	75.91	90.52	32.12	93.76	85.93	98.76	61.00	60.14	56.59
Physical	83.33	90.14	84.62	81.19	84.98	91.49	85.33	60.30	72.23	85.27
SR Textures	96.43	81.12	85.73	59.46	99.81	83.35	93.62	71.10	87.36	89.47
Proposed	94.14	96.23	97.66	97.43	88.17	95.32	90.24	96.17	96.24	93.79

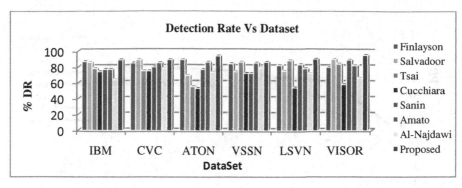

Fig. 3 Shadow Detection Accuracy

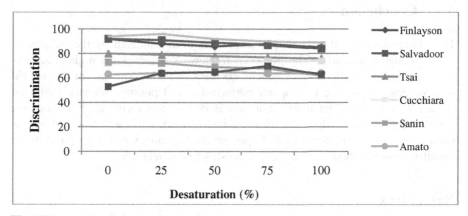

Fig. 4 Shadow Discrimination Rate

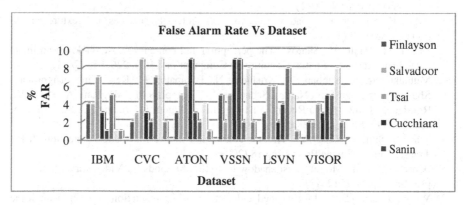

Fig. 5 False Alarm Rate

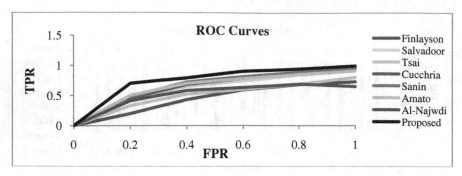

Fig. 6 ROC Curves

5   Conclusion

The effectiveness of the proposed method has been validated by the qualitative and quantitative measure. The proposed framework achieves the higher recognition rates over a collection of benchmark datasets. The outcome of the hybrid shadow detector is able to properly perform in most possible scenarios with an accuracy of 96.4%. The main limitation is due to multi colored light sources, which create multiple shadows with multicolored intensity, which would be our further direction. Perceptual studies approve the supremacy of the proposal in the visually plausible shadow-free scene reconstruction domain.

References

1. Pouli, F.T.: Statistics of image categories for computer graphics applications. Diss. University of Bristol (2011)
2. Arbel, E., Hel-Or, H.: Shadow removal using intensity surfaces and texture anchor points. PAMI 99 (2011)
3. Dee, H.M., Paulo, E.: Santos. "The perception and content of cast shadows: an interdisciplinary review". Spatial Cognition & Computation 11(3), 226–253 (2011)
4. Muthukumar, S., Subban, R., Krishnan, N., Pasupathi, P.: Real Time Insignificant Shadow Extraction from Natural Sceneries. In: Thampi, S.M., Abraham, A., Pal, S.K., Rodriguez, J.M.C. (eds.) Recent Advances in Intelligent Informatics. AISC, vol. 235, pp. 391–399. Springer, Heidelberg (2014)
5. Tian, J., Sun, J., Tang, Y.: Tricolor attenuation model for shadow detection. IEEE Transactions on Image Processing 18 (2009)
6. Amato, A., et al.: Moving Cast shadow Detection Methods for Video surveillance Application, pp. 1–25 (2013)
7. Wesolkowski, S.B.: Color image edge detection and segmentation: a comparison of the vector angle and the Euclidean distance color similarity measures. Dissertation University of Waterloo (1999)
8. Xiao, C., et al.: Fast Shadow Removal Using Adaptive Multi‐Scale Illumination Transfer. In: Computer Graphics Forum (2013)

9. Liu, F., Gleicher, M.: Texture-consistent shadow removal. In: Forsyth, D., Torr, P., Zisserman, A. (eds.) ECCV 2008, Part IV. LNCS, vol. 5305, pp. 437–450. Springer, Heidelberg (2008)
10. Scanlan, J.M., Chabries, D.M., Christiansen, R.: A shadow detection and Removal algorithm for 2-d images. In: Proc. of IEEE International Conference on Acoustics, Speech, and Signal Processing, pp. 2057–2060 (1990)
11. Jiang, H., Drew, M.S.: Shadow-resistance tracking in video. In: ICME 2003: Intl. Conf. on Multimedia and Expo, pp. 7–80 (2003)
12. Funka-Lea, G., Bajcsy, R.: Combining color and geometry for the active, visual recognition of shadows. In: Proc. of IEEE Int. Conf. on Computer Vision (ICCV), pp. 203–209 (1995)
13. Salvadoor, E., et al.: Cast Shadow Segmentation Using Invariant Color Features. Computer Vision and Image Understanding 95(2), 238–259 (2004)
14. Mikic, I., Cosman, P., Kogut, G., Trivedi, M.M.: Moving Shadow and Object Detection in Traffic Scenes. In: Proc. Int Conf. Pattern Recognition, vol. 1, pp. 321–324 (2000)
15. Horprasert, et al.: statistical approach for real-time robust background subtraction and shadow detection. In: IEEE ICCV, vol. 99, pp. 1–19 (1999)
16. Nadimi, S., et al.: Physical models for moving shadow and object detection in video. IEEE Transactions on Pattern Analysis and Machine Intelligence 26(8), 1079–1087 (2004)
17. Wu, et al.: A bayesian approach for shadow extraction from a single image. In: ICCV 2005, vol. 1, pp. 480–487. IEEE (2005)
18. Leone, A., et al.: A texture-based approach for shadow detection. In: IEEE Conference on Advanced Video and Signal Based Surveillance, pp. 371–376 (2005)
19. Withagen, P.J., Groen, F.C.A., Schutte, K.: IAS technical report IAS UVA-07-02 Shadow detection using a physical basis. Intelligent Autonomous Systems, University of Amsterdam (2007)
20. Xiao, Chunxia, et al., Fast Shadow Removal Using Adaptive Multi-Scale Illumination Transfer. In: Computer Graphics Forum (2013)
21. Ibrahim, M.M., Rajagopal, A.: Shadow detection in images. US Patent No.2007/0110309 A1 (2007)
22. Finlayson, G., Hordley, S., Drew, M.: Removing Shadows From Images. Eccv, 129–132. 2 (2006)
23. Zhu, J., Samuel, K.G.G., Masood, S., Tappen, M.F.: &ldquo, Learning to Recognize Shadows in Monochromatic Natural Images. In: Proc. IEEE Conf. Computer Vision and Pattern Recognition (2010)
24. Rita, C., et al.: Detecting moving objects, ghosts, and shadows in video streams. IEEE Transactions on Pattern Analysis and Machine Intelligence 25(10), 1337–1342 (2003)
25. Andrea, P., et al.: Detecting moving shadows: algorithms and evaluation. IEEE Transactions on Pattern Analysis and Machine Intelligence 25(7), 918–923 (2003)
26. Huerta, et al.: Detection and removal of chromatic moving shadows in surveillance scenarios. In: 12th International Conference Computer Vision. IEEE (2009)
27. Muthukumar, S., Krishnan, N., Tulasi Nachiyar, K., Pasupathi, P.: Shadow Detection in an image using Fuzzy based Approach. International Journal on Information and Communication Technology, 123–4560 (2011), doi:DOI10.5120/502-819, ISSN 0123-4560

28. Subban, R., Muthukumar, S., Pasupathi, P.: Image Restoration based on Scene Adaptive Patch In-Painting for Tampered Natural Scenes. In: Thampi, S.M., Abraham, A., Pal, S.K., Rodriguez, J.M.C. (eds.) Recent Advances in Intelligent Informatics. AISC, vol. 235, pp. 65–72. Springer, Heidelberg (2014)
29. Muthukumar, S., Krishnan, N., Tulasi Nachiyar, K., Pasupathi, P., Deepa, S.: Fuzzy information system based on image segmentation by using shadow detection. In: 2010 IEEE International Conference on Computational Intelligence and Computing Research (ICCIC), pp. 1–6. IEEE (2010)

Scrutiny of Nonlinear Adaptive Output Feedback Control for Robotic Manipulators

Davood Mohammadi Souran, Mohammad Hassan Askari,
Nima Razagh Pour, and Behrooz Razeghi

Abstract. In this paper we present a nonlinear adaptive output feedback control algorithm. The algorithm is for model reference adaptive control of robotic manipulators. This algorithm uses model signals in the regressor and the linearization law and hence, does not require an observer. We show via various simulations that this algorithm has a region of convergence. We also show that the region of convergence can be increased if a normalizing factor is used in the adaptation law.

Keywords: Adaptive control, Nonlinear System, Algorithm, Robotic Manipulators.

1   Introduction

The use of adaptive control strategies for robot manipulators is an area that has received interest from researchers over the last number of years. When exact

Davood Mohammadi Souran
School of Electrical and Computer Engineering, Shiraz University
e-mail: davood_souran@yahoo.com

Mohammad Hassan Askari
Sun Air Research Institute, Ferdowsi University of Mashhad
e-mail: mh.askari@mail.com

Nima Razagh Pour
Department of Electrical Engineering, Sadjad Institute of Higher Education
e-mail: nima.r.n@ieee.org

Behrooz Razeghi
Department of Electrical Engineering, Ferdowsi University of Mashhad
e-mail: behrooz.razeghi@gmail.com

S.M. Thampi, A. Gelbukh, and J. Mukhopadhyay (eds.), *Advances in Signal Processing*
and Intelligent Recognition Systems, Advances in Intelligent Systems and Computing 264,
DOI: 10.1007/978-3-319-04960-1_46, © Springer International Publishing Switzerland 2014

knowledge of the robot parameters is unknown, or the dynamics of a manipulator may be changing over time due to varying payload masses, an adaptive strategy is useful in estimating unknown parameters and modifying the controller to minimize tracking error. In many such adaptive control algorithms (e.g. [1] and [2]) position, velocity, and sometimes acceleration are required for the control and/or adaptation laws. However, while the position of a robot link can be measured accurately, measurement of velocity and acceleration tends to result in noisy signals [3]. In order to overcome the problem of noisy velocity and acceleration measurements, an observer can be used to estimate these values based on position measurements only. Not only can such a method help to yield velocity and acceleration estimates with less noise than their measured values, but robot manipulator setups using this sort of approach need only to be equipped with position sensors. This can help to decrease the costs of production.

Lee and Khalil [4] develop one such method of output feedback adaptive control using a high-gain observer. This observer is used to estimate position error and velocity error, and these estimated values are used in the control and adaptation laws. An acceleration signal is not required for either the control law or adaptation law. The advantage of such a high-gain observer is that the error in the observed signals tends toward zero quite rapidly, and the system is quick to recover performance similar to that achieved under full-state feedback control. A potential drawback of this method is that, due to the high gain of the observer, the presence of noise in the measurements (such as quantization error on digitized position measurements) would be amplified many times by the high observer gain.

In order to overcome difficulties with noise in a high-gain observer, we propose an addition to the adaptive controller developed in [1] that involves the use of a linear second-order observer to estimate position, velocity, and acceleration signals. When the Computed Torque Method (CTM) is used to control a non-linear robot manipulator, assuming perfect linearization occurs, each link of the closed-loop system can be modeled as a double integrator. We then use a simple second-order linear observer constructed based on the dynamics of the double integrator to estimate position, velocity, and acceleration from position measurements. The observer gains are set large enough that the observer poles are significantly faster than the error dynamics of the system, so that the effect of observer error has a minimal impact on the control of the system. However, the observer gains must be small enough to ensure that the algorithm is still effective in the presence of noise. Many adaptive controllers [5]-[6] for robot manipulators have been proposed to deal with unknown parameters. However, the use of an adaptive control algorithm may lead to a complicated implementation. On the other hand, the robust controllers developed in [7]-[8] allow for unstructured time-varying uncertainties but they are unable to guarantee asymptotic tracking (even in the absence of disturbances).

An adaptive output-feedback controller with bounded inputs was put forward by Lopez-Araujo et al. [9], achieving global convergence of position errors to zero. A proposed solution using a variable structure observer is reported by Abdessameud and Khelfi [10]. Other variations of this controller are proposed by Moreno-Valenzuela et al. [11]-[12], who prove local asymptotic stability via singular perturbations theory.

In this paper, we derive this algorithm and evaluate its region of stability. The paper is organized as follows. Section 2 develops the algorithm and Section 3 shows the simulations and analyzes the results. Finally, Section 4 concludes the paper and discusses future work.

2   The Algorithms

Figure 1 illustrates the block diagram for the nonlinear model reference adaptive controller using the reference model signals in the regressor of the adaptation law. The observer signals are used only in the linearization law.

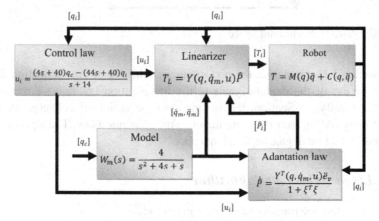

Fig. 1 Simulation Block Structure of Algorithm

The equation of the dynamics for a two degree of freedom (DOF) rigid body robot, which is illustrated in figure 2 and presented in joint space in figure 3, is given with the following equation.

$$T = M(q)\ddot{q} + C(q,\dot{q}) \tag{1}$$

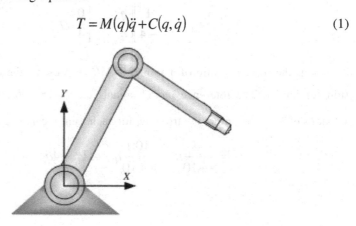

Fig. 2 Two DOF Robot Manipulator

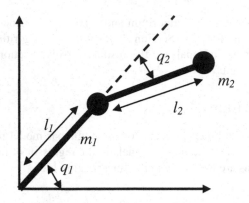

Fig. 3 Simulated Robot in Joint Space

In Section 2.1, we analyze the performance of this algorithm. In Section 2.2, we extend the use of the model signals to the linearization law and we remove the observer. Finally, in Section 2.3, we increase the region of convergence of the algorithm by adding a normalizing factor to the adaptation law. The algorithms are simulated for a two degree of freedom robot.

2.1 Linear Observer Algorithm

The model reference transfer function is given by:

$$W_M(s) = \frac{4}{S^2 + 4S + 4}$$

(2)

Which in state space form is,

$$\begin{bmatrix} \dot{q}_{mi} \\ \ddot{q}_{mi} \end{bmatrix} = \begin{bmatrix} 0 & 1 \\ -4 & -4 \end{bmatrix} \begin{bmatrix} q_{mi} \\ \dot{q}_{mi} \end{bmatrix} + \begin{bmatrix} 0 \\ 4 \end{bmatrix} q_{Ci}$$

(3)

The q_{mi} is the model position for joint i and q_{ci} represents the commanded position for joint i. The reference model signals are q_{mi} and $\dot{q}_{mi}$ which are filtered versions of q_{ci}. The virtual control law, for each joint i, can be computed by:

$$u_i = \frac{-4}{S+10} u_i + \frac{400}{S+10} q_i - 44q_i + 4q_{ci}$$

(4)

Or as it is given in Figure 1.

$$u_i = \frac{(4s+40)q_c - (44s+40)q_i}{S+14}$$

(5)

The next step is to define torques which will be applied to the each joint of the robot manipulator. We know that the robot dynamic model is given by equation (1) and can be written in linear regression form as,

$$T = Y(q,\dot{q},\ddot{q})P$$

(6)

In order to achive perfect linearization of the robot nonlinear dynamic model (6), the torques applied to the each joint of the robot manipulator should be computed according to equation (7).

$$T_L = M(q)u + C(q,\dot{q})$$

(7)

Where u is a virtual control signal given by equation (4), while TL is an output from the linearzer. Equation (7) can be written in linear regression from as,

$$T_L = Y(q,\dot{q},u)\hat{P} \quad \text{where}$$

(8)

$$Y = \begin{bmatrix} u_1 & 2u_1 + u_2 + 2u_1\cos(q_2) - 2\dot{q}_1\dot{q}_2\sin(q_2) + u_2\cos q_2 - \dot{q}_2^2\sin q_2 \\ 0 & u_1 + u_1\cos(q_2) + u_2 + \dot{q}_1^2\sin(q_2) \end{bmatrix}$$

and

$$\hat{P} = \begin{bmatrix} \hat{P}_1 \\ \hat{P}_2 \end{bmatrix}$$

The last part of the simulation block diagram illustrated in Figure 1 that has to be determined is the adaptation law. The adaptation law of algorithm for parameter P is given by equation (9).

$$\dot{\hat{P}} = \frac{Y^T(q,\dot{q}_m,u)\ddot{e}_v}{1+\varsigma_1\varsigma_1^T} = \frac{1}{1+\varsigma^T\varsigma}\begin{bmatrix} Y_{11} & Y_{21} \\ Y_{12} & Y_{22} \end{bmatrix}\begin{bmatrix} u_1 - \ddot{q}_{m1} \\ u_2 - \ddot{q}_{m2} \end{bmatrix}$$

(9)

Where $(1+\varsigma^T\varsigma)$ is a normalizing factor, $\ddot{e}_v$ is virtual error, $\hat{P}$ is an estimate of parameter P, and

$Y_{11} = u_1$

$Y_{12} = 2u_1 + u_2 + 2u_1\cos(q_2) - 2\dot{q}_1\dot{q}_2\sin(q_2) + u_2\cos q_2 - \dot{q}_2^2\sin q_2$

$Y_{21} = 0$

$Y_{22} = u_1 + u_1\cos(q_2) + u_2 + \dot{q}_1^2\sin(q_2)$

The ξ is defined as

$$\zeta = W_m(s)IY^T(q,\dot{q}_m,u) \tag{10}$$

The estimated link position $\hat{q}$ in equation (8) is calculated using the linear observer. This observer is given by:

$$\begin{bmatrix} \dot{\hat{q}}_i \\ \ddot{\hat{q}}_i \end{bmatrix} = \begin{bmatrix} 0 & 1 \\ 0 & 0 \end{bmatrix}\begin{bmatrix} \hat{q}_i \\ \dot{\hat{q}}_i \end{bmatrix} + \begin{bmatrix} 0 \\ 1 \end{bmatrix}u_i + \begin{bmatrix} 20 \\ 100 \end{bmatrix}(q_i - \hat{q}_i) \tag{11}$$

The acceleration of the robot due to the this torque is determined using the following equation:

$$\ddot{q} = M^{-1}(q)[T - C(q,\dot{q})] \tag{12}$$

Where

$$M(q) = \begin{bmatrix} m_1l^2 + 2m_2l^2 + 2m_2l^2\cos(q_2) & m_2l^2 + m_2l^2\cos(q_2) \\ m_2l^2 + m_2l^2\cos(q_2) & m_2l^2 \end{bmatrix}$$

Which is the mass matrix and

$$C(q,q) = \begin{bmatrix} -2m_2l^2\dot{q}_1\dot{q}_2\sin q_2 - m_2l^2\dot{q}_2^2\sin q_2 \\ m_2l^2\dot{q}_1^2\sin q_2 \end{bmatrix}$$

which represents the Coriolis and Centrifugal forces.Finally, the adaptation law is defined as follows:

$$\dot{\phi} = Y(q_m,\dot{q}_m,\ddot{q}_m)(t)e_v \tag{13}$$

where the vector $e_v = u - \ddot{q}_m$ and

$$Y(q_m,\dot{q}_m,\ddot{q}_m) =$$

$$\begin{bmatrix} \ddot{q}_{m1} & 2\ddot{q}_{m1} + \ddot{q}_{m2} + 2\ddot{q}_{m1}\cos(q_{m2}) - 2\dot{q}_{m1}\dot{q}_{m2}\sin(q_{m2}) + \ddot{q}_{m2}\cos q_{m2} - \dot{q}_{m2}^2\sin(q_{m2}) \\ 0 & \ddot{q}_{m1} + \ddot{q}_{m1}\cos(q_{m2}) + \ddot{q}_{m2} - \dot{\hat{q}}_{m1}^2\sin(q_{m2}) \end{bmatrix}$$

The term e_{vi} represents the virtual error for each joint. If perfect linearization has occurred then e_{vi} will equal zero and the virtual control signal u_i, will equal the desired acceleration $\ddot{q}_{mi}$, for each joint. This adaptation law uses model values only. Since, the simulations of this algorithm show good results we can extend the use of the model values further. Next we analyze what happens if we remove the observer. The estimated value, $\dot{\hat{q}}$, used in the linearization law will be replaced by the model value, $\dot{q}_m$. This algorithm is discussed next.

2.2   No Observer Algorithm

The model dynamics for the manipulator joints are given by equation (2), (3) and the virtual control law is given by equation (5). Once this control law is calculated, the torque applied to the robot can be calculated by:

$$T_L = Y(q_m, \dot{q}_m, u)\hat{P} \quad \text{where} \tag{14}$$

$$Y = \begin{bmatrix} u_1 & 2u_1 + u_2 + 2u_1\cos(q_2) - 2\dot{q}_{m1}\dot{q}_{m2}\sin(q_2) + u_2\cos q_2 - \dot{q}_{m2}^2\sin q_2 \\ 0 & u_1 + u_1\cos(q_2) + u_2 + \dot{q}_{m1}^2\sin(q_2) \end{bmatrix}$$

and

$$\hat{P} = \begin{bmatrix} \hat{P}_1 \\ \hat{P}_2 \end{bmatrix}$$

The acceleration of the robot due to this torque is determined using equation (12) and the adaptation law for the parameters is given by (13). The simulations indicate that this algorithm also has a region of convergence. But this region is small compared to the algorithm with the observer. From further analysis of this algorithm, we found that the region of convergence can be increased if a normalizing factor is used in the adaptiaton law. This algorithm is discussed next.

2.3   No Observer Normalized Algorithm

Equation (2) and (3) give the dynamics for the manipulator joints and the virtual control law is given by equation (5). Once this control law is calculated, the torque applied to the robot can be calculated by (14). The acceleration of the robot due to this torque is determined using equation (12) and the adaptation law for the parameters is given by:

$$\dot{\phi} = \frac{Y(q_m, \dot{q}_m, \ddot{q}_m)(t)e_v}{\left(1 + \zeta^T\zeta\right)} \quad \text{where}$$

$\zeta = W_m(s)IY\left(q_m, \dot{q}_m, \ddot{q}_m\right)$ and $e_v = u - \ddot{q}_m$. W_m is given by equation (2). The simulations of this algorithm clearly show that this algorithm has a bigger region of convergence than the non-normalizing algorithm.

3   Simulations and Results

We now show the results of the simulations using these algorithms. The commanded position, q_{ci}, for each manipulator joint i, is a sequence of unit steps. The period of the unit step sequence is 8 seconds. The amplitude of the unit step sequence is varied in the simulations. We consider three simulation cases as follows.

Simulation 1: Parameter initial values: 1,3
Simulation 2: Parameter initial values: 10,8
Simulation 3: Parameter initial values: 1,3

In Section 3.1 we look at the linear observer algorithm and in Section 3.2 we look at the no observer algorithm. In Section 3.3 we consider the no observer normalized algorithm. Finally in Section 3.4 we summarize the results of other simulations that were tried.

3.1   Linear Observer Algorithm

In this algorithm, the model values are used only in the regressor vector. We now analyze the region of convergence of the algorithm using Simulation 1, Simulation 2 and Simulation 3. Figure 4 below shows the trajectory error for Simulation 1.

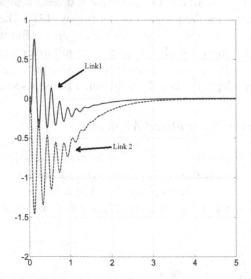

Fig. 4 Trajectory Error for Simulation 1

The algorithm converges and the error reduces to zero in about 5 seconds. Next we look at Simulation 2. Figure 5 shows the trajectory error for this simulation.

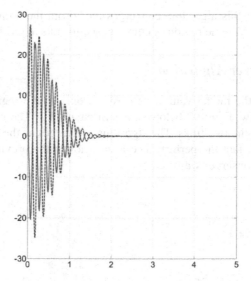

Fig. 5 Trajectory Error for Simulation 2: Joint 1(blue) and Joint 2 (Red)

This algorithm converges for Simulation 2 as well, though the time it takes for the error to reduce to zero is now 2 seconds. Next we consider Simulation 3. Figure 6 shows the graph for the trajector error

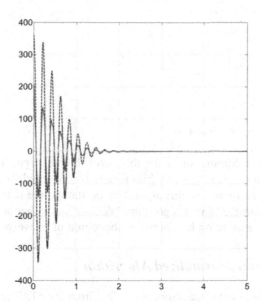

Fig. 6 Trajectory Error for Simulation 3

Simulation 3 for this algorithm converges too. From the above results we can conclude that there is some region of convergence for this algorithm.

3.2   No Observer Algorithm

In this algorithm, the model values are used not only in the regressor but also in the linearization law. Figure 7 below shows Simulation 1. The graph displays the trajectory error for the two links. The algorithm converges as the trajectory error is reduced to zero. In fact the performance is superior to the previous simulation in Figure 4 where an observer was used.

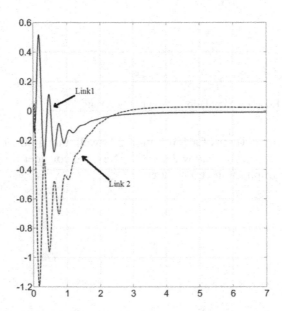

Fig. 7 Trajectory Error for Simulation 1

Simulation 2 and Simulation 3 for this algorithm diverge. These simulation graphs are not shown. The fact that Simulation 1 converged means that there is some region of convergence of this algorithm too though the region is quite small compared to the linear observer algorithm. We now show that adding the normalizing factor in the adaptation law increases the region of convergence.

3.3   No Observer Normalized Algorithm

Figure 8 below shows the trajectory error for Simulation1. The trajectory error goes to zero in about 4 seconds and hence, the algorithm converges for this case.

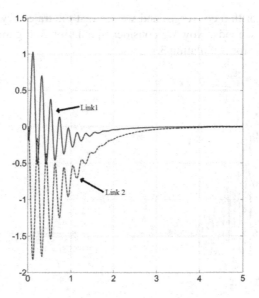

Fig. 8 Trajectory Error for Simulation 1

Next, we consider Simulation 2. Figure 9 shows the trajectory error for this simulation.

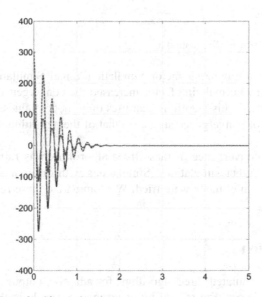

Fig. 9 Trajectory Error for Simulation 2

Again, the algorithm converges and we see that the trajectory error goes to zero in just about 2 seconds. Now we consider Simulation 3. Figure 10 below shows
the trajectory error for Simulation 3.

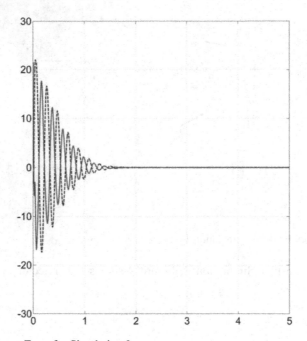

Fig. 10 Trajectory Error for Simulation 3

By adding the normalizing factor Simulation 2 and Simulation 3 have converged. Hence, the normalizing factor increased the convergence region. For the
simulation cases tried, this algorithm that uses only model values and no observer
now has a region of convergence similar to that of the algorithm that uses a linear
observer.

The region of convergence of these three algorithms was further explored by
trying out various other simulations. Simulations with different initial parameter
errors and various amplitudes were tried. We summarize these results in the next
section.

4   Conclusion

In this paper we simulated three algorithms for adaptive output feedback control
of robotic manipulators. We started by using model signals in the regressor and
found that the algorithm had a good region of convergence. We then extended the
use of the model signals to the linearization law, removing the observer from the
algorithm. We noticed that this algorithm also had a region of convergence but

this region was small compared to the previous algorithm. We further observed that the region of convergence could be increased if we used a normalizing factor. We then simulated the three algorithms and presented the results. From various simulations we can conclude though the algorithms are not globally stable, they do have some region of convergence, which decreases as the input amplitude is increased. The valuable contribution is that we now have an algorithm for nonlinear model reference adaptive control that does not require an observer and uses only model signals in the regressor. The fact that no observer is required simplifies the implementation and the use of model signals in the regressor removes the need of any measurement or estimation of the joint velocity and acceleration signals. The future work involves determining theoretically the region of convergence and proving the stability of this algorithm.

References

1. Craig, J.J., Hsu, P., Sastry, S.S.: Adaptive control of mechanical manipulators. The International Journal of Robotics Research 6(2), 16–27 (1987)
2. Slotine, J.-J.E., Li, W.: On the adaptive control of robot manipulators. The International Journal of Robotics Research 6(3), 49–59 (1987)
3. Hajjir, H., Schwartz, H.M.: An adaptive nonlinear output feedback controller for robot manipulators. In: Proceedings of the American Control Conference (June 1999)
4. Lee, K.W., Khalil, H.K.: Adaptive output feedback control of robot manipulators using high-gain observer. International Journal of Control 6, 869–886 (1997)
5. Arcak, G.M., Kokotovic, P.: Nonlinear Observers: A Circle Criterion Design D. In: Proc. of 38th Conf. on Decision and Control (December 1999)
6. Park, J., Chung, W.K.: Analytic Nonlinear H∞Inverse-Optimal Control for Euler-Lagrange System. IEEE Transactions on Robotics and Automation 16(6), 847–854 (2000)
7. Lee, G.-W., Cheng, F.-T.: Robust Control of Manipulators Using the Computed Torque Plus H∞Compensation Method. IEE Proceedings Control Theory Application 143(1), 64–72 (1996)
8. Wang, H., Wei, L., Mu, D., Li, Y.: Robust Adaptive Control of X-Y Position Table with Uncertainty. In: Proceedings of the 5th World Congress on Intelligent Control and Automation, pp. 3366–3369 (June 2004)
9. Lopez-Araujo, D., Zavala-Rio, A., Santibanez, V., Reyes, F.: Output-feedback adaptive control for the global regulation of robot manipulators with bounded inputs. International Journal of Control, Automation and Systems 11(1), 105–115 (2012)
10. Abdessameud, A., Khelfi, M.F.: A variable structure observer for the control of robot manipulators. International Journal of Applied Mathematics and Computer Science 16(2), 189–196 (2006)
11. Moreno-Valenzuela, J., Santibanez, V., Campa, R.: A class of OFT controllers for torque-saturated robot manipulators: Lyapunov stability and experimental evaluation. Journal of Intelligent & Robotic Systems 51(1), 65–88 (2008)
12. Moreno-Valenzuela, J., Santibanez, V., Campa, R.: On output feedback tracking control of robots manipulators with bounded torque input. International Journal of Control, Automation, and Systems 6(1), 76–85 (2008)

Single Channel Speech Enhancement for Mixed Non-stationary Noise Environments

Sachin Singh, Manoj Tripathy, and R.S. Anand

Abstract. Speech enhancement is very important step for improving quality and intelligibility of noisy speech signal. In practical environment more than one noise sources are present, hence it is necessary to design a technique/ algorithm that can remove mixed noises or more than one noises from single-channel speech signals. In this paper, a single channel speech enhancement method is proposed for reduction of mixed non-stationary noises. The proposed method is based on wavelet packet and ideal binary mask thresholding function for speech enhancement. Db10 mother wavelet packet transform is used for decomposition of speech signal in three levels. After decomposition of speech signal a binary mask threshold function is used to threshold the noisy coefficients from the noisy speech signal coefficients. The performance of the proposed wavelet with ideal mask method is compared with Wiener, Spectral Subtraction, p-MMSE, log-MMSE, Ideal channel selection, Ideal binary mask, hard and soft wavelet thresholding function in terms of PESQ, SNR improvement, Cepstral Distance, and frequency weighted segmental SNR. The proposed method has shown improved performance over conventional speech enhancement methods.

1 Introduction

A medium is necessary for communication among humans, if communicating medium is highly noisy than it is much difficult to transfer speech. To understanding speech, an enhancement technique is required for removing noisy environment from noisy speech signal.

Sachin Singh · Manoj Tripathy · R.S. Anand
Indian Institute of Technology Roorkee, India
e-mail: oxygen_sachin@rediff.com
    tripathy.manoj@gmail.com
    anandfee@iitr.ernet.in

S.M. Thampi, A. Gelbukh, and J. Mukhopadhyay (eds.), *Advances in Signal Processing*   545
and Intelligent Recognition Systems, Advances in Intelligent Systems and Computing 264,
DOI: 10.1007/978-3-319-04960-1_47, © Springer International Publishing Switzerland 2014

In previous literature various speech enhancement techniques are given for noise reduction [1]. Some are given as spectral subtraction [2], modified spectral subtraction [3], Wiener [4], and gain based method like MMSE STSA [5], log-MMSE [6], p-MMSE [7] etc. All methods give improvements in speech quality and not improve intelligibility upto a satisfactory levels. For improving intelligibility and quality both, some Modulation channel selection based methods [8-11] are given in literature. Further improvements in speech is also possible by using binary mask as a threshold function for removing noisy coefficients and this function improve quality and intelligibility both at appropriate levels. Some wavelet transform based methods are also given in the literature for speech enhancement [12-14]. Adaptive thresholding is much effective in comparison to hard and soft threshold function for removing noise from noisy speech signal to improve in quality and intelligibility [15].

In the real environment speech may be distorted by more than one noise source. Since most of the time it is not possible to consider only one noise in communicated speech or number of noises in speech signal. In this paper, a wavelet packet transform with binary mask threshold function is proposed for reduction of mixed non-stationary noises from speech signal.

The remaining paper is organized as follows: Section 2 brief introduction of wavelet packet transform. In section 3 a wavelet based on binary mask method is proposed for speech enhancement. Results and discussion is given in section 4 and finally conclusion is given in section 5.

2   Noisy Speech Enhancement

In this paper, a noisy single channel Hindi speech signal is modeled as the sum of the clean Hindi speech and two real non-stationary additive background noises. This noisy speech is given as:

$$y(n) = cl(n) + d_1(n) + d_2(n) \tag{1}$$

Where, $y(n)$, $cl(n)$ and $d(n)$ denote frames of noisy speech, clean speech and additive background noises, respectively.

Wavelet transform is a powerful tool for removing noises from speech signal. Hard, soft and adaptive threshold functions are used for thresholding wavelet coefficients [15] [16] [17]. The hard thresholding is defined as:

$$\hat{X} = \begin{cases} Y & if \ |Y| \geq t \\ 0 & if \ |Y| < t \end{cases} \tag{2}$$

The soft thresholding is defined as:

$$\hat{X} = \begin{cases} Y - \text{sgn}(Y) & \text{if } |Y| \geq T \\ 0 & \text{if } |Y| < T \end{cases} \tag{3}$$

Where, Y represents the wavelet coefficients and t, T are threshold limits which are given as

$$t = \frac{med(|Y|)}{0.6745} \quad \text{for hard threshold} \tag{4}$$

$$T = t\sqrt{2.\log(n)} \quad \text{for soft threshold} \tag{5}$$

For decomposition, Db10 mother wavelet packet transform is used upto 3rd levels. After getting eight band wavelet packet coefficients, binary mask threshold function is used in place of hard and soft thresholding functions.

3   Proposed Method

A speech enhancement method is proposed for reduction of mixed non-stationary noises from single channel speech that is based on Db10 wavelet packet and binary mask. To estimate the clean speech signal from the noisy speech, a Wiener gain is multiplied with noisy speech signal in each frame. This Wiener gain $G(k,t)$ is calculated as in [18]:

$$G(k,t) = \sqrt{\frac{SNR_{priori}(k,t)}{1 + SNR_{priori}(k,t)}} \tag{6}$$

In this equation, SNR_{priori} is the priori SNR, which is calculated by using the following equation [18]:

$$SNR_{priori}(k,t) = \alpha . \frac{\hat{X}(k, t-1)}{\hat{\lambda}_D(k, t)} + (1-\alpha) * \max\left[\frac{Y^2(k, t)}{\hat{\lambda}_D(k, t)} - 1, 0\right] \tag{7}$$

$$\hat{X}(k,t) = G(k,t) * Y(k,t) \tag{8}$$

$$x_{enhanced} = \begin{cases} x(k,t), & \text{if } \hat{X}(k,t) \leq cl(k,t) \\ 0 & , \quad otherwise \end{cases} \tag{9}$$

Where, $\hat{\lambda}_D$ is the background noise variance estimation and smoothing constant $\alpha = 0.98$. The background noise spectrum is estimated by using method proposed by Rangachari and Loizou in [18]. The block diagram of the proposed method is given in figure 1. The estimated speech spectrum is calculated using above eq. 6, 7 in eq. 8. After calculating the estimated spectrum of speech, to enhance the quality and intelligibility a binary mask function (eq. 9) is used for thresholding the unexpected wavelet coefficients in speech spectrum. Now inverse wavelet transform is applied for all eight band wavelet coefficients as given in fig. 1. An enhanced speech signal is obtained after inverse wavelet packet transform.

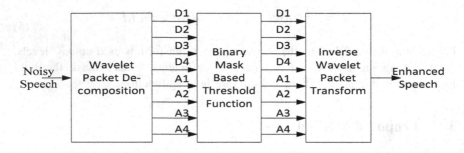

Fig. 1 Block diagram of the proposed speech enhancement method

4   Results and Discussion

For performance evaluation of the proposed method, Hindi language speech patterns have been taken from IIIT-H (International Institute of Information Technology Hyderabad) Indic speech database [19]. Hindi language speech corpus has 1000 clean speech patterns. The noise data base is taken from NOIZEX-92 database [20] for generating mixed noise patterns for analysis. The clean speech patterns of Hindi language have been added with mixed noise patterns. These mixed noise patterns are f16 + babble, machinegun + pink, and factory floor + white. The original speech patterns were sampled at 16 kHz.

The performance of the proposed method is compared in terms of parameters like PESQ, Cepstral Distance, output SNR, and Fw-SSNR. The values of the parameters are given in tables 1-4 for comparative analysis. In table 1 output SNR values are given in mixed noise environment for Hindi speech enhancement. The output SNR is maximum at +5 dB input SNR in factory floor and white noise mixed environment. The maximum improvement is given by proposed wavelet with IDBM method. This shows that the maximum quality speech can be reconstruct by proposed method.

In table 2 Frequency weighted SSNR values are given at various levels of input SNR. The higher value shows good improvement in quality of enhanced speech and lower values show fewer improvements in quality of speech and more remaining noise in the speech. For the given comparative analysis, proposed method shows fewer amounts of remaining noises in the enhanced speech signal.

The PESQ performance measure parameter gives the mean opinion score of the estimated speech from noisy speech. The range of PESQ lies from 1 to 4.5 for bad to excellent speech quality. In the table 3 PESQ values are given for Hindi speech signal in mixed noise environment.

The minimum CD value gives maximum improvement in estimated speech. in the table 4 output CD values are given for all levels of input SNR. The proposed method gives minimum value of CD measure parameter in all noise environments.

SII parameters value give intelligibility of speech pattern. The higher value gives more listening quality or intelligibility. These SII values in various mixed noises with +5dB input SNR is given in table5 for Hindi speech signals. The range of SII is from 0 to 1 scale. Where, 0 is given for not intelligible or not clear meaning and near 1 for highest meaningful speech.

In the table 5 the maximum SII value is given in machinegun + pink mixed noise environment at +5 dB input SNR. This maximum value is 0.4442 after estimating speech signals from proposed method. The maximum SII value is given by wavelet based IDBM proposed method.

Table 1 Output SNR values

Noises	SNR Input	Wiener	Spectral Sub.	p-MMSE	log-MMSE	SOFT	HARD	ICS	IDBM	WPT IDBM
F16 & Babble	-15	-2.3228	-6.3402	-2.7278	-1.4599	0.1074	0.1227	-0.341	1.7234	**1.9144**
	-10	-1.7747	-3.9115	-3.2184	-1.6779	0.3564	0.3975	0.4341	1.9081	**4.9554**
	-5	-0.4984	-1.2080	-1.2184	-1.0523	1.1027	1.1781	1.6116	2.1099	**5.0651**
	0	0.5938	0.9557	0.9455	0.3557	2.9073	3.0064	2.3550	2.2943	**8.7918**
	5	0.9135	3.6325	2.9951	1.2099	6.0987	6.2080	2.7208	2.3954	**7.4173**
Machinegun & Pink	-15	-0.7753	-5.0933	-1.7526	-1.6779	0.0951	0.1188	-0.340	1.9557	**5.1966**
	-10	-0.1275	-4.3398	-1.8198	-1.4599	0.3175	0.3849	0.4341	2.1763	**6.5033**
	-5	1.3932	-0.2685	-1.0912	-1.0523	1.0053	1.1570	1.6116	2.3215	**6.5157**
	0	3.2380	3.2606	0.7479	0.3557	2.7176	2.9725	2.3550	2.3828	**6.9647**
	5	3.2257	6.0620	4.4728	1.2099	5.8121	6.1535	2.7208	2.4372	**9.4900**
Factory floor & White	-15	-0.3544	-0.8270	-1.8356	-0.8670	0.0914	0.1186	0.2282	1.8533	**2.5929**
	-10	-0.2839	-0.6202	-1.6846	-0.7145	0.3095	0.3840	1.2293	2.0681	**3.7844**
	-5	0.3937	-0.0985	0.3967	0.2028	0.9900	1.1596	2.5925	2.2488	**9.4924**
	0	0.4275	2.1069	1.8154	0.8680	2.6975	2.9735	3.9886	2.3678	**9.6020**
	5	0.4498	4.5250	3.7515	0.9420	5.7831	6.1485	4.8562	2.4343	**13.641**

Table 2 Frequency weighted SSNR (fw-SSNR) values

Noises	SNR Input	Wiener	Spectral Sub.	p-MMSE	log-MMSE	SOFT	HARD	ICS	IDBM	WPT IDBM
	-15	5.0571	3.1652	4.0320	3.6735	-0.748	-0.147	5.5686	5.3070	**10.5472**
F16 & Babble	-10	6.0642	3.5508	4.9666	4.7690	-0.446	0.1070	6.3331	6.3098	**11.5256**
	-5	7.0349	4.3527	5.7113	5.7564	0.1922	0.7147	7.1150	7.9000	**12.6038**
	0	7.8101	5.3783	6.8051	6.5908	1.3700	1.8111	7.8795	9.6244	**13.7704**
	5	8.5420	6.2126	7.6541	7.4505	3.1060	3.4668	8.5125	11.0828	**15.2217**
	-15	5.8261	3.6454	4.5390	4.3748	0.0777	0.8930	6.1386	5.6329	**11.1857**
	-10	7.0447	3.8034	5.5766	5.4821	0.6654	1.4324	6.9527	7.0784	**12.1470**
Machinegun & Pink	-5	8.1942	4.2869	6.5798	6.5276	1.6390	2.3279	8.0529	8.6422	**13.1690**
	0	8.9267	5.3417	7.5296	7.6897	3.0707	3.7346	9.2448	10.6607	**14.5452**
	5	9.7453	6.5323	8.1045	8.4328	5.0220	5.7362	10.407	12.7112	**16.1519**
	-15	4.4029	3.4376	4.8502	4.5135	0.1634	1.0884	4.9541	5.5395	**11.0368**
	-10	5.5310	4.4487	5.6798	5.5474	0.6235	1.5122	5.9086	6.7019	**11.9830**
Factory floor & White	-5	6.6120	5.2682	6.8295	6.7192	1.3315	2.2791	6.9053	8.6436	**13.0345**
	0	7.4657	6.2192	7.5922	7.7587	2.7032	3.6208	8.1404	10.5455	**14.2575**
	5	8.4698	7.3728	8.6906	8.3784	4.5114	5.5029	9.3946	12.3724	**15.7827**

Table 3 PESQ values

Noises	SNR Input	Wiener	Spectral Sub.	p-MMSE	log-MMSE	SOFT	HARD	ICS	IDBM	WPT IDBM
	-15	1.4113	0.7360	1.1122	0.8631	0.6088	0.5178	1.6374	2.1603	**2.9461**
F16 & Babble	-10	1.7369	1.2246	1.6300	1.5443	0.6570	0.6227	1.9248	2.4629	**3.1161**
	-5	1.9502	1.5909	1.9117	1.9385	0.9095	0.8751	2.0734	2.7196	**3.3128**
	0	2.0538	1.9101	2.1188	2.0642	1.1928	1.1543	2.2435	2.9735	**3.4904**
	5	2.1666	2.1459	2.2772	2.1359	1.5488	1.5081	2.3775	3.2065	**3.6692**
	-15	1.7724	0.7781	1.0428	0.9297	0.3475	0.3362	2.0334	2.4390	**3.0693**
	-10	2.1900	1.2577	1.7752	1.7312	0.6439	0.6150	2.3538	2.6606	**3.2843**
Machinegun & Pink	-5	2.3156	1.5517	2.1287	2.0665	1.0599	1.0054	2.6068	2.9464	**3.4962**
	0	2.4644	1.9179	2.3978	2.2364	1.5037	1.4469	2.8100	3.1747	**3.6290**
	5	2.3163	2.2206	2.5526	2.4324	1.9344	1.8887	3.0038	3.4010	**3.7283**
	-15	1.2016	0.5885	1.3526	1.2638	0.5855	0.4971	1.6822	2.3729	**2.9461**
	-10	1.5249	1.0285	1.6659	1.5962	0.8421	0.7599	1.9231	2.6667	**3.1161**
Factory floor & White	-5	1.8298	1.4618	1.9585	1.9130	1.1432	1.0697	2.1347	2.9649	**3.3128**
	0	1.9031	1.8881	2.1148	2.0634	1.5136	1.4554	2.3160	3.1514	**3.4904**
	5	2.0063	2.1964	2.3665	2.0749	1.9444	1.8910	2.4765	3.4042	**3.6692**

Table 4 Cepstral Distance measure values

Noises	SNR Input	Wiener	Spectral Sub.	p-MMSE	log-MMSE	SOFT	HARD	ICS	IDBM	WPT IDBM
	-15	6.3712	8.9649	6.4998	6.8541	8.9740	8.1510	5.7062	6.6023	**3.8327**
F16 & Babble	-10	5.7469	8.3147	5.7121	6.0816	8.8841	8.0589	5.1584	6.0197	**3.5062**
	-5	5.0810	7.7327	5.2491	5.3721	8.6720	7.8635	4.5837	5.1234	**3.1802**
	0	4.5786	6.9332	4.6580	4.7810	8.3275	7.6145	4.1056	4.0709	**2.8688**
	5	4.1148	6.2579	4.2262	4.3377	7.8272	7.2667	3.7389	3.4918	**2.4457**
	-15	6.9649	9.2130	6.9907	7.2290	8.7038	7.7673	6.2689	6.7802	**3.7350**
	-10	6.3223	8.6247	6.1562	6.4585	8.6388	7.6427	5.6466	6.0449	**3.4162**
Machinegun & Pink	-5	5.6094	8.1748	5.4169	5.7169	8.4468	7.4515	4.9943	5.2407	**3.0929**
	0	4.9475	7.2596	4.8335	5.0060	8.1577	7.2002	4.3914	4.3674	**2.8114**
	5	4.3986	6.3997	4.3537	4.5193	7.6599	7.0104	3.8266	3.5898	**2.4343**
	-15	6.9740	7.4833	6.8650	7.0829	8.9829	7.6844	6.4056	6.9002	**3.8550**
	-10	6.3688	7.0698	6.2460	6.5174	8.8797	7.6085	5.8588	6.3690	**3.5749**
Factory floor & White	-5	5.7123	6.5540	5.5772	5.8053	8.6688	7.4638	5.1973	5.4502	**3.2958**
	0	5.0298	5.9057	4.9941	5.1695	8.2869	7.2162	4.5903	4.6898	**3.0416**
	5	4.4418	5.1847	4.4824	4.6432	7.8035	7.0270	3.9377	4.1103	**2.7106**

Table 5 Speech Intelligibility Index (SII) values

Noises	SNR Input	Wiener	Spect. Subt.	p-MMSE	log-MMSE	SOFT	HARD	ICS	IDBM	WPT IDBM
F16 & Babble	5dB	0.1499	0.1854	0.1676	0.1485	0.3444	0.3549	0.1779	0.4618	0.3470
Machinegun & pink	5dB	0.2391	0.2633	0.1731	0.1972	0.4118	0.4292	0.3129	0.4618	0.4442
Factory floor & White	5dB	0.1979	0.2568	0.2356	0.2069	0.4220	0.4407	0.3044	0.4618	0.4417

In the fig 2, spectrograms of (a) clean (b) noisy speech at 5dB (c) proposed method (d) log-MMSE (e) Wiener (f) Spectral Sub. (g) p-MMSE (h) IDBM (i) ICS (j) hard wavelet thresholding (k) Soft in white + factory floor mixed noise environment at +5 dB. From spectrograms comparison, it is clear that the wavelet based IDBM method gives higher quality and intelligibility.

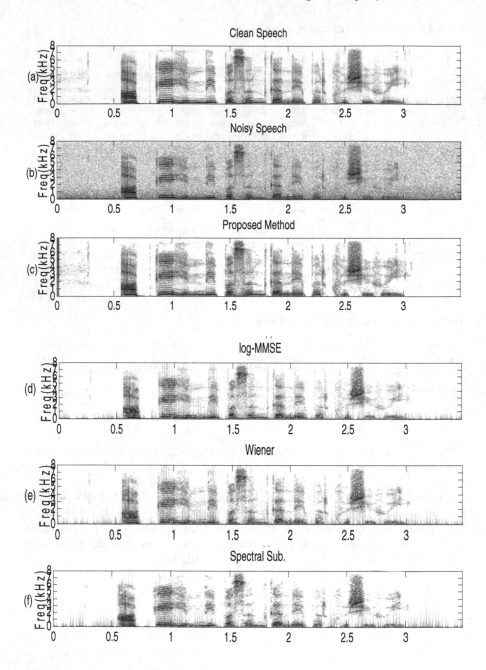

Fig. 2 Shows comparative time-frequency spectrograms of Hindi speech pattern *"apke hindi pasand karne par khusi hui"*

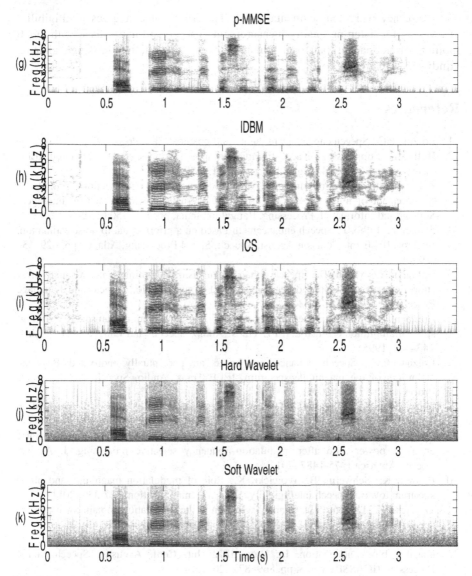

Fig. 2 (*continued*)

5   Conclusions

In this paper, a method based on wavelet packet and binary mask threshold function is proposed for mixed non-stationary noises reduction. The performance comparison has been made with Wiener, Spectral Subtraction, p-MMSE, log-MMSE, Ideal channel selection, Ideal binary mask, hard and soft wavelet thresholding function in terms of PESQ, SNR improvement, Cepstral Distance, and frequency weighted segmental SNR. The SII parameter gives intelligibility test values for listening quality measurement. The given binary mask with Db10 mother wavelet packet transform is more robust to mixed noises for single channel Hindi speech patterns.

References

1. Loizou, P.C.: Speech enhancement theory and practice. CRC Press, USA (2007)
2. Boll, S.F.: Suppression of acoustic noise in speech using spectral subtraction. IEEE Trans. Acoustic, Speech, Signal Processing 113-120 (1979)
3. Krishnamurthy, P., Prasanna, S.R.M.: Modified spectral subtraction method for enhancement of noisy speech. In: Proc. 3rd International Conference on Intelligent Sensing and Information Processing, Bangalore, India, pp. 146–150 (2005)
4. Scalart, P., Filho, J.: Speech enhancement based on a priori signal to noise estimation. In: Proc. IEEE Int. Conf. on Acoust, Speech, Signal Processing, Atlanta, pp. 629–632 (1996)
5. Ephraim, Y., Malah, D.: Speech enhancement using a minimum mean-square error short-time spectral amplitude estimator. IEEE Trans. Audio, Speech, and Language Processing, 1109–1121 (1984)
6. Ephraim, Y., Malah, D.: Speech enhancement using a minimum mean square error log-spectral amplitude estimator. IEEE Trans. Audio, Speech, and Language Processing, 443–445 (1995)
7. Loizou, P.C.: Speech enhancement based on perceptually motivated Bayesian estimators of the magnitude spectrum. IEEE Trans. Audio, Speech, and Language Processing, 857-869 (2005)
8. Dubbelboer, F., Houtgast, T.: The concept of signal-to-noise ratio in the modulation domain and speech intelligibility. J. Acoust. Sociaty America, 3937-3947 (2008)
9. Jorgensen, S., Dau, T.: Predicting speech intelligibility based on the signal-to-noise envelope power ratio after modulation-frequency selective processing. J. Acoust. Sociaty America 1475-1487 (2011)
10. Paliwal, K., Schwerin, B., Wojcicki, K.: Role of modulation magnitude and phase spectrum towards speech intelligibility. Speech Communication, 327-339 (2011)
11. Wojcicki, K., Loizou, P.C.: Channel selection in the modulation domain for improved speech intelligibility in noise. J. Acoust. Sociaty America, 2904-2913 (2012)
12. Guoshen, Y., Bacry, E., Mallat, S.: Audio signal denoising with complex wavelets and adaptive block attenuation. In: Proc. IEEE Int. Conf. Acoustic, Speech Signal Processing (ICASSP), vol. 3, pp. 869–872 (2007)
13. Zhou, B., et al.: An improved wavelet-based speech enhancement method using adaptive block thresholding. In: IEEE Conference (2010)
14. Sanam, T.F., Shahnaz, C.: Enhancement of noisy speech based on a custom thresholding function with a statistically determined threshold. Int. J. Speech Technology (April 2012)
15. Donoho, D.L.: De-noising by soft thresholding. IEEE Trans. Inform. Theory 41, 613–627 (1995)
16. Donoho, D.L., Johnstone, I.M.: Ideal spatial adaptation by wavelet shrinkage. Biometrika 81, 425–455 (1994)

17. Yasser, G., Karami, M.R.: A new approach for speech enhancement based on the adaptive thresholding of the wavelet packets. Speech Communication 48, 927–940 (2006)
18. Rangachari, S., Loizou, P.C.: A noise-estimation algorithm for highly non-stationary environments. Speech Communication 48, 220–231 (2006)
19. Prahallad, K., Kumar, E.N., Keri, V.: The IIIT-H Indic Speech Databases. In: Proceedings of Interspeech, Portland, Oregon, USA (2012), http://speech.iiit.ac.in/index.php/research-svl/69.html
20. Varga, P., Steeneken, H.J.M.: Technical report, DRA Speech Research Unit) (1992), http://www.speech.cs.cmu.edu/comp.speech/ Sect-ion1/Data/noisex.html

Channel Robust MFCCs for Continuous Speech Speaker Recognition

Sharada Vikram Chougule and Mahesh S. Chavan

Abstract. Over the years, MFCC (Mel Frequency Cepstral Coefficients), has been used as a standard acoustic feature set for speech and speaker recognition. The models derived from these features gives optimum performance in terms of recognition of speakers for the same training and testing conditions. But mismatch between training and testing conditions and type of channel used for creating speaker model, drastically drops the performance of speaker recognition system. In this experimental research, the performance of MFCCs for closed-set text independent speaker recognition is studied under different training and testing conditions. *Magnitude spectral subtraction* is used to estimate magnitude spectrum of clean speech from additive noise magnitude. The mel-warped cepstral coefficients are then normalized by taking their mean, referred as *cepstral mean normalization* used to reduce the effect of convolution noise created due to change in channel between training and testing. The performance of this modified MFCCs, have been tested using *Multi-speaker continuous (Hindi) speech database* (By Department of Information Technology, Government of India). Use of *improved MFCC* as compared to conventional MFCC perk up the speaker recognition performance drastically.

Keywords: Text independent speaker recognition, MFCC, magnitude spectral subtraction, cepstral mean normalization.

Sharada Vikram Chougule
Department of Electronics and Telecommunication Engineering,
Finolex Academy of Management and Technology, Ratnagiri
Maharashtra, India

Mahesh S. Chavan
Department of Electronics Engineering,
KIT's College of Engineering, Kolhapur
Maharashtra, India

S.M. Thampi, A. Gelbukh, and J. Mukhopadhyay (eds.), *Advances in Signal Processing*
and Intelligent Recognition Systems, Advances in Intelligent Systems and Computing 264,
DOI: 10.1007/ 978-3-319-04960-1_48, © Springer International Publishing Switzerland 2014

1   Introduction

The largest challenge to use speaker recognition technology is the channel variability, which refers to changes in channel effects between enrolment and successive recognition (verification/identification).This mismatch between training and testing, greatly degrades the performance of automatic speaker recognition systems (e.g.[1],[2]). The most widely used speech recognition features are the Mel Frequency Cepstrum Coefficients (MFCCs), which are also used for speaker recognition. The wide-spread use of the MFCCs is due to the low complexity of the estimation algorithm and their good performance for automatic speech and speaker recognition tasks under clean and matched conditions [3],[6]. However, MFCCs are easily affected by common frequency localized random perturbations, to which human perception is largely insensitive. MFCC's lack of robustness in noisy or mismatched conditions have led many researchers to investigate robust variants of MFCCs. Studies in [5] had shown that estimation of both vocal source and vocal track related features were extracted by denoising the speech and using MFCC with wavelet octave coefficients of residues (WOCOR). Furthermore, dynamic cepstral features such as delta and delta-delta cepstral have been shown to play an essential role in capturing the transitional characteristics of the speech signal. So, delta MFCC, delta-delta MFCC, and other related features such as delta cepstral energy (DCE) and delta-delta cepstral energy (DDCE) are also has been introduced into the speaker recognition systems [7]. Also several acoustic features, like MFCC, LPCC, PLP are admired and extensions of these (frequency-constrained LPCC, LFCC) and new features called PYKFEC [6],[8],[9] are evaluated over on the different conditions and measured their respective contribution to feature fusion.

In this work, we design a front-end that is motivated from auditory perception. The speech signal is improved by pre-processing, done with the help of *speech activity detection* for detecting speech portion in continuous speech and also to determine voiced and unvoiced part of the speech. M*agnitude spectral subtraction* is used to estimate magnitude spectrum of clean speech from noisy speech magnitude and *cepstral mean normalization* to reduce convolution distortion in slowly varying channel.

The organization of this paper is as follows: in Section 2, we provide the theoretical background of speech pre-processing. In Section 3, the proposed improved feature extraction algorithm using MFCC is presented. In Section 4, the performance of the improved features is evaluated under different recording conditions in terms of recognition. Conclusions are presented in Section 5.

2   Speech Pre-processing

As features related to speech as well as speaker are present in spectral content, it is desired to have input speech to the recognition system to be as clean as possible. This is especially required for different recording devices and transmission channels. Therefore we have pre-processed the speech using speech activity detection and pre-emphasis filtering. The pre-processing of speech before actual feature extraction will help to eliminate some part of noise as well as raw speech data.

2.1 Speech Activity Detection (SAD)

The fundamental problem in many speech and speaker recognition systems is to separate our speech signal from non-speech part such as silence and various types of noise and disturbances. Speech activity detection is an algorithm used in speech processing wherein, the presence or absence of human speech is detected from the audio samples. The primary function of it is to provide an indication of speech presence in order to facilitate speech processing as well as possibly providing de-limiters for the beginning and end of a speech segment. As our database for training and testing consists of continuous speech, there is possibility of voice inactive or silence segments. We have used SAD to acquire a speech segment, eliminating non-speech part such as silence and noise. Thus it is possible to distinguish speech and silence/noise and to get feature vectors better representing true speech characteristics.

The discriminative characteristics of the speech can be extracted in time domain, spectral domain or cepstral domain. As signal energy remains the basic of the feature vector, we have used Energy-based SAD [14]. Here it assumed that speech is louder than silence and background noise. Therefore we can assign high energy frames as speech, whereas low energy frames as silence or noise. Speech is detected when the energy estimation lies over the threshold. Short-time energy is used to distinguish voiced speech and zero-crossing rate is used to distinguish unvoiced part. We set two constant thresholds in SAD. If achieve the higher one, decide voiced speech. If between the higher threshold and the lower one, calculate the zero-crossing rate to decide whether it is unvoiced speech or noise. A problem here is obvious: If the input SNR is low (small speech amplitude), after the normalization of SAD, it is more likely to decide noise as speech since it has a relative large energy. Thus, to choose proper decision thresholds and keep large SNR are very important. We get the best by trying and adjusting.

As shown in figure (1), the combination of energy and zero crossing rate is used to distinguish voiced part and unvoiced part of the speech signal. It is observed that energy is high for voiced part and low for unvoiced part, whereas zero crossing rates are low for voiced part and high for unvoiced part. Thus, a burst of energy in a stipulated time is used to recognize a voiced speech whereas based on assumption that zero crossing rate of speech and noise are different, it is possible to distinguish speech and noise.

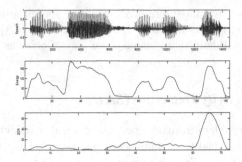

Fig. 1 Energy and zero-crossing rate of speech signal in speech activity detector

2.2 Pre-emphasis

In next step, the speech signal is emphasized using a highpass filter. Speech spectrum has more energy at low frequencies compared to high frequencies. This is due to nature of glottal pulse. This is called as *spectral tilt*. Pre-emphasis boosts the high-frequency part that was suppressed during the sound production mechanism of humans. Moreover, it can also amplify the importance of high-frequency formants.

The speech signal s(n) is sent to a high-pass filter:

$$y(n) = s(n) - a * s(n-1) \qquad (1)$$

where $s(n)$ is the output signal and the value of a is usually between 0.95 and 0.97. The z-transform of the filter is:

$$H(z) = 1 - a * z^{-1} \qquad (2)$$

which is an FIR highpass filter. Here we choose a=0.97.

3 Improved MFCC

Mel Frequency Cepstral Coefficients (MFCC) is one of the most commonly used feature extraction technique used in speech and speaker recognition systems. The technique is so-called *FFT-based*, which means that feature vectors are extracted from the frequency spectra of the windowed speech frames. MFCCs can be considered as: (a) as a filter-bank processing adapted to speech specificities and (b) as a modification of the conventional cepstrum, a well known deconvolution technique based on homomorphic processing.

The following section discusses various steps to get channel robust MFCCs.

3.1 Framing and Windowing

Though the speech signal is constantly changing, it is assumed to have quasi-stationary spectral characteristics over short time interval. Therefore, speech signal is processed in small chunks called frames. Framing divides the speech signal to get piecewise stationarity. The purpose of windowing is to limit the time interval to be analyzed so that the properties of the waveform do not change appreciably. For this, speech is usually segmented in frames of 20 to 30 msec. A typical frame overlap is around 30 to 50 % of the frame size. The purpose of the overlapping analysis is that each speech sound of the input sequence would be approximately centered at some frames. Simply cutting out speech signal into frames is equivalent to using rectangular window. But as rectangular window causes discontinuities at the edges of the segments, smooth tapers (like Hamming or Hanning) are usually used. Thus each frame is multiplied with a hamming window in order to keep the continuity of the first and the last points in the frame.

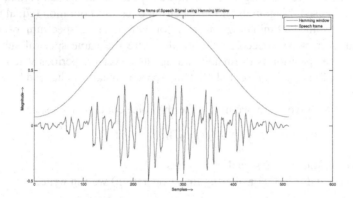

Fig. 2 One frame of voiced speech multiplied by Hamming window

3.2 Spectral Analysis

Spectral analysis is required to determine the spectral contents of the speech signal from a finite time samples obtained through framing and windowing. It gives the knowledge of distribution of power over frequency. Short time fourier transform (STFT) is a tool most widely used for speech signal analysis. Given the time series of speech signal $s(n)$, the STFT at time n is given as:

$$S(n, w) = \sum_{m=-\infty}^{\infty} s(m)w(m-n)e^{-jwm} \tag{3}$$

where $w(n)$ is the analysis window (of length N), which is assumed to be non-zero only in the interval 0 to N-1. The estimated power spectrum contains details of spectral shape as well as spectral fine components. Here the length of frame/window decides the resolution in time and frequency domain. A short frame width gives high time resolution and low frequency resolution, whereas long frame width gives low time resolution but high frequency resolution. A small window length (results in wider bandwidth) can capture fast time varying components (e.g. in rapid conversational speech), whereas a longer window length (narrow bandwidth) gives better information about sinusoidal components (e.g. harmonics of formants). Thus, we can say that short time window (approximately 5-10 msec) will represent vocal fold details (source information) whereas longer duration window (approximately 20-30 msec) gives vocal track details (filter characteristics) considering source-filter model of human speech production mechanism.

3.3 Spectral Subtraction (SS)

The magnitude or power estimate obtained with STFT is susceptible to various types of additive noise (such as background noise). To compensate for additive noise and to restore the magnitude or power spectrum of speech signal, spectral subtraction is used. Magnitude of the spectrum over short duration (equal to frame length) is obtained eliminating phase information. Here spectrum of noise is subtracted from noisy speech spectrum, therefore the name spectral subtraction. For this, noise spectrum is estimated and updated over the periods when signal is absent and only noise is present [18]. Thus speech signal is enhanced by eliminating noise.

In case of additive noise, we may write the noise contaminated speech signal as :

$$y(i) = s(i) + n(i) \tag{4}$$

where n(i) is some noise signal.

In frequency domain (considering linear operation) we can write,

$$Y(k) = S(k) + N(k) \tag{5}$$

Assuming $n(i)$ with zero mean and is uncorrelated with $x(i)$, the power spectrum of y(i) can be written as:

$$|Y(k)|^2 = |X(k)|^2 + |N(k)|^2 \tag{6}$$

Using (6), it is possible to estimate $|X(k)|^2$ as:

$$|\hat{X}(k)|^2 = |Y(k)|^2 - |\hat{N}(k)|^2 \tag{7}$$

where $|\hat{N}(k)|^2$ is some estimate of noise.

One way to estimate this noise is to average $|Y(k)|^2$ over the sequence of frames known to be non-speech (by using speech activity detection):

$$|\hat{N}(k)|^2 = \frac{1}{M}\sum_{t=0}^{M-1} |Y_t(k)|^2 \tag{8}$$

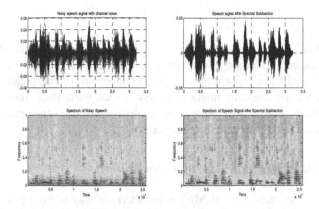

Fig. 3 Effect of Spectral Subtraction on noisy speech signal

As observed in figure (3), the noisy speech signal (contaminated by background noise) processed with magnitude spectral subtraction improves the spectral content of the speech signal by removing noise. This helps to extracts the true features of the speaker from his/her speech.

3.4 Mel-scale Bank and Cepstral Analysis

The mel scale is based on an empirical study of the human perceived pitch or frequency. The scale is divided into the units of "mel" s. The mel scale is generally speaking a linear mapping below 1000 Hz and logarithmically spaced above that frequency[3], [6]. The mel frequency warping is most conveniently done by utilizing filter bank with filters centered according to mel frequencies. The width of the triangular filters vary according to the mel scale, so that the log total energy in a critical band around the centre frequency is included. The result after warping is a number of coefficients Y(k):

$$Y(k) = \sum_{j=1}^{\frac{N}{2}} S(j) \, H_k(j) \tag{9}$$

Using N point IDFT, the cepstral coefficients are calculated by transforming log of the quefrency domain coefficients to the frequency domain as:

$$c(n) = \frac{1}{N} \sum_{k=0}^{N-1} Y(k) e^{j\frac{k2\pi}{N}n} \tag{10}$$

which can be simplified, because Y(k) is real and symmetric about N/2, by replacing the exponential by a cosine:

$$c(n) = \frac{1}{N} \sum_{k=0}^{N-1} Y(k) \cos\left(k \frac{2\pi}{N} n\right) \tag{11}$$

A reliable way of obtaining an estimate of the dominant fundamental frequency for long, clean, stationary speech signals is to use the *cepstrum*. The cepstrum is a Fourier analysis of the logarithmic amplitude spectrum of the signal. If the log amplitude spectrum contains many regularly spaced harmonics, then the Fourier analysis of the spectrum will show a peak corresponding to the spacing between the harmonics: i.e. the fundamental frequency.

3.5 Normalization of Cepstral Coefficients

Cepstral mean normalization (CMN) is an alternate way to high-pass filter cepstral coefficients. In cepstral mean normalization the mean of the cepstral vectors $c(n)$, is subtracted from the cepstral coefficients of that utterance on a sentence-by-sentence basis:

$$y(n) = c(n) - \frac{1}{N} \sum_{n=1}^{N} c(n) \tag{12}$$

Here we try to enhance the characteristics present in speech signal and reduce the channel effects on speech signal by computing the mean over finite number of frames. To compensate the channel effect, the channel cepstrum can be removed by subtraction of the cepstral mean. This temporal mean is a rough estimate of the channel response.

4 Experiments and Results

The proposed improved MFCC coefficients extracts the features on frame by frame basis. We use the standard MFCCs as the baseline features. Speech pre-processing is performed before framing. Each frame is multiplied with a 30 ms Hamming window, shifted by 20 msec. From the windowed frame, FFT is computed, and the magnitude spectrum is subtracted using MMSE (Maximum Mean Square Error) algorithm. These samples are filtered with a bank of 13 triangular filters spaced linearly on the mel-scale. The log-compressed filter outputs are converted into cepstral coefficients by DCT, and the 0th cepstral coefficient is ignored. The cepstral coefficients thus obtained are normalized by cepstral mean technique discussed earlier. Further speaker models are generated by the LBG/VQ clustering algorithm [16]. The quantization distortion with Euclidean distance is used as the matching function. The number of MFCCs and model sizes were fixed to 12 and 64, respectively. The effect of the number of MFCCs was also studied. Increasing the number of coefficients improved the identification accuracy up to 10–15 coefficients, after which the error rates stabilized. Therefore, we fixed the number of coefficients to 12.

4.1   Database

Speaker recognition (identification) experiments have been conducted to test the performance of the proposed algorithm. For performance evaluation, we have used the *multi-speaker, continuous (Hindi) speech database* generated by TIFR, Mumbai (India) and made available by Department of Information Technology, Government of India. The database contains a total of approximately 1000 Hindi sentences, a set of 10 sentences read by each of 100 speakers. These 100 sets of sentences were designed such that each set is `phonetically rich' [17]. The speech data was simultaneously recorded using two microphones: one good quality, close-talking, directional microphone and another desk-mounted Omni-directional microphone.

4.2   Database Set and Performance

(I) Database Set I- Continuous Speech Hindi Database (100 speakers, 59 Male and 41 female speakers)

Training Set- **Recorded with close-talking, directional microphone**
Testing Set - **Recorded with close-talking, directional microphone**

Table 1 Identification Results of Speaker Recognition with baseline MFCC

Feature Extraction Technique	Training Database	Testing Database	Result (%) Identification
Baseline MFCC	Phonetically Rich	Phonetically Rich	100
Baseline MFCC	Phonetically Rich	Broad Acoustic Class of phonemes in different phonetic contexts	99
Baseline MFCC	Broad Acoustic Class of phonemes in different phonetic contexts	Broad Acoustic Class of phonemes in different phonetic contexts	88
Baseline MFCC	Broad Acoustic Class of phonemes in different phonetic contexts	Phonetically Rich	98

(II) Dataset-II

Continuous Speech Hindi Database (97 speakers, 59 Male and 38 female)
Training Set- Recorded with close-talking, directional microphone
Testing Set - Recorded with desk-mounted Omni-directional microphone

Table 2 Identification Results of Speaker Recognition with modified MFCCs under phonetically rich condition

Feature Extraction Technique	Training Database	Testing Database	Result (%) Identification
Baseline MFCC	Phonetically Rich	Phonetically Rich	15.46
Baseline MFCC with CMN	Phonetically Rich	Phonetically Rich	52.57
Baseline MFCC with SS	Phonetically Rich	Phonetically Rich	60.82
Baseline MFCC with CMN & SS	Phonetically Rich	Phonetically Rich	88.65

Table 3 Identification Results of Speaker Recognition with modified MFCCs under phonetically mismatched condition

Feature Extraction Technique	Training Database	Testing Database	Result (%) Identification
Baseline MFCC	Phonetically Rich	Broad Acoustic class of phonemes	9.27
Baseline MFCC with CMN	Phonetically Rich	Broad Acoustic class of phonemes	32.61
Baseline MFCC with SS	Phonetically Rich	Broad Acoustic class of phonemes	44.33
Baseline MFCC with CMN & SS	Phonetically Rich	Broad Acoustic class of phonemes	74.22

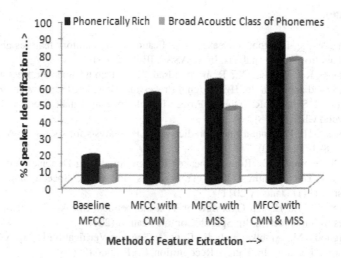

Fig. 4 Comparison of Robustness of MFCCs for Speaker Identification

5 Conclusion

In this work, we have evaluated the performance of baseline MFCCs as well as modified MFCCs under two different recording conditions, one recorded with close-talking, directional microphone and other recorded with desk-mounted Omni-directional microphone. The baseline MFCC gives optimum recognition performance for same training and testing conditions, recorded with close-tracking directional microphone (clean speech). Whereas the speaker recognition performance totally fall down (100 % to 15.46 %) under different training and testing conditions, which proves that MFCC is very susceptible to mismatched conditions.

It is observed that, percentage correct identification rate of the system is improved by modifying MFCCs with magnitude spectral subtraction and cepstral mean normalization. The combination of spectral subtraction and cepstral mean normalization compensates the adverse effect of channel mismatch. Modified MFCCs improves the performance of speaker recognition system from 15.46 % to 79.38 % when both training and testing data is phonetically rich and 9.27% to 74.22 % when under acoustically mismatched training and testing data. Experimental results demonstrate the effectiveness and robustness of the modified MFCCs as compared to regular MFCCs in different recording conditions. From results it is observed that the nature of speech (clean or noisy) and characteristics of speech (phonetical contents) also plays an important role for extracting true speaker related features in individual's speech.

Acknowledgements. The authors would like to thank Department of Information Technology, Government of India and Dr. Samudravijaya K. of TIFR, Mumbai for providing Speech Database. Special thanks to Dr. K.L.Asanare (Director, FAMT Ratnagiri), for his constant support and motivation.

References

1. Shao, Y., Wang, D.: Robust speaker identification using auditory features and computational auditory scene analysis. In: ICASSP. IEEE (2008)
2. Mammone, R.J., Zhang, X., Ramachandran, R.P.: Robust Speaker Recognition: A Feature based approach. In: IEEE Signal Processing Magazine (September 1996)
3. Rabiner, L., Schafer, R.: Digital Processing of Speech Signal. Prentice Hall, Inc., Englewood Cliffs (1978)
4. Hermansky, H.: Perceptual linear predictive (PLP) analysis for speech. J. Acoust. Soc. Am., 1738–1752 (1990)
5. Wang, N., Ching, P.C.: Robust Speaker Recognition Using Denoised Vocal Source and Vocal Tract Features. IEEE Transactions on Audio, Speech, and Language Processing 19(1) (January 2011)
6. Kinnunen, T., Li, H.: An Overview of Text-Independent Speaker Recognition: From Features to Supervectors. In: Speech Communication (2010)
7. Nosratighods, M., Ambikairajah, E., Epps, J.: Speaker Verification Using A Novel Set of Dynamic Features. In: Pattern Recognition, ICPR 2006 (2006)
8. Openshaw, J., Sun, Z., Mason, J.: A comparison of composite features under degraded speech in speaker recognition. In: IEEE Proceedings of the International Conference on Acoustics, Speech and Signal Processing, vol. 2, pp. 371–374 (1993)
9. Reynolds, D., Rose, R.: Robust text-independent speaker identification using Gaussian mixture speaker models. IEEE Trans. on Speech and Audio Processing 3 (January 1995)
10. Campbell Jr., J.P.: Speaker Recognition- A Tutorial. Proceedings of The IEEE 85(9), 1437–1462 (1997)
11. Lawson, A., Vabishchevich, P., Huggins, M., Ardis, P., Battles, B., Stauffer, A.: Survey and Evaluation of Acoustic Features for Speaker Recognition. In: ICASSP 2011. IEEE (2011)
12. Reynolds, D.A.: An Overview of Automatic Speaker Recognition Technology. In: ICASSP 2001. IEEE (2001)
13. Glsh, H., Schmidt, M.: Text Independent Speaker Identification. In: IEEE Signal Processing Magazine (1994)
14. Prasad, V., Sangwan, R., et al.: Comparison of voice activity detection algorithms for VoIP. In: Proc. of the Seventh International Symposium on Computers and Communications, Taormina, Italy, pp. 530–532 (2002)
15. Menéndez-Pidal, X., Chan, R., Wu, D., Tanaka, M.: Compensation of channel and noise distortions combining normalization and speech enhancement techniques. Speech Communication 34, 115–126 (2001)
16. Linde, Y., Buzo, A., Gray, R.M.: An algorithm for vector quantizer design. IEEE Trans. Commun. 28(1), 84–95 (1980)
17. Samudravijaya, K., Ra0, P.V.S., Agrawal, S.S.: Hindi Speech Database. In: Proceedings of International Conference on Spoken Language Processing, China (2000)
18. Vaseghi, S.V.: Advanced Digital Signal Processing and Noise Reduction, 2nd edn. John Wiley & Sons Ltd. (2000)

Novel Extension of Binary Constraint Method for Automated Frequency Assignment: Application to Military Battlefield Environment

Sandeep Kumar, Sanjay Motia, and Kuldeep Singh

Abstract. Wireless communication plays an important role in mission plan activities of various military and battlefield situations. In each of these situations, a frequency assignment problem arises with application specific characteristics. This paper presents a novel extension to binary constraint [1] method for automation of frequency assignment in battlefield scenarios. The proposed method calculates the cost function considering each node of the radio net in battlefield equally important. Two cases for calculation of cost function are considered; in the first case the average of the signal (from intra radio nodes) and interference (from inter radio nodes) is considered whereas in the second case the intra radio nodes causing minimum signal and inter radio nodes producing maximum interference are taken into consideration for cost calculation. A simulation exercise in lab environment has been done for validation of proposed methods by creating a battlefield scenario with three radio nets each consisting three-radio nodes.

Keywords: Cost Function, Binary Constraint, Battlefield, Frequency Assignment, Radio Nets.

1 Introduction

With the deployment of large wireless communication systems to support the digitization of the military battlefield, the need to manage the spectrum is paramount for their successful operation. Major area of concern is the assignment of frequencies, or channels to individual sub- networks within the system [2]. The rapid development of wireless communication systems has resulted in a scarcity of

Sandeep Kumar · Sanjay Motia · Kuldeep Singh
Central Research Lab. Bharat Electronics Ltd. Ghaziabad, India
e-mail: {sandeepkumar,sanjaymotia,kuldeepsingh}@bel.co.in

S.M. Thampi, A. Gelbukh, and J. Mukhopadhyay (eds.), *Advances in Signal Processing*
and Intelligent Recognition Systems, Advances in Intelligent Systems and Computing 264,
DOI: 10.1007/978-3-319-04960-1_49, © Springer International Publishing Switzerland 2014

the most important resource, the radio spectrum, which if not coordinated and carefully preplanned, will have an adverse effect on battlefield situations. Present day military forces may find themselves needing to move rapidly into an area with little time for familiarization [3]. Upon arrival, they have to establish the communication link quickly so that the necessary and critical information can be communicated timely to other units. The manual and conventional methods of deployment of communication resources for radio link establishment using available communication plan is very cumbersome and time consuming process for the troops in tactical battlefield. Usually the communication plan consists of predefined usable and unusable frequency list developed based on the experiences and personal knowledge of the communication operator [2] and the operator faces crunch of usable frequencies or available spectrum every time he plans the deployment of communication resources in a new scenario different from the one for which the knowledgebase was already developed. Therefore, the effective utilization of available spectrum and automation of frequency assignment methodology is the need of hour for defense forces.

This paper is organized as follows: Section 2 presents Literature Survey, Section 3 throws light on Military assignment problem and Binary Constraint method, Section 4 describes the proposed algorithms, Section 5 gives experimental results, and section 6 concludes the paper.

2   Literature Survey

The literature on frequency assignment has grown substantially over the years. The inspiration for this research has been the widespread implementation of wireless networks. Metzger [4] was the first one to publish the research on frequency assignment. Hale [5] extended the idea of Metzger by introducing the graph-coloring algorithm in the frequency assignment problems. Another effective algorithm named simulated annealing was first published by Metropolis et. al. [6] and later his work was extended by Kirkpatrick et. al. [7] who proposed use of the simulated annealing approach in optimization of frequency assignment problems. The concept of Tabu search algorithm to solve frequency assignment problems was introduced by Bouju et.al. [8]. Then ANTS algorithm [9] was proposed to solve problems on binary constraint, signal-to-interference and their combinations [10]. The idea of minimizing a cost function-using signal-to-interference calculations were introduced in [11]. The meta-heuristic algorithms, namely the simulated annealing and Tabu search, mainly provide the solution of frequency assignment problems for GSM networks. The majority of the published literature is concerned with algorithms that use binary constraints. The use of Binary Constraint method in solving frequency assignment problems in battlefield situations is still an open area for research. Salaril et. Al.[12] Proposed an Ant Colony Optimization (ACO) algorithm for the graph coloring problem which conforms to Max-Min Ant System structure and exploits a local search heuristic to improve its performance. Wu et. Al.[13] proposed a neural network based approach to tackle the frequency assignment problem (FAP) named as stochastic competitive Hopfield neural network-efficient genetic algorithm (SCH-EGA) approach . Yang et. Al.[14] proposed another hybrid approach named as SCH-ABC which is combination of stochastic competitive Hopfield neural network (SCHNN) and artificial bee colony (ABC) to cope with the frequency assignment problem (FAP). Kateros et. Al.[15] have modeled the FAP with a constraint graph represented by an interference matrix and also developed an algorithm that uses meta-heuristic techniques, in order to obtain possible solutions. They have also evaluated three versions of the proposed algorithm i.e. Tabu Search Simulated Annealing and Genetic Algorithm for frequency Bands IV/V (UHF). Audhya et. Al.[16] presented a polynomial time approximation algorithm for solving the channel assignment problems (CAPs) for cellular networks having non-homogeneous demands in a 2-band buffering system (where channel interference does not extend beyond two cells).

3   Military Frequency Assignment Problem

In battlefield situations, the communication network is formed by set of radio nets comprising of multiple radio nodes, which covers the large battlefield area. A basic radio net has a command station and a number of remote stations. Every command station in a radio net communicates on a single frequency with all remote nodes within that radio net.

Figure 1 depicts a typical military battlefield scenario where R_A, R_B and R_C are radio nets operating on frequency f_A, f_B and f_C respectively. A frequency assignment problem can be formulated as:

"Find a unique combination of frequencies assigned to all the radio nets considering optimum separation in frequencies and lowest interference among the radio nets."

The radio nodes have a certain amount of selectivity, allowing them to ignore nearby transmitters. Thus, the main requirement is a sufficient frequency separation between all radio nets so that the radio nodes of one radio nets should not cause interference with the other radio nets.The use of terrestrial battlefield communication systems such as radios is dynamic in time and space, which leads to frequency assignment problems (FAPs). The military frequency assignment problem is different from the cellular networks. In military applications, all the radio nodes can act as Transreceiver in half-duplex mode however, in cellular networks only one transmitter is taken into account.

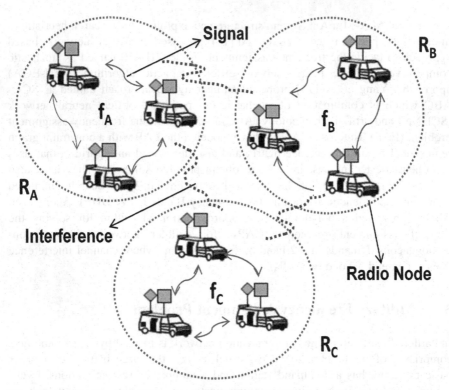

Fig. 1 Radio Nets R_A, R_B and R_C operating on frequencies f_A, f_B and f_C respectively

3.1   Binary Constraint Method for Frequency Assignment

The total possible frequency combination for assignment can be written as P(N, M), where N is the number radio nets and M is the number of frequencies. For a each frequency combination the cost function is calculated and the frequency combination with maximum cost function is chosen for the assignment. The cost function is based on the signal strength received by one transmitter from other transmitters within same radio net considering each transmitter in the radio net separately and the interference caused by the radio nodes of neighboring radio nets. This method calculates cost value for all the combinations of frequencies and suggest a frequency combination with maximum cost value. The cost function is variable of signal and interference.

Received signal strength at the ith node due to signal transmitted from k^{th} node is calculated as per eq. (1)

$$Sig_i = \frac{P_k}{D_{ik}^{\beta}} \tag{1}$$

Where P_k is the transmitted power from k^{th} node and D_{ik} is the distance between the ith node and the kth node and $2 \le \beta \le 4$

The total interference at the ith node receiver from all other transmitters is defined as eq. (2)

$$Int_i = \sum_{i=1; i \neq k}^{N} \frac{P_j}{D_{ij}^\beta} \theta \tag{2}$$

Where N is number of transmitters and θ is assumed as per eq. (3)

$$\theta = \begin{cases} 10^{\frac{-\alpha(1+\log_2 f)}{10}}, & f = 0 \ (Co - Channel) \\ 1, & f \neq 0 \ (Adj. Channel) \end{cases} \tag{3}$$

Where f is the channel separation between the transmitter and the interfering transmitter. α is the attenuation factor for adjacent channel interference having integer value.

Cost function as per eq. (1) and eq. (2) is formulated as

$$Cost = \sum_{i=1}^{m} \left(\frac{Sig_i}{Int_i}\right)^\gamma \tag{4}$$

Where γ is an integer value.

4 Novel Extension of Binary Constraint Method

Typical battlefield scenario consists of various radio nets located geographically distant apart. Each radio net consists of multiple radio nodes operating on the same frequency. While designing the algorithms for solving frequency assignment problem in military operations, following considerations are taken:

1. Every node can act as half-duplex Transreceiver. Each node shall act as signal source for intra radio net nodes and shall act as interference source for inter radio net nodes.
2. For military operations, every radio node is equally important hence optimum frequency assignment solution should be chosen which is the best fit for all the nodes.
3. Signal to noise ratio from every individual radio node inside a radio net for a particular frequency combination should be at least higher than a minimum defined threshold.
4. Transmitting power and receiver sensitivity is same across the radio nodes.

Considering above mentioned factors and requirements the traditional binary method requires modifications. The authors proposed an extension to binary constraint method to suit the battlefield frequency assignment requirements. In proposed approach, two scenarios are visualized for calculating cost function.

4.1   Average Signal and Interference Scenario

As per eq. 1 the average signal strength at i^{th} node of kth radio net can be defined as

$$S_{ki} = \frac{1}{N-1}\sum_{j=1}^{N-1} Sig_j \qquad (5)$$

Where Sig_j is the signal strength received from j^{th} node of that particular radio net.

As per eq. 2 the average interference at i^{th} node of kth radio net can be defined as

$$I_{ki} = \frac{1}{N}\sum_{l=1}^{M-1}\sum_{p=1}^{N} Int_{lp} \qquad (6)$$

Where Int_{lp} is the Interference caused by from p^{th} node of lth radio net.

Cost at i^{th} node of kth radio net can be defined as

$$Cost_{ki} = \frac{S_{ki}}{I_{ki}} \qquad (7)$$

Now the total cost of a particular frequency combination can be written as

$$Cost = \sum_{l=1}^{M-1}\sum_{p=1}^{N} Cost_{lp} \qquad (8)$$

4.2   Worst Case Scenario

As per eq. (1) the signal strength is inversely proportional to the distance between the two nodes. For worst-case scenario farthest radio node of that net shall be taken into consideration for calculating signal strength.

$$S_{ki} = \min(Sig_j) \qquad j = 1,2,.....N-1 \qquad (9)$$

For calculating interference in worst-case scenario, the maximum interference exhibited by a node in a net shall be considered.

$$I_{ki} = \sum_{l=1}^{M-1} \max (Int_{lp}) \qquad p = 1,2,........N \qquad (10)$$

The final cost is calculated as per equation (8)

A Threshold is applied to cost value calculated on every single node. The cost threshold implies that there shall not be any node having individual cost less than that threshold value corresponding to a frequency combination. Any frequency combination in which some nodes yielding individual cost less than the threshold is not considered for assignment.

5   Simulation Results and Discussion

For performance testing of algorithm three radio nets each consisting three-radio nodes are taken and results are simulated on MALAB 7.0. Battlefield scenario was created for the same configuration in desert areas. Five frequencies are taken 55.550, 59.750, 64.650, 67.250, 71.025 (MHz) for assignment. Total combinations for 5 frequencies and 3 radio nets are $P(5,3) = 60$. The transmitted power of VHF radio is 25 Watt. For the calculations following assumption are taken: $= 1$, $\gamma = 1$ and $\beta = 2$

Table 1 Distance Matrix

	A1	A2	A3	B1	B2	B3	C1	C2	C3
A1	0	11	5	17	19	7	17	15	18
A2	11	0	10	27	27	18	10	16	15
A3	5	10	0	18	17	9	12	10	13
B1	17	27	18	0	8	10	28	21	26
B2	19	27	17	8	0	13	25	17	21
B3	7	18	9	10	13	0	21	16	20
C1	17	10	12	28	25	21	0	9	6
C2	15	16	10	21	17	16	9	0	5
C3	18	15	13	26	21	20	6	5	0

Table 1 depicts the distance between various radio nodes across the radio nets. Here $A1$, $A2$, and $A3$ are radio nodes of Radio net R_A similarly $B1$, $B2$, and $B3$ are radio nodes of Radio net R_B and $C1$, $C2$, and $C3$ are radio nodes of Radio net R_C.

Table 2 Frequency assignment results

	R_A	R_B	R_C
Average Method	71.025	64.650	55.550
Worst Case without Threshold	71.025	64.650	55.550
Worst Case with Threshold= -3.036 dB	55.550	64.650	71.025
Worst Case with Threshold= -2.868 dB	55.550	67.250	71.025
Worst Case with Threshold= -2.738 dB	55.550	71.025	64.650
Worst Case with Threshold= -2.397 dB	71.025	55.550	64.650

Table 2 depicts the suggested frequency results for every radio net. For average case frequency with maximum separation are assigned to three radio nets. Four different cases for worst-case scenario are taken with different cost thresholds. From the table it can be noticed that the frequency assignment vary with threshold values.

Although the simulation results shown here seems promising in lab environment, however the authors believes that the proposed method needs validation in real-time battlefield environment. There is a tremendous future scope in the direction of automation of frequency assignment problems in battlefield scenarios.

6   Conclusion

The proposed method shall provide very important aid to military forces for automation of frequency assignment. This approach shall be very handy in the scenarios where troop movement is very frequent and there is a daily need of new frequency assignment for effective communication across the military base. Novel modification is incorporated in the traditional binary constraint method for making the method more robust and suitable for military environment. Thresholding with worst case guarantees operability of every distant node in battlefield.

References

[1]   Smith, D.H., Allen, S.M., Hurley, S., Watkins, W.J.: Frequency Assignment: Methods and Algorithms. In: Symposium, R.I. (ed.) RTO IST Symposium on Frequency Assignment, Sharing and Conservation in Systems (Aerospace), Aalborg, Denmark (1998)

[2]   Beuter, R.: Frequency Assignment and Network Planning for Digital Terrestrial Broadcasting Systems. Kluwer Academic Publishers Springer (2004)

[3]   Bradbeer, R.: Application of New Techniques to Military Frequency Assignment. In: RTO IST Symposium on Frequency Assignment, Sharing and Conservation in Systems (Aerospace), Aalborg, Denmark (1998)

[4]   Metzger, B.H.: Spectrum management technique. Presentation at 38th National ORSA Meeting, Detroit (1970)

[5]   Hale, W.K.: Frequency Assignment: Theory and Applications. Proceedings of IEEE 68(12), 1497–1514 (1980)

[6]   Metropolis, N., Rosenbluth, A., Rosenbluth, M., Teller, A., Teller, E.: Equations of state calculations by fast computing machines. Journal of Chemical Physics 21(6), 1087–1092 (1953)

[7]   Kirkpatrick, S., Gelatt, C., Vecchi, M.: Optimization by simulated annealing Science 220 (4598), pp. 671–680 (1983)

[8]   Bouju, A., Boyce, J.F., Dimitropoulos, C.H.D., vom Scheidt, G., Taylor, J.G.: Tabu search for radio links frequency assignment problem Conference on Applied Decision Technologies: Modern Heuristic Methods, England, pp. 233–250. Brunel University, Uxbridge (1995)

[9]   Montemanni, R., Smith, D.H., Allen, S.M.: An ANTS algorithm for the minimum-span frequency assignment problem with multiple interference. IEEE Transactions on Vehicular Technology 51(5), 949–953 (2002)

[10] Graham, J.S., Montemanni, R., Moon, J.N.J., Smith, D.H.: Frequency Assignment. Multiple Interference and Binary Constraints Wireless Networks Springer 14, 449–464 (2008)

[11] Watkins, W.J., Hurley, S., Smith, D.H.: Evaluation of Models for Area Coverage Radio communications Agency Agreement, RCCM070, Final Report (1998)

[12] Salaril, E., Eshghi, K.: An ACO algorithm for graph coloring problem. Int. J. Contemp. Math. Sciences 3(6), 293–304 (2008)

[13]   Wu, S., Yang, G., Xu, J., Li, X.: SCH-EGA: An Efficient Hybrid Algorithm for the Frequency Assignment Problem. Engineering Applications of Neural Networks Communications in Computer and Information Science 383, 32–41 (2013)

[14]   Yang, G., Wu, S., Xu, J., Li, X.: A Novel Hybrid SCH-ABC Approach for the Frequency Assignment Problem. In: Lee, M., Hirose, A., Hou, Z.-G., Kil, R.M. (eds.) ICONIP 2013, Part II. LNCS, vol. 8227, pp. 522–529. Springer, Heidelberg (2013)

[15]   Kateros, D.A., Georgallis, P.G., Katsigiannis, C.I., Prezerakos, G.N., Venieris, I.S.: An Algorithm for the Frequency Assignment Problem in the Case of DVB-T Allotments. In: Lirkov, I., Margenov, S., Waśniewski, J. (eds.) LSSC 2007. LNCS, vol. 4818, pp. 561–568. Springer, Heidelberg (2008)

[16]   Audhya, G.K., Sinha, K., Mandal, K., Dattagupta, R., Ghosh, S.C., Sinha, B.P.: A New Approach to Fast Near-Optimal Channel Assignment in Cellular Mobile Networks. IEEE Trans. on Mobile Computing 12(9), 1814–1827 (2013)

Two Approaches for Mobile Phone Image Insignia Identification

Nitin Mishra and Sunil Kumar Kopparapu

Abstract. Insignia identification is an important task especially as a self help application on mobile phones which can be used in museums. We propose a knowledge driven rule-based approach and a learning based approach using artificial neural network (ANN) for insignia recognition. Both the approaches are based on a common set of insignia image segmentation followed by extraction of simple, yet effective features. The features used are based on one of frugal processing and computing to suit the mobile computing power. In both the approaches we identify each extracted segment in the insignia; the correct recognition of the segment followed by post processing results in the identification of the insignia. Experimental results show that both approaches work equally well in terms of recognition accuracy of over 90% in terms of identification of the segments and 100% in terms of the actual insignia identification.

1 Introduction

Visit to a museum often requires a guide who can describe the details associated with an artifact on display. There are problems understanding the artifacts especially if the visitor is from a different country and speaking a different language. A handy mobile device that can capture the image of the artifact on display and then show details of the artifact, in the visitors own language is an usable solution. This does not only make the experience of the visitor to the museum satisfying, but also provides him the chance to see in detail the information about what he wants to see rather than depend on a human guide available at the museum.

Nitin Mishra
TCS Innovation Labs - Delhi, India
e-mail: Mishra.Nitin@TCS.Com

Sunil Kumar Kopparapu
TCS Innovation Labs - Mumbai, India
e-mail: SunilKumar.Kopparapu@TCS.Com

S.M. Thampi, A. Gelbukh, and J. Mukhopadhyay (eds.), *Advances in Signal Processing and Intelligent Recognition Systems*, Advances in Intelligent Systems and Computing 264,
DOI: 10.1007/978-3-319-04960-1_50, © Springer International Publishing Switzerland 2014

There are several mobile applications (for example, [1, 2, 3]) built for giving detailed information about an insignia, however to the best of our knowledge, there is no known mobile application that can identify an insignia, in real-time, with the camera of the phone pointed at an insignia. In this paper, we propose two approaches to identify insignia with the express intent of making them work in real-time on a mobile phone with good recognition ability. The proposed scheme is a simple, yet effective, technique that can identify the insignia (see Figure 1) from the image captured by the mobile phone camera.

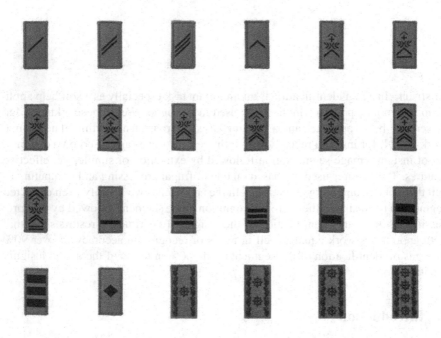

Fig. 1 Swiss insignia images. From left to right and top to bottom. Soldier, Private, Private First Class, Corporal, Sergeant, Sergeant First Class, Sergeant major, Quartermaster Sergeant, Chief sergeant major, Warrant Officer, Staff Warrant Officer, Master Warrant Officer, Chief Warrant Officer, Second Lieutenant, First Lieutenant, Captain, Major, Lieutenant Colonel, Colonel, Specialist Officer, Brigadier General, Major General, Lieutenant General, General.

It is to be noted that there is a significant variation in the quality of the images captured using a mobile phone camera. The variation in the image quality is not only due to use of different camera specifications on different commercially available mobile phones in terms of ISO, maximum supported resolution, exposure value, white balance and inbuilt image pre-processing chips but also due to (a) different external lighting conditions, (b) different light source orientations and (c) different angles of shoot in $x - y - z$ space (see Figure 2) which makes the task of facilitating real time and on-device insignia recognition more challenging.

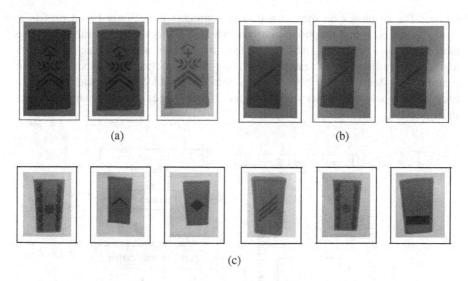

Fig. 2 Degradations in quality of images due to different: (a) External lighting conditions (Dim, Medium and Bright), (b) Light source orientations (Top, Left and Right) and (c) Angles of shoot in $x - y - z$ space

The main contribution of this paper is (a) the identification and use of a simple and effective feature set for on-device insignia identification that can work on a mobile phone captured image and (b) development of a rule-based and a learning based algorithm for insignia identification. The rest of the paper is organized as follows. We first describe both the rule-based and the ANN based approach in Section 2 in detail. In Section 3 we compare the performance of the rule-based and learning-based approach in terms of recognition of insignia for a wide variety of insignia images captured in different environmental conditions by different mobile phone cameras and we conclude in Section 4.

2   Insignia Identification

The process of identification of the insignia is performed in three steps, namely,

Step 1: Extraction of different segments in the insignia
Step 2: Assigning labels to each segment
Step 3: Post-processing, namely, associating all the labeled segments (from Step 2) in an insignia to identify the insignia (one of the 24 shown in Figure 1).

In Step 1, we used a set of some well known image pre-processing (example, [4, 5]) and segmentation techniques (example, [6, 7, 8]), to extract regions of interest in the insignia images. There are several challenges in extraction of segments from an image captured by a mobile phone camera resulting in missing known segments (false negative) and identifying non-existing segments (false alarms) in an insignia image. The emphasis on the extraction of segments is not only based on frugal computation on a mobile phone platform but also on trying to minimize the false alarms and false negatives. In Step 2, all the extracted segments from an insignia are analyzed to extract a set of features. Both rule-based and learning-based (see Figure 3) approaches were used to label all the extracted segment (Step 1) in the insignia. In Step 3, the labeled segments were analyzed to match one of the 24 insignia images in Figure 1.

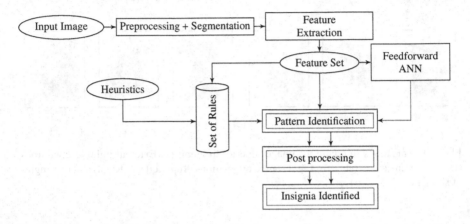

Fig. 3 Rule-Based and Learning-Based Approaches

It should be noted that the same sequence of pre-processing steps were adopted to extract segments from the insignia; the extracted segments were identified using two different approaches (rule-based and learning-based) to identify the segments; the identified segments were then post-processed to identify the insignia.

2.1 Pre-processing + Segmentation

The sequence of pre-processing steps[1] is depicted in Figure 4. The boundary of insignia is marked by identifying the periphery of a trapezoid having longest width and height. The identified trapezoid is then de-warped into a rectangular image using a warp perspective transform. The process of cropping extracts the region of interest. Light intensity correction is taken care through a process of gray-level morphological closing operation and then dividing the resulting image by a median blurred image. The binarization, morphological dilation and inversion assist in retaining regions of interest as continuous unbroken segments. The median filtering process eliminates small segments which are characterized as salt and pepper noise. The output of the pre-processing is a binary image which is the input to the feature extraction module.

---

[1] We tried several pre-processing schemes and choose the one that performed well on mobile phone images.

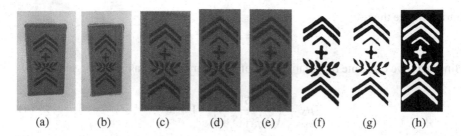

Fig. 4 Pre-processing output. (a) Original, (b) Boundary Detected, (c) Dewarp, (d) Crop, (e) Noise Blur, (f) Light intensity correction and Binarization, (g) Dilate, (h) Invert and Median Filtering.

2.2 Feature Extraction

The feature extraction process involves extraction of features (mostly low level) associated with each of the extracted segments. Let the size of the insignia image be X, Y. Assume that k segments have been extracted from an insignia image. Then for each segment $\{S_i\}_{i=1}^k$ we extract the desired segment features. Consider the segment S (we drop the subscript for ease of notation), we extract the centroid of the segment S by computing the mean of all the pixels (p_x, p_y) that belong to the S. Let $|S|$ represent the pixels which correspond to the segment S, then the centroid is computed as

$$C_S(x,y) = (C_S(x), C_S(y)) = \frac{1}{|S|} \left(\sum_{k \in S} p_x^k, \sum_{k \in S} p_y^k \right)$$

The position of the segment S, (P_S) is computed based on the location of the C_S in the insignia image as follows,

$$P_S = \begin{cases} \text{UP} & \text{if } 0 \leq C_S(y) \leq \frac{Y}{3} \\ \text{CENTER} & \text{if } \frac{1}{3}Y \leq C_S(y) \leq \frac{2}{3}Y \\ \text{BELOW} & \text{otherwise} \end{cases} \quad (1)$$

Similarly, we compute the position feature L_S as being IN (side) or OUT (side) the segment S as follows

$$L_S = \begin{cases} \text{IN} & \text{if } C_S(x,y) \in S \\ \text{OUT} & \text{otherwise} \end{cases} \quad (2)$$

Additionally, we extract the spread of the segment S in both the x and the y directions as follows,

$$\gamma_{x_S} = \sqrt{\frac{\sum_{l \in S} (p_x^l - C_S(x))^2}{|S| - 1}}, \quad \gamma_{y_S} = \sqrt{\frac{\sum_{l \in S} (p_y^l - C_S(y))^2}{|S| - 1}}$$

and compute the ratio

$$\gamma_S = \frac{\gamma_{xS}}{\gamma_{yS}}. \tag{3}$$

Similarly, we compute the relative size of the segment S, μ_S as follows

$$\mu_S = \left(\frac{|S|}{\sum_{j=1}^{k} |S_j|} \right) \times 100\% \tag{4}$$

where $\sum_{j=1}^{k} |S_j|$ represents the sum of all pixels in all the segments $\{S_i\}_{i=1}^{k}$ in an insignia image. Finally, using (1), (2), (3) and (4), each segment S in the insignia image is represented by the feature vector F_S namely,

$$F_S = \{P, L, \gamma, \mu\} \tag{5}$$

The feature set F_S was normalized for ANN training as,

$$C_x' = \frac{C_S(x)}{X}$$
$$C_y' = \frac{C_S(y)}{Y}$$
$$L' = \begin{cases} 1 & \text{if } C_S(x,y) \in S \\ 0 & \text{otherwise} \end{cases}$$

2.3 Pattern Learning and Identification

A visual inspection of the insignia images shown in Figure 1 reveals that there are essentially 11 unique patterns as seen in insignia images. These patterns can be represented by the set $P = \{/, \cap, +, U, V, \#, -, =, \approx, |, *\}$ (see Figure 5). Each of the segment extracted from the insignia image, if identified correctly, should be mapped to one of the 11 segmental patterns. We use two different approaches (rule-based and learning-based) to associate the extracted segment to one of the 11 segmental patterns.

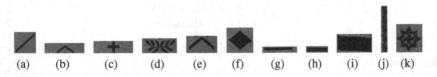

(a)   (b)   (c)   (d)   (e)   (f)   (g)   (h)   (i)   (j)   (k)

Fig. 5 The 11 unique patterns that occur in insignia image. $P = \{/, \cap, +, U, V, \#, -, =, \approx, |, *\}$

In the rule-based approach, the feature set extracted from a set of images is used to manually construct based on heuristics the rule-set base. Algorithm 4.1 shows the rules that were extracted from the insignia images. While in the case of the learning-based approach, the same features, associated with the extracted segments, were used to train the feedforward error back propagation ANN. It should be noted that a large set of training data is required for any learning to be effective. We used the process described in Algorithm 4.2 to generate the training dataset which consisted of 7500 patterns. This training data was used to learn the weights of the feedforward ANN using error back-propagation algorithm. Several ANN configurations and learning rates were experimented and the resulted presented in this paper are for the $5 - 21 - 11$ ANN[2] configuration (input, hidden and the output layer nodes) with a learning rate of 0.01.

Each segment was then labeled as one of the 11 patterns. For the rule-based approach, we used the rule base (Algorithm 4.1) to identify the given segment as one of the 11 patterns while in case of ANN, the extracted feature set was given as the input to the trained ANN to get the identified pattern.

Note 1 *All the numerical constants appearing in Algorithm 4.1 were heuristically obtained by observation.*

2.4 Post-processing

All the labeled segments corresponding to an insignia image are sequentially organized by using the centroid feature of each segment. The sequencing is based on the spatial occurance of the segment, namely, the top left segment would be first in the sequence while the segment occurring to the right bottom would come last in the sequence. Post-processing involves the following steps which are based on the observations, namely,

1. ∩, + and U always appear in same sequence
2. $\{∩ + U\}$ if present, occur together once or not at all
3. * can only co-occur with a | and another *
4. If a * occurs, its count could vary between 1 and 4
5. * can occur only if a | exist to its leftmost and rightmost

Note 2 *The post-processing rules enumerated above are based on observation of the insignia images (see Figure 1) and are data specific.*

3 Experimental Results

The original insignia images (see Figure 1) consists of 24 images of resolution 1890×3071 pixels captured by high resolution digital camera with almost zero noise, zero $x - y$, $y - z$ and $x - z$ planar rotations and negligible intensity variations. The experimental dataset consists of 216 images ($24 \times 3 \times 3$) captured by three colleagues in the lab who were not part of this work. Each of them captured 24 image using their personal mobile handset cameras in three different resolution namely,

---

[2] Details about Artificial Neural Networks have been omitted because it is well known in literature.

low (0.3 MP), medium (3 MP) and high (5 MP) with noise, light variations and $x - y$, $y - z$ and $x - z$ planar rotations of their choice.

The test insignia images were first pre-processed to extract segments (as shown in Figure 4); for each of the segment extracted from the insignia image, features F_S (see 5) was extracted. A pattern ($\in P$) was associated with the feature F_S using Algorithm 4.1 for the rule-based approach and by passing the feature set through the ANN for the learning based approach. We now discuss the experimental results.

We conducted experiments to measure the performance of both the rule-based and learning-based approaches. In the initial set of experiments, we computed the segment level accuracy in terms of associating the extracted segment with one of the 11 patterns and in the second set of results, we computed the actual insignia recognition accuracy. The experimental results for the rule-based approach is shown in Table 1 and the results of the learning-based approach (ANN with best configuration of $5 - 21 - 11$ and learning rate 0.01) is shown in Table 2.

Table 1 Experimental results of the rule-based approach

Segment Level Recognition			
With Pattern U (1155)		Without Pattern U (761)	
Correct	Accuracy(%)	Correct	Accuracy(%)
763	66.06	721	94.74

Insignia Level Recognition (216)	
Correct	Accuracy(%)
216	100.00

Table 2 Experimental Results for the learning-based approach

Segment Level Recognition					
		With Pattern U (1155)		Without Pattern U (761)	
L	H	Correct	Accuracy(%)	Correct	Accuracy(%)
0.01	21	1128	97.66	753	98.95

Insignia Level Recognition (216)			
L	H	Correct	Accuracy(%)
0.01	21	216	100.00

Observation 1. *Most of the errors occurred during the recognition of the pattern U. This is primarily due to the incorrect segmentation of the pattern into smaller fragments (see Figure 6). For this reason, in all our experimental results, we show experimental results for segment level recognition with and without pattern U.*

It should be noted that though the learning-based approach returns an accuracy of 97.66% and looks far more promising than the rule-based approach with an accuracy of 66.06% for segment level recognition with pattern U. But as mentioned

(a)   (b)

Fig. 6 Incorrect segmentation of pattern U into smaller fragments. (a) Before segmentation and (b) After segmentation

in Observation 1, segment level recognition was considered without taking pattern U into account. Experimental results show that both rule-based and learning-based approach, is able to work with an accuracy of above 90% in terms of identification of the patterns while the recognition of the insignia itself is 100% for both the approaches.

4 Conclusions

We proposed two different different approaches to insignia identification based on a common set of features extracted from a segmented image. We focused on low-level feature extraction to enable computation on the mobile device and established that the choice of features does not sacrifice the final recognition ability, both at the segment level and the insignia level. Both the rule-based approach and the learning-based approach perform well at the insignia level. Clearly the learning-based approach out performs the rule-based approach (97% compared to 66%). However, it should be noted that the generation of the rule-base for the rule-based approach is manual, time consuming and prone to error, while learning-based ANN approach has the benefit of learning the patterns based on the training data without any manual support.

References

1. Skytrait-Mobile, US Military Ranks (July 2013), https://play.google.com/store/apps/details?id=com.brandao.militaryranks&hl=en
2. CC-Intelligent-Solutions, Military Ranks XRef (July 2013), https://play.google.com/store/apps/details?id=com.ccis.mobile.MilitaryRanksXRef&hl=en
3. Zentrum-Elektronische-Medien, iSoldat (July 2013), https://play.google.com/store/apps/details?id=com.tis.isoldat&hl=en
4. Engel, J., Gerretzen, J., Szymańska, E., Jansen, J.J., Downey, G., Blanchet, L., Buydens, L.: Breaking with trends in pre-processing. TrAC Trends in Analytical Chemistry 50, 96–106 (2013)
5. Rinnan, S., Berg, F.V.D., Engelsen, S.B.: Review of the most common pre-processing techniques for near-infrared spectra. TrAC Trends in Analytical Chemistry 28(10), 1201–1222 (2009)

6. Zhang, D., Islam, M.M., Lu, G.: A review on automatic image annotation techniques. Pattern Recognition 45(1), 346–362 (2012)
7. Tavakoli, V., Amini, A.A.: A survey of shaped-based registration and segmentation techniques for cardiac images. Computer Vision and Image Understanding (2013)
8. Moaveni, M., Wang, S., Hart, J.M., Tutumluer, E., Ahuja, N.: Evaluation of aggregate size and shape by means of segmentation techniques and aggregate image processing algorithms. Transportation Research Record: Journal of the Transportation Research Board 2335(1), 50–59 (2013)

Appendix

Algorithm 4.1: HEURISTICIDENTIFYPATTERNS(F_S)

---

comment: $F_S = \{P, L, \gamma, \mu\}$

if $P = \{\text{BELOW}\}$

then $\begin{cases} \textbf{if } L = \{\text{OUT}\} \textbf{ then } z \leftarrow \{V\} \\ \textbf{else if } L = \{\text{IN}\} \\ \quad \textbf{then} \begin{cases} \textbf{if } 0 \le \gamma \le 1.57 \textbf{ then } z \leftarrow \{*\} \\ \textbf{else if } 1.57 \le \gamma \le 3.0 \textbf{ then } z \leftarrow \{\approx\} \\ \textbf{else if } 3.0 \le \gamma \le 6.5 \textbf{ then } z \leftarrow \{=\} \\ \textbf{else if } \gamma > 6.5 \textbf{ then } z \leftarrow \{-\} \end{cases} \end{cases}$

else if $P = \{\text{CENTER}\}$

then $\begin{cases} \textbf{if } L = \{\text{IN}\} \\ \quad \textbf{then} \begin{cases} \textbf{if } 0 \le \mu \le 20 \textbf{ then } z \leftarrow \{+\} \\ \textbf{else if } \mu > 20 \\ \quad \textbf{then} \begin{cases} \textbf{if } 0 \le \gamma \le 0.07 \textbf{ then } z \leftarrow \{|\} \\ \textbf{else if } \gamma \le 1.15 \textbf{ then } z \leftarrow \{/\} \\ \textbf{else if } \gamma \le 1.50 \textbf{ then } z \leftarrow \{\#\} \\ \textbf{else if } \gamma \le 3.0 \textbf{ then } z \leftarrow \{\approx\} \\ \textbf{else if } \gamma > 3.0 \textbf{ then } z \leftarrow \{=\} \end{cases} \end{cases} \\ \textbf{else if } L = \{\text{OUT}\} \\ \quad \textbf{then} \begin{cases} \textbf{if } \mu \le 15\% \textbf{ then } z \leftarrow \{U\} \\ \textbf{else if } \mu > 15\% \textbf{ then } z \leftarrow \{V\} \end{cases} \end{cases}$

else if $P = \{\text{UP}\}$

then $\begin{cases} \textbf{if } L = \{\text{IN}\} \\ \quad \textbf{then} \begin{cases} \textbf{if } \mu \le 20\% \textbf{ then } z \leftarrow \{+\} \\ \textbf{else if } \mu > 20\% \textbf{ then } z \leftarrow \{\approx\} \end{cases} \\ \textbf{else if } L = \{\text{OUT}\} \\ \quad \textbf{then} \begin{cases} \textbf{if } \mu \le 10\% \textbf{ then } z \leftarrow \{\cap\} \\ \textbf{else if } \mu > 10\% \textbf{ then } z \leftarrow \{V\} \end{cases} \end{cases}$

return (z)

Algorithm 4.2: ANNTRAININGDATAGENERATE(D)

---

comment: $D=\{d_{/},d_{\cap},d_{+},d_{U},d_{V},d_{\#},d_{-},d_{=},d_{\approx},d_{|},d_{*}\}$

comment: $RangeDataset, d_S \in D=\{min_{L'}, min_{C_{x'}}, min_{C_{y'}},$

$min_{\gamma}, min_{\mu}, max_{L'}, max_{C_{x'}}, max_{C_{y'}}, max_{\gamma}, max_{\mu}, S\}$

$TrainingData \leftarrow \{\}$
for each $d_S \in D$
  do for $i,j,k,l \leftarrow 0$ **to** 4

$$\text{do}\begin{cases} \textbf{if } (S! = \{U\}) \\ \quad \textbf{then}\begin{cases} x \leftarrow min_{L'} \\ y \leftarrow min_{C_{x'}} + i \times \left(\frac{max_{C_{x'}} - min_{C_{x'}}}{4}\right) \\ z \leftarrow min_{C_{y'}} + j \times \left(\frac{max_{C_{y'}} - min_{C_{y'}}}{4}\right) \\ w \leftarrow min_{\gamma} + k \times \left(\frac{max_{\gamma} - min_{\gamma}}{4}\right) \\ h \leftarrow min_{\mu} + l \times \left(\frac{max_{\mu} - min_{\mu}}{4}\right) \\ TrainingData \leftarrow TrainingData \\ \quad \cup\{x,y,z,w,h\} \end{cases} \\ \\ \quad \textbf{else}\begin{cases} x_1 \leftarrow 0 \\ y \leftarrow min_{C_{x'}} + i \times \left(\frac{max_{C_{x'}} - min_{C_{x'}}}{4}\right) \\ z \leftarrow min_{C_{y'}} + j \times \left(\frac{max_{C_{y'}} - min_{C_{y'}}}{4}\right) \\ w \leftarrow min_{\gamma} + k \times \left(\frac{max_{\gamma} - min_{\gamma}}{4}\right) \\ h \leftarrow min_{\mu} + l \times \left(\frac{max_{\mu} - min_{\mu}}{4}\right) \\ TrainingData \leftarrow TrainingData \\ \quad \cup\{x_1,y,z,w,h\} \\ \\ x_2 \leftarrow 1 \\ TrainingData \leftarrow TrainingData \\ \quad \cup\{x_2,y,z,w,h\} \end{cases} \end{cases}$$

return ($TrainingData$)

Color Image Segmentation Using Semi-supervised Self-organization Feature Map

Amiya Halder, Shruti Dalmiya, and Tanmana Sadhu

Abstract. Image segmentation is one of the fundamental steps in digital image processing, and is an essential part of image analysis. This paper presents an image segmentation of color images by semi-supervised clustering method based on modal analysis and mutational agglomeration algorithm in combination with the self-organization feature map (SOM) neural network. The modal analysis and mutational agglomeration is used for initial segmentation of the images. Subsequently, the sampled image pixels of the segmented image are used to train the network through SOM. Results are compared with four different state of the art image segmentation algorithms and are found to be encouraging for a set of natural images.

Keywords: Clustering, Image segmentation, K-means Algorithm, Modal analysis, Mutational agglomeration, Neural network, Self-organizing feature map.

1 Introduction

Image segmentation is the process of division of the image into regions with similar attributes. Colors are one of the most important distinguishing features in biological visual systems, used for complex pattern and object recognition. The segmentation of color images is a recent research as the presence of color information lends greater complexity as compared to greyscale images. In problems of pattern recognition for artificial vision, color image segmentation is a vital requirement.

Every pixel of the color image consists of red (R), green (G) and blue (B) components, each with varying measures of brightness, hue and saturation.

Amiya Halder · Shruti Dalmiya · Tanmana Sadhu
St. Thomas' College of Engineering and Technology, Kolkata, West Bengal, India
e-mail: {amiya_halder77,shruti.dalmiya.30,tanmana5}@gmail.com

S.M. Thampi, A. Gelbukh, and J. Mukhopadhyay (eds.), *Advances in Signal Processing and Intelligent Recognition Systems*, Advances in Intelligent Systems and Computing 264, DOI: 10.1007/978-3-319-04960-1_51, © Springer International Publishing Switzerland 2014

Segmentation of a color image promotes greater understanding of the details represented by it, and is implemented by using different methods such as histogram thresholding, edge and line oriented segmentation, region growing, and clustering [1, 2]. Specifically, segmentation of natural color images without previously acquired information is considered to be a challenging task.

A "cluster" is essentially a collection of data values (intensity values) that share similar characteristics. Clustering methods generally aim to partition an input data space into "k" number of regions, based on a similarity or dissimilarity metric. One of the most popular algorithms used is the K-means algorithm [3, 4, 5], which uses an iterative cluster analysis technique in which each data belongs to the cluster with the nearest distance. Here, the number of clusters k has to be known beforehand, and this constrains it to a maximum value. Also, optimal clusters may not be obtained.

A solution of the segmentation problem may be obtained by employing the artificial neural network based self-organizing feature map method described in [6, 7]. The intensity values of each of the red, green and blue components of color image are taken as features to be trained by the SOM network. A number of prototype vectors of the input image are used for training. Segmentation using classification by SOM to group pixels into a priori unknown number of classes has been stated in [8], whereas the cluster number is predetermined using K-means method. However, the drawback of this technique is that the number of clusters determined by the clustering procedure might be overestimated. In such a case, some homogeneous regions can be classified as more than one class, consequently increasing the inter cluster error measure. To overcome this drawback various methods have been proposed [9, 10].

Another approach in which the over-segmentation problem has been dealt with in the clustering process has been proposed in [11]. A thresholding technique has been employed to eliminate extra clusters by combining two or more clusters if they are at a distance less than a set threshold. Here, the determination of the cluster centers is done through SOM, which is dependent on the initial selected parameters for learning and number of iterations. The method of using an ensemble of SOM networks to implement more reliable and robust clustering has been stated in [6]. A major limitation of this method is that the number of clusters has to be specified before the network can be trained.

In proposed method, initially determine the preliminary cluster centers by modal analysis technique. The statistical mode does not take into account largely anomalous values in a set of data and is thus more representative of all the cluster data points as compared to the mean, which is used in the K-means clustering. Mutational agglomeration then combines similar clusters, thus removing the chances of over or under-clustering problem occurrence. Finally SOM has been employed as a semi-supervised algorithm in which unsupervised feature extraction through competitive learning rule has been implemented, along with a supervised classification process for clustering achieved by training of the neural network. Comparison of proposed algorithm with those of K-means, SOM, a combination of K-means and SOM, and only modal analysis and mutational agglomeration methods has been done by means of a validity index, proposed by Turi [12], which provides a measure of intra- and inter-cluster dependencies.

The paper is organized into seven section. Section 2 addresses the concept of modal analysis. Section 3 describes the mutational agglomeration. Section 4 provides a brief discussion of Kohonen's self-organizing feature map neural network. Section 5 describes the proposed algorithm. Experimental results are reported in Section 6. Finally, in Section 7, conclusion are drawn.

2 Modal Analysis

Initial cluster centers are determined by modal analysis. Each pixel belongs to a cluster centroid if it satisfies a predetermined threshold condition; otherwise it is treated as an independent centroid. The threshold condition is based on a Euclidean distance measure between the pixel intensity and the centroid intensity values. Consider any pixel p, the membership to a cluster centroid, Z_i is defined as

$$Z_i = \{p : p \in \psi(x,y) \ \ and \ \ |I(Z_i) - I(p)| \leq T\} \tag{1}$$

$I(Z_i)$ = Intensity value of ith centroid,
$I(p)$ = Intensity value of the pixel p,
T = Threshold value,
$\psi(x,y)$ = Input image.
Modal analysis is applied once the clusters have been defined after thresholding. Modal value of each cluster is the pixel value with the highest frequency in the cluster, which replaces its original cluster centroid [10].

3 Mutational Agglomeration

Initial cluster groups are formed by modal analysis. Then reduce the initial cluster groups by mutational agglomeration. Once the set of cluster centers have been determined, each is converted into its binary equivalent. The two groups are same if the two centroid's at least four Most Significant Bits (MSB) are same, then the two clusters can be merged into one and the centroid of the new cluster thus formed will be the modal value of all the pixels belonging to the new group. Equality measure is checked by modulo-2 (XOR) operation. This iterative process is continued, till the deviation between previously established groups and new groups formed is least or negligible. Then the image pixel value can be replaced by the group pixel value to obtain a segmented image [10].

4 Self-organization Feature Map

Kohonen's self-organizing feature map network [13, 14] consists of an input
and an output layer. Let z be the dimension of the input space and an input
pattern be denoted by $\overrightarrow{I_i} = [i_1, i_2, ..i_z]$. Let synaptic weight vector of an
output neuron j be denoted by $\overrightarrow{W_j} = [w_{j1}, w_{j2}, ..w_{jz}]$, j=1,2,... ,$l$ where l is
the total number of neurons in the output layer. If synaptic weight vector
$\overrightarrow{W_k}$ of neuron k is the best matched one with input vector $\overrightarrow{I}$ then the error
(denoted as, e_k) is as follows:

$$e_k = arg\ min_j \|\overrightarrow{I} - \overrightarrow{W_j}\| \tag{2}$$

The neuron k is then called the wining neuron for the input vector $\overrightarrow{I}$.
Let $h_{j,k}(n)$ denote the influence of the wining neuron k to its topological
neighbor j at epoch number n; and similarly, $d_{j,k}$ denote the lateral distance
between wining neuron k and its neighbor j. Now $h_{j,k}(n)$ is defined in such
a way that it attains the maximum value at the wining neuron k for which
$d_{j,k}$ is zero and decreases monotonically with increasing lateral distance $d_{j,k}$
. A typical choice of $h_{j,k}(n)$ is

$$h_{j,k}(n) = exp(-\frac{d_{j,k}^2}{2\sigma^2(n)}) \tag{3}$$

where $\sigma(n)$ represents the size of the topological neighborhood at epoch n.
The size of the topological neighborhood shrinks with n. The synaptic weight
updating process can be simulated by the following equation:

$$\overrightarrow{W_j}(n+1) = \overrightarrow{W_j}(n)\eta(n)h_{j,k}(n)(\overrightarrow{I} - \overrightarrow{W_j}(n)) \tag{4}$$

The updating procedure moves the weight vector $\overrightarrow{W_j}$ towards the input
vector $\overrightarrow{I}$. So repeated presentation of training patterns tries to make the
synaptic weight vectors tend to follow the distribution of the input vectors.

5 Proposed Method

In this paper, modal analysis and mutational agglomeration based segmentation in combination with the Kohonen's neural network based iterative
self-organizing feature map data analysis technique to achieve a more stable
and homogeneous segmentation. The modal analysis technique of assigning
modal values as the preliminary cluster centroids contributes to robustness of
the method by eliminating the influence of highly deviating values in the data
set in their computation. The mode values overcome a major shortcoming of
assigning mean as the cluster centroids. Mutational agglomeration plays an
important role in abolishing cluster redundancy by combining comparable
centroid values. Also, the number of clusters becomes known automatically

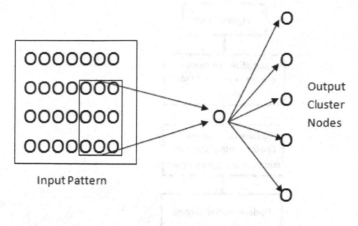

Fig. 1 Self-organizing feature map

for further realization of the lattice of output neurons in SOM. Further, it is to be noted that the characteristic dimensionality reduction for the color images is achieved with the application of SOM. In this paper, one-dimensional self-organizing feature map technique is used, where initialize weight vectors W_j of $j_t h$ neuron with random values in the red, green and blue planes are considered.

$$W_j(r, g, b) = [w_j(r), w_j(g), w_j(b)] \tag{5}$$

Where j=1,2,..m,m being the total number of clusters. The input features are generated corresponding to each pixel of the color components taking the intensity value of that pixel and its neighboring pixels. Here 2nd order (N^2) neighborhood is considered. So the dimension of the input pattern is nine.

A block diagram of the proposed algorithm is given in Fig.2.

6   Experimental Results

The algorithm has been simulated on MATLAB software using .tif and .jpg images. A set of input vectors, approximately 30 percent from the image data set, have been picked randomly. Each input vector taken as a 3×3 neighboring pixels, for each of the color components, red, green and blue. The threshold for modal analysis has been taken as 5. The initial learning rate is chosen to be 0.1. The total number of iterations for the adaptation phase is taken as 1000 as an optimal value. These values can be varied as suitability of the clustering.

The results are evaluated with respect to compactness measure as validity index [12]. The average validity indices (ten simulation) of the proposed algorithm and other algorithms, have been shown in Fig.5 to facilitate its

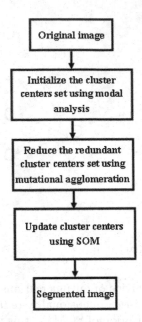

Fig. 2 Block diagram of the proposed algorithm

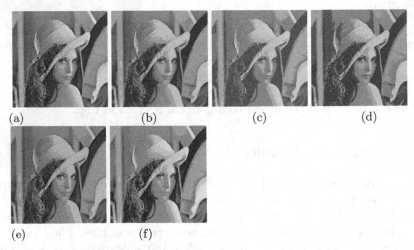

(a)   (b)   (c)   (d)

(e)   (f)

Fig. 3 Segmented image using different techniques:(a)Original image(b) K-means(c) SOM (d) K-means-SOM (e) Modal analysis and mutational agglomeration (f) Proposed method

comparative analysis between K-means, SOM, K-means with SOM, and only modal and mutational Agglomeration methods. The resultant segmented images of Lena and Peppers by implementation of the above different methods

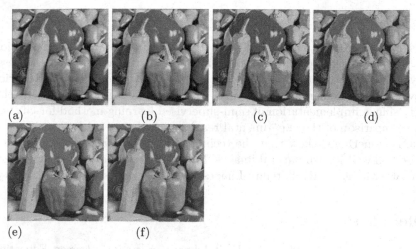

(a)   (b)   (c)   (d)

(e)   (f)

Fig. 4 Segmented image using different techniques:(a)Original image(b) K-means(c) SOM (d) K-means-SOM (e) Modal analysis and mutational agglomeration (f) Proposed method

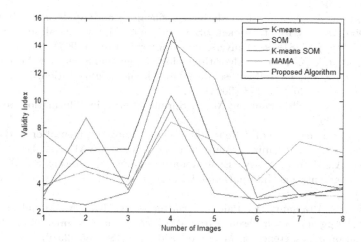

Fig. 5 Compactness value(VI) obtained using K-means, SOM, K-means-SOM, MAMA (Modal Analysis and mutational agglomeration) and Proposed method for eight natural images

have been shown for comparison with proposed method in Fig.3 and Fig.4. Fig.5 clearly shows that the proposed algorithm gives better validity index than the other image segmentation algorithms.

7 Conclusions

This paper presents a new approach for semi-supervised self-organization feature map based segmentation technique for color image that can successfully segment the images. The advantages of the method include no prior knowledge requirement of the number of clusters for segmentation and relatively simple implementation of semi-supervised learning method for clustering. Comparison of the experimental results with that of other unsupervised clustering methods, show that the technique gives satisfactory results when applied on well known natural images. Moreover results of its use on images from other fields (MRI, Satellite Images) demonstrate its wide applicability.

References

1. Gonzalez, R.C., Woods, R.E.: Digital Image processing. Pearson Education (2002)
2. Pratt, W.K.: Digital Image Processing. A Wiley-Interscience Publication (2003)
3. Yerpude, A., Dubey, S.: Colour image segmentation using K - Medoids Clustering. International Journal of Computer Techology and Applications 3(1), 152–154 (2012)
4. MacQueen, J.B.: Some Methods for classification and Analysis of Multivariate Observations. In: 5th Berkeley Symposium on Mathematical Statistics and Probability, pp. 281–297. University of California Press (1967)
5. Hartigan, J.A.: Clustering Algorithms. John Wiley and Sons, New York (1975)
6. Jiang, Y., Zhou, Z.: SOM Ensemble-Based Image Segmentation. Neural Processing Letters 20, 171–178 (2004)
7. Kohonen, T.: Self-Organizing Maps. Springer Series in Information Sciences, vol. 30 (1995)
8. Moreira, J., Costa, L.D.F.: Neural-based color image segmentation and classification using self-organizing maps. Anais do IX SIBGRAPI, pp. 47–54 (1996)
9. Halder, A., Pathak, N.: An Evolutionary Dynamic Clustering Based Colour Image Segmentation. International Journal of Image Processing 4(6), 549–556 (2011)
10. Halder, A., Pramanik, S., Pal, S., Chatterjee, N., Kar, A.: Modal and Mutational Agglomeration based Automatic Colour Image Segmentation. In: 3rd International Conference on Machine Vision, pp. 481–485 (2010)
11. Awada, M.: An Unsupervised Artificial Neural Network Method for Satellite Image Segmentation. The International Arab Journal of Information Technology 7(2), 199–205 (2010)
12. Turi, R.H.: Clustering-Based Color Image Segmentation, PhD Thesis, Monash University, Australia (2001)
13. Haykin, S.: Neural Networks. Pearson Education (2001)
14. Fausett, L.V.: Fundamentals of Neural Networks. Prentice Hall (1994)

A Review on Localization in Wireless Sensor Networks

Jeril Kuriakose, Sandeep Joshi, R. Vikram Raju, and Aravind Kilaru

Abstract. Localization is extensively used in Wireless Sensor Networks (WSNs) to identify the current location of the sensor nodes. A WSN consist of thousands of nodes that make the installation of GPS on each sensor node expensive and moreover GPS will not provide exact localization results in an indoor environment. Manually configuring location reference on each sensor node is also not possible in the case of dense network. This gives rise to a problem where the sensor nodes must identify its current location without using any special hardware like GPS and without the help of manual configuration. Localization techniques makes the deployment of WSNs economical. Most of the localization techniques are carried out with the help of anchor node or beacon node, which knows its present location. Based on the location information provided by the anchor node or beacon node, other nodes localize themselves. In this paper we present a succinct survey on the localization techniques used in wireless sensor networks covering its problems and research gap.

1 Introduction

Wireless sensor devices have a wide range of application in surveillance and monitoring. Most of the devices or nodes in wireless sensor network are made up of off-the-shelf materials and deployed in the area of surveillance and monitoring. The responsibility of each sensor node is to identify the changes in its particular region or area. The changes are like movement of animals, increase or decrease in temperature or rainfall and these changes are periodically reported to the aggregation point or the central server. The central server or the

Jeril Kuriakose · Sandeep Joshi · R. Vikram Raju · Aravind Kilaru
School of Computing and Information Technology (SCIT),
Manipal University Jaipur, Jaipur
e-mail: jeril@muj.manipal.edu

S.M. Thampi, A. Gelbukh, and J. Mukhopadhyay (eds.), *Advances in Signal Processing* 599
and Intelligent Recognition Systems, Advances in Intelligent Systems and Computing 264,
DOI: 10.1007/978-3-319-04960-1_52, © Springer International Publishing Switzerland 2014

aggregation server identifies the area with the help of the location reference
sent by the sensor node.

Initially during deployment each sensor nodes are given their location reference. This is done either manually or the sensor nodes automatically calculate
the distance with the help of GPS devices attached to it. Installing a GPS device or manually calculating the location may not be possible in the context
of large network because of the excessive cost and workforce involved, respectively. To overcome this sensor nodes are made to identify their locations with
the help of neighbouring nodes. This paper focuses on the localization techniques used by the sensor nodes to identify their location. Several researches
are going on in the field of localization to identify the exact location.

The location of the nodes plays a significant role in many areas like routing,
surveillance and monitoring, and military. The sensor nodes must know their
location reference in order to carry-out Location-based routing (LR) [1 -
4]. So as to find out the shortest route, the Location Aided Routing (LAR)
protocol [5 - 7] makes use of the locality reference of the sensor nodes. In some
industries the sensor nodes are used to identify minute changes like pressure,
temperature and gas leak, and in military, robots are used to detect land-
mines, where in both the cases location information plays a key part.

Organization of the paper - Section 2 and 3 covers the concepts and
problems in localization. Section 4 covers the localization techniques. Section
5 and 6 presents the performance, discussion and future events. Section 7
concludes the paper.

2   Concepts and Properties of Localization

In most of the localization techniques, localization is carried out with the
help of neighbouring nodes. Initially few nodes are made available with their
location reference either by manual configuration or using GPS devices. Several localization techniques are discussed in this paper. Fig. 1 illustrates the
different techniques or methods used to identify the location of the nodes.

The localization can be classified as known location based localization,
proximity based localization, angle based localization, range and distance
based localization. In fig. 1 the range and distance based localization are
categorized separately, though both are same. For range based localization,
special hardware is required to find out the range, however it is not required
for distance based localization.

2.1   Known Location Based Localization

In this type of localization the sensor nodes know their location in prior.
This is done either by manually configuring or using a GPS [8 - 12] device.
Manual configuration of the sensor node is done with the help of GPS. The
GPS device can be effective where there are no reference nodes available to

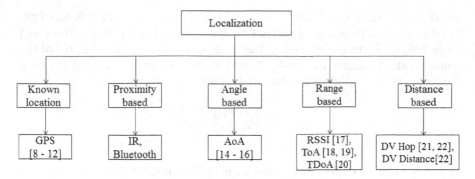

Fig. 1 Overview of localization

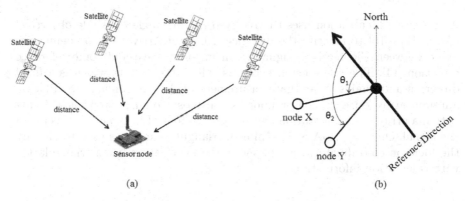

Fig. 2 (a) Working of GPS receiver installed in a sensor node. (b) Localization using triangulation with orientation information.

get localized. The location of the sensor node is calculated with the help of GPS satellites. A minimum of four satellites are required to calculate the location of the GPS receiver. Fig. 2a shows the working of GPS receiver.

The distance between the GPS receiver and the GPS satellites are calculated using the time taken for the signal to reach the device. Once the distances are known, the GPS receiver uses triangulation [24] or trilateration [29] technique to determine its location. It has a good accuracy with a standard deviation of 4 to 10 meters.

2.2 Proximity Based Localization

In this type of localization the wireless sensor network is divided into several clusters. Each cluster has a cluster head which is equipped with a GPS device. Using Infrared (IR) or Bluetooth, the nodes find out the nearness or proximity location. When comparing proximity based localization with range based localization, proximity based localization does not suffer fading [30]. Proximity based localization will not be applicable when the power threshold drops below a threshold value i.e., the range or power (signal strength) of the central node. Consider two nodes X and Y with a power threshold P_1, then the localization eligibility is calculated as follows: [30]

$$Q_{x,y} = \begin{cases} 1, \ P_{x,y} \geq P_1 \\ 0, \ P_{x,y} < P_1 \end{cases}$$

where,
$P_{x,y}$ is the measured received power at node X transmitted by node Y.

2.3 Angle Based Localization

Angle based localization uses the received signals angle or Angle of Arrival (AoA) [14 - 16] to identify the distance. Angle of Arrival can be defined as angle between the received signal of an incident wave and some reference direction [14]. The reference direction is called orientation, which is a fixed direction and against that the measurement of AoA is carried out. Placing antenna array on each sensor node is the most common approach. Using antenna array with orientation can be used to identify the angular sector of the signal. Once the AoA is determined, triangulation [24] is used to identify the location co-ordinates. Fig. 2b shows the localization using triangulation with orientation information.

2.4 Range Based Localization

This localization is carried out based on the range. The range is calculated using the Received Signal Strength Indicator (RSSI) [17] or Time of Arrival (ToA) [18, 19] or Time Difference of Arrival (TDoA) [20]. In RSSI based localization the receiver sends the signal strength with reference to the sender, and sender calculates the distance based on the signal strength. ToA and TDoA use timing to calculate the range. Time synchronization is an important factor when using ToA and TDoA.

2.5 Distance Based Localization

Distance based localization technique uses hop distance between the sender and receiver node to identify the location reference. It uses DV-hop propagation method [21, 22] or DV-distance [22] propagation method for localization.

3 Problems in Localization Techniques

There are few limitations encountered during localization. The problems are listed as follows,

3.1 Known Location Based Localization

The cost of a GPS device is around $500 - $1000, and installing a GPS for all the nodes is a dense network is not recommended. There are some situations where the GPS fail to find out the exact location reference such as in an underground, underwater or indoor environment. A small amount of accuracy can be reduced due to multipath propagation. Satellite availability also plays a key role in location estimation.

3.2 Proximity Based Localization

In this type of localization the nodes estimate the proximity location with the assistance of a central node. Larger the range of central node smaller is the accuracy. Localization is not achievable when the central server is down. This technique is an economical one with less accuracy.

3.3 Angle Based Localization

Special antennas are required for Angle based localization. The node that wants to localize does it with the help of three or more nodes, which have the special antenna installed in them. The angle measurement error can vary from $1°$ to $25°$ as an effect of noise. As angle based localization technique requires special antenna's that are expensive, AoA is generally used in Base Station's (BS) or cell phone tower's.

3.4 Range Based localization

Link reliability and noise interference plays a significant role in reducing the accuracy of range measurements. Other factors like multipath fading and environmental changes also decrease the accuracy in range measurement.

3.5 Distance Based Localization

In this type of localization the accuracy can be improved only if the network is dense. For a sparse network the localization accuracy is reduced considerably.

4   Localization Techniques

The localization techniques can be grouped into two types namely range based and range free approach. Fig. 3 shows the localization techniques grouped into different types.

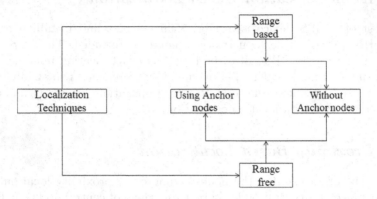

Fig. 3 Categorizing localization techniques

4.1   Range Based Approach

This method uses the range information to calculate the distance between each node. The localization can be carried out with or without the anchor nodes.

4.1.1   Using Anchor Nodes

During the deployment of wireless sensor network, few sensor nodes are made configured with their location reference either manually or using GPS. These nodes act as the anchor nodes. Other nodes localize themselves with the support of anchor nodes.

Localization is carried out using the range or angle based techniques discussed in the previous section. Each sensor node must be equipped with special hardware to achieve localization. In RSSI based distance measurement, the distance is calculated with respect to the sender's signal strength. A node can calculate the distance using the signal strength measurement received from the sender. The signal strength gradually decreases as the node moves away. Fig. 4a shows the typical signal strength or coverage area of a node.

Next type of localization technique that uses range to identify the distance is ToA based localization. The nodes that use ToA for localization must be time synchronized. The transmitted and received time are used by the sender to calculate the distance. Fig. 5a shows the working of range estimation using

ToA. The distance between the sender and receiver are calculated as follows: [19]

$$d_{xy} = \frac{1}{2} \left[(T^x_{recv} - T^x_{trans}) - (T^y_{trans} - T^y_{recv}) \right]$$

where,
d_{xy} is the distance between node X and node Y,
T^x_{recv} is the received power of node X,
T^x_{trans} is the transmitted power of node X,
T^y_{recv} is the received power of node Y,
T^y_{trans} is the transmitted power of node Y.

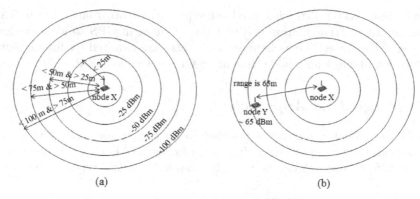

Fig. 4 (a) Typical signal strength or coverage area of a node. (b) Range estimation using RSSI.

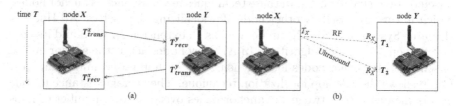

Fig. 5 (a) Range estimation using ToA. (b) Range estimation using TDoA.

Once the distances are discovered, multilateration or trilateration [29] technique is implemented to find out the location reference of the node. RF signal travel at the speed of light, this make the RF propagation to get varied in indoor environments and gives rise to a high localization overhead. In order to overcome the RF propagation in indoor environment, in [13], a combination of RF signals with Ultrasound was proposed. The speed of Ultrasound is less when compared to the speed of light. Based on the TDoA of the two signals, the distance is calculated. Fig. 5b shows the working of range estimation using TDoA. Another method for locating a node using TDoA is done by observing the time for a signal to reach two or more receivers. It is made sure that all the receiver nodes are time synchronized. The TDoA is calculated as follows: [23]

$$\tau = \frac{(r_2 - r_1)}{c}$$

where,
τ is the TDoA,
$r_1 \& r_2$ are the range from the transmitter to the two receivers,
c is the speed of propagation.

4.1.2 Without Using Anchor Nodes

A device that has GPS attached need not require a support from anchor nodes for localization. Triangulation [24] technique is used in GPS to identify the location of the node. The assistance of satellites are required for finding out the location of the sensor node that has a GPS device installed.

4.2 Range Free Approach

There are few localization techniques that do not require special hardware for localization, they compute their distance using on DV hop or DV distance. The range free approach can be broadly classified into two types as follows:

4.2.1 Using Anchor Nodes

Techniques, namely Probability Grid [21] and Kcdlocation [24] works on DV based distance localization. In these techniques few nodes act as anchor nodes, which in turn are used by other nodes for localizing themselves. Ad Hoc Positioning System (APS) [22] can be used as an extension for GPS and its uses hop by hop positioning algorithm. In APS few nodes act as the anchor nodes, based on which other nodes localize using hop by hop positioning technique. Fig. 6 shows the DV hop localization technique. The average distance for one hop is the distance between the anchor nodes over the total number of hops between the anchor nodes. The equation to compute average distance for one hop is as follows: [22]

$$H_d = \frac{\sum \sqrt{(A_i - A_j)^2 + (B_i - B_j)^2}}{\sum h}$$

where,
H_d is average distance for one hop,
$\sum h$ is the total number of hops,
$(A_i, B_i) \& (A_j, B_j)$ are the location co-ordinates of the anchor nodes.

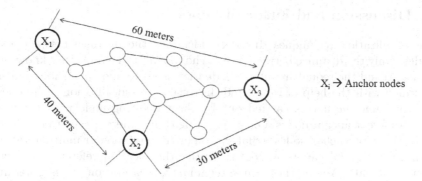

Fig. 6 Hop by hop localization technique

4.2.2 Without Using Anchor Nodes

Convex Position Estimation technique [28] works without an anchor node. The network is modelled by a central sever giving equations for relating the distance between the nodes. It uses a good optimization technique to find out the location of the nodes based on the equations.

5 Performance of Localization Schemes

Table 1 shows the performance comparison of different localization schemes. Each localization techniques serve different purposes. More the number of anchor nodes, less the localization error. In a closed environment with more obstructions, the localization error tends to increase. This can be controlled by making the network dense.

Table 1 Comparison of localization techniques

Localization Techniques used	Accuracy (in meters)	Limitations
GPS	2 to 15	Indoor localization is not possible in many cases
Proximity based	1 to 30	Depends on the range of the signal used
Angle based approach	1 to 8	Require special antenna
Range based approach	4 to 10	Require special hardware and time synchronization
DV based approach	10 to 20	Accuracy can be improved in a dense network

6 Discussion and Future Events

The localization techniques discussed identifies the distance between two nodes. Only the distance between the anchor node (or the node that knows its location) and the requesting node (node that needs to find out its location) is obtained with the help of range and distance based localization techniques. So, the same method is carried out by the requesting node with three or more different anchor nodes. Once obtaining three different range's i.e., from three different anchor nodes, trilateration or triangulation or multilateration technique [24, 29] can be used to identify the location reference or location coordinate. Among these three techniques, trilateration [29] is generally preferred. In the present scenario there is a trade-off between localization accuracy and algorithm runtime. The localization techniques discussed here are based in the consideration of a static network. Monitoring and surveillance has become painless because of wireless sensor network.

Several researches are being carried out in the field of localization. Few future events still remain unaddressed in the 3-D localization, security, mobility and energy conservation. 2-D scenario is generally used for localization and there will be a requirement for new localization techniques for 3-D scenario. As most of the localization techniques require the help of neighbouring nodes to identify their localization, there must be a surety that the neighbouring nodes are valid. Localization fails in presence of a mole in the network. Several security measures are required to secure the network as well as the sensor node from attacks. Localization can be easily carried out for fixed nodes. There exist a dire fall in localization accuracy, for mobile sensor nodes. Designing an energy efficient localization technique is an essential consideration for wireless sensor network.

7 Conclusion

There is a considerable rise in the use of wireless sensor network because of their cost and size. The localization techniques discussed in this paper, help in reducing the deployment cost of dense wireless sensor networks. Several techniques with abridged hardware which can identify their current location has been discussed in this paper along with their key features and drawbacks. Since a wireless sensor nodes are provided with limited resources there is a need in designing an intelligent power aware and secure localization approach in both 2D and 3D scenario.

References

1. Camp, T., Boleng, J., et al.: Performance Comparison of Two Location Based Routing Protocols for Ad Hoc Networks. IEEE Infocom. (2002)

2. Blazevic, L., Boudec, J.-Y.L., Giordano, S.: A Location-Based Routing Method for Mobile Ad Hoc Networks. IEEE Transactions on Mobile Computing (2005)
3. Fubler, H., Mauve, M., et al.: Location Based Routing for Vehicular AdHoc Networks. In: Proceedings of ACM MOBICOM (2002)
4. Qu, H., Wicke, S.B.: Co-designed anchor-free localization and location-based routing algorithm for rapidly-deployed wireless sensor networks. Information Fusion (2008)
5. Kuhn, F., et al.: Worst-case optimal and average-case efficient geometric ad-hoc routing. In: Proc. of the 4th ACM International Symposium on Mobile Ad Hoc Networking & Computing (2003)
6. El Defrawy, K., Tsudik, G.: Alarm: Anonymous location aided routing in suspicious Manets. In: IEEE International Conference on Network Protocols, pp. 304–313 (2007)
7. Ko, Y.-B., Vaidya, N.H.: Location-Aided Routing (LAR) in mobile ad hoc networks. Wireless Networks (2000)
8. Whitehouse, H.J., Leese de Escobar, A.M., et al.: A GPS Sonobuoy Localization System. In: Position Location and Navigation Symposium (2004)
9. Lita, I., et al.: A New Approach of Automobile Localization System Using GPS and GSM/GPRS Transmission. International Spring Seminar on Electronics Technology (2006)
10. Stoleru, R., et al.: Walking GPS: A Practical Solution for Localization in Manually Deployed Wireless Sensor Networks. In: IEEE International Conference on Local Computer Networks (2004)
11. Stefano, P., Pascucci, F., Ulivi, G.: An outdoor navigation system using GPS and inertial platform. In: IEEE/ASME Transactions on Mechatronics (2002)
12. Parkinson, B., et al.: Global Positioning System: Theory and Application. Progress in Astronautics and Aeronautics, vol. I (1996)
13. Priyantha, N., Chakraborty, A., Balakrishnan, H.: The Cricket Location-Support System. In: Proc. ACM MobiCom. (2000)
14. Peng, R., Sichitiu, M.L.: Angle of Arrival Localization for Wireless Sensor Networks. In: Third Annual IEEE Comm. Society Conference on Sensor and Ad Hoc Comm. and Networks (2006)
15. Niculescu, D., Nath, B.: Ad hoc positioning system (APS) using AOA. In: IEEE INFOCOM (2003)
16. Nasipuri, A., Li, K.: A directionality based location discovery scheme for wireless sensor networks. In: ACM International Workshop on Wireless Sensor Networks and Applications (2002)
17. Mao, G., Anderson, B.D.O., Fidan, B.: Path Loss Exponent Estimation for Wireless Sensor Network Localization. Computer Networks (2007)
18. Moses, R., Krishnamurthy, D., Patterson, R.: A Self-Localization Method for Wireless Sensor Networks. In: Eurasip J. Applied Signal Processing, Special Issue on Sensor Networks (2003)
19. Sarigiannidis, G.: Localization for Ad Hoc Wireless Sensor Networks. M.S. thesis, Technical University Delft, The Netherlands (2006)
20. Xiao, J., Ren, L., Tan, J.: Research of TDOA Based Self-Localization Approach in Wireless Sensor Network. In: Proceedings IEEE/RSJ International Conference on Intelligent Robots and Systems (2006)
21. Stoleru, R., Stankovic, J.: Probability grid: A location estimation scheme for wireless sensor networks. In: Proceedings of Sensor and Ad-Hoc Comm. and Networks Conference (SECON) (2004)

22. Niculescu, D., Nath, B.: Ad hoc positioning system (APS). In: Proceedings of the IEEE Global Telecommunications Conference (2001)
23. Chestnut, P.C.: Emitter location accuracy using TDOA and differential Doppler. IEEE Transactions in Aerospace and Electronic System (1982)
24. Fang, Z., Zhao, Z., et al.: Localization in Wireless Sensor Networks with Known Coordinate Database. EURASIP Journal on Wireless Communications and Networking (2010)
25. Suresh, P., et al.: Mobile Information Retrieval using Topic-Sensitive PageRank and Page Freshness. International Journal of Advanced Research in Computer and Communication Engineering (2013)
26. Suresh, P., Shekeela, N., et al.: Detection and Elimination of Malicious Beacon Nodes in Broadband Based Wireless Networks. International Journal of Advanced and Innovative Research (2013)
27. Shivprasad, B.J., Amruth, V., et al.: Feature Level Image Fusion. In: Elsevier/Emerging Research in Computing, Information, Communication and Applications, ERCICA 2013 (2013)
28. Doherty, L., Pister, K., Ghaoui, L.: Convex Position Estimation in Wireless Sensor Networks. In: Proceedings of the IEEE Computer and Communication Societies (INFOCOM 2001) (2001)
29. Manolakis, D.: Efficient Solution and Performance Analysis of 3-D Position Estimation by Trilateration. IEEE Transactions on Aerospace and Electronic Systems (1996)
30. Patwari, N., et al.: Using Proximity and Quantized RSS for Sensor Location in Wireless Location in Wireless Networks. In: Proceedings Workshop Wireless Sensor Networks and Applications (2003)
31. Bal, M., Liu, M., Shen, W., Ghenniwa, H.: Localization in cooperative Wireless Sensor Networks: A review. In: 13th International Conference on Computer Supported Cooperative Work in Design, CSCWD 2009 (2009)
32. http://www.libelium.com/products/waspmote/

Author Index